WITHDRAWN

LIVERPOOL JMU LIBRARY

VOLUME FOUR HUNDRED AND NINETY-ONE

METHODS IN ENZYMOLOGY

The Unfolded Protein Response and Cellular Stress, Part C

METHODS IN ENZYMOLOGY

Editors-in-Chief

JOHN N. ABELSON AND MELVIN I. SIMON

Division of Biology
California Institute of Technology
Pasadena, California

Founding Editors

SIDNEY P. COLOWICK AND NATHAN O. KAPLAN

VOLUME FOUR HUNDRED AND NINETY-ONE

Methods in
ENZYMOLOGY

The Unfolded Protein Response and Cellular Stress, Part C

EDITED BY

P. MICHAEL CONN
*Divisions of Reproductive Sciences and Neuroscience (ONPRC)
Departments of Pharmacology and Physiology
Cell and Developmental Biology, and
Obstetrics and Gynecology (OHSU)
Beaverton, OR, USA*

AMSTERDAM • BOSTON • HEIDELBERG • LONDON
NEW YORK • OXFORD • PARIS • SAN DIEGO
SAN FRANCISCO • SINGAPORE • SYDNEY • TOKYO
Academic Press is an imprint of Elsevier

Academic Press is an imprint of Elsevier
525 B Street, Suite 1900, San Diego, CA 92101-4495, USA
30 Corporate Drive, Suite 400, Burlington, MA 01803, USA
32 Jamestown Road, London NW1 7BY, UK

First edition 2011

Copyright © 2011, Elsevier Inc. All Rights Reserved.

No part of this publication may be reproduced, stored in a retrieval system or transmitted in any form or by any means electronic, mechanical, photocopying, recording or otherwise without the prior written permission of the publisher

Permissions may be sought directly from Elsevier's Science & Technology Rights Department in Oxford, UK: phone (+44) (0) 1865 843830; fax (+44) (0) 1865 853333; email: permissions@elsevier.com. Alternatively you can submit your request online by visiting the Elsevier web site at http://elsevier.com/locate/permissions, and selecting *Obtaining permission to use Elsevier material*

Notice
No responsibility is assumed by the publisher for any injury and/or damage to persons or property as a matter of products liability, negligence or otherwise, or from any use or operation of any methods, products, instructions or ideas contained in the material herein. Because of rapid advances in the medical sciences, in particular, independent verification of diagnoses and drug dosages should be made

For information on all Academic Press publications
visit our website at elsevierdirect.com

ISBN: 978-0-12-385928-0
ISSN: 0076-6879

Printed and bound in United States of America
11 12 13 10 9 8 7 6 5 4 3 2 1

**Working together to grow
libraries in developing countries**

www.elsevier.com | www.bookaid.org | www.sabre.org

ELSEVIER   BOOK AID International   Sabre Foundation

Contents

Contributors	xiii
Preface	xix
Volumes in Series	xxi

Section I. New Approaches to Studying UPR and Cell Stress   1

1. CFTR Expression Regulation by the Unfolded Protein Response   3
Rafal Bartoszewski, Andras Rab, Lianwu Fu, Sylwia Bartoszewska, James Collawn, and Zsuzsa Bebok

1. Introduction	4
2. Methods	6
Acknowledgments	22
References	22

2. GRP78/BiP Chapter: Modulation of GRP78/BiP in Altering Sensitivity to Chemotherapy   25
Thomas C. Chen

1. Introduction	26
2. Measurement of ER Stress *In Vitro* and *In Vivo*	27
3. Conclusion	36
References	36

3. Targeting the Unfolded Protein Response in Cancer Therapy   37
Marina V. Backer, Joseph M. Backer, and Prakash Chinnaiyan

1. Introduction	38
2. The UPR and Cancer	38
3. Targeting the UPR	40
4. Combination of EGF–SubA with Other UPR-Targeting Drugs	50
Acknowledgments	54
References	54

4. Large-Scale Analysis of UPR-Mediated Apoptosis in Human Cells   57

Andrew M. Fribley, Justin R. Miller, Tyler E. Reist, Michael U. Callaghan, and Randal J. Kaufman

1. Introduction	58
2. Monitoring Proliferation and Caspase Activation Following UPR Activation	60
3. Monitoring the Expression of UPR and Cell Death Target Genes	61
4. DNA Fragmentation Analysis	68
Acknowledgments	69
References	69

5. Quantitative Analysis of Amino Acid Oxidation Markers by Tandem Mass Spectrometry   73

Anuradha Vivekanandan-Giri, Jaeman Byun, and Subramaniam Pennathur

1. Introduction	75
2. Experimental Procedures	76
3. Results	82
4. Conclusions	86
Acknowledgments	86
References	87

6. Animal Models in the Study of the Unfolded Protein Response   91

Hemamalini Bommiasamy and Brian Popko

1. Introduction	92
2. Activating Transcription Factor 6	93
3. IRE1/X-Box-Binding Protein-1	94
4. PKR-Like ER Kinase	97
5. eIF2α	98
6. ATF4	99
7. CHOP	100
8. GADD34	101
9. P58IPK	102
10. Transgenic Mouse Models for Monitoring ER Stress	102
11. UPR and Lipid Metabolism	103
12. UPR, Hypoxia, and Cancer	104
13. UPR and Inflammatory-Mediated Demyelination	105
14. Future Challenges	105
References	106

7. **Measurement of Fluoride-Induced Endoplasmic Reticulum Stress Using *Gaussia* Luciferase** — 111

Ramaswamy Sharma, Masahiro Tsuchiya, Bakhos A. Tannous, and John D. Bartlett

1. Introduction — 112
2. Properties of *Gaussia* Luciferase — 114
3. Materials — 115
4. Procedure — 118
5. Conclusions — 123
References — 124

8. **Analysis of Nelfinavir-Induced Endoplasmic Reticulum Stress** — 127

Ansgar Brüning

1. Introduction — 128
2. Obtaining Nelfinavir — 129
3. Using Nelfinavir — 129
4. Analysis of Nelfinavir-Induced ER Stress — 130
5. Immunoblot Analysis — 134
6. RT-PCR Analysis — 135
7. Conclusions — 141
Acknowledgment — 142
References — 142

9. **Using Temporal Genetic Switches to Synchronize the Unfolded Protein Response in Cell Populations *In Vivo*** — 143

Alexander Gow

1. Introduction — 144
2. Variable Functions of CHOP in the PERK Signaling Pathway — 146
3. Mutations in the *PLP1* Gene: A Model UPR Disease — 147
4. Problems with Current Paradigms Used to Characterize the UPR — 149
5. A Novel *In Vivo* Genetic Switch Solves Many of These Problems — 150
6. Generalizing GST to Study Other UPR Diseases — 157
Acknowledgments — 157
References — 158

10. **Glycoprotein Maturation and the UPR** — 163

Andreas J. Hülsmeier, Michael Welti, and Thierry Hennet

1. Introduction — 163
2. N-glycosylation — 165
3. O-glycosylation — 175
Acknowledgment — 181
References — 181

11. **Monitoring and Manipulating Mammalian Unfolded Protein Response** — 183

Nobuhiko Hiramatsu, Victory T. Joseph, and Jonathan H. Lin

1. Introduction — 184
2. Monitoring Mammalian UPR — 185
3. Chemical–Genetic Manipulation of Mammalian UPR — 189
4. Concluding Remarks — 194
Acknowledgments — 195
References — 195

12. **A Screen for Mutants Requiring Activation of the Unfolded Protein Response for Viability** — 199

Guillaume Thibault and Davis T. W. Ng

1. Introduction — 200
2. Screening for Mutants That Require UPR Activation for Viability — 202
3. Cloning and Sequencing *per* Genes — 209
4. Monitoring the UPR Activity — 212
5. SGA Database — 213
6. Media Recipes — 214
7. Closing Remarks — 214
Acknowledgment — 215
References — 215

13. **Signaling Pathways of Proteostasis Network Unrevealed by Proteomic Approaches on the Understanding of Misfolded Protein Rescue** — 217

Patrícia Gomes-Alves, Sofia Neves, and Deborah Penque

1. Introduction — 218
2. 2DE-Based Proteomics Approach Analysis to Study Protein Expression Profile — 219
3. Experimental Design — 220
4. 2D Gel Electrophoresis — 221
5. Protein Identification by MS — 223
6. Biochemical Validation — 225
7. Data Mining and Publication — 226
Acknowledgments — 231
References — 231

14. Decreased Secretion and Unfolded Protein Response Upregulation   235

Carissa L. Young, Theresa Yuraszeck, and Anne S. Robinson

1. Introduction — 236
2. Heterologous Protein Expression — 237
3. Quality Control Mechanisms of the Secretory Pathway — 238
4. Endoplasmic Reticulum Export and Trafficking — 239
5. Experimental Systems to Evaluate Expression, UPR, and Secretory Processing — 239
6. *S. cerevisiae* Strains Used for Optimal Expression — 240
7. Plasmid Design — 241
8. Evaluation of Heterologous Protein Expression — 243
9. Statistical Analysis of Microarray Results — 248

Acknowledgment — 257
References — 257

15. Measuring Signaling by the Unfolded Protein Response   261

David J. Cox, Natalie Strudwick, Ahmed A. Ali, Adrienne W. Paton, James C. Paton, and Martin Schröder

1. Introduction — 262
2. Methods to Induce the UPR — 266
3. Measuring Signaling by the UPR in *S. cerevisiae* — 266
4. Measuring Signaling by the UPR in Mammalian Cells — 276

Acknowledgments — 288
References — 289

16. Quantitative Measurement of Events in the Mammalian Unfolded Protein Response   293

Jie Shang

1. Introduction — 294
2. Cell Culture and ER Stress Inducers — 295
3. Northern Blots for GRP78/BiP and EDEM — 296
4. RT-PCR Analysis of XBP1 mRNA Splicing — 299
5. Immunoblotting and Measurement of ATF6 — 302
6. Measurement of Inhibition of Protein Synthesis — 302
7. Measurement of Stimulation of LLO Extension — 303
8. Calculations of Transcript Accumulation, Signal Activation, and Correlation — 304
9. Conclusion — 305

Acknowledgments — 306
References — 306

17. **Regulation of Immunoglobulin Synthesis, Modification, and Trafficking by the Unfolded Protein Response: A Quantitative Approach**   309

    Adi Drori and Boaz Tirosh

    1. Introduction   310
    2. Isolation of Splenic B Cells and Infection Thereof by Retroviruses   312
    3. Measurement of Protein Synthesis in Plasma Cells   316
    4. Measurement of Ig Mislocalization in Primary B Cells   320
    Acknowledgments   324
    References   324

18. **Use of Chemical Genomics in Assessment of the UPR**   327

    Sakae Saito and Akihiro Tomida

    1. Introduction   328
    2. Assessment of the Activation of UPR Transcriptional Program in Cancer Cell   329
    3. Gene Expression Signature-Based Identification of UPR Modulators   334
    4. Future Perspective of Chemical Genomics in UPR Research   338
    Acknowledgments   339
    References   339

19. **Small GTPase Signaling and the Unfolded Protein Response**   343

    Marion Bouchecareilh, Esther Marza, Marie-Elaine Caruso, and Eric Chevet

    1. Introduction   344
    2. Materials and Methods   346
    3. Monitoring the Unfolded Protein Response   348
    4. Monitoring Small GTPase Activity   353
    5. Pharmacological and Genetic Modulation of UPR and GTPase Signaling   356
    Acknowledgments   358
    References   358

20. **Inhibitors of Advanced Glycation and Endoplasmic Reticulum Stress**   361

    Reiko Inagi

    1. Introduction   362
    2. Advanced Glycation and Its Pathophysiology   363
    3. Measurement of Advanced Glycation   364
    4. Link Between Advanced Glycation and ER Stress   369

5.	Effects of Advanced Glycation Inhibitor Against ER Stress	373
6.	Conclusion	377
	Acknowledgments	377
	References	377

Author Index   *381*

Subject Index   *399*

Contributors

Ahmed A. Ali
School of Biological and Biomedical Sciences, Durham University, Durham, United Kingdom, and Department of Molecular Biology, National Research Centre, Cairo, Egypt

Joseph M. Backer
Sibtech, Inc., Brookfield, Connecticut, USA

Marina V. Backer
Sibtech, Inc., Brookfield, Connecticut, USA

John D. Bartlett
Department of Cytokine Biology, Forsyth Institute, Cambridge Massachusetts, and Department of Developmental Biology, Harvard School of Dental Medicine, Boston, Massachusetts, USA

Sylwia Bartoszewska
Department of Cell Biology, University of Alabama at Birmingham, Birmingham, Alabama, USA

Rafal Bartoszewski
Department of Cell Biology, University of Alabama at Birmingham, Birmingham, Alabama, USA

Zsuzsa Bebok
Department of Cell Biology, University of Alabama at Birmingham, Birmingham, Alabama, USA

Hemamalini Bommiasamy
Department of Neurology, Center for Peripheral Neuropathy, The University of Chicago, Chicago, Illinois, USA

Marion Bouchecareilh
Avenir, Inserm U889, and Université Bordeaux 2, Bordeaux, France

Ansgar Brüning
Department of OB/GYN, Molecular Biology Laboratory, University Hospital Munich, Campus Innenstadt and Grosshadern, Munich, Germany

Jaeman Byun
Division of Nephrology, Department of Internal Medicine, University of Michigan, Ann Arbor, Michigan, USA

Michael U. Callaghan
Division of Pediatric Hematology/Oncology, Department of Pediatrics, Wayne State School of Medicine, Detriot, Michigan, USA

Marie-Elaine Caruso
Avenir, Inserm U889, and Université Bordeaux 2, Bordeaux, France

Thomas C. Chen
Department of Neurosurgery and Pathology, University of Southern California, Los Angeles, USA

Eric Chevet
Avenir, Inserm U889, and Université Bordeaux 2, Bordeaux, France

Prakash Chinnaiyan
Department of Experimental Therapeutics and Radiation Oncology, H. Lee Moffitt Cancer Center and Research Institute, Tampa, Florida, USA

James Collawn
Department of Cell Biology, University of Alabama at Birmingham, Birmingham, Alabama, USA

David J. Cox
School of Biological and Biomedical Sciences, Durham University, Durham, United Kingdom

Adi Drori
Institute for Drug Research, School of Pharmacy, Faculty of Medicine, The Hebrew University, Jerusalem, Israel

Andrew M. Fribley
Department of Biological Chemistry, University of Michigan School of Medicine, Ann Arbor, Michigan, USA

Lianwu Fu
Department of Cell Biology, University of Alabama at Birmingham, Birmingham, Alabama, USA

Patrícia Gomes-Alves
Laboratório de Proteómica, Departamento de Genética, Instituto Nacional de Saúde Dr Ricardo Jorge (INSA, I.P.), Av. Padre Cruz, Lisboa, Portugal

Alexander Gow
Center for Molecular Medicine and Genetics, Carman and Ann Adams Department of Pediatrics, Department of Neurology, Wayne State University School of Medicine, Detroit, Michigan, USA

Andreas J. Hülsmeier
Institute of Physiology, University of Zürich, Winterthurerstrasse, Zürich, Switzerland

Thierry Hennet
Institute of Physiology, University of Zürich, Winterthurerstrasse, Zürich, Switzerland

Nobuhiko Hiramatsu
Department of Pathology, School of Medicine, University of California, San Diego, La Jolla, California, USA

Reiko Inagi
Division of Nephrology and Endocrinology, University of Tokyo School of Medicine, Tokyo, Japan

Victory T. Joseph
Department of Pathology, and Department of Neuroscience, School of Medicine, University of California, San Diego, La Jolla, California, USA

Randal J. Kaufman
Department of Biological Chemistry, University of Michigan School of Medicine, Ann Arbor, Michigan, USA

Jonathan H. Lin
Department of Pathology, School of Medicine, University of California, San Diego, La Jolla, California, USA

Esther Marza
Avenir, Inserm U889, and Université Bordeaux 2, Bordeaux, France

Justin R. Miller
Division of Pediatric Hematology/Oncology, Department of Pediatrics, Wayne State School of Medicine, Detriot, Michigan, USA

Sofia Neves
Laboratório de Proteómica, Departamento de Genética, Instituto Nacional de Saúde Dr Ricardo Jorge (INSA, I.P.), Av. Padre Cruz, Lisboa, Portugal

Davis T.W. Ng
Temasek Life Sciences Laboratory, and Department of Biological Sciences, National University of Singapore, Singapore

Adrienne W. Paton
Research Centre for Infectious Diseases, School of Molecular and Biomedical Science, University of Adelaide, Adelaide, Australia

James C. Paton
Research Centre for Infectious Diseases, School of Molecular and Biomedical Science, University of Adelaide, Adelaide, Australia

Subramaniam Pennathur
Division of Nephrology, Department of Internal Medicine, University of Michigan, Ann Arbor, Michigan, USA

Deborah Penque
Laboratório de Proteómica, Departamento de Genética, Instituto Nacional de Saúde Dr Ricardo Jorge (INSA, I.P.), Av. Padre Cruz, Lisboa, Portugal

Brian Popko
Department of Neurology, Center for Peripheral Neuropathy, The University of Chicago, Chicago, Illinois, USA

Andras Rab
Department of Cell Biology, University of Alabama at Birmingham, Birmingham, Alabama, USA

Tyler E. Reist
Department of Biological Chemistry, University of Michigan School of Medicine, Ann Arbor, Michigan, USA

Anne S. Robinson
Department of Chemical Engineering, University of Delaware, Newark, Delaware, USA

Sakae Saito
Cancer Chemotherapy Center, Japanese Foundation for Cancer Research, Tokyo, Japan

Martin Schröder
School of Biological and Biomedical Sciences, Durham University, Durham, United Kingdom

Jie Shang
Department of Pharmacology, University of Texas Southwestern Medical Center, Dallas, Texas, USA

Ramaswamy Sharma
Department of Cytokine Biology, Forsyth Institute, Cambridge Massachusetts, and Department of Developmental Biology, Harvard School of Dental Medicine, Boston, Massachusetts, USA

Natalie Strudwick
School of Biological and Biomedical Sciences, Durham University, Durham, United Kingdom

Bakhos A. Tannous
Departments of Neurology and Radiology, Massachusetts General Hospital, and Program in Neuroscience, Harvard Medical School, Boston, Massachusetts, USA

Guillaume Thibault
Temasek Life Sciences Laboratory, National University of Singapore, Singapore

Boaz Tirosh
Institute for Drug Research, School of Pharmacy, Faculty of Medicine, The Hebrew University, Jerusalem, Israel

Akihiro Tomida
Cancer Chemotherapy Center, Japanese Foundation for Cancer Research, Tokyo, Japan

Masahiro Tsuchiya
Division of Aging and Geriatric Dentistry, Tohoku University, Japan

Anuradha Vivekanandan-Giri
Division of Nephrology, Department of Internal Medicine, University of Michigan, Ann Arbor, Michigan, USA

Michael Welti
Institute of Physiology, University of Zürich, Winterthurerstrasse, Zürich, Switzerland

Carissa L. Young
Department of Chemical Engineering, University of Delaware, Newark, Delaware, USA

Theresa Yuraszeck
Department of Chemical Engineering, University of California Santa Barbara, Santa Barbara, California, USA

Preface

The observation that the living cell contains a mechanism to sense and correct the accumulation of unfolded (or incorrectly folded) proteins in the endoplasmic reticulum was formidable in organizing thoughts about cellular integration. This mechanism both halts further protein synthesis and promotes the production of chaperone proteins that act to relieve this problem. If this problem cannot be corrected, the mechanism can initiate programmed cell death. Aspects of this unfolded protein response (UPR) are conserved from yeast to man, an observation that suggests a key role in the process of maintaining a living cell.

The UPR presents a way of understanding cellular regulation, a mechanism for disease, and a therapeutic opportunity.

The present volume provides descriptions of the occurrence of the UPR, the methods used to assess it, and the pharmacological tools and other methodological approaches to analyze its impact on cellular regulation. The authors explain how these methods are able to provide important biological insights.

Authors were selected based on research contributions in the area about which they have written and based on their ability to describe their methodological contribution in a clear and reproducible way. They have been encouraged to make use of graphics, comparisons to other methods, and to provide tricks and approaches not revealed in prior publications that make it possible to adapt methods to other systems.

The editor wants to express appreciation to the contributors for providing their contributions in a timely fashion, to the senior editors for guidance, and to the staff at Academic Press for helpful input.

P. Michael Conn
Portland, Oregon, USA

Methods in Enzymology

VOLUME I. Preparation and Assay of Enzymes
Edited by SIDNEY P. COLOWICK AND NATHAN O. KAPLAN

VOLUME II. Preparation and Assay of Enzymes
Edited by SIDNEY P. COLOWICK AND NATHAN O. KAPLAN

VOLUME III. Preparation and Assay of Substrates
Edited by SIDNEY P. COLOWICK AND NATHAN O. KAPLAN

VOLUME IV. Special Techniques for the Enzymologist
Edited by SIDNEY P. COLOWICK AND NATHAN O. KAPLAN

VOLUME V. Preparation and Assay of Enzymes
Edited by SIDNEY P. COLOWICK AND NATHAN O. KAPLAN

VOLUME VI. Preparation and Assay of Enzymes *(Continued)*
Preparation and Assay of Substrates
Special Techniques
Edited by SIDNEY P. COLOWICK AND NATHAN O. KAPLAN

VOLUME VII. Cumulative Subject Index
Edited by SIDNEY P. COLOWICK AND NATHAN O. KAPLAN

VOLUME VIII. Complex Carbohydrates
Edited by ELIZABETH F. NEUFELD AND VICTOR GINSBURG

VOLUME IX. Carbohydrate Metabolism
Edited by WILLIS A. WOOD

VOLUME X. Oxidation and Phosphorylation
Edited by RONALD W. ESTABROOK AND MAYNARD E. PULLMAN

VOLUME XI. Enzyme Structure
Edited by C. H. W. HIRS

VOLUME XII. Nucleic Acids (Parts A and B)
Edited by LAWRENCE GROSSMAN AND KIVIE MOLDAVE

VOLUME XIII. Citric Acid Cycle
Edited by J. M. LOWENSTEIN

VOLUME XIV. Lipids
Edited by J. M. LOWENSTEIN

VOLUME XV. Steroids and Terpenoids
Edited by RAYMOND B. CLAYTON

VOLUME XVI. Fast Reactions
Edited by KENNETH KUSTIN

VOLUME XVII. Metabolism of Amino Acids and Amines (Parts A and B)
Edited by HERBERT TABOR AND CELIA WHITE TABOR

VOLUME XVIII. Vitamins and Coenzymes (Parts A, B, and C)
Edited by DONALD B. MCCORMICK AND LEMUEL D. WRIGHT

VOLUME XIX. Proteolytic Enzymes
Edited by GERTRUDE E. PERLMANN AND LASZLO LORAND

VOLUME XX. Nucleic Acids and Protein Synthesis (Part C)
Edited by KIVIE MOLDAVE AND LAWRENCE GROSSMAN

VOLUME XXI. Nucleic Acids (Part D)
Edited by LAWRENCE GROSSMAN AND KIVIE MOLDAVE

VOLUME XXII. Enzyme Purification and Related Techniques
Edited by WILLIAM B. JAKOBY

VOLUME XXIII. Photosynthesis (Part A)
Edited by ANTHONY SAN PIETRO

VOLUME XXIV. Photosynthesis and Nitrogen Fixation (Part B)
Edited by ANTHONY SAN PIETRO

VOLUME XXV. Enzyme Structure (Part B)
Edited by C. H. W. HIRS AND SERGE N. TIMASHEFF

VOLUME XXVI. Enzyme Structure (Part C)
Edited by C. H. W. HIRS AND SERGE N. TIMASHEFF

VOLUME XXVII. Enzyme Structure (Part D)
Edited by C. H. W. HIRS AND SERGE N. TIMASHEFF

VOLUME XXVIII. Complex Carbohydrates (Part B)
Edited by VICTOR GINSBURG

VOLUME XXIX. Nucleic Acids and Protein Synthesis (Part E)
Edited by LAWRENCE GROSSMAN AND KIVIE MOLDAVE

VOLUME XXX. Nucleic Acids and Protein Synthesis (Part F)
Edited by KIVIE MOLDAVE AND LAWRENCE GROSSMAN

VOLUME XXXI. Biomembranes (Part A)
Edited by SIDNEY FLEISCHER AND LESTER PACKER

VOLUME XXXII. Biomembranes (Part B)
Edited by SIDNEY FLEISCHER AND LESTER PACKER

VOLUME XXXIII. Cumulative Subject Index Volumes I-XXX
Edited by MARTHA G. DENNIS AND EDWARD A. DENNIS

VOLUME XXXIV. Affinity Techniques (Enzyme Purification: Part B)
Edited by WILLIAM B. JAKOBY AND MEIR WILCHEK

VOLUME XXXV. Lipids (Part B)
Edited by JOHN M. LOWENSTEIN

VOLUME XXXVI. Hormone Action (Part A: Steroid Hormones)
Edited by BERT W. O'MALLEY AND JOEL G. HARDMAN

VOLUME XXXVII. Hormone Action (Part B: Peptide Hormones)
Edited by BERT W. O'MALLEY AND JOEL G. HARDMAN

VOLUME XXXVIII. Hormone Action (Part C: Cyclic Nucleotides)
Edited by JOEL G. HARDMAN AND BERT W. O'MALLEY

VOLUME XXXIX. Hormone Action (Part D: Isolated Cells, Tissues, and Organ Systems)
Edited by JOEL G. HARDMAN AND BERT W. O'MALLEY

VOLUME XL. Hormone Action (Part E: Nuclear Structure and Function)
Edited by BERT W. O'MALLEY AND JOEL G. HARDMAN

VOLUME XLI. Carbohydrate Metabolism (Part B)
Edited by W. A. WOOD

VOLUME XLII. Carbohydrate Metabolism (Part C)
Edited by W. A. WOOD

VOLUME XLIII. Antibiotics
Edited by JOHN H. HASH

VOLUME XLIV. Immobilized Enzymes
Edited by KLAUS MOSBACH

VOLUME XLV. Proteolytic Enzymes (Part B)
Edited by LASZLO LORAND

VOLUME XLVI. Affinity Labeling
Edited by WILLIAM B. JAKOBY AND MEIR WILCHEK

VOLUME XLVII. Enzyme Structure (Part E)
Edited by C. H. W. HIRS AND SERGE N. TIMASHEFF

VOLUME XLVIII. Enzyme Structure (Part F)
Edited by C. H. W. HIRS AND SERGE N. TIMASHEFF

VOLUME XLIX. Enzyme Structure (Part G)
Edited by C. H. W. HIRS AND SERGE N. TIMASHEFF

VOLUME L. Complex Carbohydrates (Part C)
Edited by VICTOR GINSBURG

VOLUME LI. Purine and Pyrimidine Nucleotide Metabolism
Edited by PATRICIA A. HOFFEE AND MARY ELLEN JONES

VOLUME LII. Biomembranes (Part C: Biological Oxidations)
Edited by SIDNEY FLEISCHER AND LESTER PACKER

VOLUME LIII. Biomembranes (Part D: Biological Oxidations)
Edited by SIDNEY FLEISCHER AND LESTER PACKER

VOLUME LIV. Biomembranes (Part E: Biological Oxidations)
Edited by SIDNEY FLEISCHER AND LESTER PACKER

VOLUME LV. Biomembranes (Part F: Bioenergetics)
Edited by SIDNEY FLEISCHER AND LESTER PACKER

VOLUME LVI. Biomembranes (Part G: Bioenergetics)
Edited by SIDNEY FLEISCHER AND LESTER PACKER

VOLUME LVII. Bioluminescence and Chemiluminescence
Edited by MARLENE A. DELUCA

VOLUME LVIII. Cell Culture
Edited by WILLIAM B. JAKOBY AND IRA PASTAN

VOLUME LIX. Nucleic Acids and Protein Synthesis (Part G)
Edited by KIVIE MOLDAVE AND LAWRENCE GROSSMAN

VOLUME LX. Nucleic Acids and Protein Synthesis (Part H)
Edited by KIVIE MOLDAVE AND LAWRENCE GROSSMAN

VOLUME 61. Enzyme Structure (Part H)
Edited by C. H. W. HIRS AND SERGE N. TIMASHEFF

VOLUME 62. Vitamins and Coenzymes (Part D)
Edited by DONALD B. MCCORMICK AND LEMUEL D. WRIGHT

VOLUME 63. Enzyme Kinetics and Mechanism (Part A: Initial Rate and Inhibitor Methods)
Edited by DANIEL L. PURICH

VOLUME 64. Enzyme Kinetics and Mechanism
(Part B: Isotopic Probes and Complex Enzyme Systems)
Edited by DANIEL L. PURICH

VOLUME 65. Nucleic Acids (Part I)
Edited by LAWRENCE GROSSMAN AND KIVIE MOLDAVE

VOLUME 66. Vitamins and Coenzymes (Part E)
Edited by DONALD B. MCCORMICK AND LEMUEL D. WRIGHT

VOLUME 67. Vitamins and Coenzymes (Part F)
Edited by DONALD B. MCCORMICK AND LEMUEL D. WRIGHT

VOLUME 68. Recombinant DNA
Edited by RAY WU

VOLUME 69. Photosynthesis and Nitrogen Fixation (Part C)
Edited by ANTHONY SAN PIETRO

VOLUME 70. Immunochemical Techniques (Part A)
Edited by HELEN VAN VUNAKIS AND JOHN J. LANGONE

VOLUME 71. Lipids (Part C)
Edited by JOHN M. LOWENSTEIN

VOLUME 72. Lipids (Part D)
Edited by JOHN M. LOWENSTEIN

VOLUME 73. Immunochemical Techniques (Part B)
Edited by JOHN J. LANGONE AND HELEN VAN VUNAKIS

VOLUME 74. Immunochemical Techniques (Part C)
Edited by JOHN J. LANGONE AND HELEN VAN VUNAKIS

VOLUME 75. Cumulative Subject Index Volumes XXXI, XXXII, XXXIV–LX
Edited by EDWARD A. DENNIS AND MARTHA G. DENNIS

VOLUME 76. Hemoglobins
Edited by ERALDO ANTONINI, LUIGI ROSSI-BERNARDI, AND EMILIA CHIANCONE

VOLUME 77. Detoxication and Drug Metabolism
Edited by WILLIAM B. JAKOBY

VOLUME 78. Interferons (Part A)
Edited by SIDNEY PESTKA

VOLUME 79. Interferons (Part B)
Edited by SIDNEY PESTKA

VOLUME 80. Proteolytic Enzymes (Part C)
Edited by LASZLO LORAND

VOLUME 81. Biomembranes (Part H: Visual Pigments and Purple Membranes, I)
Edited by LESTER PACKER

VOLUME 82. Structural and Contractile Proteins (Part A: Extracellular Matrix)
Edited by LEON W. CUNNINGHAM AND DIXIE W. FREDERIKSEN

VOLUME 83. Complex Carbohydrates (Part D)
Edited by VICTOR GINSBURG

VOLUME 84. Immunochemical Techniques (Part D: Selected Immunoassays)
Edited by JOHN J. LANGONE AND HELEN VAN VUNAKIS

VOLUME 85. Structural and Contractile Proteins (Part B: The Contractile Apparatus and the Cytoskeleton)
Edited by DIXIE W. FREDERIKSEN AND LEON W. CUNNINGHAM

VOLUME 86. Prostaglandins and Arachidonate Metabolites
Edited by WILLIAM E. M. LANDS AND WILLIAM L. SMITH

VOLUME 87. Enzyme Kinetics and Mechanism (Part C: Intermediates, Stereo-chemistry, and Rate Studies)
Edited by DANIEL L. PURICH

VOLUME 88. Biomembranes (Part I: Visual Pigments and Purple Membranes, II)
Edited by LESTER PACKER

VOLUME 89. Carbohydrate Metabolism (Part D)
Edited by WILLIS A. WOOD

VOLUME 90. Carbohydrate Metabolism (Part E)
Edited by WILLIS A. WOOD

VOLUME 91. Enzyme Structure (Part I)
Edited by C. H. W. HIRS AND SERGE N. TIMASHEFF

VOLUME 92. Immunochemical Techniques (Part E: Monoclonal Antibodies and General Immunoassay Methods)
Edited by JOHN J. LANGONE AND HELEN VAN VUNAKIS

VOLUME 93. Immunochemical Techniques (Part F: Conventional Antibodies, Fc Receptors, and Cytotoxicity)
Edited by JOHN J. LANGONE AND HELEN VAN VUNAKIS

VOLUME 94. Polyamines
Edited by HERBERT TABOR AND CELIA WHITE TABOR

VOLUME 95. Cumulative Subject Index Volumes 61–74, 76–80
Edited by EDWARD A. DENNIS AND MARTHA G. DENNIS

VOLUME 96. Biomembranes [Part J: Membrane Biogenesis: Assembly and Targeting (General Methods; Eukaryotes)]
Edited by SIDNEY FLEISCHER AND BECCA FLEISCHER

VOLUME 97. Biomembranes [Part K: Membrane Biogenesis: Assembly and Targeting (Prokaryotes, Mitochondria, and Chloroplasts)]
Edited by SIDNEY FLEISCHER AND BECCA FLEISCHER

VOLUME 98. Biomembranes (Part L: Membrane Biogenesis: Processing and Recycling)
Edited by SIDNEY FLEISCHER AND BECCA FLEISCHER

VOLUME 99. Hormone Action (Part F: Protein Kinases)
Edited by JACKIE D. CORBIN AND JOEL G. HARDMAN

VOLUME 100. Recombinant DNA (Part B)
Edited by RAY WU, LAWRENCE GROSSMAN, AND KIVIE MOLDAVE

VOLUME 101. Recombinant DNA (Part C)
Edited by RAY WU, LAWRENCE GROSSMAN, AND KIVIE MOLDAVE

VOLUME 102. Hormone Action (Part G: Calmodulin and Calcium-Binding Proteins)
Edited by ANTHONY R. MEANS AND BERT W. O'MALLEY

VOLUME 103. Hormone Action (Part H: Neuroendocrine Peptides)
Edited by P. MICHAEL CONN

VOLUME 104. Enzyme Purification and Related Techniques (Part C)
Edited by WILLIAM B. JAKOBY

VOLUME 105. Oxygen Radicals in Biological Systems
Edited by LESTER PACKER

VOLUME 106. Posttranslational Modifications (Part A)
Edited by FINN WOLD AND KIVIE MOLDAVE

VOLUME 107. Posttranslational Modifications (Part B)
Edited by FINN WOLD AND KIVIE MOLDAVE

VOLUME 108. Immunochemical Techniques (Part G: Separation and Characterization of Lymphoid Cells)
Edited by GIOVANNI DI SABATO, JOHN J. LANGONE, AND HELEN VAN VUNAKIS

VOLUME 109. Hormone Action (Part I: Peptide Hormones)
Edited by LUTZ BIRNBAUMER AND BERT W. O'MALLEY

VOLUME 110. Steroids and Isoprenoids (Part A)
Edited by JOHN H. LAW AND HANS C. RILLING

VOLUME 111. Steroids and Isoprenoids (Part B)
Edited by JOHN H. LAW AND HANS C. RILLING

VOLUME 112. Drug and Enzyme Targeting (Part A)
Edited by KENNETH J. WIDDER AND RALPH GREEN

VOLUME 113. Glutamate, Glutamine, Glutathione, and Related Compounds
Edited by ALTON MEISTER

VOLUME 114. Diffraction Methods for Biological Macromolecules (Part A)
Edited by HAROLD W. WYCKOFF, C. H. W. HIRS, AND SERGE N. TIMASHEFF

VOLUME 115. Diffraction Methods for Biological Macromolecules (Part B)
Edited by HAROLD W. WYCKOFF, C. H. W. HIRS, AND SERGE N. TIMASHEFF

VOLUME 116. Immunochemical Techniques
(Part H: Effectors and Mediators of Lymphoid Cell Functions)
Edited by GIOVANNI DI SABATO, JOHN J. LANGONE, AND HELEN VAN VUNAKIS

VOLUME 117. Enzyme Structure (Part J)
Edited by C. H. W. HIRS AND SERGE N. TIMASHEFF

VOLUME 118. Plant Molecular Biology
Edited by ARTHUR WEISSBACH AND HERBERT WEISSBACH

VOLUME 119. Interferons (Part C)
Edited by SIDNEY PESTKA

VOLUME 120. Cumulative Subject Index Volumes 81–94, 96–101

VOLUME 121. Immunochemical Techniques (Part I: Hybridoma Technology and Monoclonal Antibodies)
Edited by JOHN J. LANGONE AND HELEN VAN VUNAKIS

VOLUME 122. Vitamins and Coenzymes (Part G)
Edited by FRANK CHYTIL AND DONALD B. MCCORMICK

VOLUME 123. Vitamins and Coenzymes (Part H)
Edited by FRANK CHYTIL AND DONALD B. MCCORMICK

VOLUME 124. Hormone Action (Part J: Neuroendocrine Peptides)
Edited by P. MICHAEL CONN

VOLUME 125. Biomembranes (Part M: Transport in Bacteria, Mitochondria, and Chloroplasts: General Approaches and Transport Systems)
Edited by SIDNEY FLEISCHER AND BECCA FLEISCHER

VOLUME 126. Biomembranes (Part N: Transport in Bacteria, Mitochondria, and Chloroplasts: Protonmotive Force)
Edited by SIDNEY FLEISCHER AND BECCA FLEISCHER

VOLUME 127. Biomembranes (Part O: Protons and Water: Structure and Translocation)
Edited by LESTER PACKER

VOLUME 128. Plasma Lipoproteins (Part A: Preparation, Structure, and Molecular Biology)
Edited by JERE P. SEGREST AND JOHN J. ALBERS

VOLUME 129. Plasma Lipoproteins (Part B: Characterization, Cell Biology, and Metabolism)
Edited by JOHN J. ALBERS AND JERE P. SEGREST

VOLUME 130. Enzyme Structure (Part K)
Edited by C. H. W. HIRS AND SERGE N. TIMASHEFF

VOLUME 131. Enzyme Structure (Part L)
Edited by C. H. W. HIRS AND SERGE N. TIMASHEFF

VOLUME 132. Immunochemical Techniques (Part J: Phagocytosis and Cell-Mediated Cytotoxicity)
Edited by GIOVANNI DI SABATO AND JOHANNES EVERSE

VOLUME 133. Bioluminescence and Chemiluminescence (Part B)
Edited by MARLENE DELUCA AND WILLIAM D. MCELROY

VOLUME 134. Structural and Contractile Proteins (Part C: The Contractile Apparatus and the Cytoskeleton)
Edited by RICHARD B. VALLEE

VOLUME 135. Immobilized Enzymes and Cells (Part B)
Edited by KLAUS MOSBACH

VOLUME 136. Immobilized Enzymes and Cells (Part C)
Edited by KLAUS MOSBACH

VOLUME 137. Immobilized Enzymes and Cells (Part D)
Edited by KLAUS MOSBACH

VOLUME 138. Complex Carbohydrates (Part E)
Edited by VICTOR GINSBURG

VOLUME 139. Cellular Regulators (Part A: Calcium- and Calmodulin-Binding Proteins)
Edited by ANTHONY R. MEANS AND P. MICHAEL CONN

VOLUME 140. Cumulative Subject Index Volumes 102–119, 121–134

VOLUME 141. Cellular Regulators (Part B: Calcium and Lipids)
Edited by P. MICHAEL CONN AND ANTHONY R. MEANS

VOLUME 142. Metabolism of Aromatic Amino Acids and Amines
Edited by SEYMOUR KAUFMAN

VOLUME 143. Sulfur and Sulfur Amino Acids
Edited by WILLIAM B. JAKOBY AND OWEN GRIFFITH

VOLUME 144. Structural and Contractile Proteins (Part D: Extracellular Matrix)
Edited by LEON W. CUNNINGHAM

VOLUME 145. Structural and Contractile Proteins (Part E: Extracellular Matrix)
Edited by LEON W. CUNNINGHAM

VOLUME 146. Peptide Growth Factors (Part A)
Edited by DAVID BARNES AND DAVID A. SIRBASKU

VOLUME 147. Peptide Growth Factors (Part B)
Edited by DAVID BARNES AND DAVID A. SIRBASKU

VOLUME 148. Plant Cell Membranes
Edited by LESTER PACKER AND ROLAND DOUCE

VOLUME 149. Drug and Enzyme Targeting (Part B)
Edited by RALPH GREEN AND KENNETH J. WIDDER

VOLUME 150. Immunochemical Techniques (Part K: *In Vitro* Models of B and T Cell Functions and Lymphoid Cell Receptors)
Edited by GIOVANNI DI SABATO

VOLUME 151. Molecular Genetics of Mammalian Cells
Edited by MICHAEL M. GOTTESMAN

VOLUME 152. Guide to Molecular Cloning Techniques
Edited by SHELBY L. BERGER AND ALAN R. KIMMEL

VOLUME 153. Recombinant DNA (Part D)
Edited by RAY WU AND LAWRENCE GROSSMAN

VOLUME 154. Recombinant DNA (Part E)
Edited by RAY WU AND LAWRENCE GROSSMAN

VOLUME 155. Recombinant DNA (Part F)
Edited by RAY WU

VOLUME 156. Biomembranes (Part P: ATP-Driven Pumps and Related Transport: The Na, K-Pump)
Edited by SIDNEY FLEISCHER AND BECCA FLEISCHER

VOLUME 157. Biomembranes (Part Q: ATP-Driven Pumps and Related Transport: Calcium, Proton, and Potassium Pumps)
Edited by SIDNEY FLEISCHER AND BECCA FLEISCHER

VOLUME 158. Metalloproteins (Part A)
Edited by JAMES F. RIORDAN AND BERT L. VALLEE

VOLUME 159. Initiation and Termination of Cyclic Nucleotide Action
Edited by JACKIE D. CORBIN AND ROGER A. JOHNSON

VOLUME 160. Biomass (Part A: Cellulose and Hemicellulose)
Edited by WILLIS A. WOOD AND SCOTT T. KELLOGG

VOLUME 161. Biomass (Part B: Lignin, Pectin, and Chitin)
Edited by WILLIS A. WOOD AND SCOTT T. KELLOGG

VOLUME 162. Immunochemical Techniques (Part L: Chemotaxis and Inflammation)
Edited by GIOVANNI DI SABATO

VOLUME 163. Immunochemical Techniques (Part M: Chemotaxis and Inflammation)
Edited by GIOVANNI DI SABATO

VOLUME 164. Ribosomes
Edited by HARRY F. NOLLER, JR., AND KIVIE MOLDAVE

VOLUME 165. Microbial Toxins: Tools for Enzymology
Edited by SIDNEY HARSHMAN

VOLUME 166. Branched-Chain Amino Acids
Edited by ROBERT HARRIS AND JOHN R. SOKATCH

VOLUME 167. Cyanobacteria
Edited by LESTER PACKER AND ALEXANDER N. GLAZER

VOLUME 168. Hormone Action (Part K: Neuroendocrine Peptides)
Edited by P. MICHAEL CONN

VOLUME 169. Platelets: Receptors, Adhesion, Secretion (Part A)
Edited by JACEK HAWIGER

VOLUME 170. Nucleosomes
Edited by PAUL M. WASSARMAN AND ROGER D. KORNBERG

VOLUME 171. Biomembranes (Part R: Transport Theory: Cells and Model Membranes)
Edited by SIDNEY FLEISCHER AND BECCA FLEISCHER

VOLUME 172. Biomembranes (Part S: Transport: Membrane Isolation and Characterization)
Edited by SIDNEY FLEISCHER AND BECCA FLEISCHER

VOLUME 173. Biomembranes [Part T: Cellular and Subcellular Transport: Eukaryotic (Nonepithelial) Cells]
Edited by SIDNEY FLEISCHER AND BECCA FLEISCHER

VOLUME 174. Biomembranes [Part U: Cellular and Subcellular Transport: Eukaryotic (Nonepithelial) Cells]
Edited by SIDNEY FLEISCHER AND BECCA FLEISCHER

VOLUME 175. Cumulative Subject Index Volumes 135–139, 141–167

VOLUME 176. Nuclear Magnetic Resonance (Part A: Spectral Techniques and Dynamics)
Edited by NORMAN J. OPPENHEIMER AND THOMAS L. JAMES

VOLUME 177. Nuclear Magnetic Resonance (Part B: Structure and Mechanism)
Edited by NORMAN J. OPPENHEIMER AND THOMAS L. JAMES

VOLUME 178. Antibodies, Antigens, and Molecular Mimicry
Edited by JOHN J. LANGONE

VOLUME 179. Complex Carbohydrates (Part F)
Edited by VICTOR GINSBURG

VOLUME 180. RNA Processing (Part A: General Methods)
Edited by JAMES E. DAHLBERG AND JOHN N. ABELSON

VOLUME 181. RNA Processing (Part B: Specific Methods)
Edited by JAMES E. DAHLBERG AND JOHN N. ABELSON

VOLUME 182. Guide to Protein Purification
Edited by MURRAY P. DEUTSCHER

VOLUME 183. Molecular Evolution: Computer Analysis of Protein and Nucleic Acid Sequences
Edited by RUSSELL F. DOOLITTLE

VOLUME 184. Avidin-Biotin Technology
Edited by MEIR WILCHEK AND EDWARD A. BAYER

VOLUME 185. Gene Expression Technology
Edited by DAVID V. GOEDDEL

VOLUME 186. Oxygen Radicals in Biological Systems (Part B: Oxygen Radicals and Antioxidants)
Edited by LESTER PACKER AND ALEXANDER N. GLAZER

VOLUME 187. Arachidonate Related Lipid Mediators
Edited by ROBERT C. MURPHY AND FRANK A. FITZPATRICK

VOLUME 188. Hydrocarbons and Methylotrophy
Edited by MARY E. LIDSTROM

VOLUME 189. Retinoids (Part A: Molecular and Metabolic Aspects)
Edited by LESTER PACKER

VOLUME 190. Retinoids (Part B: Cell Differentiation and Clinical Applications)
Edited by LESTER PACKER

VOLUME 191. Biomembranes (Part V: Cellular and Subcellular Transport: Epithelial Cells)
Edited by SIDNEY FLEISCHER AND BECCA FLEISCHER

VOLUME 192. Biomembranes (Part W: Cellular and Subcellular Transport: Epithelial Cells)
Edited by SIDNEY FLEISCHER AND BECCA FLEISCHER

VOLUME 193. Mass Spectrometry
Edited by JAMES A. MCCLOSKEY

VOLUME 194. Guide to Yeast Genetics and Molecular Biology
Edited by CHRISTINE GUTHRIE AND GERALD R. FINK

VOLUME 195. Adenylyl Cyclase, G Proteins, and Guanylyl Cyclase
Edited by ROGER A. JOHNSON AND JACKIE D. CORBIN

VOLUME 196. Molecular Motors and the Cytoskeleton
Edited by RICHARD B. VALLEE

VOLUME 197. Phospholipases
Edited by EDWARD A. DENNIS

VOLUME 198. Peptide Growth Factors (Part C)
Edited by DAVID BARNES, J. P. MATHER, AND GORDON H. SATO

VOLUME 199. Cumulative Subject Index Volumes 168–174, 176–194

VOLUME 200. Protein Phosphorylation (Part A: Protein Kinases: Assays, Purification, Antibodies, Functional Analysis, Cloning, and Expression)
Edited by TONY HUNTER AND BARTHOLOMEW M. SEFTON

VOLUME 201. Protein Phosphorylation (Part B: Analysis of Protein Phosphorylation, Protein Kinase Inhibitors, and Protein Phosphatases)
Edited by TONY HUNTER AND BARTHOLOMEW M. SEFTON

VOLUME 202. Molecular Design and Modeling: Concepts and Applications (Part A: Proteins, Peptides, and Enzymes)
Edited by JOHN J. LANGONE

VOLUME 203. Molecular Design and Modeling: Concepts and Applications (Part B: Antibodies and Antigens, Nucleic Acids, Polysaccharides, and Drugs)
Edited by JOHN J. LANGONE

VOLUME 204. Bacterial Genetic Systems
Edited by JEFFREY H. MILLER

VOLUME 205. Metallobiochemistry (Part B: Metallothionein and Related Molecules)
Edited by JAMES F. RIORDAN AND BERT L. VALLEE

VOLUME 206. Cytochrome P450
Edited by MICHAEL R. WATERMAN AND ERIC F. JOHNSON

VOLUME 207. Ion Channels
Edited by BERNARDO RUDY AND LINDA E. IVERSON

VOLUME 208. Protein–DNA Interactions
Edited by ROBERT T. SAUER

VOLUME 209. Phospholipid Biosynthesis
Edited by EDWARD A. DENNIS AND DENNIS E. VANCE

VOLUME 210. Numerical Computer Methods
Edited by LUDWIG BRAND AND MICHAEL L. JOHNSON

VOLUME 211. DNA Structures (Part A: Synthesis and Physical Analysis of DNA)
Edited by DAVID M. J. LILLEY AND JAMES E. DAHLBERG

VOLUME 212. DNA Structures (Part B: Chemical and Electrophoretic Analysis of DNA)
Edited by DAVID M. J. LILLEY AND JAMES E. DAHLBERG

VOLUME 213. Carotenoids (Part A: Chemistry, Separation, Quantitation, and Antioxidation)
Edited by LESTER PACKER

VOLUME 214. Carotenoids (Part B: Metabolism, Genetics, and Biosynthesis)
Edited by LESTER PACKER

VOLUME 215. Platelets: Receptors, Adhesion, Secretion (Part B)
Edited by JACEK J. HAWIGER

VOLUME 216. Recombinant DNA (Part G)
Edited by RAY WU

VOLUME 217. Recombinant DNA (Part H)
Edited by RAY WU

VOLUME 218. Recombinant DNA (Part I)
Edited by RAY WU

VOLUME 219. Reconstitution of Intracellular Transport
Edited by JAMES E. ROTHMAN

VOLUME 220. Membrane Fusion Techniques (Part A)
Edited by NEJAT DÜZGÜNEŞ

VOLUME 221. Membrane Fusion Techniques (Part B)
Edited by NEJAT DÜZGÜNEŞ

VOLUME 222. Proteolytic Enzymes in Coagulation, Fibrinolysis, and Complement Activation (Part A: Mammalian Blood Coagulation Factors and Inhibitors)
Edited by LASZLO LORAND AND KENNETH G. MANN

VOLUME 223. Proteolytic Enzymes in Coagulation, Fibrinolysis, and Complement Activation (Part B: Complement Activation, Fibrinolysis, and Nonmammalian Blood Coagulation Factors)
Edited by LASZLO LORAND AND KENNETH G. MANN

VOLUME 224. Molecular Evolution: Producing the Biochemical Data
Edited by ELIZABETH ANNE ZIMMER, THOMAS J. WHITE, REBECCA L. CANN, AND ALLAN C. WILSON

VOLUME 225. Guide to Techniques in Mouse Development
Edited by PAUL M. WASSARMAN AND MELVIN L. DEPAMPHILIS

VOLUME 226. Metallobiochemistry (Part C: Spectroscopic and Physical Methods for Probing Metal Ion Environments in Metalloenzymes and Metalloproteins)
Edited by JAMES F. RIORDAN AND BERT L. VALLEE

VOLUME 227. Metallobiochemistry (Part D: Physical and Spectroscopic Methods for Probing Metal Ion Environments in Metalloproteins)
Edited by JAMES F. RIORDAN AND BERT L. VALLEE

VOLUME 228. Aqueous Two-Phase Systems
Edited by HARRY WALTER AND GÖTE JOHANSSON

VOLUME 229. Cumulative Subject Index Volumes 195–198, 200–227

VOLUME 230. Guide to Techniques in Glycobiology
Edited by WILLIAM J. LENNARZ AND GERALD W. HART

VOLUME 231. Hemoglobins (Part B: Biochemical and Analytical Methods)
Edited by JOHANNES EVERSE, KIM D. VANDEGRIFF, AND ROBERT M. WINSLOW

VOLUME 232. Hemoglobins (Part C: Biophysical Methods)
Edited by JOHANNES EVERSE, KIM D. VANDEGRIFF, AND ROBERT M. WINSLOW

VOLUME 233. Oxygen Radicals in Biological Systems (Part C)
Edited by LESTER PACKER

VOLUME 234. Oxygen Radicals in Biological Systems (Part D)
Edited by LESTER PACKER

VOLUME 235. Bacterial Pathogenesis (Part A: Identification and Regulation of Virulence Factors)
Edited by VIRGINIA L. CLARK AND PATRIK M. BAVOIL

VOLUME 236. Bacterial Pathogenesis (Part B: Integration of Pathogenic Bacteria with Host Cells)
Edited by VIRGINIA L. CLARK AND PATRIK M. BAVOIL

VOLUME 237. Heterotrimeric G Proteins
Edited by RAVI IYENGAR

VOLUME 238. Heterotrimeric G-Protein Effectors
Edited by RAVI IYENGAR

VOLUME 239. Nuclear Magnetic Resonance (Part C)
Edited by THOMAS L. JAMES AND NORMAN J. OPPENHEIMER

VOLUME 240. Numerical Computer Methods (Part B)
Edited by MICHAEL L. JOHNSON AND LUDWIG BRAND

VOLUME 241. Retroviral Proteases
Edited by LAWRENCE C. KUO AND JULES A. SHAFER

VOLUME 242. Neoglycoconjugates (Part A)
Edited by Y. C. LEE AND REIKO T. LEE

VOLUME 243. Inorganic Microbial Sulfur Metabolism
Edited by HARRY D. PECK, JR., AND JEAN LEGALL

VOLUME 244. Proteolytic Enzymes: Serine and Cysteine Peptidases
Edited by ALAN J. BARRETT

VOLUME 245. Extracellular Matrix Components
Edited by E. RUOSLAHTI AND E. ENGVALL

VOLUME 246. Biochemical Spectroscopy
Edited by KENNETH SAUER

VOLUME 247. Neoglycoconjugates (Part B: Biomedical Applications)
Edited by Y. C. LEE AND REIKO T. LEE

VOLUME 248. Proteolytic Enzymes: Aspartic and Metallo Peptidases
Edited by ALAN J. BARRETT

VOLUME 249. Enzyme Kinetics and Mechanism (Part D: Developments in Enzyme Dynamics)
Edited by DANIEL L. PURICH

VOLUME 250. Lipid Modifications of Proteins
Edited by PATRICK J. CASEY AND JANICE E. BUSS

VOLUME 251. Biothiols (Part A: Monothiols and Dithiols, Protein Thiols, and Thiyl Radicals)
Edited by LESTER PACKER

VOLUME 252. Biothiols (Part B: Glutathione and Thioredoxin; Thiols in Signal Transduction and Gene Regulation)
Edited by LESTER PACKER

VOLUME 253. Adhesion of Microbial Pathogens
Edited by RON J. DOYLE AND ITZHAK OFEK

VOLUME 254. Oncogene Techniques
Edited by PETER K. VOGT AND INDER M. VERMA

VOLUME 255. Small GTPases and Their Regulators (Part A: Ras Family)
Edited by W. E. BALCH, CHANNING J. DER, AND ALAN HALL

VOLUME 256. Small GTPases and Their Regulators (Part B: Rho Family)
Edited by W. E. BALCH, CHANNING J. DER, AND ALAN HALL

VOLUME 257. Small GTPases and Their Regulators (Part C: Proteins Involved in Transport)
Edited by W. E. BALCH, CHANNING J. DER, AND ALAN HALL

VOLUME 258. Redox-Active Amino Acids in Biology
Edited by JUDITH P. KLINMAN

VOLUME 259. Energetics of Biological Macromolecules
Edited by MICHAEL L. JOHNSON AND GARY K. ACKERS

VOLUME 260. Mitochondrial Biogenesis and Genetics (Part A)
Edited by GIUSEPPE M. ATTARDI AND ANNE CHOMYN

VOLUME 261. Nuclear Magnetic Resonance and Nucleic Acids
Edited by THOMAS L. JAMES

VOLUME 262. DNA Replication
Edited by JUDITH L. CAMPBELL

VOLUME 263. Plasma Lipoproteins (Part C: Quantitation)
Edited by WILLIAM A. BRADLEY, SANDRA H. GIANTURCO, AND JERE P. SEGREST

VOLUME 264. Mitochondrial Biogenesis and Genetics (Part B)
Edited by GIUSEPPE M. ATTARDI AND ANNE CHOMYN

VOLUME 265. Cumulative Subject Index Volumes 228, 230–262

VOLUME 266. Computer Methods for Macromolecular Sequence Analysis
Edited by RUSSELL F. DOOLITTLE

VOLUME 267. Combinatorial Chemistry
Edited by JOHN N. ABELSON

VOLUME 268. Nitric Oxide (Part A: Sources and Detection of NO; NO Synthase)
Edited by LESTER PACKER

VOLUME 269. Nitric Oxide (Part B: Physiological and Pathological Processes)
Edited by LESTER PACKER

VOLUME 270. High Resolution Separation and Analysis of Biological Macromolecules (Part A: Fundamentals)
Edited by BARRY L. KARGER AND WILLIAM S. HANCOCK

VOLUME 271. High Resolution Separation and Analysis of Biological Macromolecules (Part B: Applications)
Edited by BARRY L. KARGER AND WILLIAM S. HANCOCK

VOLUME 272. Cytochrome P450 (Part B)
Edited by ERIC F. JOHNSON AND MICHAEL R. WATERMAN

VOLUME 273. RNA Polymerase and Associated Factors (Part A)
Edited by SANKAR ADHYA

VOLUME 274. RNA Polymerase and Associated Factors (Part B)
Edited by SANKAR ADHYA

VOLUME 275. Viral Polymerases and Related Proteins
Edited by LAWRENCE C. KUO, DAVID B. OLSEN, AND STEVEN S. CARROLL

VOLUME 276. Macromolecular Crystallography (Part A)
Edited by CHARLES W. CARTER, JR., AND ROBERT M. SWEET

VOLUME 277. Macromolecular Crystallography (Part B)
Edited by CHARLES W. CARTER, JR., AND ROBERT M. SWEET

VOLUME 278. Fluorescence Spectroscopy
Edited by LUDWIG BRAND AND MICHAEL L. JOHNSON

VOLUME 279. Vitamins and Coenzymes (Part I)
Edited by DONALD B. MCCORMICK, JOHN W. SUTTIE, AND CONRAD WAGNER

VOLUME 280. Vitamins and Coenzymes (Part J)
Edited by DONALD B. MCCORMICK, JOHN W. SUTTIE, AND CONRAD WAGNER

VOLUME 281. Vitamins and Coenzymes (Part K)
Edited by DONALD B. MCCORMICK, JOHN W. SUTTIE, AND CONRAD WAGNER

VOLUME 282. Vitamins and Coenzymes (Part L)
Edited by DONALD B. MCCORMICK, JOHN W. SUTTIE, AND CONRAD WAGNER

VOLUME 283. Cell Cycle Control
Edited by WILLIAM G. DUNPHY

VOLUME 284. Lipases (Part A: Biotechnology)
Edited by BYRON RUBIN AND EDWARD A. DENNIS

VOLUME 285. Cumulative Subject Index Volumes 263, 264, 266–284, 286–289

VOLUME 286. Lipases (Part B: Enzyme Characterization and Utilization)
Edited by BYRON RUBIN AND EDWARD A. DENNIS

VOLUME 287. Chemokines
Edited by RICHARD HORUK

VOLUME 288. Chemokine Receptors
Edited by RICHARD HORUK

VOLUME 289. Solid Phase Peptide Synthesis
Edited by GREGG B. FIELDS

VOLUME 290. Molecular Chaperones
Edited by GEORGE H. LORIMER AND THOMAS BALDWIN

VOLUME 291. Caged Compounds
Edited by GERARD MARRIOTT

VOLUME 292. ABC Transporters: Biochemical, Cellular, and Molecular Aspects
Edited by SURESH V. AMBUDKAR AND MICHAEL M. GOTTESMAN

VOLUME 293. Ion Channels (Part B)
Edited by P. MICHAEL CONN

VOLUME 294. Ion Channels (Part C)
Edited by P. MICHAEL CONN

VOLUME 295. Energetics of Biological Macromolecules (Part B)
Edited by GARY K. ACKERS AND MICHAEL L. JOHNSON

VOLUME 296. Neurotransmitter Transporters
Edited by SUSAN G. AMARA

VOLUME 297. Photosynthesis: Molecular Biology of Energy Capture
Edited by LEE MCINTOSH

VOLUME 298. Molecular Motors and the Cytoskeleton (Part B)
Edited by RICHARD B. VALLEE

VOLUME 299. Oxidants and Antioxidants (Part A)
Edited by LESTER PACKER

VOLUME 300. Oxidants and Antioxidants (Part B)
Edited by LESTER PACKER

VOLUME 301. Nitric Oxide: Biological and Antioxidant Activities (Part C)
Edited by LESTER PACKER

VOLUME 302. Green Fluorescent Protein
Edited by P. MICHAEL CONN

VOLUME 303. cDNA Preparation and Display
Edited by SHERMAN M. WEISSMAN

VOLUME 304. Chromatin
Edited by PAUL M. WASSARMAN AND ALAN P. WOLFFE

VOLUME 305. Bioluminescence and Chemiluminescence (Part C)
Edited by THOMAS O. BALDWIN AND MIRIAM M. ZIEGLER

VOLUME 306. Expression of Recombinant Genes in Eukaryotic Systems
Edited by JOSEPH C. GLORIOSO AND MARTIN C. SCHMIDT

VOLUME 307. Confocal Microscopy
Edited by P. MICHAEL CONN

VOLUME 308. Enzyme Kinetics and Mechanism (Part E: Energetics of Enzyme Catalysis)
Edited by DANIEL L. PURICH AND VERN L. SCHRAMM

VOLUME 309. Amyloid, Prions, and Other Protein Aggregates
Edited by RONALD WETZEL

VOLUME 310. Biofilms
Edited by RON J. DOYLE

VOLUME 311. Sphingolipid Metabolism and Cell Signaling (Part A)
Edited by ALFRED H. MERRILL, JR., AND YUSUF A. HANNUN

VOLUME 312. Sphingolipid Metabolism and Cell Signaling (Part B)
Edited by ALFRED H. MERRILL, JR., AND YUSUF A. HANNUN

VOLUME 313. Antisense Technology
(Part A: General Methods, Methods of Delivery, and RNA Studies)
Edited by M. IAN PHILLIPS

VOLUME 314. Antisense Technology (Part B: Applications)
Edited by M. IAN PHILLIPS

VOLUME 315. Vertebrate Phototransduction and the Visual Cycle (Part A)
Edited by KRZYSZTOF PALCZEWSKI

VOLUME 316. Vertebrate Phototransduction and the Visual Cycle (Part B)
Edited by KRZYSZTOF PALCZEWSKI

VOLUME 317. RNA–Ligand Interactions (Part A: Structural Biology Methods)
Edited by DANIEL W. CELANDER AND JOHN N. ABELSON

VOLUME 318. RNA–Ligand Interactions (Part B: Molecular Biology Methods)
Edited by DANIEL W. CELANDER AND JOHN N. ABELSON

VOLUME 319. Singlet Oxygen, UV-A, and Ozone
Edited by LESTER PACKER AND HELMUT SIES

VOLUME 320. Cumulative Subject Index Volumes 290–319

VOLUME 321. Numerical Computer Methods (Part C)
Edited by MICHAEL L. JOHNSON AND LUDWIG BRAND

VOLUME 322. Apoptosis
Edited by JOHN C. REED

VOLUME 323. Energetics of Biological Macromolecules (Part C)
Edited by MICHAEL L. JOHNSON AND GARY K. ACKERS

VOLUME 324. Branched-Chain Amino Acids (Part B)
Edited by ROBERT A. HARRIS AND JOHN R. SOKATCH

VOLUME 325. Regulators and Effectors of Small GTPases
(Part D: Rho Family)
Edited by W. E. BALCH, CHANNING J. DER, AND ALAN HALL

VOLUME 326. Applications of Chimeric Genes and Hybrid Proteins
(Part A: Gene Expression and Protein Purification)
Edited by JEREMY THORNER, SCOTT D. EMR, AND JOHN N. ABELSON

VOLUME 327. Applications of Chimeric Genes and Hybrid Proteins
(Part B: Cell Biology and Physiology)
Edited by JEREMY THORNER, SCOTT D. EMR, AND JOHN N. ABELSON

VOLUME 328. Applications of Chimeric Genes and Hybrid Proteins (Part C: Protein–Protein Interactions and Genomics)
Edited by JEREMY THORNER, SCOTT D. EMR, AND JOHN N. ABELSON

VOLUME 329. Regulators and Effectors of Small GTPases (Part E: GTPases Involved in Vesicular Traffic)
Edited by W. E. BALCH, CHANNING J. DER, AND ALAN HALL

VOLUME 330. Hyperthermophilic Enzymes (Part A)
Edited by MICHAEL W. W. ADAMS AND ROBERT M. KELLY

VOLUME 331. Hyperthermophilic Enzymes (Part B)
Edited by MICHAEL W. W. ADAMS AND ROBERT M. KELLY

VOLUME 332. Regulators and Effectors of Small GTPases (Part F: Ras Family I)
Edited by W. E. BALCH, CHANNING J. DER, AND ALAN HALL

VOLUME 333. Regulators and Effectors of Small GTPases (Part G: Ras Family II)
Edited by W. E. BALCH, CHANNING J. DER, AND ALAN HALL

VOLUME 334. Hyperthermophilic Enzymes (Part C)
Edited by MICHAEL W. W. ADAMS AND ROBERT M. KELLY

VOLUME 335. Flavonoids and Other Polyphenols
Edited by LESTER PACKER

VOLUME 336. Microbial Growth in Biofilms (Part A: Developmental and Molecular Biological Aspects)
Edited by RON J. DOYLE

VOLUME 337. Microbial Growth in Biofilms (Part B: Special Environments and Physicochemical Aspects)
Edited by RON J. DOYLE

VOLUME 338. Nuclear Magnetic Resonance of Biological Macromolecules (Part A)
Edited by THOMAS L. JAMES, VOLKER DÖTSCH, AND ULI SCHMITZ

VOLUME 339. Nuclear Magnetic Resonance of Biological Macromolecules (Part B)
Edited by THOMAS L. JAMES, VOLKER DÖTSCH, AND ULI SCHMITZ

VOLUME 340. Drug–Nucleic Acid Interactions
Edited by JONATHAN B. CHAIRES AND MICHAEL J. WARING

VOLUME 341. Ribonucleases (Part A)
Edited by ALLEN W. NICHOLSON

VOLUME 342. Ribonucleases (Part B)
Edited by ALLEN W. NICHOLSON

VOLUME 343. G Protein Pathways (Part A: Receptors)
Edited by RAVI IYENGAR AND JOHN D. HILDEBRANDT

VOLUME 344. G Protein Pathways (Part B: G Proteins and Their Regulators)
Edited by RAVI IYENGAR AND JOHN D. HILDEBRANDT

VOLUME 345. G Protein Pathways (Part C: Effector Mechanisms)
Edited by RAVI IYENGAR AND JOHN D. HILDEBRANDT

VOLUME 346. Gene Therapy Methods
Edited by M. IAN PHILLIPS

VOLUME 347. Protein Sensors and Reactive Oxygen Species (Part A: Selenoproteins and Thioredoxin)
Edited by HELMUT SIES AND LESTER PACKER

VOLUME 348. Protein Sensors and Reactive Oxygen Species (Part B: Thiol Enzymes and Proteins)
Edited by HELMUT SIES AND LESTER PACKER

VOLUME 349. Superoxide Dismutase
Edited by LESTER PACKER

VOLUME 350. Guide to Yeast Genetics and Molecular and Cell Biology (Part B)
Edited by CHRISTINE GUTHRIE AND GERALD R. FINK

VOLUME 351. Guide to Yeast Genetics and Molecular and Cell Biology (Part C)
Edited by CHRISTINE GUTHRIE AND GERALD R. FINK

VOLUME 352. Redox Cell Biology and Genetics (Part A)
Edited by CHANDAN K. SEN AND LESTER PACKER

VOLUME 353. Redox Cell Biology and Genetics (Part B)
Edited by CHANDAN K. SEN AND LESTER PACKER

VOLUME 354. Enzyme Kinetics and Mechanisms (Part F: Detection and Characterization of Enzyme Reaction Intermediates)
Edited by DANIEL L. PURICH

VOLUME 355. Cumulative Subject Index Volumes 321–354

VOLUME 356. Laser Capture Microscopy and Microdissection
Edited by P. MICHAEL CONN

VOLUME 357. Cytochrome P450, Part C
Edited by ERIC F. JOHNSON AND MICHAEL R. WATERMAN

VOLUME 358. Bacterial Pathogenesis (Part C: Identification, Regulation, and Function of Virulence Factors)
Edited by VIRGINIA L. CLARK AND PATRIK M. BAVOIL

VOLUME 359. Nitric Oxide (Part D)
Edited by ENRIQUE CADENAS AND LESTER PACKER

VOLUME 360. Biophotonics (Part A)
Edited by GERARD MARRIOTT AND IAN PARKER

VOLUME 361. Biophotonics (Part B)
Edited by GERARD MARRIOTT AND IAN PARKER

VOLUME 362. Recognition of Carbohydrates in Biological Systems (Part A)
Edited by YUAN C. LEE AND REIKO T. LEE

VOLUME 363. Recognition of Carbohydrates in Biological Systems (Part B)
Edited by YUAN C. LEE AND REIKO T. LEE

VOLUME 364. Nuclear Receptors
Edited by DAVID W. RUSSELL AND DAVID J. MANGELSDORF

VOLUME 365. Differentiation of Embryonic Stem Cells
Edited by PAUL M. WASSAUMAN AND GORDON M. KELLER

VOLUME 366. Protein Phosphatases
Edited by SUSANNE KLUMPP AND JOSEF KRIEGLSTEIN

VOLUME 367. Liposomes (Part A)
Edited by NEJAT DÜZGÜNEŞ

VOLUME 368. Macromolecular Crystallography (Part C)
Edited by CHARLES W. CARTER, JR., AND ROBERT M. SWEET

VOLUME 369. Combinational Chemistry (Part B)
Edited by GUILLERMO A. MORALES AND BARRY A. BUNIN

VOLUME 370. RNA Polymerases and Associated Factors (Part C)
Edited by SANKAR L. ADHYA AND SUSAN GARGES

VOLUME 371. RNA Polymerases and Associated Factors (Part D)
Edited by SANKAR L. ADHYA AND SUSAN GARGES

VOLUME 372. Liposomes (Part B)
Edited by NEJAT DÜZGÜNEŞ

VOLUME 373. Liposomes (Part C)
Edited by NEJAT DÜZGÜNEŞ

VOLUME 374. Macromolecular Crystallography (Part D)
Edited by CHARLES W. CARTER, JR., AND ROBERT W. SWEET

VOLUME 375. Chromatin and Chromatin Remodeling Enzymes (Part A)
Edited by C. DAVID ALLIS AND CARL WU

VOLUME 376. Chromatin and Chromatin Remodeling Enzymes (Part B)
Edited by C. DAVID ALLIS AND CARL WU

VOLUME 377. Chromatin and Chromatin Remodeling Enzymes (Part C)
Edited by C. DAVID ALLIS AND CARL WU

VOLUME 378. Quinones and Quinone Enzymes (Part A)
Edited by HELMUT SIES AND LESTER PACKER

VOLUME 379. Energetics of Biological Macromolecules (Part D)
Edited by JO M. HOLT, MICHAEL L. JOHNSON, AND GARY K. ACKERS

VOLUME 380. Energetics of Biological Macromolecules (Part E)
Edited by JO M. HOLT, MICHAEL L. JOHNSON, AND GARY K. ACKERS

VOLUME 381. Oxygen Sensing
Edited by CHANDAN K. SEN AND GREGG L. SEMENZA

VOLUME 382. Quinones and Quinone Enzymes (Part B)
Edited by HELMUT SIES AND LESTER PACKER

VOLUME 383. Numerical Computer Methods (Part D)
Edited by LUDWIG BRAND AND MICHAEL L. JOHNSON

VOLUME 384. Numerical Computer Methods (Part E)
Edited by LUDWIG BRAND AND MICHAEL L. JOHNSON

VOLUME 385. Imaging in Biological Research (Part A)
Edited by P. MICHAEL CONN

VOLUME 386. Imaging in Biological Research (Part B)
Edited by P. MICHAEL CONN

VOLUME 387. Liposomes (Part D)
Edited by NEJAT DÜZGÜNEŞ

VOLUME 388. Protein Engineering
Edited by DAN E. ROBERTSON AND JOSEPH P. NOEL

VOLUME 389. Regulators of G-Protein Signaling (Part A)
Edited by DAVID P. SIDEROVSKI

VOLUME 390. Regulators of G-Protein Signaling (Part B)
Edited by DAVID P. SIDEROVSKI

VOLUME 391. Liposomes (Part E)
Edited by NEJAT DÜZGÜNEŞ

VOLUME 392. RNA Interference
Edited by ENGELKE ROSSI

VOLUME 393. Circadian Rhythms
Edited by MICHAEL W. YOUNG

VOLUME 394. Nuclear Magnetic Resonance of Biological Macromolecules (Part C)
Edited by THOMAS L. JAMES

VOLUME 395. Producing the Biochemical Data (Part B)
Edited by ELIZABETH A. ZIMMER AND ERIC H. ROALSON

VOLUME 396. Nitric Oxide (Part E)
Edited by LESTER PACKER AND ENRIQUE CADENAS

VOLUME 397. Environmental Microbiology
Edited by JARED R. LEADBETTER

VOLUME 398. Ubiquitin and Protein Degradation (Part A)
Edited by RAYMOND J. DESHAIES

VOLUME 399. Ubiquitin and Protein Degradation (Part B)
Edited by RAYMOND J. DESHAIES

VOLUME 400. Phase II Conjugation Enzymes and Transport Systems
Edited by HELMUT SIES AND LESTER PACKER

VOLUME 401. Glutathione Transferases and Gamma Glutamyl Transpeptidases
Edited by HELMUT SIES AND LESTER PACKER

VOLUME 402. Biological Mass Spectrometry
Edited by A. L. BURLINGAME

VOLUME 403. GTPases Regulating Membrane Targeting and Fusion
Edited by WILLIAM E. BALCH, CHANNING J. DER, AND ALAN HALL

VOLUME 404. GTPases Regulating Membrane Dynamics
Edited by WILLIAM E. BALCH, CHANNING J. DER, AND ALAN HALL

VOLUME 405. Mass Spectrometry: Modified Proteins and Glycoconjugates
Edited by A. L. BURLINGAME

VOLUME 406. Regulators and Effectors of Small GTPases: Rho Family
Edited by WILLIAM E. BALCH, CHANNING J. DER, AND ALAN HALL

VOLUME 407. Regulators and Effectors of Small GTPases: Ras Family
Edited by WILLIAM E. BALCH, CHANNING J. DER, AND ALAN HALL

VOLUME 408. DNA Repair (Part A)
Edited by JUDITH L. CAMPBELL AND PAUL MODRICH

VOLUME 409. DNA Repair (Part B)
Edited by JUDITH L. CAMPBELL AND PAUL MODRICH

VOLUME 410. DNA Microarrays (Part A: Array Platforms and Web-Bench Protocols)
Edited by ALAN KIMMEL AND BRIAN OLIVER

VOLUME 411. DNA Microarrays (Part B: Databases and Statistics)
Edited by ALAN KIMMEL AND BRIAN OLIVER

VOLUME 412. Amyloid, Prions, and Other Protein Aggregates (Part B)
Edited by INDU KHETERPAL AND RONALD WETZEL

VOLUME 413. Amyloid, Prions, and Other Protein Aggregates (Part C)
Edited by INDU KHETERPAL AND RONALD WETZEL

VOLUME 414. Measuring Biological Responses with Automated Microscopy
Edited by JAMES INGLESE

VOLUME 415. Glycobiology
Edited by MINORU FUKUDA

VOLUME 416. Glycomics
Edited by MINORU FUKUDA

VOLUME 417. Functional Glycomics
Edited by MINORU FUKUDA

VOLUME 418. Embryonic Stem Cells
Edited by IRINA KLIMANSKAYA AND ROBERT LANZA

VOLUME 419. Adult Stem Cells
Edited by IRINA KLIMANSKAYA AND ROBERT LANZA

VOLUME 420. Stem Cell Tools and Other Experimental Protocols
Edited by IRINA KLIMANSKAYA AND ROBERT LANZA

VOLUME 421. Advanced Bacterial Genetics: Use of Transposons and Phage for Genomic Engineering
Edited by KELLY T. HUGHES

VOLUME 422. Two-Component Signaling Systems, Part A
Edited by MELVIN I. SIMON, BRIAN R. CRANE, AND ALEXANDRINE CRANE

VOLUME 423. Two-Component Signaling Systems, Part B
Edited by MELVIN I. SIMON, BRIAN R. CRANE, AND ALEXANDRINE CRANE

VOLUME 424. RNA Editing
Edited by JONATHA M. GOTT

VOLUME 425. RNA Modification
Edited by JONATHA M. GOTT

VOLUME 426. Integrins
Edited by DAVID CHERESH

VOLUME 427. MicroRNA Methods
Edited by JOHN J. ROSSI

VOLUME 428. Osmosensing and Osmosignaling
Edited by HELMUT SIES AND DIETER HAUSSINGER

VOLUME 429. Translation Initiation: Extract Systems and Molecular Genetics
Edited by JON LORSCH

VOLUME 430. Translation Initiation: Reconstituted Systems and Biophysical Methods
Edited by JON LORSCH

VOLUME 431. Translation Initiation: Cell Biology, High-Throughput and Chemical-Based Approaches
Edited by JON LORSCH

VOLUME 432. Lipidomics and Bioactive Lipids: Mass-Spectrometry–Based Lipid Analysis
Edited by H. ALEX BROWN

VOLUME 433. Lipidomics and Bioactive Lipids: Specialized Analytical Methods and Lipids in Disease
Edited by H. ALEX BROWN

VOLUME 434. Lipidomics and Bioactive Lipids: Lipids and Cell Signaling
Edited by H. ALEX BROWN

VOLUME 435. Oxygen Biology and Hypoxia
Edited by HELMUT SIES AND BERNHARD BRÜNE

VOLUME 436. Globins and Other Nitric Oxide-Reactive Protiens (Part A)
Edited by ROBERT K. POOLE

VOLUME 437. Globins and Other Nitric Oxide-Reactive Protiens (Part B)
Edited by ROBERT K. POOLE

VOLUME 438. Small GTPases in Disease (Part A)
Edited by WILLIAM E. BALCH, CHANNING J. DER, AND ALAN HALL

VOLUME 439. Small GTPases in Disease (Part B)
Edited by WILLIAM E. BALCH, CHANNING J. DER, AND ALAN HALL

VOLUME 440. Nitric Oxide, Part F Oxidative and Nitrosative Stress in Redox Regulation of Cell Signaling
Edited by ENRIQUE CADENAS AND LESTER PACKER

VOLUME 441. Nitric Oxide, Part G Oxidative and Nitrosative Stress in Redox Regulation of Cell Signaling
Edited by ENRIQUE CADENAS AND LESTER PACKER

VOLUME 442. Programmed Cell Death, General Principles for Studying Cell Death (Part A)
Edited by ROYA KHOSRAVI-FAR, ZAHRA ZAKERI, RICHARD A. LOCKSHIN, AND MAURO PIACENTINI

VOLUME 443. Angiogenesis: *In Vitro* Systems
Edited by DAVID A. CHERESH

VOLUME 444. Angiogenesis: *In Vivo* Systems (Part A)
Edited by DAVID A. CHERESH

VOLUME 445. Angiogenesis: *In Vivo* Systems (Part B)
Edited by DAVID A. CHERESH

VOLUME 446. Programmed Cell Death, The Biology and Therapeutic Implications of Cell Death (Part B)
Edited by ROYA KHOSRAVI-FAR, ZAHRA ZAKERI, RICHARD A. LOCKSHIN, AND MAURO PIACENTINI

VOLUME 447. RNA Turnover in Bacteria, Archaea and Organelles
Edited by LYNNE E. MAQUAT AND CECILIA M. ARRAIANO

VOLUME 448. RNA Turnover in Eukaryotes: Nucleases, Pathways and Analysis of mRNA Decay
Edited by LYNNE E. MAQUAT AND MEGERDITCH KILEDJIAN

VOLUME 449. RNA Turnover in Eukaryotes: Analysis of Specialized and Quality Control RNA Decay Pathways
Edited by LYNNE E. MAQUAT AND MEGERDITCH KILEDJIAN

VOLUME 450. Fluorescence Spectroscopy
Edited by LUDWIG BRAND AND MICHAEL L. JOHNSON

VOLUME 451. Autophagy: Lower Eukaryotes and Non-Mammalian Systems (Part A)
Edited by DANIEL J. KLIONSKY

VOLUME 452. Autophagy in Mammalian Systems (Part B)
Edited by DANIEL J. KLIONSKY

VOLUME 453. Autophagy in Disease and Clinical Applications (Part C)
Edited by DANIEL J. KLIONSKY

VOLUME 454. Computer Methods (Part A)
Edited by MICHAEL L. JOHNSON AND LUDWIG BRAND

VOLUME 455. Biothermodynamics (Part A)
Edited by MICHAEL L. JOHNSON, JO M. HOLT, AND GARY K. ACKERS (RETIRED)

VOLUME 456. Mitochondrial Function, Part A: Mitochondrial Electron Transport Complexes and Reactive Oxygen Species
Edited by WILLIAM S. ALLISON AND IMMO E. SCHEFFLER

VOLUME 457. Mitochondrial Function, Part B: Mitochondrial Protein Kinases, Protein Phosphatases and Mitochondrial Diseases
Edited by WILLIAM S. ALLISON AND ANNE N. MURPHY

VOLUME 458. Complex Enzymes in Microbial Natural Product Biosynthesis, Part A: Overview Articles and Peptides
Edited by DAVID A. HOPWOOD

VOLUME 459. Complex Enzymes in Microbial Natural Product Biosynthesis, Part B: Polyketides, Aminocoumarins and Carbohydrates
Edited by DAVID A. HOPWOOD

VOLUME 460. Chemokines, Part A
Edited by TRACY M. HANDEL AND DAMON J. HAMEL

VOLUME 461. Chemokines, Part B
Edited by TRACY M. HANDEL AND DAMON J. HAMEL

VOLUME 462. Non-Natural Amino Acids
Edited by TOM W. MUIR AND JOHN N. ABELSON

VOLUME 463. Guide to Protein Purification, 2nd Edition
Edited by RICHARD R. BURGESS AND MURRAY P. DEUTSCHER

VOLUME 464. Liposomes, Part F
Edited by NEJAT DÜZGÜNEŞ

VOLUME 465. Liposomes, Part G
Edited by NEJAT DÜZGÜNEŞ

VOLUME 466. Biothermodynamics, Part B
Edited by MICHAEL L. JOHNSON, GARY K. ACKERS, AND JO M. HOLT

VOLUME 467. Computer Methods Part B
Edited by MICHAEL L. JOHNSON AND LUDWIG BRAND

VOLUME 468. Biophysical, Chemical, and Functional Probes of RNA Structure, Interactions and Folding: Part A
Edited by DANIEL HERSCHLAG

VOLUME 469. Biophysical, Chemical, and Functional Probes of RNA Structure, Interactions and Folding: Part B
Edited by DANIEL HERSCHLAG

VOLUME 470. Guide to Yeast Genetics: Functional Genomics, Proteomics, and Other Systems Analysis, 2nd Edition
Edited by GERALD FINK, JONATHAN WEISSMAN, AND CHRISTINE GUTHRIE

VOLUME 471. Two-Component Signaling Systems, Part C
Edited by MELVIN I. SIMON, BRIAN R. CRANE, AND ALEXANDRINE CRANE

VOLUME 472. Single Molecule Tools, Part A: Fluorescence Based Approaches
Edited by NILS G. WALTER

VOLUME 473. Thiol Redox Transitions in Cell Signaling, Part A Chemistry and Biochemistry of Low Molecular Weight and Protein Thiols
Edited by ENRIQUE CADENAS AND LESTER PACKER

VOLUME 474. Thiol Redox Transitions in Cell Signaling, Part B Cellular Localization and Signaling
Edited by ENRIQUE CADENAS AND LESTER PACKER

VOLUME 475. Single Molecule Tools, Part B: Super-Resolution, Particle Tracking, Multiparameter, and Force Based Methods
Edited by NILS G. WALTER

VOLUME 476. Guide to Techniques in Mouse Development, Part A Mice, Embryos, and Cells, 2nd Edition
Edited by PAUL M. WASSARMAN AND PHILIPPE M. SORIANO

VOLUME 477. Guide to Techniques in Mouse Development, Part B Mouse Molecular Genetics, 2nd Edition
Edited by PAUL M. WASSARMAN AND PHILIPPE M. SORIANO

VOLUME 478. Glycomics
Edited by MINORU FUKUDA

VOLUME 479. Functional Glycomics
Edited by MINORU FUKUDA

VOLUME 480. Glycobiology
Edited by MINORU FUKUDA

VOLUME 481. Cryo-EM, Part A: Sample Preparation and Data Collection
Edited by GRANT J. JENSEN

VOLUME 482. Cryo-EM, Part B: 3-D Reconstruction
Edited by GRANT J. JENSEN

VOLUME 483. Cryo-EM, Part C: Analyses, Interpretation, and Case Studies
Edited by GRANT J. JENSEN

VOLUME 484. Constitutive Activity in Receptors and Other Proteins, Part A
Edited by P. MICHAEL CONN

VOLUME 485. Constitutive Activity in Receptors and Other Proteins, Part B
Edited by P. MICHAEL CONN

VOLUME 486. Research on Nitrification and Related Processes, Part A
Edited by MARTIN G. KLOTZ

VOLUME 487. Computer Methods, Part C
Edited by MICHAEL L. JOHNSON AND LUDWIG BRAND

VOLUME 488. Biothermodynamics, Part C
Edited by MICHAEL L. JOHNSON, AND JO M. HOLT AND GARY K. ACKERS

VOLUME 489. The Unfolded Protein Response and Cellular Stress, Part A
Edited by P. MICHAEL CONN

VOLUME 490. The Unfolded Protein Response and Cellular Stress, Part B
Edited by P. MICHAEL CONN

VOLUME 491. The Unfolded Protein Response and Cellular Stress, Part C
Edited by P. MICHAEL CONN

SECTION ONE

NEW APPROACHES TO STUDYING UPR AND CELL STRESS

CHAPTER ONE

CFTR Expression Regulation by the Unfolded Protein Response

Rafal Bartoszewski, Andras Rab, Lianwu Fu,
Sylwia Bartoszewska, James Collawn, *and* Zsuzsa Bebok

Contents

1. Introduction	4
2. Methods	6
2.1. ER stress induction and the assessment of UPR activity	6
2.2. The effects of the UPR on CFTR mRNA levels	10
2.3. Assays for *CFTR* transcriptional regulation	13
2.4. Assessment of CFTR protein levels	18
Acknowledgments	22
References	22

Abstract

The cystic fibrosis transmembrane conductance regulator (CFTR) is a chloride channel and key regulator of epithelial functions. Mutations in the *CFTR* gene lead to reduced or dysfunctional CFTR protein and cause cystic fibrosis (CF), a generalized exocrinopathy affecting multiple organs. In the airways, loss of CFTR function leads to thickened mucus, reduced mucociliary clearance, chronic infections, and respiratory failure. Common airway disorders such as bronchitis and chronic obstructive pulmonary disease (COPD) also present CF-like symptoms such as mucus congestion and chronic inflammation without mutations in *CFTR*. The primary risk factors for COPD and chronic bronchitis include environmental stress insults such as pollutants and infections that often result in hypoxic conditions. Furthermore, environmental factors such as cigarette smoke and reactive oxygen species have been implicated in reduced CFTR function. Activation of cellular stress responses by these factors promotes differential, stress-associated gene expression regulation. During our investigations on the mechanisms of CFTR expression regulation, we have shown that the ER stress response, the unfolded protein response (UPR), decreases CFTR

Department of Cell Biology, University of Alabama at Birmingham, Birmingham, Alabama, USA expression at the transcriptional, translational, and maturational levels. Here, we provide a detailed description of the methods we employ to study CFTR expression regulation by the UPR. Similar approaches are applicable in studies on other genes and how they are affected by the UPR.

1. INTRODUCTION

Mutations in the cystic fibrosis transmembrane conductance regulator (*CFTR*) gene result in reduced or dysfunctional CFTR protein and cause cystic fibrosis (CF). CFTR is a multidomain, membrane glycoprotein that functions as a chloride channel at the cell surface (Riordan and Chang, 1992; Schwiebert *et al.*, 1999). CFTR has also been shown to regulate the epithelial sodium channel (ENaC; Berdiev *et al.*, 2009; Ismailov *et al.*, 1996), plasma membrane recycling (Bradbury *et al.*, 1992), macrophage function (Di *et al.*, 2006), and plays a significant role in the assembly of cell surface signaling complexes in epithelial cells (Naren *et al.*, 2003). Since the discovery of the *CFTR* gene in 1989, it has become evident that proper CFTR function is crucial for the physiology of epithelial tissues in multiple organs with the most significant CFTR function in the excretory glands (for review, Bebok and Collawn, 2006; Quinton, 2007; Riordan, 2008).

The majority of CF (>75%) is caused by the deletion of three nucleotides resulting in the loss of phenylalanine at the 508 position of the CFTR protein (ΔF508 CFTR; Kerem *et al.*, 1989). ΔF508 CFTR misfolds during biogenesis and is eliminated by the endoplasmic reticulum–associated degradation (ERAD) machinery (Cheng *et al.*, 1990; Ward *et al.*, 1995). Premature termination codons (PTC), UAA, UAG, or UGA are responsible for ∼10% of CF cases (Collawn *et al.*, 2010; Kerem *et al.*, 2008). PTCs generate unstable, truncated mRNA, from which nonfunctional or toxic proteins are translated (Frischmeyer and Dietz, 1999). A small fraction of CF cases (<10%) is caused by a variety of more than 1700 rare mutations in *CFTR* (http://www.genet.sickkids.on.ca/cftr). Because the outcome of the mutations is reduced CFTR function, determining the level of CFTR necessary to ameliorate the symptoms of CF and understanding CFTR expression regulation during pathological conditions will be crucial for the development of any potential therapy.

The most prevalent complications of CF develop in the lungs and gastrointestinal tract. While symptomatic treatments have significantly improved the gastrointestinal complications, in the airways, mucus congestion and reduced mucociliary clearance promote the development of chronic bacterial infections that lead to respiratory failure (for review,

Collawn et al., 2010; Rowe et al., 2005). Chronic airway disorders such as chronic bronchitis and COPD present CF-like symptoms that are characterized by thick mucus, reduced mucociliary clearance, and inflammation (Liu et al., 2005; O'Byrne and Postma, 1999). Therefore, *CFTR* has become one of the central genetic risk factors associated with asthma and COPD (Dahl and Nordestgaard, 2009; van der Deen et al., 2005). However, most of the previous studies have concentrated on mutations and genetic variations without considering the effects of environmental factors and their role in epigenetic gene expression regulation.

Our initial studies suggested that CFTR expression is regulated by cell culture conditions (Bebok et al., 2001) and epithelial cell differentiation (Bebok et al., 1998). It is now clear that cigarette smoke (Cantin et al., 2006b), reactive oxygen nitrogen species (RONS; Bebok et al., 2002), oxidant stress (Cantin et al., 2006a), and hypoxia (Guimbellot et al., 2008) promote COPD and reduce CFTR expression and function. Furthermore, the environmental factors that reduce CFTR expression also activate cellular stress responses that promote cell recovery and reestablish physiological function (Kultz, 2005). Therefore, these studies emphasized the importance for understanding the mechanism of endogenous CFTR expression regulation by cellular stress responses in epithelial cells.

Based on the central role of the ER in membrane protein synthesis, we focused our efforts on the ER and the ER stress response, the UPR (Schroder and Kaufman, 2005). Our studies indicated that the UPR reduced endogenous, but not recombinant CFTR mRNA levels, and this effect was not related to mRNA stability (Rab et al., 2007). Subsequent studies confirmed that CFTR transcription is inhibited through epigenetic mechanisms triggered by the UPR and involve the binding of ATF6 to the minimal promoter of *CFTR*, as well as histone deacetylation and promoter methylation. We now know that transcriptional repression as well as enhanced ERAD (reduced maturation) minimize cell surface CFTR levels and function (Bartoszewski et al., 2008b). Surprisingly, the UPR-associated transcriptional repression of genes is not a generalized process since only a small fraction of genes is repressed during ER stress (Bartoszewski and Bebok, unpublished observations).

Considering the fact that asthma, chronic bronchitis, and COPD are the most frequent chronic airway disorders associated with cellular stress, understanding how the expression of key transporters such as CFTR is regulated during cellular stress responses is an important consideration in any potential therapeutic intervention. Here, we provide a detailed technical roadmap for investigating CFTR expression regulation by the UPR. A model illustrating UPR-associated CFTR expression regulation is presented in Fig. 1.1.

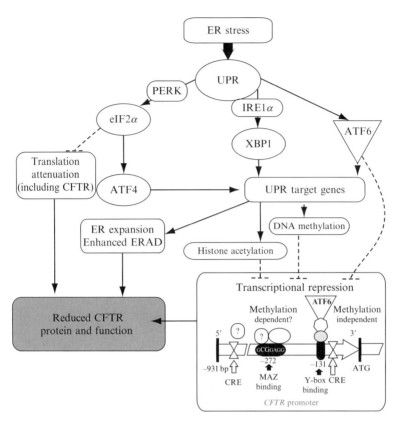

Figure 1.1 Regulation of CFTR expression by the UPR. The three main pathways of the mammalian UPR and their identified associations with endogenous CFTR expression regulation are shown (based on Bartoszewski et al., 2008b; Rab et al., 2007). CRE, cAMP response element; MAZ, Myc-associated zinc finger protein binding; Y-box binding, Y-box protein binding; ATG, transcription start.

2. Methods

2.1. ER stress induction and the assessment of UPR activity

Several methods that interfere with normal cellular processes have been shown to induce ER stress and activate the UPR. Here, we describe the use of proteasome and glycosylation inhibitors as ER stress inducers.

2.1.1. ER stress induction

Proteasome inhibition (ALLN or epoxomycin; Hong et al., 2004; Rab et al., 2007) or glycosylation inhibition (tunicamycin (TM); Bartoszewski et al., 2008a; Hung et al., 2004) can be used to induce ER stress. Cells are grown under standard tissue culture conditions. The number of cells used for each experiment should be determined based on RNA content (see below). Time course studies in several cell lines indicate that 100 μM ALLN (Sigma) or 20 μM epoxomycin (BIOMOL) treatment for 8–12 h is sufficient to induce ER stress, activate the UPR and cause significant reduction in CFTR mRNA levels (Bartoszewski et al., 2008a,b; Rab et al., 2007). TM (Sigma) treatment for 12–14 h at 5 µg/ml (final concentration) is sufficient in most cases (Bartoszewski et al., 2008a,b; Nadanaka et al., 2004; Rab et al., 2007).

2.1.2. RNA isolation from human airway epithelial cells

For all mRNA measurements, total RNA is isolated from control and ER-stressed cells. UPR reporters, internal controls, and CFTR mRNA are amplified using specific probes described below. Here, we describe the steps of total RNA isolation.

2.1.2.1. Materials and equipment RNeasy Mini Kit (Qiagen), buffers (RLT, RW1, RPE are supplied with kit), PBS (KCl 2.7 mM, KH_2PO_4 1.5 mM, NaCl 138 mM, $Na_2HPO_4 \cdot 7H_2O$ 8 mM, pH 7.4, Gibco), RNase-free DNase (Qiagen), 14.3 M β-mercaptoethanol, sterile, RNase-free pipette tips (ART), microcentrifuge (with rotor for 2 ml tubes), 96–100% ethanol (molecular biology grade, Sigma), 70% ethanol (molecular biology grade, Sigma), 10-cm cell culture dishes (Corning), tissue culture media (cell type specific). All materials are sterile and RNase free.

2.1.2.2. Before starting Determine the sufficient number of cells for each experiment. Based on the binding capacity of the RNeasy spin column (100 µg of RNA), the minimum number is $\sim$100 cells and the maximum number of cells depends on the RNA content. Examples: (1) high RNA content (COS-7 cells, $\sim$35 µg of RNA/10^6 cells); (2) average RNA content (HeLa and Calu-3 cells, $\sim$15 µg of RNA/10^6 cells); (3) low RNA content (NIH/3 T3, $\sim$10 µg of RNA/10^6 cells). The starting cell number should not exceed 3–4 × 10^6 cells. The cell number can be determined in pilot experiments. The optimal confluency of cells for the experiments should also be determined based on the time required for treatments (e.g., ER stress induction). Cell confluency should not exceed 90% at the time of RNA isolation in order to maximize the RNA content.

2.1.2.3. RNA isolation Isolate RNA as described in the Qiagen RNeasy Mini handbook (04/2006, pp. 27–30). The individual steps are described below. RDD, RPE, RLT, and RW1 buffers are supplied with the kit. All experimental steps are performed at room temperature unless stated otherwise.

Remove media from cells (~90% confluent, grown on 100-mm cell culture dishes, ~4 × 10^6 cells or 90–100 μg of RNA); wash the cells twice with ice-cold PBS (2 ml/wash). Lyse the cells in 600 μl of RLT buffer (prior to use add 10 μl of β-mercaptoethanol/ml of RLT buffer) and collect the lysate with a pipette. The cell lysate can be stored at −70 °C for several months. Prior to further processing, samples should be thawed quickly at 37 °C and centrifuged for 5 min at 5000×g to remove the insoluble material. Add an equal volume of 70% ethanol (~600 μl) to the cell lysate, mix by pipetting, transfer the resulting mixture in 700 μl aliquots onto RNeasy spin columns and centrifuge for 30 s at 8000×g. Discard the flow-through material. Wash the columns with 350 μl RW1 buffer (centrifuge for 30 s at 8000×g and discard the flow-through material). Add 80 μl, freshly prepared DNase I incubation mix (10 μl DNase I stock solution, 3 K units/μl mixed with 70 μl RDD buffer, Qiagen) to the columns. Incubate the columns with DNase I-containing solution for 20 min at room temperature. Wash the columns with another 350 μl of RW1 buffer (Qiagen) and place the columns in new collection tubes (supplied with the kit) and wash with 500 μl of RPE (Qiagen) buffer. Centrifuge 15 s at 8000×g and discard the flow-through material. Repeat the 500 μl of RPE buffer wash step, centrifuge for 2 min at maximum speed, and discard the flow-through material. To ensure removal of residual ethanol from the column, spin the columns for additional 2 min at maximum speed. Place the columns in fresh collection tubes and add 30 μl of RNase-free water to the center. Incubate at room temperature for 1 min and centrifuge for 1 min at maximum speed to elute RNA. Discard the columns and measure RNA concentration and purity in the eluted solution by measuring absorbance at 260 (A_{260}) and 280 nm (A_{280}). At neutral pH, an A_{260} reading of 1 equals 44 μg/ml RNA. The A_{260}/A_{280} ratio for pure RNA is between 1.9 and 2.1.

2.1.3. Semiquantitative RT-PCR using ABI 7500 Real-Time PCR System

Equipment and materials

ABI 7500 Real-Time PCR System, RNase-free sterile pipette tips, RNase-free water, 96-well reaction plate (ABI), TaqMan One-Step RT-PCR Master Mix Reagents (ABI), 10-cm cell culture dish (Corning). All steps are performed on ice unless stated otherwise. The optimum RNA concentration in the samples is 50 ng/μl. RNase-free water can be added to adjust RNA concentration. Using 50 ng/μl RNA stock solutions will result in 10 ng/μl final RNA concentration in the reaction mixture. This total RNA concentration allows efficient *CFTR* mRNA quantification from a broad spectrum of epithelial cell lines such as

Calu-3, T-84, CFPAC-1 (Bartoszewski et al., 2008b; Rab et al., 2007). All samples, including internal controls, should be run in triplicate. Experiments include a 5-point, logarithmic standard sample dilution row (10^0–10^{-5}) for both *CFTR* and the endogenous control. When cells with minimal CFTR expression are tested, a standard dilution row of 10^0–$10^{-2.5}$ should be added. The standard curves aid in determining primer efficiency and calculation of *CFTR* mRNA levels. Controls without mRNA (no template controls, NTC) should also be included.

The reaction mixture consists of 5 µl total mRNA template; 12.5 µl, 2× Master Mix (ABI); 0.625 µl, 40× Multiscribe RNase Inhibitor Mix (ABI); 1.25 µl, CFTR or endogenous control primer probes (20×); and 5.625 µl RNase-free water. Calculate the volume of RT-PCR reactions and prepare stock solution for each primer (UPR reporter, CFTR, and internal control). Mix all compounds except the RNA template and dispense 20 µl of the mix into individual wells of the PCR plate. Add 5 µl of the total RNA templates. Centrifuge the plate briefly and perform one-step RT-PCR under the following conditions: reverse transcription at 48 °C for 30 min, DNA polymerase activation at 95 °C for 10 min, followed by 40 cycles of denaturation at 95 °C for 15 s and annealing extension at 60 °C for 1 min. Fluorescence data collection is performed during the 60 °C step. Calculate the *CFTR* mRNA levels relative to endogenous control using the relative standard curve method (ABI 2004 Guide to Performing Relative Quantification of Gene Expression Using Real-Time Quantitative PCR).

2.1.4. Assessment of UPR reporter (spliced XBP1 and BiP/HSPA5) mRNA levels by semiquantitative RT-PCR

To determine the effects of ER stress on CFTR expression, we first monitor changes in UPR reporters to confirm that ER stress has occurred, the UPR is activated, and then assay for changes in CFTR mRNA levels (Rab et al., 2007). We routinely measure spliced XBP1 (sXBP) and BiP mRNA levels as reporters of UPR activity (Bartoszewski et al., 2008a,b; Rab et al., 2007). The mammalian UPR is activated through the ER-resident reporters PERK, ATF6, and IRE1α (Schroder and Kaufman, 2005). When the UPR is activated, IRE1α splices the XBP1 mRNA to produce the XBP transcription factor (Yoshida et al., 2001). Activation of all three pathways leads to increased chaperone levels including BiP (Lee, 2005).

Primer probes

HSPA5/BiP (assay ID: Hs00607129_gH) and sXBP1 (assay ID: Hs00231936_m1) are used as UPR reporters. GAPDH (ABI, assay ID: Hs99999905_m1) or 18S rRNA (ABI, assay ID: HS_99999901_m1) are used as internal controls.

2.2. The effects of the UPR on CFTR mRNA levels

Stable, recombinant CFTR-expressing cell lines are widely used to study different aspects of CFTR biogenesis and proteomics (Bebok et al., 2005; Carvalho-Oliveira et al., 2004; Gomes-Alves et al., 2009; Sheppard et al., 1994). However, these models are not suitable for studying transcriptional regulation of CFTR since the recombinant construct lacks the native promoter and regulatory regions. In our studies, we have tested a combination of recombinant and endogenous CFTR-expressing cell lines in order to dissect the different aspects of CFTR biogenesis (Bartoszewski et al., 2008b; Bebok et al., 2005; Rab et al., 2007; Varga et al., 2004). Furthermore, we developed cell lines expressing both endogenous and recombinant CFTR to study the differences in the regulation of the two variants (Bartoszewski et al., 2008a).

2.2.1. Measurement of CFTR mRNA levels

Epithelial cell lines from airways (Calu-3, 16HBE14o-), intestines (T 84, HT29), or primary airway epithelial cell cultures can be used for studies on endogenous CFTR expression regulation. Stable, recombinant CFTR-expressing cell lines such as CFBE41o-ΔF, CFBE41o-WT (Bebok et al., 2005), BHK (Carvalho-Oliveira et al., 2004; Gomes-Alves et al., 2009), or other model cell line such as HeLa-WT (Bartoszewski et al., 2008a; Bebok et al., 2005; Rab et al., 2007) can be used for recombinant CFTR mRNA measurements.

Probes

CFTR mRNA levels can be evaluated using assay on demand primer mix (ABI Assay ID: Hs00357011_m1). GAPDH (assay ID: Hs99999905_m1) or 18S RNA (ID: Hs99999901_s1) can be amplified as internal controls from the samples. We also tested Transferrin receptor (TR) mRNA levels following UPR induction and found no changes in message levels (Rab et al., 2007). TR mRNA levels can be determined using assay ID: Hs99999911_m1.

2.2.2. Determination of CFTR mRNA stability

Reduced CFTR mRNA levels in response to UPR activation could result from inhibited transcription or from enhanced mRNA decay. Therefore, it is necessary to determine possible changes in mRNA stability as the result of ER stress. RNA degradation, processing, and quality control are regulated mainly by the $3' \rightarrow 5'$ribonucleolytic complex called exosome (Belostotsky, 2009; Schmid and Jensen, 2008). While it is unlikely that reduced endogenous CFTR mRNA levels result from an increase in exosome activity, since there is no reduction in recombinant CFTR mRNA levels or control mRNAs, a posttranscriptional modification of the endogenous CFTR mRNA during ER stress may cause a specific CFTR mRNA decay. The following assays help to determine whether changes in RNase activity and/or mRNA half-life contribute to reduced CFTR mRNA levels during ER stress.

2.2.2.1. Assessment of cytoplasmic RNase activity

Equipment and materials

Microcentrifuge (with rotor for 2 ml tubes), ultracentrifuge (Beckman TL-100 or equivalent), and Potter–Elveheim homogenizers. Cells are grown on 10-cm tissue culture dishes ($\sim$80% confluent). Hypotonic homogenization buffer (10 mM KCl, 10 mM HEPES, pH 7.4, and 1 mM MgCl$_2$). Total RNA isolated from untreated cells as described in Section 2.1.2. RNase digestion buffer (2×): 0.6 M NaCl, 20 mM Tris, 10 mM EDTA, pH 7.4.

Cytoplasmic protein extraction

All steps are performed in an RNase-free environment. Following induction of ER stress (Section 2.1.1), cells are washed three times with ice-cold PBS, scraped into homogenization buffer (600 µl) with protease (Protease Inhibitor Complete Mini, Roche) and phosphatase inhibitors (Sigma), and incubated for 30 min at 4 °C. Cells are then homogenized using Potter homogenizer and unbroken cells are removed by centrifugation at 2500×g for 15 min. The supernatant is then separated into membrane and cytosolic components by a 1 h centrifugation at 100,000×g (Beckman TL-100 or equivalent). Following determination and normalization of protein concentrations, cytoplasm extracts should be used immediately for RNA incubation experiments. The total amount of cytoplasmic protein obtained from Calu-3 cells grown to 80% confluency on 10-cm tissue culture dishes is $\sim$700–800 µg. Each RNA incubation reaction contains 20 µg cytoplasmic protein extract in a total of 30 µl reaction mixture. RNA incubation reactions are performed at least in triplicate.

Incubation of RNA samples with cytoplasmic proteins

Incubate total RNA isolated from control, untreated cells (20 µg RNA/sample in 30 µl 2× RNase digestion buffer) with an equal volume of cytosolic protein extract (20 µg total proteins/sample) under gentle stirring at 25 °C for 20 min. At the end of incubation, stop the RNase reaction with 300 µl TRIzol (Invitrogen). Important controls include samples without RNA in the RNase digestion buffer, and RNA solution without cytoplasmic protein extracts. Following the incubation, RNA is extracted from samples to determine the remaining CFTR mRNA levels.

Extraction of remaining RNA from samples

RNA extraction is performed using the protocol described in Section 2.1.2.

Compare CFTR mRNA levels following cytoplasmic RNase digestion

CFTR mRNA content following RNase digestion is determined by real-time RT-PCR using the protocol described above for total CFTR mRNA (Section 2.2.1). Enhanced RNase activity in the cytoplasmic fraction following induction of ER stress would result in reduced CFTR mRNA levels compared to controls. In our experience, there is a ~20% reduction in the CFTR mRNA levels as the result of baseline RNase activity in the control, untreated samples (Baudouin-Legros *et al.*, 2005; Rab *et al.*, 2007).

2.2.2.2. CFTR mRNA half-life

Equipment and materials

Actinomycin D (Sigma A1410), TM from *Streptomyces* sp. (Sigma T7765), 18S primer probe mix—assay ID: HS_99999901_m1 (ABI), CFTR primer probe mix—assay ID: Hs00357011_m1 (ABI). ABI 7500 Real-Time PCR System and supplies (see above).

Protocol

Unless stated otherwise, all steps are performed on ice. Parallel experiments are performed on mRNA samples isolated from control (untreated) and ER-stressed cells (Section 2.1.1). Cells are cultured on 10-cm plastic dishes to 80% confluency. Actinomycin D is added at 5 µg/ml final concentration after 8 h of ER stress induction to stop transcription. Isolate RNA (according to protocol in Section 2.1.2) 2, 4, 6, and 8 h following actinomycin D treatment. Total CFTR mRNA levels at each time point are assessed using one-step RT-PCR and normalized to endogenous 18S rRNA levels as described in Section 2.1.2. CFTR mRNA levels for each time point are calculated from three individual samples. Relative CFTR mRNA levels at the time points indicated are plotted as percent differences from CFTR mRNA levels at the initial time point (t_0). The mRNA half-lives can be calculated from the exponential decay using the trend line equation $C/C_0 = e^{-k_d t}$ (where C and C_0 are mRNA amounts at time t and at time 0 (t_0), respectively, and k_d is the mRNA decay constant; Bartoszewski *et al.*, 2008b). A typical plot of the results is shown in Fig. 1.2.

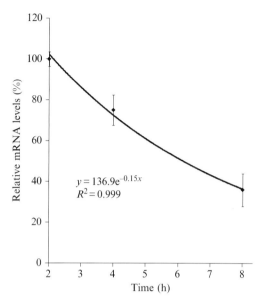

Figure 1.2 CFTR mRNA half-life measurements. A representative plot of CFTR mRNA half-life measurements is shown. mRNA half-life is calculated based on the equation described in the text (Bartoszewski et al., 2008b).

2.3. Assays for *CFTR* transcriptional regulation

Our studies indicated no differences in endogenous CFTR mRNA stability following UPR activation (Bartoszewski et al., 2008a,b; Rab et al., 2007). Here, we provide detailed description of methods that were developed to test CFTR transcriptional activity and the mechanism of CFTR transcriptional repression during ER stress.

2.3.1. *CFTR* promoter reporter assay

Principle

Promoter reporter vectors contain a specified section of the promoter from the gene of interest upstream of a reporter gene such as Firefly luciferase. Panomics Inc. pioneered the development of basic promoter reporter vectors. Variants of these vectors can be developed to answer specific questions regarding transcriptional regulation. The CFTR promoter reporter vector from Panomics contains a ~ 1000 bp fragment of the human CFTR 5′ regulatory region upstream of the firefly luciferase gene (Bartoszewski et al., 2008b). Expression of firefly luciferase depends on the activity of the *CFTR* promoter. The promoter contains a cAMP responsive element (CRE) that can be activated by forskolin to increase luciferase expression by $\sim 30\%$ (Bartoszewski et al., 2008b). This provides the option for testing the functionality of the regulatory region. As an internal control for transfection efficiency, a *Renilla* luciferase vector can be used. The expression of Renilla luciferase is driven by a viral promoter and therefore, like the recombinant CFTR expression vectors, is not regulated by the UPR. The schematic of the CFTR promoter reporter vector is shown in Fig. 1.3.

Materials and equipment

Cell lines without endogenous CFTR expression (HeLa, Cos-7), transfection reagent—Fugene™ 6 (Roche), Opti-MEM™ reduced serum media (Invitrogen), 6-well cell culture plates (Costar), Luciferase assay reagent (Promega), *CFTR* promoter reporter vector (Firefly luciferase, Panomics Inc.), Renilla luciferase control vector (Panomics Inc.), dual luciferase assay reagent kit (Promega), Luminometer.

Protocol

HeLa cells grown on 6-well tissue culture plates are transfected with the promoter reporter constructs and Renilla luciferase as described in the manufacturer's protocol. At 12 h posttransfection, cells are treated with ER stress inducers (Section 2.1.1) or forskolin (to activate the CRE) for an additional 12 h. Control wells do not receive any treatment. Cells are washed in PBS, harvested, and lysed using luciferase assay lysis buffer (provided in Luciferase assay reagent kit). Luciferase activity is measured using dual luciferase assay reagent. To reduce error, at least four independent replicates should be measured. Results are presented as relative firefly luciferase/Renilla luciferase light units or percentage changes in light units compared to the untreated control.

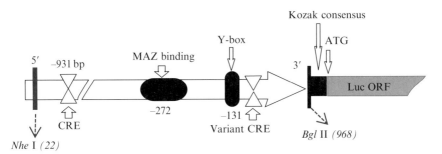

Figure 1.3 Schematic diagram of the CFTR promoter reporter vector. A CFTR promoter containing regulatory elements is shown. Restriction sites for cloning (*Nhe*I and *Bgl*II) are also labeled. CRE, cAMP response element; MAZ, Myc-associated zinc finger protein binding; Y-box, Y-box protein binding; ATG, transcription start (Bartoszewski *et al.*, 2008b).

2.3.2. Chromatin immunoprecipitation assays

Principle

Chromatin immunoprecipitation (ChIP) assays were developed to identify proteins (histones and transcription factors) that are associated with specific regions of genes (Rodriguez and Huang, 2005; Weinmann and Farnham, 2002). The initial step is the cross-linking of protein–protein and protein–DNA complexes in cells with formaldehyde. After cross-linking, the cells are lysed and sonicated to shear the DNA with the proteins attached. Proteins of interest such as transcription factors that are cross-linked to DNA fragments are subsequently immunoprecipitated. The immunoprecipitates contain a specific fragment of the gene, the transcription factor of interest, and very likely, other components of the transcriptional complex. After removal of the proteins from the DNA, the specific fragment of the gene is purified and identified using PCR amplification. Our first approach was to analyze the 1-kb fragment of the human *CFTR* 5′ regulatory region to find putative direct or indirect binding sites for the UPR-induced transcription factors XBP1 and ATF6. Then, we performed ChIP experiments to test the role of these transcriptional factors in CFTR transcriptional repression (Bartoszewski *et al.*, 2008b).

Materials and equipment

Sonicator with microtip, water bath with shaker, or heat block with shaker capable of 65 °C heating and 450 rpm shakes, microcentrifuge for 1.5 ml tubes. Cell type specific tissue culture media (Invitrogen), TM from *Streptomyces* sp., TM (Sigma), DMSO (Sigma), NaCl (Sigma), Proteinase K (Sigma), EDTA (Sigma), 1 M Tris–HCl, pH 6.5, phenol–chloroform–isoamyl alcohol (25:24:1), pH 8.0 (Sigma).

Buffers

2× RIPA: 100 mM Tris–HCl, pH 7.4; 2% NP-40 (Sigma); 0.5% SDS (Sigma); 300 mM NaCl; 2 mM EDTA; 2 mM PMSF (Sigma); 2 μg/ml of aprotinin (Sigma); leupeptin (Sigma); pepstatin A (Sigma). Hypotonic buffer: 10 mM HEPES, pH 7.9 (Sigma); 1.5 mM MgCl$_2$ (Sigma); 10 mM KCl (Sigma). LiCl buffer (10 mM Tris–HCl, pH 8.1; 0.25 M LiCl; Sigma); 1% NP-40; 1% Na-deoxycholate; 1 mM EDTA. Low salt: 0.1% SDS; 1% Triton-X-100; 2 mM EDTA; 20 mM Tris–HCl, pH 8.1; 150 mM NaCl. High salt: 0.1% SDS; 1% Triton-X-100; 2 mM EDTA; 20 mM Tris–HCl, pH 8.1; 500 mM NaCl. Qiagen MiniElute PCR purification kit (Qiagen). Elution buffer: 10.8 ml sterile, nuclease-free water; 1% SDS; 100 mM NaHCO$_3$ (S5761 Sigma). TE buffer (Qiagen). PBS buffer (Invitrogen).

Protocol

Sample preparation and cross-linking: Epithelial cells (e.g., Calu-3; T-84) are grown on 20-cm tissue culture dishes to 80% confluence. Induce ER stress as described in Section 2.1.1. Nontreated cells serve as the controls. Exchange regular media with cross-linking solution that contains 275 µl of 37% formaldehyde in 10 ml of serum-free tissue culture media. Cover the cells with cross-linking solution and incubate at room temperature for 15–30 min. Aspirate the cross-linking solution, wash cells on ice with cold PBS, and scrape cells in 3 ml PBS/plate. Centrifuge cells at $200 \times g$ 10 min at 4 °C, discard supernatant, and resuspend cells in hypotonic buffer with fresh protease inhibitors added (950 µl/sample). Incubate cells on ice for 5 min and add 50 µl of 10% NP-40/sample (the final concentration of NP-40 is 0.5%), followed by an additional 5 min incubation on ice. Centrifuge samples at $1000 \times g$ for 10 min at 4 °C, discard supernatant, and resuspend the pellet (containing nuclei) in 2 ml/sample TE buffer with protease inhibitors. Samples can be stored at -80 °C for up to 2 weeks.

Sonication and determination of DNA content: Sonicate nuclei in 15-ml conical tubes on ice. Sonication times and settings need to be determined for each sonicator and cell line tested. Calu-3 cells require twenty 20-s pulses at 40% power using VibraCell sonicator with microtip (Sonics Materials). Chromatin fragmentation needs to be verified using agarose gel electrophoresis. The main DNA fraction after fragmentation should be in size range from 150 to 400 bp. Transfer samples into 2 ml Eppendorf tubes and centrifuge at $10,000 \times g$ for 15 min at 4 °C and measure DNA concentration in supernatant at A_{280nm}. Aliquot samples into 5–10 units (1 unit = 100 µg DNA). Samples can be stored at -20 °C for several months.

Input preparation (to determine the quality of the samples): Remove ~50 µg DNA from each sample and dilute with TE buffer to 500 µl followed by the addition of 20 µl of 5 M NaCl to each sample. Mix samples by vortexing and centrifuge at maximum speed in microcentrifuge for 1 min. Place samples into the shaker (450 rpm) incubator (65 °C) and shake overnight. Add protein digestion mix to the samples (10 µl 0.5 M EDTA, 20 µl 1 M Tris–HCl, pH 6.5, 2 µg Proteinase K) and incubate for 2 h at 37 °C in the shaker. To extract DNA, add 500 µl phenol–chloroform–isoamyl alcohol (25:24:1) pH 8.0 to each sample, mix well, and centrifuge at maximum speed (14,000 rpm) at 4 °C for 15 min. The upper aqueous phase contains the DNA. Transfer the DNA fraction to new tubes and purify by Qiagen MiniElute PCR purification kit as described in manufacturer's instructions. Elute each sample with 50 µl of nuclease-free water (stored at -20 °C). These samples are used to test the quality of the shredded DNA (Fig. 1.4).

CFTR Expression Regulation by the Unfolded Protein Response

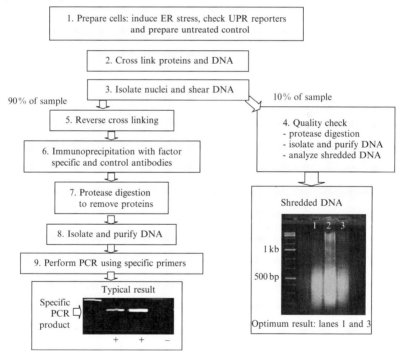

Figure 1.4 Flow chart of the ChIP assay. The *CFTR* 5' regulatory region is shown as an example with some identified and some putative binding factors. The shredded DNA sample is from Calu-3 cells. The PCR products represent the regions of the CFTR promoter with ATF6 binding (Bartoszewski *et al.*, 2008b).

Perform PCR reactions using the shredded DNA samples as a template with primers designed to amplify the region of interest. A positive PCR reaction from the input samples indicates that samples are appropriate for immunoprecipitation.

Immunoprecipitation and DNA recovery: Use 500 μg of DNA/sample and adjust the volume to 500 μl with TE buffer. Add 500 μl of 2× RIPA buffer with protease inhibitors and 40 μl of protein A and G agarose beads preincubated with salmon sperm (SS) DNA to block nonspecific binding. Rotate samples for 30 min at 4 °C to remove proteins that bind nonspecifically to the beads. Centrifuge at 14,000 rpm for 1 min at 4 °C and transfer supernatants to fresh chilled tubes. Add 5 μg of ChiP grade antibody to each sample and rotate overnight at 4 °C. Controls with nonspecific antibody or nonimmune purified IgG should be included. After incubation with antibodies, add 60 μl of SS DNA preincubated Protein A or G Agarose beads (depending on the antibody) and rotate for 2 h at 4 °C to capture the protein–DNA complexes. Centrifuge at 14,000 rpm for 3 min at 4 °C, wash pellets in 1 ml buffer in the following order: (1) low salt buffer, (2) high salt buffer, (3) LiCl buffer, and (4) twice in TE buffer. Between washes the samples are centrifuged at 14,000 rpm for 3 min at RT and supernatants discarded. Resuspend the final pellet in 275 μl of fresh Elution buffer and nutate for 15 min at RT, followed by centrifugation (14,000 rpm for 3 min). Remove 250 μl of the supernatant and save on ice. Add 250 μl fresh elution buffer, nutate for 15 min at RT, and centrifuge for 1 min at maximum speed at RT. Remove and add the second 250 μl supernatant to the first so the total eluted volume will be 500 μl. Add 20 μl of 5 M NaCl to each sample and recover DNA as described for input preparation. The purified DNA samples can now be used as templates for PCR reactions. Use primer probes designed to amplify DNA sequences to which the expected transcription factor binds either directly or indirectly and amplify the expect sequence by PCR. Immunoprecipitation with nonspecific antibodies or using nonspecific primer probes should not result in PCR product. It is crucial to test all controls in order to eliminate nonspecific effects. PCR products can be analyzed on 2% EtBr-stained agarose gels (Fig. 1.4). Alternatively, real-time PCR can be performed.

2.4. Assessment of CFTR protein levels

CFTR protein levels decrease during ER stress (Bartoszewski *et al.*, 2008a,b; Rab *et al.*, 2007). Reduced CFTR protein levels may result from (1) reduced mRNA levels, (2) decreased translation, (3) enhanced ERAD, and/or (4) reduced CFTR protein half-life. Here, we describe the methodology for assessing CFTR protein expression.

2.4.1. CFTR maturation efficiency measurements

Metabolic pulse-chase assays are designed to investigate the synthesis, intracellular processing, and half-life of newly synthesized proteins and provide important information regarding the effects of ER stress on these processes (Rab et al., 2007; Varga et al., 2004). Endogenous CFTR expression levels and translational rates are low and therefore CFTR maturation efficiency studies are difficult in cell lines that endogenously express CFTR. Basic information regarding metabolic labeling of proteins is available from other sources (Diaz and Varki, 2009). Here, we describe important points for metabolic pulse-chase studies designed to investigate endogenous CFTR synthesis and processing. Epithelial cells (e.g., Calu-3, T-84) must be grown under optimal conditions for these experiments. For example, it is important to change the tissue culture media every day in order to avoid nutrient starvation that induces ER stress and reduces CFTR transcription (Rab et al., 2007). Furthermore, use low volumes of media to reduce the possibility of hypoxia, since hypoxia has been shown to reduce CFTR levels (Bebok et al., 2001; Guimbellot et al., 2008; Varga et al., 2004).

Materials and equipment

Six-well tissue culture dishes, methionine/cysteine-free MEM (Invitrogen), EasyTag Protein Labeling Mixture (^{35}S-methionine/cysteine, NEN), RIPA buffer (150 mM NaCl, 1% NP-40, 0.5% sodium deoxycholate, 0.1% SDS, 50 mM Tris–HCl, pH 8.0), Protease Inhibitor Complete Mini (Roche). CFTR-specific antibodies appropriate for immunoprecipitation (24-1; Varga et al., 2004; M3A7 (R&D Systems)), Protein A or G Agarose beads (depending on antibody binding efficiency), microcentrifuge, 2× concentrate Laemmli sample buffer (Sigma), 6% or 8% SDS–polyacrylamide gels (Invitrogen). Gel running apparatus, Phosphor Imager screen, Phosphor Imager (Molecular Dynamics), Image analysis software (IPLab or equivalent).

Protocol

Cells are grown on 6-well tissue culture dishes to ~70% confluency and ER stress induced as described in Section 2.1.1. Untreated control samples are also tested. Remove tissue culture media, rinse cells with PBS (CaCl$_2$ 0.9 mM, MgCl$_2$ 0.5 mM, pH 7.4; Gibco), and replace regular growth medium with methionine/cysteine-free MEM. Incubate cells at 37 °C with 5% CO$_2$ for 30 min (methionine starvation). Protein synthesis is paused in the absence of methionine. At the end of methionine starvation, replace the media with methionine-free media supplemented with 300 μCi/ml of EasyTag Protein Labeling Mixture and place the cells in the incubator. (Samples are radioactive and therefore handled accordingly during the entire experiment.) Labeling time (pulse) should be determined for each cell line. For the cell lines listed and based on the time required for CFTR synthesis, a 30-min pulse period is usually sufficient (Varga et al., 2004). Radioactive labeling is stopped by removing the EasyTag Protein Labeling Mixture, washing the wells, and adding fresh, regular tissue culture media to the cells. Cells are lysed in RIPA buffer supplemented with protease inhibitors at the time points specified (chase period). Samples are rotated at 4 °C for at least 1 h, vortexed several times, and centrifuged at 14,000 rpm for 30 min. Supernatants are collected in fresh tubes. CFTR can be immunoprecipitated from the samples using 24-1 (or M3A7) anti-CFTR monoclonal antibody preloaded Protein G Agarose beads. The immunoprecipitation reaction is for 2 h or longer at 4 °C with constant rotation. Following immunoprecipitation, samples are centrifuged for 1 min at max speed and supernatants are removed. Immunoprecipitated samples (bead–antibodies–CFTR complexes) are then washed four times using 1 ml/sample ice-cold RIPA buffer by mixing the beads with RIPA buffer followed centrifugation and removal of the washing buffer using vacuum. The beads should not be disturbed in order minimize variability between samples. Immunoprecipitated CFTR is released from the beads with the addition of 2× Laemmli sample buffer supplemented with 5% β-mercaptoethanol. The volume of sample buffer should be equal to the volume of beads and remaining buffer (∼30 μl if 20 μl of the beads were used with the sample). Samples are incubated at 37 °C for 40–45 min (do not boil immunoprecipitated CFTR samples). Spin down samples before loading for PAGE. Samples are analyzed by SDS-PAGE (6% gels) and detected using autoradiography (Molecular Dynamics, PhosphorImager). CFTR maturation efficiency is measured by calculating the percentage of newly synthesized CFTR converted into fully glycosylated CFTR based on densitometry using IPLab software (Scanalytics, Inc.) as described previously (Varga et al., 2004). A typical CFTR maturation efficiency experiment is shown in Fig. 1.5.

2.4.2. Detection of steady-state CFTR levels by Western blot

CFTR-expressing cells are lysed with RIPA buffer supplemented with protease inhibitors (Complete Mini, Roche) on ice for 15 min. The cell lysates are rotated at 4 °C for 30 min and the insoluble material is removed by centrifugation at 14,000 rpm for 15 min. Protein concentrations are determined by BCA^{TM} protein assay (Pierce) using bovine serum albumin (BSA) as the standard. Following normalization of protein concentrations, lysates are mixed with equal volume of 2× Laemmli sample buffer and incubated for 30–45 min at 37 °C prior to separation by SDS-PAGE. Using epithelial cell lines endogenously expressing CFTR (Calu-

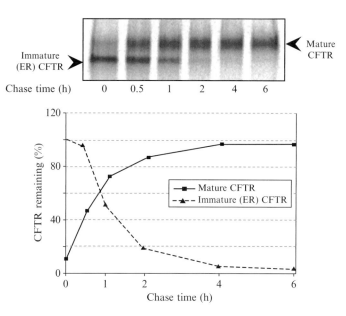

Figure 1.5 Pulse-chase assay to measure CFTR maturation efficiency. A representative gel and the results of densitometry measurements are shown. Calu-3 cells expressing endogenous wild-type CFTR were labeled for 30 min with 300 µCi/ml Easy Tag protein labeling mixture and chased for the time periods specified (immature CFTR, the core glycosylated ER form of CFTR; mature CFTR, the fully glycosylated, post-ER form of CFTR).

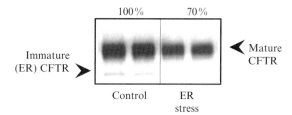

Figure 1.6 CFTR detection by Western blot. Representative results of CFTR protein detection in whole cell lysates by Western blot. Total cell lysates (15 µg) were separated by 8% PAGE, Western transferred and CFTR was detected using MM13-4 anti-CFTR monoclonal antibody, anti-mouse IgG-HRP (Pierce), and ECL (Pierce). Samples were analyzed in duplicates. Densitometry (Scion Image) was performed to assess CFTR expression levels in control and ER-stressed Calu-3 cells (immature CFTR, the core glycosylated ER form of CFTR; mature CFTR, the fully glycosylated, post-ER form of CFTR).

3, T-84), 15 µg of total protein in the cell lysate contains sufficient CFTR to detect using the MM13-4 antibody. Following SDS-PAGE, the proteins from the gel are transferred to polyvinylidene difluoride (PVDF) membranes (300 mA for 90 min at 4 °C). The membranes are then blocked with milk proteins dissolved in PBS/Tween-20 (5% milk, 0.5% Tween-20 for 1–12 h), followed by immunoblotting with CFTR-specific antibody (mouse monoclonal antibody, MM13-4, Millipore, 2 μg/ml), HRP-conjugated anti-mouse IgG antibody (Pierce), and detected using ECL (Pierce). A typical Western blot result of wild-type CFTR is shown in Fig. 1.6.

ACKNOWLEDGMENTS

Supported by NIH, HL076587 (Bebok), DK060065 (Collawn).

REFERENCES

Bartoszewski, R., et al. (2008a). Activation of the unfolded protein response by deltaF508 CFTR. Am. J. Respir. Cell Mol. Biol. **39,** 448–457.

Bartoszewski, R., et al. (2008b). The mechanism of cystic fibrosis transmembrane conductance regulator transcriptional repression during the unfolded protein response. J. Biol. Chem. **283,** 12154–12165.

Baudouin-Legros, M., et al. (2005). Cell-specific posttranscriptional regulation of CFTR gene expression via influence of MAPK cascades on 3′UTR part of transcripts. Am. J. Physiol. Cell Physiol. **289,** C1240–C1250.

Bebok, Z., et al. (2005). Failure of cAMP agonists to activate rescued {Delta}F508 CFTR in CFBE41o- airway epithelial monolayers. J. Physiol. **569,** 601–615.

Bebok, Z., and Collawn, J. F. (2006). Cystic fibrosis. In "The Encyclopedia of Respiratory Medicine," (S. Shapiro, ed.), pp. 599–609. Elsevier, Oxford, UK.

Bebok, Z., et al. (2001). Improved oxygenation promotes CFTR maturation and trafficking in MDCK monolayers. Am. J. Physiol. Cell Physiol. **280,** C135–C145.

Bebok, Z., et al. (2002). Reactive oxygen nitrogen species decrease cystic fibrosis transmembrane conductance regulator expression and cAMP-mediated Cl-secretion in airway epithelia. J. Biol. Chem. **277,** 43041–43049.

Bebok, Z., et al. (1998). Activation of DeltaF508 CFTR in an epithelial monolayer. Am. J. Physiol. **275,** C599–C607.

Belostotsky, D. (2009). Exosome complex and pervasive transcription in eukaryotic genomes. Curr. Opin. Cell Biol. **21,** 352–358.

Berdiev, B. K., et al. (2009). Assessment of the CFTR and ENaC association. Mol. Biosyst. **5,** 123–127.

Bradbury, N. A., et al. (1992). Regulation of plasma membrane recycling by CFTR. Science **256,** 530–532.

Cantin, A. M., et al. (2006a). Oxidant stress suppresses CFTR expression. Am. J. Physiol. Cell Physiol. **290,** C262–C270.

Cantin, A. M., et al. (2006b). Cystic fibrosis transmembrane conductance regulator function is suppressed in cigarette smokers. Am. J. Respir. Crit. Care Med. **173,** 1139–1144.

Carvalho-Oliveira, I., et al. (2004). CFTR localization in native airway cells and cell lines expressing wild-type or F508del-CFTR by a panel of different antibodies. J. Histochem. Cytochem. **52,** 193–203.

Cheng, S. H., et al. (1990). Defective intracellular transport and processing of CFTR is the molecular basis of most cystic fibrosis. Cell **63,** 827–834.

Collawn, J. F., Fu, L., and Bebok, Z. (2010). Therapeutic targets in cystic fibrosis: The role of pharmaceutical chaperones. *Expert Rev. Proteomics* **7**, 495–506.

Dahl, M., and Nordestgaard, B. G. (2009). Markers of early disease and prognosis in COPD. *Int. J. Chron. Obstruct. Pulmon. Dis.* **4**, 157–167.

Di, A., *et al.* (2006). CFTR regulates phagosome acidification in macrophages and alters bactericidal activity. *Nat. Cell Biol.* **8**, 933–944.

Diaz, S., and Varki, A. (2009). Metabolic radiolabeling of animal cell glycoconjugates. *Curr. Protoc. Protein Sci.* Chapter 12, Unit 12 2 12 2 1-55.

Frischmeyer, P. A., and Dietz, H. C. (1999). Nonsense-mediated mRNA decay in health and disease. *Hum. Mol. Genet.* **8**, 1893–1900.

Gomes-Alves, P., *et al.* (2009). Low temperature restoring effect on F508del-CFTR misprocessing: A proteomic approach. *J. Proteomics* **73**, 218–230.

Guimbellot, J. S., *et al.* (2008). Role of oxygen availability in CFTR expression and function. *Am. J. Respir. Cell Mol. Biol.* **39**, 514–521.

Hong, M., *et al.* (2004). Endoplasmic reticulum stress triggers an acute proteasome-dependent degradation of ATF6. *J. Cell. Biochem.* **92**, 723–732.

Hung, J. H., *et al.* (2004). Endoplasmic reticulum stress stimulates the expression of cyclooxygenase-2 through activation of NF-kappaB and pp 38 mitogen-activated protein kinase. *J. Biol. Chem.* **279**, 46384–46392.

Ismailov, I. I., *et al.* (1996). Regulation of epithelial sodium channels by the cystic fibrosis transmembrane conductance regulator. *J. Biol. Chem.* **271**, 4725–4732.

Kerem, B., *et al.* (1989). Identification of the cystic fibrosis gene: Genetic analysis. *Science* **245**, 1073–1080.

Kerem, E., *et al.* (2008). Effectiveness of PTC124 treatment of cystic fibrosis caused by nonsense mutations: A prospective phase II trial. *Lancet* **372**, 719–727.

Kultz, D. (2005). Molecular and evolutionary basis of the cellular stress response. *Annu. Rev. Physiol.* **67**, 225–257.

Lee, A. S. (2005). The ER chaperone and signaling regulator GRP78/BiP as a monitor of endoplasmic reticulum stress. *Methods* **35**, 373–381.

Liu, S., *et al.* (2005). Dynamic activation of cystic fibrosis transmembrane conductance regulator by type 3 and type 4D phosphodiesterase inhibitors. *J. Pharmacol. Exp. Ther.* **314**, 846–854.

Nadanaka, S., *et al.* (2004). Activation of mammalian unfolded protein response is compatible with the quality control system operating in the endoplasmic reticulum. *Mol. Biol. Cell* **15**, 2537–2548.

Naren, A. P., *et al.* (2003). A macromolecular complex of beta 2 adrenergic receptor, CFTR, and ezrin/radixin/moesin-binding phosphoprotein 50 is regulated by PKA. *Proc. Natl. Acad. Sci. USA* **100**, 342–346.

O'Byrne, P. M., and Postma, D. S. (1999). The many faces of airway inflammation. Asthma and chronic obstructive pulmonary disease. Asthma Research Group. *Am. J. Respir. Crit. Care Med.* **159**, S41–S63.

Quinton, P. M. (2007). Cystic fibrosis: Lessons from the sweat gland. *Physiology (Bethesda)* **22**, 212–225.

Rab, A., *et al.* (2007). Endoplasmic reticulum stress and the unfolded protein response regulate genomic cystic fibrosis transmembrane conductance regulator expression. *Am. J. Physiol. Cell Physiol.* **292**, C756–C766.

Riordan, J. R. (2008). CFTR function and prospects for therapy. *Annu. Rev. Biochem.* **77**, 701–726.

Riordan, J. R., and Chang, X. B. (1992). CFTR, a channel with the structure of a transporter. *Biochim. Biophys. Acta* **1101**, 221–222.

Rodriguez, B. A., and Huang, T. H. (2005). Tilling the chromatin landscape: Emerging methods for the discovery and profiling of protein-DNA interactions. *Biochem. Cell Biol.* **83,** 525–534.

Rowe, S. M., *et al.* (2005). Cystic fibrosis. *N. Engl. J. Med.* **352,** 1992–2001.

Schmid, M., and Jensen, T. H. (2008). The exosome: A multipurpose RNA-decay machine. *Trends Biochem. Sci.* **33,** 501–510.

Schroder, M., and Kaufman, R. J. (2005). The mammalian unfolded protein response. *Annu. Rev. Biochem.* **74,** 739–789.

Schwiebert, E. M., *et al.* (1999). CFTR is a conductance regulator as well as a chloride channel. *Physiol. Rev.* **79,** S145–S166.

Sheppard, D. N., *et al.* (1994). Expression of cystic fibrosis transmembrane conductance regulator in a model epithelium. *Am. J. Physiol.* **266,** L405–L413.

van der Deen, M., *et al.* (2005). ATP-binding cassette (ABC) transporters in normal and pathological lung. *Respir. Res.* **6,** 59.

Varga, K., *et al.* (2004). Efficient intracellular processing of the endogenous cystic fibrosis transmembrane conductance regulator in epithelial cell lines. *J. Biol. Chem.* **279,** 22578–22584.

Ward, C. L., *et al.* (1995). Degradation of CFTR by the ubiquitin-proteasome pathway. *Cell* **83,** 121–127.

Weinmann, A. S., and Farnham, P. J. (2002). Identification of unknown target genes of human transcription factors using chromatin immunoprecipitation. *Methods* **26,** 37–47.

Yoshida, H., *et al.* (2001). XBP1 mRNA is induced by ATF6 and spliced by IRE1 in response to ER stress to produce a highly active transcription factor. *Cell* **107,** 881–891.

CHAPTER TWO

GRP78/BiP Chapter: Modulation of GRP78/BiP in Altering Sensitivity to Chemotherapy

Thomas C. Chen

Contents

1. Introduction	26
2. Measurement of ER Stress *In Vitro* and *In Vivo*	27
2.1. *In vitro* measurements of GRP78 and ER stress	27
2.2. Early markers of ER stress	29
2.3. Intermediate markers of ER stress	30
2.4. Late markers of ER stress	31
2.5. Modulation of GRP78/BiP levels *in vitro*	31
2.6. *In vivo* modulation of GRP78	34
3. Conclusion	36
References	36

Abstract

The glucose-regulated protein 78 (GRP78) is a key regulator of the endoplasmic reticulum (ER) stress response. This chapter discusses how GRP78 may be measured *in vitro* and *in vivo*. Because of the authors' expertise and experience with brain tumors, most of our work on GRP78 has focused on malignant gliomas. Here, we present our methodology for determining GRP78 *in vitro* using glioma cell lines and primary cell cultures, and in measuring GRP78 expression in tissue sections from rodent glioma models and human brain tumors. After discussion of measurement of GRP78 levels, modulation of GRP78 *in vitro* is demonstrated using glioma cell lines. Lastly, *in vivo* modulation of GRP78 levels via chemotherapy is determined using intracranial rodent models.

Department of Neurosurgery and Pathology, University of Southern California, Los Angeles, USA

1. INTRODUCTION

The endoplasmic reticulum (ER) is responsible for the synthesis and folding of membrane and secretory proteins. When the protein load exceeds the capacity of the ER, it may induce ER stress, leading to the unfolded protein response (UPR). The UPR is a protective mechanism by which the cell cuts down protein translation, upregulates chaperone proteins to guide the misfolded proteins out of the cell, and increases the breakdown of the misfolded proteins already in the ER. The stakes are high, because if the cell is unsuccessful in responding to the ER stress, activation of the apoptotic pathways ensues, leading to cell death. One of the key regulators of this UPR is an ER chaperone called glucose-regulated protein 78 (GRP78) or BiP (immunoglobulin heavy-chain binding protein). In the normal cell, GRP78 is responsible for binding three transmembrane ER stress sensors PERK (PKR (double-stranded-RNA-dependent protein kinase)-like ER kinase), IRE1 (inositol requiring enzyme 1), and ATF6. Both PERK and IRE1 are transmembrane proteins which sense unfolded protein levels within the ER. ATF6 is a transcription factor which is essential for a prosurvival response to ER perturbation from the external microenvironment. All three transmembrane proteins are regulated by GRP78, which in an unstressed ER binds to them and inhibits their activity. Elevation of GRP78/BiP is an early marker of ER stress, and correlates with cellular protection in response to the external stimuli, that is, from hypoxia, hypoglycemia, chemotherapy, radiation therapy. In response to stressful stimuli, GRP78 dissociates from these transmembrane sensors, allowing ATF6 to be activated in the Golgi apparatus, leading to induction of the eukaryotic translation inhibition factor 2 (eIF2a), and inhibition of global protein synthesis. This response induces the protein translation shutdown that is seen secondary to ER stress. The unbound GRP78 then acts to chaperone some of the excess malfolded protein in the ER out of the cell. If GRP78 translation or activity is impeded, it will not be able to act as a transporter, and misfolded proteins will accumulate in the ER. Too much misfolded proteins will lead to activation of the apoptotic pathway apoptosis with induction of CHOP (CCAAT/enhancer-binding protein transcription factor), caspase 4, and caspase 7, critical executioners of the proapoptotic arm of the ER stress response (Schönthal, 2008).

The ER is the main site for intracellular calcium storage. As a calcium binding protein, GRP78 is responsible for maintaining and preserving ER calcium homeostasis. Calcium is normally regulated by three transmembrane IP3 proteins which control its release, and the sarco/endoplasm reticulum Ca^{2+}-ATPase (SERCA) which controls its return back to the

ER. ER stress stimulates Ca release. In order to achieve Ca homeostasis, SERCA must remove the excess calcium out of the cytoplasm (Hovnanian, 2007).

Lastly, the breakdown of the misfolded proteins is facilitated by the ubiquitin–proteasome proteins. Inhibition of proteasome leads to inhibition of breakdown of misfolded proteins already in the ER, leading to accumulation of "garbage" in the cell, resulting in apoptosis as well. Apoptosis is a well-defined energy dependent process which involves activation of caspase 4, caspase 7, and CHOP (Pyrko et al., 2007a).

This chapter will focus on GRP78/BiP. The protocols detailed are primarily based on our previous work and publications on glioma cells.

2. MEASUREMENT OF ER STRESS *IN VITRO* AND *IN VIVO*

2.1. *In vitro* measurements of GRP78 and ER stress

As ER stress is a continuum of intracellular changes that occur in response to noxious stimuli, timing and sequence in the measurement of the various ER stress markers are essential to understand the significance of changes that occur. In our laboratory, we typically measure early, intermediate, and late ER stress markers. Early markers of ER stress include GRP78 and ATF6; intermediate signs of ER stress include translation shutdown, and changes in calcium levels and SERCA. Late signs of ER stress include induction of CHOP, caspase 4, caspase 7, and TUNEL (terminal deoxynucleotidyl transferase (TdT)-mediated dUTP nick end labeling) assay. *In vitro* experiments are performed using glioma cell lines or primary glioma cell cultures, that is, primary cell cultures of brain tumor endothelial cells (TuBEC), or tissue sections (fresh frozen or formalin fixed).

2.1.1. Glioma cell lines and primary cell culture

Glioma cell lines are purchased from American Tissue Culture (ATCC), whereas primary cell cultures are obtained from glioma specimens obtained at the time of surgery. Cell lines (A-172, U-87, U-373, LN-229, U-251) are grown in DMEM or RPMI media (GIBCO BRL, Grand Island, NY), supplemented with fetal calf serum (FCS; 2.5–10%), 100 U/mL penicillin, 0.1 mg/mL streptomycin. Cells are plated in subconfluent conditions, and usually reach confluence 2–4 days after plating.

Primary cell cultures, consisting of primary glioma cells and TuBEC, are obtained from newly diagnosed GBM specimens under Institutional Review Board (IRB) approval for discarded tissue. No identification is made between the tissue and patient aside from the fact that it is a newly diagnosed GBM sample. After sterile acquisition in the operating room, the tissue is washed (3×) and then cut into small pieces in 2% FCS medium. Brain microvessels are isolated via centrifuge (1000 rpm, 10 min) using a 15% dextran solution. The microvessel pellet is then resuspended in 1 mg/mL collagenase–dispase in RPMI-1640 medium supplemented with 2% FCS in a 37 °C water bath with a shaker for 1 h. The digested specimen is resuspended in 2% FCS–RPMI and centrifuged at 1200 rpm for 5 min twice. The final pellet is resuspended in RPMI-1640 medium supplemented with 10% FCS, 100 ng/mL ECGS, and 10% Nu-serum culture supplement (BD BioCoat; BD Biosciences, Bedford, MA), and plated on precoated gelatin flasks. The medium is changed every 3 or 4 days until the cell cultures become 80% confluent. Endothelial cells are separated from glioma cells by their uptake of diacetylated low-density lipoprotein (LDL) as only endothelial cells contain LDL receptors. FACS sorting allows for separation of endothelial cells from other cells. Following sorting the purity of the EC population is confirmed by immunostaining with von Willebrand Factor (vWF; DAKO, Carpinteria, CA), CD31/PECAM-1 (Santa Cruz Biotechnology, Santa Cruz, CA), and CD105/endoglin (Santa Cruz Biotechnology). The cells that are not LDL positive are then stained for GFAP (glial fibrillary acidic protein) to confirm glioma tumor cells. Macrophages and other inflammatory cells are usually lost during the duration of the cell culture and media change. Standard media (DMEM, 10% FCS, HEPES, sodium bicarbonate, penicillin/streptomycin) is used for the cell culture; TuBEC are supplemented with 100 ng/mL Endogro. All experiments are performed on subconfluent (60–80%) TuBEC cultures, and cells between passages 4 and 6 are only used.

2.1.2. Glioma tissue

Glioma tissue is obtained from two sources: rodent models and human patients. In our rodent models, we use either immune-deficient mice or rats implanted with human glioma cells (U-87, U-251) or we use immune-competent mice or rats implanted with syngeneic mice (GL261) or rat glioma cell lines (RG2, CNS1). Not all human glioma cell lines or rodent glioma cell lines are tumorigenic, and only the tumorigenic ones are listed. Tumorigenic glioma cell lines are able to grow tumors in both subcutaneous and intracranial models. In all animals where a tumor is harvested, animals are euthanized using sodium pentobarbital (100 mg/kg) injected intraperitoneally. The brains are then removed, embedded in O.C.T. medium, and frozen. Fresh frozen tumor samples are then sectioned at a thickness of 5 μm with a cryostat. All tumors are sectioned through their largest diameter, and sections prepared for immunohistochemistry.

Malignant glioma tissues are obtained from glioblastoma patients undergoing primary resection or biopsy of their tumor. No previous exposure to radiation or chemotherapy had been experienced by this group of patients.

The peritumoral tissue is obtained from patients undergoing a partial or full lobectomy for their GBM. Consent for use of tissue specimens is obtained from the University of Southern California (USC) IRB. Fresh frozen or formalin-fixed paraffin-embedded samples are used.

2.2. Early markers of ER stress

2.2.1. Measurement of GRP78 levels

Measurement of GRP78 levels is typically performed via semiquantitative measurements such as Western blots and immunohistochemistry. Western blots are prepared from total cell lysates by lysis of cells with RIPA buffer, and protein concentrations are determined using the bicinchoninic acid (BCA) protein assay reagent (Pierce, Rockford, IL). For each lane, 50 μg of total cell lysates are loaded into cell lane, and immunostained for the primary antibody against GRP78 (Santa Cruz, Biotechnology, Inc.) according to manufacturer's recommendations. The secondary antibodies are coupled to horseradish peroxidase and detected by chemiluminescence using the Super Signal West substrate (Pierce). All immunoblots are performed at least twice to confirm the results. The blots are semiquantified using Image J software.

2.2.2. Measurement of ATF6 levels

ATF6 levels are also elevated in ER stress; protein levels are increased after ER stress as an acute response. ATF6 levels are also detected using Western blot analysis. Primary antibody is obtained from Santa Cruz Biotechnology.

2.2.3. Immunohistochemistry

In order to determine GRP78 changes in the cell or tissue, immunostaining is the most direct route. Immunostaining may be performed using cells made into cytopreps or via tissue sections. Cytopreps are made by plating detached cells ($1-2.5 \times 10^4$ cells/cytoprep) onto a glass coverslip (Fisherbrand Microscope Cover Glass, Fisher Scientific, Tustin, CA). Cytopreps made with glioma cells do not need gelatin; cytopreps made from TuBEC are plated onto coverslips coated with 1% gelatin. Cytopreps are then fixed in acetone, and blocked with 5% normal horse serum. Similarly, tissue sections obtained from rodent glioma model or human tissue (as described above) are fixed in acetone for 5 min, dried for 10 min, quenched in 0.3% H_2O_2 for 5 min, and washed three times in cold phosphate-buffered saline (PBS). Nonspecific tissue binding is preblocked for 30 min using 5% normal horse serum at room temperature. In both cytopreps and tissue sections, the primary antibody (anti-GRP78, BD Clontech, Palo Alto, CA; rabbit anti-human) is incubated for 1 h. Goat anti-rabbit secondary antibody at a 1:200 dilution at room temperature is then added after three washes in PBS. Immunoreactivity is visualized with a Vectastain ABC kit (Vector Laboratories, Burlingame, CA), followed by AEC (3-amino-9-ethylcarbazole) substrate kit according to the manufacturer's instructions. This procedure employs biotinylated secondary antibodies and a preformed avidin:biotinylated enzyme complex that has been termed the ABC technique. A red precipitate identifies positive staining. Pictures are taken using an Olympus BH2 microscope (Fluor 10× objective; numerical aperture 0.25) and a Nikon Coolpix 990 digital camera.

2.3. Intermediate markers of ER stress

Intermediate markers of ER stress include detection of translational shutdown and calcium release from the SERCA. Translational shutdown is indicative of the cells undergoing severe ER stress, which are now shutting down in order to prevent cell self-destruction or apoptosis. Release of calcium via SERCA is also another indication, that is, there is loss of active pump homeostasis, resulting in calcium release into the cytoplasm.

2.3.1. Meaurement of intermediate ER stress

In order to determine impairment of cellular translation, we determine incorporation of ^{35}S-methionine into newly translated proteins. Cells are treated with drugs that trigger ER stress (i.e., dimethylcelecoxib, DMC) and protein translation shutdown (Schonthal et al., 2008).

As controls, cells remained untreated or are exposed to the potent protein synthesis inhibitor cycloheximide. During the final 30 min of each treatment condition, the culture medium is replaced with methionine-free growth medium supplemented with ^{35}S-methionine (20 µCi/mL) in the continued presence of the respective drug. Then, total cellular lysates are prepared and equal amounts of each sample (50 µL) are separated by polyacrylamide gel electrophoresis. An autoradiograph of the gel is prepared demonstrating relative ^{35}S-methionine levels; the resulting magnitude of incorporated radioactivity per milligram of total protein is shown as counts per minute. Treated samples are standardized relative to control untreated cells which are set to 100%. The bottom of the same gel is stained with Coomassie blue to demonstrate that equal amounts of protein were loaded into each lane. In order to determine the time course of protein translation inhibition, cells are treated with varying time intervals of ER stress agents such as DMC for 2 up to 18 h.

Another measurement of intermediate ER stress is release of intracytoplasmic calcium. We had previously demonstrated that ER stress agents such as DMC induce release of calcium via SERCA inhibition. In order to determine release of intracellular calcium, the cells are loaded by incubating them with 4 µM Fura-2/AM (Invitrogen, Carlsbad, CA) for 30 min at room temperature in external solution containing 138 mM NaCl, 5.6 mM KCl, 1.2 mM MgCl$_2$, 2.6 mM CaCl$_2$, 10 mM HEPES, and 4 mM glucose (pH = 7.4). After loading, the cells are rinsed and transferred to the imaging setup. Cells are then treated with individual drugs for 10 s; fluorescence is elicited with the excitation wavelength alternating between 350 and 380 nm, using a Polychromator V (TILL Photonics GmbH) to provide illumination via a Zeiss Axiovert 100 microscope with a Zeiss Fluor 40× oil objective (Carl Zeiss). Images are captured using a Cascade 512B CCD camera (Photometrics) controlled with MetaFluor software (Molecular Devices) at 0.5 Hz acquisition frequency. Ratios of the images obtained at 350 and 380 nm excitation are used to illustrate changes in the cytoplasmic calcium concentration.

2.4. Late markers of ER stress

Late markers of ER stress include CHOP, a marker for ERS-induced apoptosis, and caspases 4 and 7. Apoptosis, not specific to ER stress induction, may be determined using techniques such as TUNEL staining. This section will only cover the ER stress markers for apoptosis. CHOP, caspase 4, and caspase 7 assays are typically induced within 48 h of ER stress. They are typically detected via Western blot or immunohistochemistry on cytopreps or tissue sections.

2.4.1. Apoptotic markers specific to ER stress

As a transcription factor, CHOP sensitizes cells to ER stress-induced apoptosis through downregulation of BCL-2 and activation of GADD34 and ER01, an ER oxidase. Antibodies to CHOP, and caspases 4, and 7 were purchased from Santa Cruz Biotechnology, Inc. and were used according to manufacturer's recommendations. Caspase 4 was also purchased from StressGen Biotechnology Corporation. The secondary antibodies were coupled to horseradish peroxidase and were detected by chemiluminescence using the SuperSignal West substrate from Pierce. All immunoblots were repeated at least once to confirm the results. Western blot semiquantitation was performed using Image J software.

2.5. Modulation of GRP78/BiP levels *in vitro*

Modulation of GRP78/BiP levels *in vitro* is most useful for evaluation of chemotherapeutic drugs and for evaluation of agents that induce ER stress. Most of our work has concentrated on the antiglioma chemotherapeutic agent temozolomide (TMZ), and on an unique ER stress agent called DMC. TMZ and DMC induce ER stress by 'induction of GRP78. In evaluating modulation of GRP78 levels, cell death assays are useful to evaluate if induction of ER stress induces apoptosis. Moreover, direct siGRP78 can be used as a positive control to modulate GRP78 levels. The green tea extract epigallocatechin 3-gallate (EGCG) has been used to chemically modulate GRP78 levels in the presence of TMZ.

2.5.1. TMZ and EGCG

TMZ is obtained from the pharmacy at the USC and dissolved in PBS to a concentration of 20 mg/mL. Immediately before use, this stock solution is diluted with 0.9% sodium chloride to 2 mg/mL. EGCG is purchased from Sigma-Aldrich (St. Louis, MO) and is dissolved in ddH$_2$O to 10 mg/mL immediately before use. We also use TEAVIGO-EGCG, which are capsules with caffeine-free, highly concentrated green tea extract ($\geq 94\%$ EGCG) that are widely available at health food stores; the capsule contents are dissolved in ultrapure water and used fresh (or stored at $-80\,^\circ$C).

2.5.2. Cell death assays

The two cell death assays that are most commonly used in our laboratory are the methylthiazoletetrazolium (MTT) assay and the cell death enzyme-linked immunosorbent assay (ELISA). MTT assays determine the percentage of viable cells present and are usually performed at 24–72 h after drug addition. Glioma cells are seeded onto 96-well plates in a volume of 50 μl per well (2.0–5.0 × 10^5 cells/mL). The next day, an additional 50 μL of medium containing various concentrations of drug is added and the cells are incubated for 48–72 h. This is followed by the addition of 10 μL thiazolyl blue tetrazolium bromide (MTT; Sigma-Aldrich) for 4 h (stock solution of MTT is 5 mg/mL in PBS). The reaction is stopped, and the cell cultures are lysed by the addition of 100 μL solubilization solution (10% sodium dodecyl sulfate (SDS) in 0.01 M hydrochloric acid (HCl)). The 96-well plate is left in the cell culture incubator overnight for complete solubilization of the MTT crystals, and the optical density (OD) of each well is determined in an ELISA reader at 490 nm. The background value (equal to the OD of control well containing medium without cells MTT solubilization solution) is subtracted from all measured values. All experiments are performed at least in quadruplicate.

Apoptosis in cell cultures *in vitro* is determined by the Cell Death Detection ELISA kit (Roche Diagnostics, Indianapolis, IN) according to manufacturer's instructions. This immunoassay specifically detects the histone region (H1, H2A, H2B, H3, and H4) of mono- and oligonucleosomes that are released during apoptosis. Ninety-six-well plates are seeded with 1000 cells per well, and read at 405 nm in a Microplate Autoreader (Model EL 311SX; Bio-Tek Instruments, Inc., Winooski, VT).

Apoptosis in cytopreps and tumor sections are measured semiquantitatively with the use of the TUNEL assays. All components for this procedure are from the ApopTag *In Situ* Apoptosis Detection kit (Chemicon, Temecula, CA), and used according to the manufacturer's instructions. Positive cells are usually marked purple or brown and read out visually.

2.5.3. siRNA modulation of GRP78

In order to modulate GRP8 expression using siGRP78, we have cloned the siGRP78 cDNA sequence into a replication incompetent lentiviral vector. The sequences of the siRNA against human GRP78 are siGRP78 A: 5′-AAGGAGCGCATTGATACTAGA-3′ and siGRP78 B: 5′-AAGAAAAGCTGGGAGGTAAAC-3′. The sequences of the control siRNA (nonsense sequence) are siControl A: 5′-GGAGAAGAATAG-CAACGGTAA-3′ and siControl B: 5′-AAGGTGGTTGTTTTGTT-CATT-3′. For construction of lentivirus expressing GRP78, full-length human GRP78 is prepared by the USC Norris Core Lab. Briefly, reverse transcription of total HEK293T RNA followed by a two-step PCR amplification and subcloning into the BamH1/XhoI sites of pcDNA3 (Invitrogen) is used to yield pcDNA3-hGRP78. Nonreplicating lentiviral vectors coexpressing 14 biscistronic human GRP78 and enhanced green fluorescent protein (EGFP) linked via the encephalomyocarditis virus (EMCV) internal ribosomal entry site (IRES) are produced using the pLenti6/V5-D-TOPO and ViraPower Lentivirus Expression system supplied by Invitrogen. Initially, EGFP (Clonetech, Mountain View, CA) is inserted into the CMV-driven expression cassette by TOPO cloning as a marker. After the viability of this construct is established, the parent construct is modified. The human GRP78 gene is then digested from pcDNA3/h78 using Xba I and Xho I ligated into pLenti6/EGFP between the CMV promoter and the EGFP gene. Next, an insert encoding the EMCV IRES is generated by PCR from pIRES–EGFP (Clonetech) and subcloned into the Xho I/Age I sites of pLenti6/huGRP78 EGFP. The IRES sequence is inserted between the human GRP78 gene and the EGFP allowing GRP78 expression to be monitored by EGFP fluorescence. The manufacturer's manual is followed for TOPO cloning and production of viral particles. For infection, 104 cells are plated into 6-well dishes and infected with lentivirus at titers of 5×10^6 units/mL (TU/mL). Infected cells are monitored for green fluorescent protein (GFP) under the fluorescent microscope. Cells are used when cultures are 100% GFP positive.

2.5.4. EGCG modulation of GRP78

The major polyphenolic green tea component EGCG is being intensively investigated for its potential chemopreventive and chemotherapeutic activity. In our laboratory, we had previously demonstrated its efficacy in inhibition of GRP78 elevation from TMZ (Pyrko et al., 2007b). Among its many biological effects is EGCG's ability to bind to GRP78 and inactivate its antiapoptotic function, enabling it to increase the chemosensitivity of tumor cells. For example, EGCG has been shown to sensitize different types of tumor cells to a variety of proapoptotic agents, such as 5-fluorouracil, taxol, vinblastine, gemcitabine, or tumor necrosis factor (TNF)-related apoptosis-inducing ligand (TRAIL) *in vitro* and to doxorubicin, paclitaxel, or interferon in mouse tumor models *in vivo* (Siddiqui *et al.*, 2008).

2.5.5. Treatment with TMZ *in vitro*: Colony formation assays

Treatment with TMZ is best performed using colony formation assays (CFA) for *in vitro* assays. TMZ used with MTT assays as described above are usually not that accurate. We generally find that a much higher concentration of TMZ is needed for MTT assays than for CFA. For CFA, 200 cells are seeded into each well of a 6-well plate. After cells have fully attached to the surface of the culture plate, they are exposed to drug treatment for 48 h. Thereafter, the drugs are removed, fresh growth medium is added, and the cells are kept in culture undisturbed for 12–14 days, during which time the surviving cells spawn a colony of proliferating cells. Colonies are visualized by staining for 4 h with 1% methylene blue (in methanol), and then counted for the number of surviving colonies.

2.6. *In vivo* modulation of GRP78

In order to determine the effect of GRP78 modulation *in vivo*, an animal model is crucial. We currently use a subcutaneous and intracranial malignant glioma model. Most of our studies are performed with immune-deficient mice secondary to the easier handling of the animals, and the ability to use immune-deficient mice for human glioma cells. Syngeneic models are usually performed with rat glioma cell lines because of the ease of obtaining syngeneic rat glioma cell lines compared to mice. All of our animal protocols are approved by the Institutional Animal Care and Use Committee (IACUC) of the USC, and all applicable policies are strictly observed during the course of this study. If nude mice are used, we generally employ 4- to 6-week-old male athymic nu/nu mice obtained from Harlan (Indianapolis, IN) and kept in a pathogen-free environment. To support more consistent tumor take and uniform growth, the animals are given whole-body irradiation with 300 cGy of ionizing radiation (Cesium 137) 4 days prior to tumor implantation by using a low dose-rate laboratory irradiator (Gammacell 40; Atomic Energy of Canada Limited, Canada).

2.6.1. Rodent glioma models—Subcutaneous and intracranial

For subcutaneous tumor inoculation, 5×10^5 U-87 glioblastoma cells are injected subcutaneously into the right flank. Once palpable tumors have developed, the animals are randomly divided into treatment groups. The subcutaneous tumor model is convenient not only from the standpoint of ease of implantation, but also with direct visualization of tumor growth, and demonstration of direct measurement of tumor size. We determine tumor size using the following calculations: Tumor size is calculated by the following formula: volume $(mm^3) = L \times W \times H \times 0.5$ (L, length;

W, width; H, height). Student's t-test is used for statistical analysis, and a P-value of <0.05 was considered significant.

For the intracranial model, athymic mice (Harlan, Inc.) are anesthetized and fixed into a stereotactic head frame. A paramedian incision is made and a 1.5-mm burr hole drilled 1 mm anterior to the coronal suture on the right hemisphere and 2 mm lateral from the midline. 2×10^5 cells in a volume of 10 µL is injected into the right frontal lobe of the brain, and thereafter the skin incision is sutured with silk thread. Seven days after tumor cell implantation, drug treatment is initiated by gavage. In the case of TMZ and EGCG, the following regimen is used: TMZ (5 mg/kg) is administered via oral gavage daily for 7 consecutive days, which is followed by 7 days without TMZ; EGCG (50 mg/kg) is given oral daily in parallel to TMZ for the same 7 days on, 7 days off cycle. The entire 14-day cycle is repeated until the animals are sacrificed or have died.

2.6.2. Progression and survival models and GRP78 modulation

In order to determine the efficacy of GRP78 modulation, progression and survival models are used. In general, we use six animals per arm in order to achieve statistical significance with a power of 80%, and alpha of 0.05. Progression models determine the increase in tumor size during treatment. Progression models may be performed using subcutaneous glioma models or by using luciferase labeled glioma cells implanted intracranially. The easiest method is to use subcutaneous glioma models in which the tumor may be measured and assessed for change in tumor size. We generally use the MacDonald criteria for determination of tumor progression (Wen et al., 2010). If there is tumor progression, there is an increase of size by greater than 25%. The intracranial models are generally used to construct Kaplan–Meier survival curves. The animals are often difficult to image directly with rodent MRI scans because of the skull. Therefore, these animals are treated and followed for neurological status. If the animals demonstrate evidence of increased intracranial pressure (bulging eyes) or neurological deficits (i.e., hemiparesis), they are deemed safety to self, sacrificed, and their brains removed. Although we do have access to a rodent MRI scan that can be used for brain imaging, we generally reserve the MRI scan to special situations in which it is crucial to directly monitor tumor size.

2.6.3. Tissue analysis of GRP78 levels

Tissue analysis of the tumor and peritumoral brain is essential to determine the effects of GRP78 modulation. The brain is harvested immediately after animal sacrifice and frozen. After freezing, it is cut into sections. Basic studies include H&E for determination of tumor size, and evaluation of tumor spread within the normal brain. The brain tumor itself is stained for GFAP, CD31 (for endothelial cells), and Ki-67 (a proliferation marker). Moreover, TUNEL assay is performed as described above for evidence of apoptosis. Staining for GRP78 and CHOP may be performed in both frozen and paraffin-embedded tissue to determine for evidence of ER stress. The peritumoral brain is just as important as the normal brain, because it will often demonstrate evidence of increased ER stress in the normal tissue.

3. Conclusion

This chapter has focused on GRP78 and its effect on malignant gliomas. The techniques outlined are applicable for GRP78 studies in all tumors. There are two main differences in glioma studies compared to systemic tumors. First, the brain tumor is more difficult to visualize because it is intracranial. Modulation of GRP78 and its effects on tumor growth is more difficult to assess than a subcutaneous systemic tumor model. Secondly, the blood–brain barrier, although "leaky" in brain tumors, still forms a significant functional barrier preventing free egression or digression of chemotherapeutic agents to the brain. In spite of these difficulties, we have demonstrated that GRP78 modulation is crucial to changing chemosensitivity in malignant brain tumors.

REFERENCES

Hovnanian, A. (2007). SERCA pumps and human diseases. *Subcell. Biochem.* **45,** 337–363.
Pyrko, P., Kardosh, A., Wang, W., Xiong, W., Schonthal, A. H., and Chen, T. C. (2007a). HIV-1 protease inhibitors nelfinavir and atazanavir induce malignant glioma death by triggering endoplasmic reticulum stress. *Cancer Res.* **67**(22), 10920–10928.
Pyrko, P., Schönthal, A. H., Hofman, F. M., Chen, T. C., and Lee, A. H. (2007b). The unfolded protein response regulator GRP78/BiP as a novel target for increasing chemosensitivity in malignant gliomas. *Cancer Res.* **67,** 9809–9816.
Schönthal, A. H. (2008). Endoplasmic reticulum stress and autophagy as targets for cancer therapy. *Cancer Lett.* **275,** 163–169.
Schonthal, A. H., Chen, T. C., Hofman, F. M., Louie, S. G., and Petasis, N. A. (2008). Celecoxib analogs that lack COX-2 inhibitory function: Preclinical development of novel anticancer drugs. *Expert Opin. Investig. Drugs* **17**(2), 197–208.
Siddiqui, I. A., Malik, A., Adhami, V. M., Asim, M., Hafeez, B. B., Sarfaraz, S., and Mukhtar, H. (2008). Green tea polyphenol EGCG sensitizes human prostate carcinoma LNCaP cells to TRAIL-mediated apoptosis and synergistically inhibits biomarkers associated with angiogenesis and metastasis. *Oncogene* **27,** 2055–2063.
Wen, P. Y., Norden, A. D., Drappatz, J., and Quant, E. (2010). Response assessment challenges in clinical trials of gliomas. *Curr. Oncol. Rep.* **12**(1), 68–75.

CHAPTER THREE

Targeting the Unfolded Protein Response in Cancer Therapy

Marina V. Backer,* Joseph M. Backer,* *and* Prakash Chinnaiyan[†]

Contents

1. Introduction	38
2. The UPR and Cancer	38
3. Targeting the UPR	40
3.1. (−)-Epigallocatechin gallate	40
3.2. Targeting the UPR in cancer cells via selective cleavage of GRP78	40
4. Combination of EGF–SubA with Other UPR-Targeting Drugs	50
4.1. Combination of EGF–SubA with thapsigargin for treatment of MDA231luc cells	50
4.2. Combination of EGF–SubA with histone deacetylase inhibitors in the glioblastoma cell line T98G	52
Acknowledgments	54
References	54

Abstract

Rapid growth of tumor cells coupled with inadequate vascularization leads to shortage of oxygen and nutrients. The unfolded protein response (UPR), a defense cellular mechanism activated during such stress conditions, is a complex process that includes upregulation of the endoplasmic reticulum chaperones, such as glucose-regulated protein 78 (GRP78). Due to its central role in UPR, GRP78 is overexpressed in many cancers; it is implicated in cancer cell survival through supporting of drug- and radioresistance as well as metastatic dissemination, and is generally associated with poor outcome. This is the reason why selective destruction of GRP78 could become a novel anticancer strategy. GRP78 is the only known substrate of the proteolytic A subunit (SubA) of a bacterial AB_5 toxin, and the selective SubA-induced cleavage of GRP78 leads to massive cell death. Targeted delivery of SubA into cancer cells via specific receptor-mediated endocytosis could be a suitable strategy for assaulting tumor cells. We fused SubA to

* Sibtech, Inc., Brookfield, Connecticut, USA
[†] Department of Experimental Therapeutics and Radiation Oncology, H. Lee Moffitt Cancer Center and Research Institute, Tampa, Florida, USA

Methods in Enzymology, Volume 491
ISSN 0076-6879, DOI: 10.1016/B978-0-12-385928-0.00003-1

© 2011 Elsevier Inc.
All rights reserved.

epidermal growth factor (EGF), whose receptor (EGFR) is frequently overexpressed in tumor cells, and demonstrated that the resulting EGF–SubA immunotoxin is an effective killer of EGFR-positive tumor cells. Furthermore, because of its unique mechanism of action, EGF–SubA synergizes with UPR-inducing drugs, which opens a possibility for the development of mechanism-based combination regimens for effective anticancer therapy. In this chapter, we provide experimental protocols for the assessment of the effects of EGF–SubA on EGFR-positive cancer cells, either alone or in combination with UPR-inducing drugs.

1. INTRODUCTION

During normal growth and differentiation, cells encounter variations in their microenvironment that can adversely affect their homeostasis and physiological functions. This involves a variety of biochemical, physiologic, and pathologic stimuli, including nutrient and oxygen deprivation, viral infection, and local tissue inflammation, all of which can place significant "stresses" on cellular function. The endoplasmic reticulum (ER) is exquisitely sensitive to these alterations in intracellular homeostasis and plays an important role in allowing cells to adapt to these environmental stresses. This is primarily accomplished through activation of the unfolded protein response (UPR), which represents a defense mechanism that allow cells to respond to adverse conditions threatening their survival (Jamora et al., 1996; Kaufman, 1999). As an organelle is primarily recognized as a compartment for protein folding and assembly, UPR activation initiates a series of adaptive processes, including decreasing the arrival of newly synthesized proteins in the ER, augmenting the folding capacity of the organelle by increasing synthesis of ER chaperones, and enhancing the extrusion of irreversibly misfolded protein (Kaufman, 1999). The chaperone protein glucose-regulated protein 78 (GRP78), also referred to as the immunoglobulin-binding protein BiP, serves as a master UPR regulator that plays a central role in modulating its downstream signaling. Under nonstressed conditions, Grp78 binds to its client proteins PERK, ATF6, and IRE1, preventing their signaling. However, when the ER is overloaded by newly synthesized proteins or is "stressed" by agents that cause accumulation of unfolded proteins, Grp78 binds to the accumulating unfolded proteins in the ER, freeing its specific client proteins, leading to pathway activation (Lee et al., 2008; Schroder and Kaufman, 2005).

2. THE UPR AND CANCER

Environmental fluctuations and cellular stresses are particularly heightened during tumorigenesis. Rapid growth of tumor cells coupled with inadequate vascularization lead to significant changes in the microenvironment, including decreased levels of oxygen, nutrients, and glucose, all of which can be detrimental to both continued cell growth and survival. Therefore, cancer cells may be particularly reliant upon the adaptive mechanisms offered by the UPR for continued growth and survival in these otherwise cytotoxic conditions (Lee, 2007). One of the first studies demonstrating this potential reliance of cancer cells on the UPR was presented by Jamora et al. (1996), in which Grp78-knockdown fibrosarcoma cells demonstrated similar *in vitro* growth characteristics as their parental line, however, were not able to sustain growth *in vivo* in a mouse model. Similar work has been demonstrated involving a downstream mediator of UPR signaling, XBP-1, which was shown to be directly implicated in the growth of a tumor under hypoxic conditions (Romero-Ramirez et al., 2004).

With data suggesting a specific reliance of tumors on the UPR to sustain continued growth, several investigators have explored the potential for aberrant expression and the subsequent prognostic implications of UPR-related proteins in cancer. Pyrko et al. (2007) demonstrated that Grp78 is expressed at low levels in normal adult brain, however significantly elevated in malignant glioma specimens and human malignant glioma cell lines. Using microarray analysis, Lee et al. (2008) similarly found that Grp78 expression was upregulated in glioma specimens and that its expression correlated with tumor grade. Further, Grp78 expression had prognostic implications in glioblastoma, with increased expression portending poor survival. Similar findings were reported in breast cancer, including increased expression of Grp78 in malignant but not benign breast lesions (Fernandez et al., 2000). In addition, Grp78 expression (Lee et al., 2006; Scriven et al., 2009) and its downstream mediator XBP-1 (Scriven et al., 2009) were predictive of clinical response to therapy. In prostate cancer, Grp78 expression was found to be upregulated, correlating with both tumor recurrence and patient survival (Daneshmand et al., 2007), and associated with hormone-resistant disease (Pootrakul et al., 2006). In addition to solid tumors, aberrant activation of the UPR has been recently identified in hematological malignancies. The formation of the spliced variant of XBP1 was detected in about 16% of acute myeloid leukemia (AML) patients. Interestingly, unlike solid tumors, UPR activation predicted a more favorable outcome in this subgroup of patients (Schardt et al., 2009).

As described above, overexpression of UPR-related proteins appears to be a negative prognostic factor in several solid tumors. Based on these findings, the potential for this pathway to contribute toward tumor aggressiveness and/or therapeutic resistance has been studied in several preclinical models. In glioblastoma models, Grp78 was shown to contribute toward resistance to a variety of chemotherapeutics, including temozolomide, 5-fluorouracil, CPT-11, etoposide, cisplatin, and ionizing radiation (Lee et al., 2008; Pyrko et al., 2007). In breast cancer models, UPR activation induced resistance to both doxorubicin and 5-fluorouracil (Scriven et al., 2009). In addition to influencing response of the primary tumor, data also suggest Grp78 confers chemoresistance to tumor-associated endothelial cells. In these studies, it was shown that Grp78 was generally highly elevated in the vasculature derived from human glioma specimens both *in situ* in tissue and *in vitro* in primary cell culture compared with normal brain tissue and vasculature. In addition, their relative resistance to standard chemotherapeutic agents was overcome following Grp78 targeting (Virrey *et al.*, 2008). Beyond therapeutic resistance, the UPR may also directly influence tumor angiogenesis. Ghosh *et al.* (2010) demonstrated that the three downstream mediators of the UPR, PERK, Ire1, and ATF6 powerfully regulate VEGF expression under stressed conditions. These findings are supported by studies using a breast cancer mouse model of Grp78 heterozygotes, where Grp78 expression was reduced by approximately one-half. The tumor microvessel density was substantially reduced in the Grp78 heterozygous mice when compared to wild type, with no effect on vasculature of normal organs (Dong *et al.*, 2008). Collectively, these findings support the role the UPR plays in both tumor growth and therapeutic resistance, suggesting its potential to serve as a therapeutic target in cancer therapy.

3. Targeting the UPR

3.1. (−)-Epigallocatechin gallate

With emerging data suggesting the UPR contributes toward tumor growth and therapeutic resistance, identifying ways to target this pathway as a form of molecularly targeted cancer therapy represents an active area of research. Although this line of investigation has only recently gained traction, some very interesting leads are actively being explored. It was recently discovered that (−)-epigallocatechin gallate (EGCG), which is the major polyphenol component of green tea and the most active single anticancer factor found in tea, directly interacts with the ATP-binding site of Grp78, thereby regulating protein function through competitive inhibition. Binding of EGCG caused the conversion of Grp78 from its active monomer to the inactive dimer and oligomer forms. Treatment with EGCG interfered with the formation of the antiapoptotic Grp78–caspase-7 complex, resulting in increased chemotherapy-induced apoptosis (Ermakova *et al.*, 2006). Similar findings were observed using EGCG in glioblastoma cell lines (Pyrko *et al.*, 2007). Trials are currently underway testing the potential of EGCG to serve as a strategy for both cancer treatment and prevention (Khan *et al.*, 2009).

3.2. Targeting the UPR in cancer cells via selective cleavage of GRP78

Selective destruction of GRP78 became possible with the discovery of a novel bacterial toxin SubAB (Morinaga *et al.*, 2007; Paton *et al.*, 2004, 2006). SubAB, the first member of a new family of potent AB_5 toxins produced by Shiga toxigenic *Escherichia coli*, comprises a single A subunit (SubA) with subtilisin-like proteolytic activity and a pentamer of targeting B subunits. Targeting B subunits bind to cellular receptors and deliver SubA into the cell, where it selectively cleaves only one protein, GRP78, at a single site, di-leucine motif (L416L417) in the hinge region connecting the ATPase and protein-binding domains of the molecule (Paton et al., 2006).

Under physiologically normal conditions, the majority of GRP78 is located within the ER lumen. Therefore, SubAB toxin must be internalized by the cell and delivered to the ER through a retrograde transport via the Golgi system (Chrong et al., 2007). Nevertheless, GRP78 cleavage is remarkably rapid; it starts after 15 min of the exposure of susceptible Vero cells to SubAB, and virtually all intact GRP78 in the cell is degraded by 1–2 h of exposure, even at the toxin doses as low as 10 ng/mL, suggesting a highly potent catalytic activity (Paton et al., 2006). Remarkably, selective GRP78 cleavage provides SubAB with 10–100 times higher cytotoxicity than that of the "conventional" Shiga toxins 1 and 2 (Stx1 and Stx2) that are produced by the same Shiga toxigenic *E. coli*, but have a different highly specific target; Stx1 and Stx2 are both site-specific N-glycosidases that cleave off A_{4324} in the 5′ terminus of 28S rRNA thereby inhibiting protein synthesis and leading to cell death (Saxena et al., 1989).

Selective GRP78 cleavage by SubAB leads to the release of PERK, IRE1, and ATF6, and hence activates the corresponding three ER stress signaling pathways (Wolfson et al., 2008), which lead to massive apoptosis (Paton et al., 2006). It should be noted that the cytotoxicity resulted from the direct cleavage of GRP78 with SubA is in contrast to the outcome of inhibition or downregulation of GRP78 via antisense or RNA interference approaches, neither of which led to cell death, or even significantly affected cell growth (Dong et al., 2005; Tsutsumi et al., 2006; Zu et al., 2006). Given the high potency and the unique cancer-relevant target, SubA appears to be well suited for targeted delivery into tumor cells.

To achieve specific antitumor activity, without significant systemic toxicity, SubA should be delivered selectively into cancer cells. For this, typically, toxins are fused to targeting moieties, so that resulting constructs, generically named immunotoxins, can bind to and be internalized by proteins selectively overexpressed on the surface of tumor cells. Epidermal growth factor rece1ptor (EGFR) is one such protein and it has been used previously as a target for immunotoxins (Kreitman, 2006). EGFR (also known as erbB1 or HER1) is a 170-kDa member of the erbB tyrosine kinase receptor family, which plays an important role in cell differentiation, proliferation, and migration and is frequently overexpressed in tumor cells, making it an attractive drug target (Rusch et al., 1996). Ligand binding induces EGFR dimerization and autophosphorylation that activates downstream signal transduction pathways (Scaltriti and Baselga, 2006; Yarden, 2001). Ligand/receptor complex is then internalized and trafficked to lysosomes for degradation, but a fraction of EGFR may escape degradation and be translocated into the nucleus, where it functions as a transcription factor or as a cofactor for STAT3 and E2F1 transcription factors. Nuclear EGFR is implicated in resistance to therapy and may contribute to poor survival of cancer patients through not well-understood mechanisms (Scaltriti and Baselga, 2006).

We cloned cDNA encoding an EGFR ligand EGF in-frame with cDNA encoding SubA, and expressed the resulting immunotoxin, named EGF–SubA, with functionally active EGF and SubA moieties (Backer et al., 2009). We found that EGF–SubA rapidly accumulates in cancer cells via EGFR-mediated endocytosis and is cytotoxic to both growing and confluent EGFR-expressing cells in the picomolar concentration range, and that this toxicity is mediated by cleavage of GRP78 (Backer et al., 2009). In this respect, despite different routes of intracellular delivery, EGF–SubA is similar to the bacterial SubAB holotoxin (Paton et al., 2006). In this chapter, we provide the optimal protocols for assessment of EGF–SubA toxicity for EGFR-positive cancer cells in tissue culture and *in vivo*. The tissue culture protocols include evaluation of overall cell viability and Western blot analysis of GRP78 cleavage in cells exposed to EGF–SubA. To characterize cellular response to EGF–SubA, we recommend performing both assays, because together they provide mechanistic insights into how cell viability depends on dynamics of GRP78 destruction and upregulation. In any case, understanding how tumor cells respond to EGF–SubA in tissue culture is an important step that should be accomplished before progressing to *in vivo* studies.

3.2.1. Viability of EGFR-positive cancer cells exposed to EGF–SubA

This assay provides a quantitative measurement of the cellular response to EGF–SubA, an IC_{50} value, which is the concentration of EGF–SubA sufficient to inhibit cell growth to 50%. Under standardized conditions, the IC_{50} is constant for any given cell line, and could be used for comparing different cell lines in terms of their overall susceptibility to EGF–SubA.

Cell viability assay takes several days of exposure to EGF–SubA, followed by 4-h incubation with CellTiter assay reagents, a tetrazolium compound MTS and an electron coupling reagent. MTS is bioreduced by cells into a formazan product that is soluble in tissue culture medium, which then is measured by absorbance at 490 nm, directly from 96-well assay plates without additional processing. The quantity of formazan product is directly proportional to the number of living cells in culture, and is a measure of cell viability.

The protocol provided here includes two cancer cell lines: human prostate cancer cells PC3 and human breast cancer cells MDA231luc. PC3 cells are available from American Type Culture Collection (ATCC), while MDA231luc are a luciferase-expressing derivative of human breast carcinoma MDA-MB-231 developed in SibTech, Inc. (Brookfield, CT).

Parental MDA-MB-231 breast cancer cells are available from ATCC. We selected these two cell lines because they express similarly moderate EGFR ($\sim 2 \times 10^5$ EGFR per cell) and similar levels of GRP78 (Backer *et al.*, 2009, in Supplementary Material), yet display different susceptibility to EGF–SubA. Hence, these two cell lines are an excellent example of the complex molecular mechanisms underlying the cellular response to EGF–SubA, indicating that by no means is it determined solely by the levels of EGFR and GRP78.

3.2.1.1. Required materials

Standard equipment for tissue culture

Plate reader

A plate reader that reads optical density in a 96-well plate format at $\lambda = 490$ nm, for example, Synergy 4 Plate Reader (BioTek Instruments, Inc., Winooski, VT)

Additional materials

- *Cells of interest*: any EGFR-expressing cancer cells, like PC3 or MDA231-luc. The growth medium for the cells contains F-12 (for PC3) or DMEM (for MDA231luc) supplemented with 10% FCS and 2% glutamine. Antibiotics (e.g., penicillin, streptomycin, nystatin) could be used during the experiment. In our experience, there is no interference of the above antibiotics with the binding, internalization, or intracellular events induced by EGF–SubA.
- EGF–SubA is available from SibTech, Inc. EGF–SubA is supplied as a sterile solution at a concentration of 1 mg/mL (22 µ*M*). EGF–SubA is shipped at ambient temperature; it is stable at $+4\,^\circ$C for at least 7 days, but for long-term storage it should be frozen at $-20\,^\circ$C in small, single-use aliquots. To ensure long-term activity, we recommend avoiding repetitive freezing–thawing cycles.

Other reagents

- CellTiter 96® Aqueous Non-Radioactive Proliferation Assay (Promega, Madison WI).
- Trypsin–EDTA solution (trypsin 0.25%, EDTA 0.05%).
- Ca^{2+} and Mg^{2+}-free PBS.

Disposables

- T-75 tissue culture flasks for routine cell propagation.
- 96-well assay plates.

3.2.1.2. Tumor cell preparation

One day prior to exposure

Grow cells in T-75 flasks under normal culture conditions (5% CO_2, 37 °C), until 70–80% confluence, then trypsinize, and seed onto a 96-well plate, at a plating density of 2000 cells per well, in a volume of 50 μL per well. We recommend using triplicate wells per every experimental point.

Note that this assay takes at least 72 h (for slow growing cell lines this time could be 96 h, or even 120 h). Keep in mind that water evaporation from the wells of edge rows is faster than that from the center of the plate; therefore, by the end of the assay, the volume of medium in the edge wells will be significantly lower than that in the wells in the plate center. To avoid uneven growth conditions in the assay wells, we recommend not to seed cells in the edge rows. Simply fill these wells with sterile PBS or water, to maintain humidity within the plate.

On the day of exposure

Approximately 20 h after cell plating, dilute EGF–SubA in complete culture medium prewarmed to 37 °C, to make 2 nM EGF–SubA. Note that this is an 11,000-fold dilution from the initial 22 μM EGF–SubA; therefore, to make an accurate concentration of 2 nM EGF–SubA, we recommend making several consecutive dilution steps using complete medium. Note also that 2 nM is a starting (the highest) concentration of EGF–SubA, which should be diluted further. In a fresh sterile 96-well plate, make eight 1:2 serial dilutions of 2 nM EGF–SubA in complete culture medium prewarmed to 37 °C. The resulting EGF–SubA concentrations range from 8 pM to 2 nM.

3.2.1.3. Assessment of cell susceptibility to EGF–SubA

Once the EGF–SubA dilutions are made, take plate with cells from CO_2 incubator. Using multichannel pipette, transfer all diluted EGF–SubA solutions to triplicate wells with cells, adding 50 μL per well. Add complete culture medium to control wells. The final volume is 100 μL per well, and the final concentrations of EGF–SubA vary from 0 (control untreated cells) to 1 nM, due to the final twofold dilution. Place the plate back in CO_2 incubator for at least 72 h.

Note that by the end of exposure, cells in control wells should reach 70–90% confluence. In our experience, it takes 72 h for the majority of cancer cell lines. However, there are slowly growing cells that would reach this point only after 96–120 h of incubation.

When cells in control wells reach 70–90% confluence, add CellTiter reagent solution to all wells with cells, 20 μL per well, according to the manufacturer's protocol. Place the plate back in CO_2 incubator for exactly 4 h, and then read optical density in the plate reader set at 490 nm. Plot the optical density as percentage of control cells versus the concentration of EGF–SubA, and determine the IC_{50} values (Fig. 3.1).

3.2.2. Analysis of selective GRP78 cleavage in cancer cells

Cleavage of GRP78 is a short-term assay; it reflects an early (within several hours) cell response, before any changes in cell growth could be detected. Notice that usually there are two distinct GRP78 responses to EGF–SubA, cleavage of GRP78 and upregulation of GRP78 expression (Backer et al., 2009). Typically, EGF–SubA-induced cleavage reflects the general susceptibility of cells to the immunotoxin. Western blot analysis of GRP78 cleavage might be instrumental for a fast screening of a number of EGFR-positive cancer cells or even cell subtypes, for example, tumor cells that

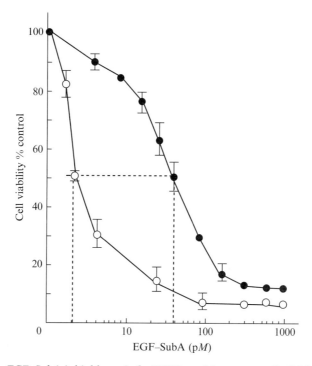

Figure 3.1 EGF–SubA is highly toxic for EGFR-positive tumor cells. PC3 cells (open circles) or MDA231 cells (closed circles) were plated in triplicate wells of 96-well plates, 2000 cells per well, and exposed to EGF–SubA 20 h later. Cell viability was determined as described in Section 2.1. Viability of control untreated cells was defined as 100%. For these cell lines, the IC_{50} values differ 20-fold, 2 pM for PC3 cells and 40 pM for MDA231luc cells, despite similar levels of EGFR and GRP78 in these cells (Backer et al., 2009).

escape conventional chemo- and radiation therapy. Such cells often overexpress GRP78 that helps them to cope with therapy-induced insults (Lee, 2007). From this perspective, fast and efficient EGF–SubA-induced cleavage of GRP78, as detected by Western blot analysis, could indicate a promising strategy for treatment of therapy-resistant tumor cells with EGF–SubA.

The effective concentration of EGF–SubA for this assay depends on the IC_{50} for given cells (should be at least 20 times the IC_{50} value; determined as described in Section 3.2.1). For example, for MDA321luc cells (IC_{50} of 40 pM), well-detectable GRP78 cleavage within the first hour of exposure is achieved at a concentration of 1 nM EGF–SubA.

3.2.2.1. Required materials

Equipment for tissue culture, SDS-PAGE, and protein transfer

Additional materials

- Cancer cells and EGF–SubA immunotoxin: as described in Section 3.2.1.1.
- GRP78-specific antibody for Western blot analysis. Note that antibodies raised against N- or C-terminal portions of GRP78 could work equally well in this assay, but keep in mind the difference in the molecular weights of the detectable GRP78 fragments. We provide here an example of the use of two GRP78-specific antibodies from different suppliers. The first is rabbit Sc-13968 GRP78/BiP antibody H-129 (Santa Cruz Biotechnology, Santa Cruz, CA) which reacts with the intact GRP78 and its 28-kDa fragment. This antibody allows for a simultaneous detection of the intact GRP878 (78 kDa), its 28-kDa fragment, and is compatible with the use of antibodies for detection of β-actin (40–44 kDa), for normalization purposes, on the same blot without stripping (Fig. 3.2A). The second is rabbit MAB4846 GRP78/BiP clone #474421antibody from R&D Systems (Minneapolis, MN) which recognizes the intact GRP78 and its 44-kDa fragment. Note that this antibody would interfere with β-actin detection (40–44-kDa bands, for the majority of commercially available β-actin antibodies), and therefore you should strip the blot probed with this GRP78-specific antibody and reprobe it for β-actin. Alternatively, you can run the same cell lysates separately, specifically for Western blot analysis for probing with β-actin.
- β-actin-specific antibody for GRP78 band normalization analysis. We used rabbit β-actin antibody from Sigma (St. Louis, MO) that reacts with a 44-kDa β-actin band.
- Secondary anti-rabbit whole IgG antibody conjugated to horseradish peroxidase (HRP) and the ECL Plus detection kit were from GE Healthcare.

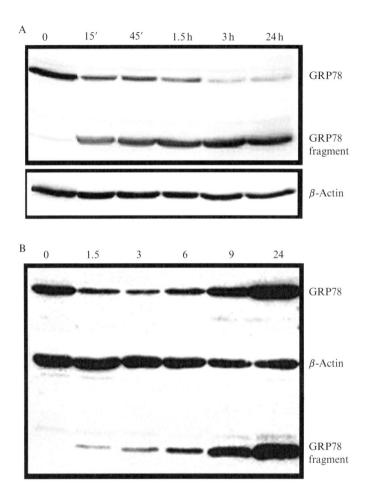

Figure 3.2 EGF–SubA induces rapid cleavage of GRP78 in EGFR-positive cancer cells. Selective cleavage of GRP78 in PC3 cells (A) and MDA231luc cells (B). Cells were plated in 6-well plates, 0.5 million cells per well, and exposed to 1 nM EGF–SubA 20 h later. Equal amounts of clarified cytosols (40 μg of total cellular protein) were analyzed by Western blotting using GRP78-specific antibodies from R&D Systems (A) or Santa Cruz (B), as described in Section 3.2.2.1. Reproduced from Backer *et al.* (2009), with permission.

Other reagents

- Trypsin–EDTA solution (trypsin 0.25%, EDTA 0.05%).
- Ca^{2+} and Mg^{2+}-free PBS supplemented with 5 mM EDTA.
- 10% (v/v) NP-40 (Sigma).
- MicroBCA assay (Pierce, Rockford, IL).

- Supported 0.22 μm nitrocellulose for protein transfer (BioRad, Hercules CA).
- X-ray film or gel documentation system (e.g., Gel Logic 1500 Imaging System from Carestream, Rochester, NY).

Disposables

- 1.5-mL (Eppendorf) tubes.
- 6-well tissue culture plates.

3.2.2.2. Tumor cell preparation

One day prior to exposure to EGF–SubA

Determine how many time points of exposure you need per cell line. Six time points may be a convenient number for an initial assessment, because it provides for critical exposure times, for example, 0.5, 1, 3, 6, 24 h, and an untreated control, and you can conveniently seed one 6-well plate per cell line. Trypsinize cells and seed onto a 6-well plate at a plating density of 500,000–700,000 cells per well, so that six individual wells of every plate have cells of one cell line at about 80% confluence on the exposure day.

On the day of exposure

Approximately 20 h after cell plating, dilute EGF–SubA in complete culture medium prewarmed to 37 °C, to make a final of 20–100× the corresponding IC_{50} value. In our protocol, we use the EGF–SubA concentration of 1 nM.

3.2.2.3. Tumor cell exposure and Western blot analysis

Aspirate medium from one well designated for the longest exposure time (e.g., 24-h exposure). Add 2 mL of 1 nM EGF–SubA in culture medium prepared as described above, and place the plate back to CO_2 incubator. Continue with 6, 3, 1, and 0.5-h exposure time points, so that by the end of the shortest time point (e.g., 0.5-h), you have all cells exposed to EGF–SubA for the corresponding times and ready for harvesting.

To harvest cells, aspirate medium from all wells, wash cells once with Ca^{2+} and Mg^{2+}-free PBS supplemented with 5 mM EDTA. Add fresh Ca^{2+} and Mg^{2+}-free PBS with 5 mM EDTA, 1.2 mL per well, and let sit at room temperature. Monitor cell shape under microscope. Cells are ready for harvesting when they become round shaped and start to detach from plastic. At this point, gently resuspend cells in each well using a 1-mL automatic pipette, and collect cell suspensions from each well into a prelabeled 1.5-mL microcentrifuge tubes. Keep tubes with cell suspensions on ice, until all cells from the 6-well plate are collected.

Centrifuge all tubes at $5000 \times g$ for 5 min; carefully aspirate the supernatants, trying not to disturb cell pellets, and place tubes on ice. Add ice-cold PBS to cell pellets, 90 μL per tube, and resuspend the cell pellets by pipetting. Add 10 μL 10% NP-40 to each tube, to make a final concentration of 1% (v/v) of NP-40, mix by pipetting, and let sit on ice for 5 min, or until cell suspension is clarified. Centrifuge, now at $12,000 \times g$ for 10 min, and collect the clarified supernatants into prelabeled ice-cold 1.5-mL tubes. Take out 2 μL of each cell lysate for microBCA analysis of total protein content. Run microBCA according to the manufacturer's protocol.

Note that in order to preserve cellular proteins from degradation, cell lysates should be frozen at $-80\,^\circ\text{C}$ immediately after taking the aliquots for microBCA assay. Keep cell lysates frozen at $-80\,^\circ\text{C}$ for the time required to accomplish microBCA assay, to calculate protein concentration in cell lysates, and to prepare for SDS-PAGE. Alternatively, add any commercially available protease inhibitor cocktail and keep the cell lysates on ice until everything is ready for SDS-PAGE.

Load cell lysates on a 15% polyacrylamide gel, 40 μg of total cellular protein per lane, separate by reducing SDS-PAGE at 200 V for 60 min, transfer to nitrocellulose, and probe with a GRP78-specific antibody according to the manufacturer's protocol. Probing with β-actin antibody can be done simultaneously, or consecutively, as discussed in Section 3.2.2.1 and shown in Fig. 3.2A and B. After development with ECL Plus, blot images could be obtained either by exposure X-ray film, or by capturing by a gel documentation system. Even though equal amounts of total cellular protein have been loaded on the gel, to account for potential sample loading and protein transfer errors, GRP78 band intensities should be normalized by the corresponding β-actin signals on the same blots.

As shown in Fig. 3.2, in PC3 and MDA231luc cells, EGF–SubA-induced changes in GRP78 progress through a rapid and significant decline of intact protein followed by upregulation by *de novo* protein synthesis, which is especially prominent in MDA231luc cells. We found that the rapid decline pattern of the intact GRP78 is typical for the majority of EGFR-positive cells, however, the percentage of uncleaved GRP78 and the levels of *de novo* synthesized GRP78 vary between cell lines.

Even though the ultimate answer on the therapeutic potential of EGF–SubA could be obtained only in *in vivo* studies, the susceptibility of tumor cells to EGF–SubA detected as described above usually correlates well with the response of the corresponding tumors to EGF–SubA treatment in animal tumor models. Indeed, after systemic administration, EGF–SubA significantly inhibited growth of MDA231luc and PC3 tumor xenografts grown in immunocompromised mice (Backer *et al.*, 2009).

4. Combination of EGF–SubA with Other UPR-Targeting Drugs

Nearly all bacterial or plant toxins explored for cancer therapy so far work via inhibition of protein synthesis (Holzman, 2009; Kreitman, 2009; Pastan et al., 2006; Potala et al., 2008), and because they target such a vital cellular process, even small amounts of these immunotoxins delivered into a physiologically normal, noncancerous cell could cause a serious damage or even cell death. From a clinical point of view, most immunotoxins display an unacceptably high level of nonspecific toxicity, providing for a too narrow therapeutic window (Kreitman, 2006, 2009).

Unlike many other bacterial or plant toxins, SubAB possesses a subtilisin-like proteolytic activity, and its unique mechanism of action, the selective cleavage of GRP78, opens new possibilities for cancer therapy. One advantage is that a 50% decrease in the level of GRP78 does not affect normal cell, but significantly impedes tumor growth and angiogenesis (Dong et al., 2008). These findings suggest that EGF–SubA might be less toxic for noncancerous cells that also express EGFR, even though at lower levels than cancerous cells. Another distinct advantage of EGF–SubA is the potential for mechanism-based combination with other drugs targeting UPR. Typically, combination of two anticancer drugs that aim at the same molecular target via two independent mechanisms calls for less amount of either drug for therapy, while resulting in a more severe insult to cancer cells. We provide here protocols on treatment of cancer cells with combination of EGF–SubA and UPR-targeting drugs.

4.1. Combination of EGF–SubA with thapsigargin for treatment of MDA231luc cells

Thapsigargin inhibits ER Ca^{2+}-dependent ATPase, leading to a depletion of ER Ca^{2+} storage, which, in turn, decreases the activity of Ca-dependent chaperones leading to an increase in unfolded proteins and the corresponding induction of UPR signaling (Denmeade and Isaacs, 2005). Based on this mechanism, thapsigargin might be a good candidate for combination with EGF–SubA to kill EGFR-positive cancer cells. Indeed, using the protocols described in Section 3.2, we demonstrated that treatment of MDA231luc with thapsigargin resulted in a time-dependent increase of GRP78 (Fig. 3.3A); combination of EGF–SubA and thapsigargin led to accumulation of the GRP78 fragment, but not to upregulation of the intact protein, as was observed for either thapsigargin (Fig. 3.3A) or EGF–SubA alone (Fig. 3.2B). Thus, the combination of EGF–SubA and thapsigargin effectively sabotaged the UPR defense mechanism.

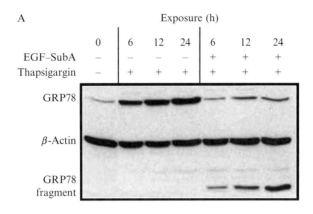
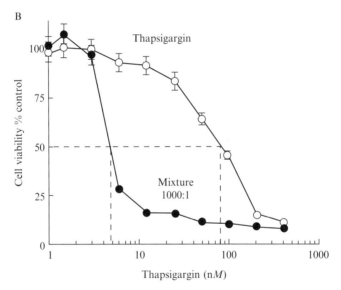

Figure 3.3 EGF–SubA enhances cytotoxic effects of thapsigargin. (A) MDA231luc cells plated in 6-well plates, 0.5 million cells per well, and 20 h later were exposed for indicated times to the following conditions: 1 nM EGF–SubA, 2.5 μM thapsigargin, or their combination. Clarified cytosol fractions were analyzed as described in Section 2.2, using GRP78-specific antibody from Santa Cruz. (B) MDA231luc cells were plated in 96-well plates 2000 cells per well. Twenty hours later, cells in triplicate wells were exposed to varying amounts of thapsigargin alone or its mixture with EGF–SubA at a molar ratio of 1000:1. Viability was determined 72 h later, as described in Section 2.1. Reproduced from Backer et al. (2009), with permission.

Here, we provide a protocol how to analyze the interactions of EGF–SubA and thapsigargin, based on the viability assay described above (Section 3.2.1).

4.1.1. Required materials
This is described in Section 3.2.1.

Additional materials

Thapsigargin and dimethyl sulfoxide (DMSO) are available from Sigma. Make a stock solution of 2 mM thapsigargin in DMSO and store it in small aliquots at $-20\ °C$ for up to 6 months.

4.1.2. MDA231luc cell preparation
This is described in Section 3.2.

4.1.3. Assessment of the impact of EGF–SubA and thapsigargin combination on MDA321luc cells

The method described by Berenbaum (1978) could be used for evaluation of the relationships between the effects of EGF–SubA and an UPR-inducing drug (in this example, thapsigargin) in terms of synergism, additivity, or antagonism. The protocol provided here is based on the cellular viability analysis described in Section 3.2.1.

First, determine the IC_{50} values for each drug alone, EGF–SubA and thapsigargin, as described in Section 3.2. For EGF–SubA, the IC_{50} is 40 pM; while for thapsigargin, it is 3 orders of magnitude higher, 80 nM (Backer et al., 2009). Make a mixture of EGF–SubA and thapsigargin at a molar ratio based on their corresponding IC_{50} values (e.g., 1:1000). The starting mixture for serial dilution should contain 2 nM EGF–SubA and 20 µM thapsigargin in prewarmed complete culture medium. Serially dilute this mixture, using the dilution protocol provided in Section 3.2.1.

If the drugs are additive, it will lead to an approximate twofold decrease in the IC_{50} value for each drug in combination, relative to the drugs alone. The sum (S) of fractional IC_{50} (a ratio of IC_{50} for a drug in a combination to IC_{50} for drug alone) serves as a quantitative criterion, indicating synergism when $S < 1$, additivity when $S \sim 1$, and antagonism when $S > 1$. The combination of EGF–SubA and thapsigargin is dramatically more effective than either drug alone ($S = 0.15$), with fractional IC_{50} values of 0.056 for thapsigargin and 0.094 for EGF–SubA, respectively. As shown in Fig. 3.3, synergism between EGF–SubA and thapsigargin was particularly obvious at the essentially nontoxic concentrations of each compound alone.

4.2. Combination of EGF–SubA with histone deacetylase inhibitors in the glioblastoma cell line T98G

Histone deacetylase (HDAC) inhibitors are a class of targeted anticancer agents that are progressing through clinical trials. It has recently been demonstrated that HDAC inhibitors can activate the UPR (either indirectly, by impairing the clearance of misfolded proteins (Kawaguchi et al., 2003), or directly, through transcriptional control (Baumeister et al., 2005) and/or Grp78 acetylation (Kahali et al., 2010; Rao et al., 2010)) and that activation may contribute toward the antitumor activity of HDAC inhibitors (Kahali et al., 2010). Based on these findings, we sought to determine if targeting the UPR with the GRP78 selective agent EGF–SubA could enhance the antitumor activity of HDAC inhibitors in glioblastoma cell lines. Using the clonogenic assay, T98G cells were seeded as single cells in 6-well culture plates. After allowing to seed, cells were exposed to the HDAC inhibitor vorinostat (SAHA) alone or combined with EGF–SubA (1 pM) for 48 h. Cultures were then replaced with fresh medium during the colony-forming period (10–14 days). Survival curves were generated after normalizing for the cell killing induced by EGF–SubA alone. As demonstrated in Fig. 3.4, combining the HDAC inhibitor vorinostat with the GRP78 selective agent EGF–SubA enhanced cytotoxicity.

4.2.1. Required materials
This is described in Section 3.2.1.

Additional materials

- Vorinostat was provided by Merck Research Laboratories (Whitehouse Station, NJ) through the NCI-CTEP and was dissolved in DMSO at

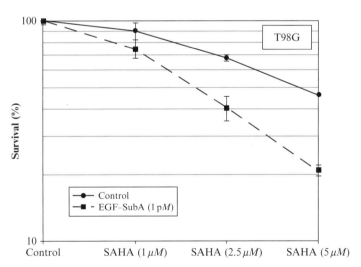

Figure 3.4 EGF–SubA enhances the cytotoxic effects of the HDAC inhibitor vorinostat (SAHA). The glioblastoma cell line T98G was seeded in 6-well tissue culture plates and exposed to vorinostat ± EGF–SubA (1 pM) 6 h later. After 48-h drug exposure, culture plates were washed and replaced with fresh media. Colonies were determined 10–14 days later, and survival curves were generated after normalizing for cell killing of EGF–SubA alone.

1 mM and stored in aliquots at $-80\,^{\circ}\mathrm{C}$. Vorinostat was added to the cell treatment at 1 mL final concentration.
- Isotonic buffer for cell counting when using a Coulter counter.
- 6-well plates (Corning).
- Crystal violet (0.5% in v/v in methanol) solution for staining.
- Colony counting pen.

4.2.2. Clonogenic assay

The clonogenic assay is an *in vitro* cell survival assay that evaluates all modalities of cell death based on the ability of a single cell to grow into a colony. The colony is defined to consist of at least 50 cells. Different cell lines have different plating efficiencies (PE). When untreated cells are plated as a single-cell suspension at low densities, they will grow to colonies in $\sim$10–14 days. PE is the ratio of the number of colonies to the number of cells seeded. The number of colonies that arise after treatment of cells, expressed relative to the individual cells PE, is called the surviving fraction (SF). Cells were seeded as single cells in 6-well plates and allowed to adhere for 6 h, treated as described above for 48 h, then medium was replaced with drug-free medium, and cells were allowed to incubate for 10–14 days. Plates were then stained with crystal violet, and colonies consisting of 50 or more cells were manually counted. Results were normalized to the colony-forming efficiency of the vehicle control. The survival fraction was calculated based on the number of colonies formed in drug-treated cells relative to that of the untreated control.

ACKNOWLEDGMENTS

The research relevant to this chapter was supported by an NIH grant 1R43CA132349-01A2.

REFERENCES

Backer, J. M., Krivoshein, A., Hamby, C. V., Pizzonia, J., Gilbert, K., Ray, J. S., Brand, H., Paton, A. W., Paton, J. C., and Backer, M. V. (2009). Chaperone-targeting cytotoxin and ER stress-inducing drug synergize to kill cancer cells. *Neoplasia* **11,** 1165–1173.

Baumeister, P., *et al.* (2005). Endoplasmic reticulum stress induction of the Grp78/BiP promoter: Activating mechanisms mediated by YY1 and its interactive chromatin modifiers. *Mol. Cell Biol.* **25,** 4529–4540.

Berenbaum, M. C. (1978). A method for testing synergy with any number of agents. *J. Infect. Dis.* **137,** 122–130.

Chrong, D. C., Paton, J. C., Thorpe, C. M., and Paton, A. W. (2007). Clathrin-dependent trafficking of subtilase cytotoxin, a novel AB5 toxin that targets the ER chaperone BiP. *Cell. Microbiol.* **10,** 795–806.

Daneshmand, S., et al. (2007). Glucose-regulated protein GRP78 is up-regulated in prostate cancer and correlates with recurrence and survival. *Hum. Pathol.* **38,** 1547–1552.

Denmeade, S. R., and Isaacs, J. T. (2005). The SERCA pump as a therapeutic target: Making a "smart bomb" for prostate cancer. *Cancer Biol. Ther.* **4,** 14–22.

Dong, D., Ko, B., Baumeister, P., Swenson, S., Costa, F., Markland, F., Stiles, C., Patterson, J. B., Bates, S. E., and Lee, A. S. (2005). Vascular targeting and antiangiogenesis agents induce drug resistance effector GRP78 within the tumor microenvironment. *Cancer Res.* **65,** 5785–5791.

Dong, D., Ni, M., Li, J., Xiong, S., Ye, W., Virrey, J. J., Mao, C., Ye, R., Wang, M., Pen, L., Dubeau, L., Groshen, S., et al. (2008). Critical role of the stress chaperone GRP78/BiP in tumor proliferation, survival, and tumor angiogenesis in transgene-induced mammary tumor development. *Cancer Res.* **68,** 498–505.

Ermakova, S. P., et al. (2006). (-)-Epigallocatechin gallate overcomes resistance to etoposide-induced cell death by targeting the molecular chaperone glucose-regulated protein 78. *Cancer Res.* **66,** 9260–9269.

Fernandez, P. M., et al. (2000). Overexpression of the glucose-regulated stress gene GRP78 in malignant but not benign human breast lesions. *Breast Cancer Res. Treat.* **59,** 15–26.

Ghosh, R., et al. (2010). Transcriptional regulation of VEGF-A by the unfolded protein response pathway. *PLoS ONE* **5,** e9575.

Holzman, C. (2009). Whatever Happened to Immunotoxins? Research and hope are still alive. *J. Natl. Cancer Inst.* **101,** 624–625.

Jamora, C., et al. (1996). Inhibition of tumor progression by suppression of stress protein GRP78/BiP induction in fibrosarcoma B/C10ME. *Proc. Natl. Acad. Sci. USA* **93,** 7690–7694.

Kahali, S., et al. (2010). Activation of the unfolded protein response contributes toward the antitumor activity of vorinostat. *Neoplasia* **12,** 80–86.

Kaufman, R. J. (1999). Stress signaling from the lumen of the endoplasmic reticulum: Coordination of gene transcriptional and translational controls. *Genes Dev.* **13,** 1211–1233.

Kawaguchi, Y., et al. (2003). The deacetylase HDAC6 regulates aggresome formation and cell viability in response to misfolded protein stress. *Cell* **115,** 727–738.

Khan, N., et al. (2009). Review: Green tea polyphenols in chemoprevention of prostate cancer: Preclinical and clinical studies. *Nutr. Cancer* **61,** 836–841.

Kreitman, R. J. (2006). Immunotoxins for targeted cancer therapy. *AAPS J.* **8,** E532–E551.

Kreitman, R. J. (2009). Recombinant immunotoxins containing truncated bacterial toxins for the treatment of hematologic malignancies. *Biodrugs* **23,** 1–13.

Lee, A. S. (2007). GRP78 induction in cancer: Therapeutic and prognostic implications. *Cancer Res.* **67,** 3496–3499.

Lee, E., et al. (2006). GRP78 as a novel predictor of responsiveness to chemotherapy in breast cancer. *Cancer Res.* **66,** 7849–7853.

Lee, H. K., et al. (2008). GRP78 is overexpressed in glioblastomas and regulates glioma cell growth and apoptosis. *Neuro Oncol.* **10,** 236–243.

Morinaga, N., Yahiro, K., Matsuura, G., Watanabe, M., Nomura, F., Moss, J., and Noda, M. (2007). Two distinct cytotoxic activities of subtilase cytotoxin produced by shiga-toxigenic *Escherichia coli*. *Infect. Immun.* **75,** 488–496.

Pastan, I., Hassan, R., Fitzgerald, D. J., and Kreitman, R. J. (2006). Immunotoxin therapy of cancer. *Nat. Rev. Cancer* **6,** 559–565.

Paton, A. W., Srimanote, P., Talbot, U. M., Wang, H., and Paton, J. C. (2004). A new family of potent AB(5) cytotoxins produced by Shiga toxigenic *Escherichia coli*. *J. Exp. Med.* **200,** 35–46.

Paton, A. W., Beddoe, T., Thorpe, C. M., Whisstock, J. C., Wilce, M. C. J., Rossjohn, J., Talbot, U. M., and Paton, J. C. (2006). AB5 subtilase cytotoxin inactivates the endoplasmic reticulum chaperone BiP. *Nature* **443,** 548–552.

Pootrakul, L., et al. (2006). Expression of stress response protein Grp78 is associated with the development of castration-resistant prostate cancer. *Clin. Cancer Res.* **12,** 5987–5993.

Potala, S., Sahoo, S. K., and Verma, R. S. (2008). Targeted therapy of cancer using diphtheria toxin-derived immunotoxins. *Drug Discov. Today* **13,** 807–815.

Pyrko, P., et al. (2007). The unfolded protein response regulator GRP78/BiP as a novel target for increasing chemosensitivity in malignant gliomas. *Cancer Res.* **67,** 9809–9816.

Rao, R., et al. (2010). Treatment with panobinostat induces glucose-regulated protein 78 acetylation and endoplasmic reticulum stress in breast cancer cells. *Mol. Cancer Ther.* **9,** 942–952.

Romero-Ramirez, L., et al. (2004). XBP1 is essential for survival under hypoxic conditions and is required for tumor growth. *Cancer Res.* **64,** 5943–5947.

Rusch, V., Mendelsohn, J., and Dmitrovsky, E. (1996). The epidermal growth factor receptor and its ligands as therapeutic targets in human tumors. *Cytokine Growth Factor Rev.* **7,** 133–141.

Saxena, S. K., O'Brien, A. D., and Ackerman, E. J. (1989). Shiga toxin, Shiga-like toxin II variant, and ricin are all single-site RNA N-glycosidases of 28S RNA when microinjected into Xenopus oocytes. *J. Biol. Chem.* **264,** 596–601.

Scaltriti, M., and Baselga, J. (2006). The epidermal growth factor receptor pathway: A model for targeted therapy. *Cancer Res.* **12,** 5268–5272.

Schardt, J. A., et al. (2009). Activation of the unfolded protein response is associated with favorable prognosis in acute myeloid leukemia. *Clin. Cancer Res.* **15,** 3834–3841.

Schroder, M., and Kaufman, R. J. (2005). The mammalian unfolded protein response. *Annu. Rev. Biochem.* **74,** 739–789.

Scriven, P., et al. (2009). Activation and clinical significance of the unfolded protein response in breast cancer. *Br. J. Cancer* **101,** 1692–1698.

Tsutsumi, S., Namba, T., Tanaka, K. I., Arai, Y., Ishihara, T., Aburaya, M., Mima, S., Hóshino, T., and Mizushima, T. (2006). Celecoxib upregulates endoplasmic reticulum chaperones that inhibit celecoxib-induced apoptosis in human gastric cells. *Oncogene* **25,** 1018–1029.

Virrey, J. J., et al. (2008). Stress chaperone GRP78/BiP confers chemoresistance to tumor-associated endothelial cells. *Mol. Cancer Res.* **6,** 1268–1275.

Wolfson, J. J., May, K. L., Thorpe, C. M., Jandhyyala, D. M., Paton, J. C., and Paton, A. W. (2008). Subtilase cytotoxin activates PERK, IRE and ATF6 endoplasmic reticulum stress-signaling pathways. *Cell. Microbiol.* **10,** 1775–1786.

Yarden, Y. (2001). The EGFR family and its ligands in human cancer: Signaling mechanisms and therapeutics opportunities. *Eur. J. Cancer* **37,** S3–S8.

Zu, K., Bihani, T., Lin, A., Park, Y. M., Mori, C., and Ip, C. (2006). Enhanced selenium effect on growth arrest by BiP/GRP78 knockdown in p53-null human prostate cancer cells. *Oncogene* **25,** 546–554.

CHAPTER FOUR

LARGE-SCALE ANALYSIS OF UPR-MEDIATED APOPTOSIS IN HUMAN CELLS

Andrew M. Fribley,[*,1] Justin R. Miller,[†,1] Tyler E. Reist,[*] Michael U. Callaghan,[†] and Randal J. Kaufman[*,2]

Contents

1. Introduction	58
2. Monitoring Proliferation and Caspase Activation Following UPR Activation	60
3. Monitoring the Expression of UPR and Cell Death Target Genes	61
3.1. Cell culture	62
3.2. Cells-to-CT™ (Applied Biosystems/Ambion)	63
3.3. Semiquantitative PCR analysis of *XBP1u* and *XBP1s* with Cells-to-CT™-derived cDNA	66
4. DNA Fragmentation Analysis	68
Acknowledgments	69
References	69

Abstract

The historic distinction between academic- and industry-driven drug discovery, whereby academicians worked to identify therapeutic targets and pharmaceutical companies advanced probe discovery, has been blurred by an academic high-throughput chemical genomic revolution. It is now common for academic labs to use biochemical or cell-based high-throughput screening (HTS) to investigate the effects of thousands or even hundreds of thousands of chemical probes on one or more targets over a period of days or weeks. To support the efforts of individual investigators, many universities have established core facilities where screening can be performed collaboratively with large chemical libraries managed by highly trained HTS personnel and guided by the experience of computational, medicinal, and synthetic organic chemists. The identification of large numbers of promising hits from such screens has driven the need

---

[*] Department of Biological Chemistry, University of Michigan School of Medicine, Ann Arbor, Michigan, USA
[†] Division of Pediatric Hematology/Oncology, Department of Pediatrics, Wayne State School of Medicine, Detroit, Michigan, USA
[1] These authors contributed equally to this work
[2] Corresponding author

for independent labs to scale down secondary *in vitro* assays in the hit to lead identification process. In this chapter, we will describe the use of luminescent and quantitative reverse transcription real-time PCR (qRT-PCR) technologies that permit evaluation of the expression patterns of multiple unfolded protein response (UPR) and apoptosis-related genes, and simultaneously evaluate proliferation and cell death in 96- or 384-well format.

1. INTRODUCTION

The ability of the unfolded protein response (UPR) to modulate cell death, following an unsuccessful attempt to restore homeostatic protein folding in the ER lumen, remains an incomplete story. Recently, some of the key molecular players have been identified at the transcriptional level and it has become clear that multiple proteins interacting in the nucleus to coordinately shut off survival genes and activate prodeath genes is a common theme. The ATF4-mediated induction of CHOP, following PERK activation and eIF2α phosphorylation, is a key event in the switch, under stress, from adaptation toward death and has received the most attention in the literature. Initial clues implicating CHOP as a participant in the UPR-mediated cell death program came to light when it was reported that overexpression of CHOP could induce cell cycle arrest and apoptosis (Barone *et al.*, 1994; Matsumoto *et al.*, 1996) and that *Chop* null mice were partially resistant to ER stress-mediated apoptosis (Oyadomari *et al.*, 2002; Zinszner *et al.*, 1998). Though it is clear that CHOP has an important role in ER stress-induced apoptosis, a comprehensive analysis of its target (downstream of *CHOP* or DOC) genes has not revealed a smoking gun (Wang *et al.*, 1998; and our unpublished observation), suggesting that this effect might be indirect.

Though specific target genes capable of directly inducing apoptosis have not been identified, CHOP can induce the expression of death receptor 5 (DR5) and *tribbles*-related protein 3 (TRB3) in a stress-dependent fashion to modulate the UPR death response. DR5 is a member of the TNFR family and can mediate cell death via the FADD signaling complex (Chaudhary *et al.*, 1997). Thapsigargin (Tg) enhanced DR5 expression was found to be mediated by CHOP in human cancer cell lines and sensitized them to TRAIL-induced cell death (Hetschko *et al.*, 2008). Increased expression of DR5 enhanced ligand binding and led to the recruitment of adaptor proteins at the intracellular DR5 death domain and initiated a signaling cascade that culminated in the cleavage and activation of caspase 8 similar to TNFR1, Fas, and DR3 and DR4. The discovery that CHOP could modulate DR5 expression linked the UPR to the "extrinsic" DR-mediated apoptosis pathway which, following caspase 8 cleavage, culminates in the activation of executioner caspases 3 and 7 to target substrates in the nucleus such as lamins and PARP immediately prior to DNA fragmentation. It should be noted that additional *in vitro* experiments revealed that siRNA knockdown of DR5 could interfere with the conformational change of Bax and caspase 3 activation required for apoptotic cell death following stress (Yamaguchi and Wang, 2004). TRB3 has also been identified as an ER stress-inducible target of CHOP/ATF4 signaling that can modulate UPR-dependent cell death induced by various ER stressors (Ohoka *et al.*, 2005; Ord and Ord, 2005; Ord *et al.*, 2007). Though there is a report that siRNA knockdown of TRB3 could reduce ER stress-dependent cell death in 293 cells (Ohoka *et al.*, 2005), most studies have reported that TRB3 antagonizes the antiproliferative and cytotoxic effects of the UPR by downregulating ATF4 transcriptional activity thereby lowering the level of intracellular reactive oxygen species (ROS; Ord *et al.*, 2007).

The UPR utilizes the BCL2 family during the cell death process via distinct and complementary mechanisms. CHOP induction can dramatically reduce cellular levels of BCL2 to directly potentiate the release of cytochrome *c* and initiate the mitochondrial or "intrinsic" cell death pathway (McCullough *et al.*, 2001). The subset of BCL2 family members that possess only the BCL2 homology domain 3 (BH3 domain), in stark contrast to BCL2, are all known to be proapoptotic. In general, this small group of proteins have a similar *modus operandi* in the apoptotic push toward death which is characterized by their ability to interact with BCL2 impeding its ability to keep Bax and Bak in an inactive conformation. Activation of Bax or Bak precipitates the release of cytochrome *c* from mitochondria and Ca^{2+} from the ER, thus setting in motion the process of apoptosis. Though currently nine members of the BH3-only protein family have been identified, only NBK/BIK, BIM, NOXA, and PUMA have been closely associated with the UPR-mediated cell death (Fribley *et al.*, 2006; Kieran *et al.*, 2007; Morishima *et al.*, 2004; Shimazu *et al.*, 2007; Zou *et al.*, 2009).

A number of molecules in addition to CHOP, ATF4, Bax/Bak, and caspase 12 are known to be involved with UPR-mediated cell death. It has been known for over a decade that Tg can activate the c-Jun NH(2)-terminal kinase cascade and apoptosis in an oxidative stress-dependent fashion (Srivastava *et al.*, 1999). Several years later it was reported that the activation of IRE1α led to the formation of a tripartite complex at the cell membrane with TRAF2 and ASK1 prior to the activation of the JNK cell death program (Matsuzawa *et al.*, 2002; Nishitoh *et al.*, 2002; Urano *et al.*, 2000). It is clear that JNK plays an important role in UPR-mediated cell death. Since we will not describe any large-scale methods focused to identify the activation of JNK signaling, further discussion has been omitted. (For thorough and recent reviews of stress-mediated activation of JNK signaling, see Nagai *et al.*, 2007; Rincon and Davis, 2009.)

2. Monitoring Proliferation and Caspase Activation Following UPR Activation

When cells undergo apoptosis, many distinct biochemical changes occur that can be readily detected to identify early or late stages of the death process. Recent advances in fluorescent and luminescent technology has made it possible to detect these changes with very limited numbers of cells and reagents in a 96- or 384-well format in a relatively cost-effective fashion. Importantly, we will describe how to monitor cell viability/proliferation and caspase activation in parallel to clearly establish the kinetics of cell growth/death and caspase activation. Members of the caspase family of cysteine proteases are necessary for nearly all apoptotic responses. Caspase $3^{-/-}$, caspase $7^{-/-}$, and caspase $9^{-/-}$ murine embryonic fibroblasts were found to be resistant to Tg, tunicamycin (Tm), brefeldin A, and calcium ionophore-induced ER stress, suggesting an essential role for mitochondria-mediated or intrinsic cell death following unresolved protein folding defect (Masud *et al.*, 2007). Although this group demonstrated that caspase $8^{-/-}$ MEFs were not protected from these stresses, it has been reported that in murine cells, caspase 12 can activate caspase 8 following UPR activation (Morishima *et al.*, 2002; Rao *et al.*, 2002). The caspase 12 gene in humans is inactive due to a single nucleotide polymorphism (Saleh *et al.*, 2004).

Required materials

Standard tissue culture facility and materials.
Electronic multichannel pipets (Matrix).
Luminescent plate reader (M5 Molecular Devices, GloMax-Multi+ or similar).
Cell Titer-Glo Luminescent Cell Viability Assay (Promega G7570).
Caspase-Glo 3/7 Assay (Promega G8091).

Luminescent assays for simultaneous detection of proliferation and caspase 3/7 activation.

1. 1.25×10^4 cells (~75% confluent) in 50 µl of appropriate medium are plated in white opaque 96-well tissue culture plates the day before addition of ER stress-inducing agent (or other compound of interest). Tg used at 250 nM–1.0 µM is a good positive control for most cell types for 24 h time course experiments. A clear observation plate should be treated in parallel to determine cell density and to visually observe the experimental wells. Measurement of caspase activation and proliferation should be performed at 4, 8, 16, and 24 h after treatment. Two plates of cells (one for 4 and 8 h and one for 16 and 24 h) should be plated with

triplicates for each time point. A 500 μM Tg stock can be prepared in DMSO; positive control culture wells should contain 250 nM–1 μM Tg. It is imperative that vehicle and Tg-treated controls are included for each plate.
2. Assuming ∼10% evaporation of the medium after overnight plating, the controls and compound/s of interest should be added to cell cultures in a volume of 5 μl. Six point dose–response assays performed in triplicate are sufficient to attain significance. Proliferation and caspase activity assays should be performed on the same plate; therefore, six wells must be treated for each time point and concentration. If serial dilutions are performed to achieve desired concentrations, the vehicle-treated wells should contain the highest concentration of vehicle to which any condition is exposed.
3. For the simultaneous measurement of proliferation and caspase 3/7 activation, both luminescent reagents should be thawed in a water bath until they have equilibrated completely with room temperature. Thirty to fifty microliters of each reagent should be added to the appropriate wells and incubated with shaking at room temperature for 10–15 min. When beginning this assay, the plates should be read every 15 min over a 45 min period to determine when the optimal signal occurs. Data may be represented graphically in a dual y-axis-fashion to compare the reduction in proliferation and activation of caspases following exacerbation of the adaptive capacity of the UPR (Fig. 4.1).

Special note on "edge effect." Unequal distribution of attached cells in 96- or 384-well plates can be caused by heat gradient fluctuations or levelness of incubators and can lead to a phenomenon known as edge effect. Edge effect is characterized by a ring or crescent-shaped pattern of adherent cells at the periphery of a well and can have dramatic effects on inter-well reproducibility. Several solutions including not using the outermost wells of a plate have been proposed. Another simple technique has been described whereby plates are allowed to sit in the tissue culture hood at room temperature for 1 h before placing cultures in an incubator (Lundholt et al., 2003).

3. Monitoring the Expression of UPR and Cell Death Target Genes

The use of 96- or 384-well thermocyclers for quantitative reverse transcription real-time PCR (qRT-PCR) has moved academic laboratories' ability to analyze gene expression light years beyond the Northern blot; however, most protocols still rely on the use of cumbersome and time-consuming phenol-based extractions from relatively large numbers of cells for RNA isolation. The recent introduction of Cells-to-CT^{TM} (Applied

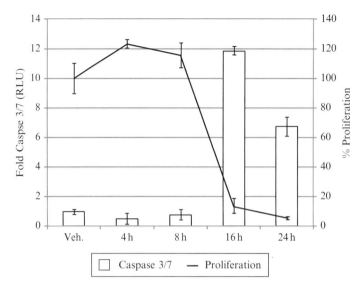

Figure 4.1 Dual *y*-axis representation of simultaneously obtained luminescent proliferation (Cell Titer-Glo) and caspase activation (Caspase-Glo 3/7) data in human squamous cell carcinoma cell lines treated with the UPR inducing inhibitor of the 26S proteasome Velcade, as indicated. Experimental samples were analyzed in triplicate and error bars represent standard deviation.

Biosystems/Ambion) has provided a phenol-free system that we have found can provide enough high-quality RNA template for the production of cDNA from as few as 500 cells in less than 10 min. In this section, we will describe a slight protocol modification that increases by twofold the number of cDNA reactions that can be performed from the manufacturer's indication. We will then describe how the cDNA can be diluted and interrogated with TaqMan (Applied Biosystems) Gene Expression Assays to evaluate the expression of UPR and apoptosis genes in stressed cells.

Required materials

Standard tissue culture facility and materials.
Electronic multichannel pipets (Matrix).
Cells-to-CT^{TM} (Applied Biosystems/Ambion).
TaqMan Gene Expression Assay (Applied Biosystems).
Quantitative Real-Time PCR thermocycler.

3.1. Cell culture

1. 1.25×10^4 cells (75–90% confluent) are plated in a 96-well tissue culture plate 16–24 h prior to stress induction. If working with an experimental

system that requires the use of very few cells, we have successfully performed this protocol with 500 cells. Note: reduced numbers of cells may require a lower (e.g., 1:10 or 1:25) dilution of cDNA before measurement of gene expression as described in Section 3.2.3.
2. ER stress and apoptosis can be induced by treating cultures with 1.0–2.5 μg/ml Tm or 0.25–1.0 μM Tg for 6–24 h.

3.2. Cells-to-CT™ (Applied Biosystems/Ambion)

3.2.1. Cell lysis

1. Aspirate and discard culture medium (1–6 h after treatment for the measurement of UPR and early apoptosis genes) from the 96-well plate.
2. Wash cells briefly with 50 μl cold (4 °C) 1× PBS.
3. Aspirate PBS from the wells and add 25 μl of lysis solution containing 1:100 DNase I to remove genomic DNA.
4. Mix well by pipetting up and down five times; incubate the lysis reactions for 5 min at room temperature (19–25 °C).
5. Add 2.5 μl (1:10) stop solution to each lysis reaction and mix well by pipetting up and down five times; incubate for 2 min at room temperature.
6. Lysate may be used immediately for cDNA synthesis or stored promptly at −20 °C.

3.2.2. Reverse transcription

1. Preparation of Master Mix:

Component	Per Rxn (μl)	96 Rxns[a] (μl)	384 Rxns[a] (μl)
2× RT buffer	12.5	1320	5280
20× RT enzyme mix	1.25	132	528
Nuclease-free water	1.25	132	528
Final volume	15	1584	6336

[a] 96 and 384 Rxn amounts reflect 10% overage.

2. Add 10 μl of each Cells-to-CT™ lysate to a 15 μl aliquot of RT Master Mix and mix gently.
3. *Important*: centrifuge tubes or plates prior to thermocycling to assure mixing.
4. Thermocycling:

	Stage	Repeats	(°C)	Time (min)
Reverse transcription	1	1	37	60
RT inactivation	2	1	95	5
Hold	3	1	4	∞

3.2.3. Quantitative reverse transcription real-time PCR

The TaqMan Gene Expression Assay and Master Mix cocktail and cDNA templates are added to 96- or 384-well plates separately, as described.

Reaction summary

Component	Per Rxn (μl)
20× TaqMan Gene Expression Assay	0.25
2× TaqMan Gene Expression Master Mix	2.5
cDNA template (diluted 1:50)	2.25
Total volume	5[a]

[a] 5 μl reaction volumes will dramatically reduce reagent use.

(A) Preparation of TaqMan Gene Expression Assay/Master Mix cocktail and plate setup

Calculate the volume of TaqMan Gene Expression Assay and Master Mix cocktail required to measure each cDNA in triplicate. For example, to measure the expression of 18S (internal control) in eight cDNA samples would require TaqMan Gene Expression Assay and Master Mix cocktail for 24 wells (Fig. 4.2, gray-shaded wells). Calculations include a 10% overage to account for errors in pipetting.

TaqMan Gene Expression Assay/Master Mix cocktail for eight samples (24 Rxns)

Component	Per Rxn (μl)	24 Rxns (μl)
20× Gene Expression Assay★	0.25	6.6
2× Gene Expression Master Mix	2.5	66
Final volume	2.75	72.6

★ For UPR, oxidative stress and cell death gene expression assays commonly used by our lab, see Table 4.1.

(B) Dilution of cDNA

Each cDNA should initially be diluted 1:50 in nuclease-free water; calculate the volume of diluted cDNA needed for triplicate samples

Large-Scale Analysis of UPR-Mediated Apoptosis in Human Cells 65

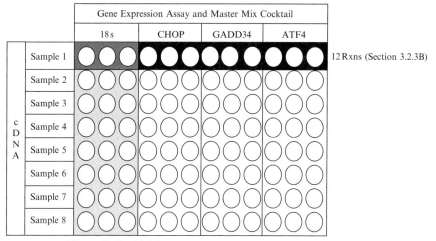

Figure 4.2 Model of 96-well plate set up for quantitative real-time PCR.

(Fig. 4.2, black-shaded wells). Calculations include a 10% overage to account for errors in pipetting.

Diluted cDNA template

Component	Per Rxn (µl)	12 Rxns (µl)
cDNA Template (1:50)	2.25	29.7

(C) Assay plate setup
  1. Transfer 2.75 µl of the TaqMan Gene Expression Assay/Master Mix cocktail (from Section 3.2.3A) to the bottom of the appropriate wells of a 96- or 384-well plate.
  2. Transfer 2.25 µl of the diluted cDNA template (made in Section 3.2.3B) to the bottom of the appropriate wells of the 384-well plate.
  3. Seal the plate with the appropriate cover, and *important*: centrifuge briefly.

(D) Thermocycling

	Stage	Step	Repeats	(°C)	Time
Hold	1	1	1	50	2 min
Hold	2	1	1	95	10 min
Cycle	3	1	40	95	15 s
		2		60^a	1 min

[a] TaqMan primer/probes are optimized for annealing at 60 °C.

Table 4.1 TaqMan gene expression assays

	Human	
	Gene	Assay ID
Internal control	18s	Hs99999901_s1
UPR	CHOP/DDIT3	Hs01090850_m1
	GADD34	Hs00169585_m1
	BIP	Hs99999174_m1
	ATF3	Hs00910173_m1
	ATF4	Hs00909569_g1
	ATF5	Hs01119208_m1
	HRI	Hs00205264_m1
	PERK	Hs00178128_m1
	P58IPK	Hs00534483_m1
Bcl-2 family	BAX	Hs99999001_m1
	BAK	Hs00832876_g1
	NOXA	Hs00560402_m1
	PUMA	Hs00248075_m1
	BIM	Hs00197982_m1
	NBK/BIK	Hs00609635_m1
	BID	Hs00609632_m1
	BCL2	Hs00153350_m1
Oxidative stress	HMOX1	Hs00157965_m1
	SOD1	Hs00916176_m1
	SOD2	Hs00167309_m1
	GPX1	Hs00829989_g1
	HIF1-α	Hs00153153_m1
	ERO1-α	Hs00205880_m1
Death	DR5	Hs00366278_m1
	TRIB3	Hs00221754_m1

(E) Data analysis: methods of calculating changes in gene expression vary by investigator and instrument and are therefore omitted.

3.3. Semiquantitative PCR analysis of *XBP1u* and *XBP1s* with Cells-to-CT™-derived cDNA

For semiquantitative PCR analysis of *XBP1* unspliced (*XBP1u*) and spliced (*XBP1s*) with Cells-to-CT™-derived cDNA, the Taq PCR Core Kit (Qiagen) is routinely utilized, according to the manufacture's protocol.

1. Human primers for *XBP1u* and *XBP1s* (Park et al., 2007):
   hXBP1 Forward:CCTTGTAGTTGAGAACCAGG.
   hXBP1 Reverse:GGGGCTTGGTATATATGTGG.
2. Preparation of Master Mix

Component	Volume/Rxn (µl)	Final Conc.
Forward primer	1 (10 pmol/µl)	200 nM
Reverse primer	1 (10 pmol/µl)	200 nM
dNTP mix	1 (10 mM each dNTP)	200 µM each dNTP
10× Buffer (w/15 mM MgCl$_2$)	5	1×
Taq DNA polymerase	0.25	2.5 U
Nuclease-free water	36.75	
Final volume Master Mix	45	

3. Mix PCR Master Mix thoroughly and add 45 µl to PCR tubes or plates.
4. Add 5 µl of undiluted cDNA (from Section 3.2.2) to the tubes/plates containing the Master Mix and centrifuge briefly.
5. Thermocycling

	Stage	Step	Repeats	(°C)	Time
Initial denaturation	1	1	1	94	3 min
Denaturation	2	1	30	94	30 s
Annealing		2		54	30 s
Extension		3		72	1 min
Final extension	3	1	1	72	10 min
Hold	4	1	40	4	∞

6. Amplicons can be resolved electrophoretically using a 1.8% agarose gel: *XBP1u* is 442 bp and *XBP1s* is 416 bp.
   Note: We have improved throughput of XBP1 amplicon analysis with the QIAxcel System (Qiagen). For amplicon analysis with this machine, the appropriate cartridge is the QIAxcel DNA High Resolution Gel Cartridge (Cat. # 929002), and the analysis can be performed using the OM500 method.

4. DNA Fragmentation Analysis

The hallmark of late-stage apoptosis has come to be the endonucleolytic cleavage (laddering) of DNA between nucleosomes into different sized fragments (Schwartzman and Cidlowski, 1993).

Required materials

DNA lysis buffer prepared by mixing:
  0.5 ml 1 M Tris–HCl (pH 7.4).
  0.1 ml 5 M NaCl, 1.0 ml 0.5 M EDTA (pH 8.0).
  2.5 ml 10% sodium dodecyl sulfate (or sodium laurel sarkosinate).
  45.9 ml ddH$_2$O.
Proteinase K (cat# P6656, Sigma).
Standard agarose gel electrophoresis equipment and reagents.

1. Approximately 2×10^6 cells are plated the day before the addition of an apoptotic stimulus in 10-cm tissue culture dishes. This assay can be performed with fewer cells if necessary; however, the reduction of cells from death and extensive organic extractions will reduce the yield of fragmented genomic DNA considerably.
2. Following treatment (16–24 h), the attached and floating cells are pooled by scraping and gently pelleted at 1200–1400 rpm in a table-top swinging bucket centrifuge.
3. The cell pellet is solubilized in 0.5 ml DNA lysis buffer. Histone proteins need to be digested from the genomic DNA by supplementing the lysis buffer at the time of use with 100 μg/ml of proteinase K. The reaction is incubated for 2 h in a 50 °C water bath. *Note*: for this and all subsequent steps, the use of blunt-ended pipet tips is recommended to avoid shearing of genomic DNA; alternatively, scissors may be used to trim a few millimeters off the ends of conventional tips.
4. DNA is extracted twice with phenol–chloroform mixed 1:1, and twice with chloroform alone. Following the final chloroform extraction, the aqueous phase is transferred to a clean Eppendorf tube and a 1/10 volume of 3 M sodium acetate (pH 5.2) is added along with either a 0.8 volume of isopropanol or 2.5 volumes of 100% ethanol to precipitate the DNA.
5. Following centrifugation at high speed, the precipitated DNA is washed and more completely pelleted by the addition of 0.5 ml of cold 70% ethanol, centrifuging at high speed for 10–15 min, and drying briefly at room temperature. The washed pellet is resuspended in 20–30 μl of TE (pH 8) supplemented with 0.25 mg/ml RNAse A (AM2270, Applied Biosystems). Fifteen to 30 μg of DNA (A_{260} determined) is resolved on a 1.5% agarose gel (Fig. 4.3).

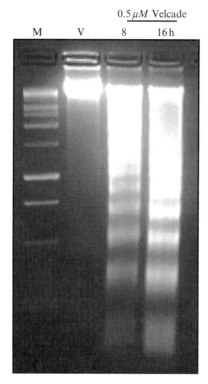

Figure 4.3 Resolution of fragmented genomic DNA harvested from human squamous cell carcinoma cells treated with Velcade, as indicated, on a 1.5% agarose gel.

Note: A laddered appearance of the DNA is indicative of apoptosis; a smeared appearance suggests that the observed cell death is the result of some other form of cell death such as necrosis.

ACKNOWLEDGMENTS

The authors appreciate the critical review of this chapter by Dr. Harmeet Malhi. Portions of this work were supported by NIH grants DK042394, HL052173, and HL057346, as well as a MH084182 and MH089782 (R. J. K.). Additionally, A. M. F. is supported by DE019678 and M.U.C is supported by Hemophilia of Georgia Clinician Scientist Award.

REFERENCES

Barone, M. V., Crozat, A., et al. (1994). CHOP (GADD153) and its oncogenic variant, TLS-CHOP, have opposing effects on the induction of G1/S arrest. *Genes Dev.* **8**(4), 453–464.

Chaudhary, P. M., Eby, M., et al. (1997). Death receptor 5, a new member of the TNFR family, and DR4 induce FADD-dependent apoptosis and activate the NF-kappaB pathway. *Immunity* **7**(6), 821–830.

Fribley, A. M., Evenchik, B., et al. (2006). Proteasome inhibitor PS-341 induces apoptosis in cisplatin-resistant squamous cell carcinoma cells by induction of Noxa. *J. Biol. Chem.* **281**(42), 31440–31447.

Hetschko, H., Voss, V., et al. (2008). Upregulation of DR5 by proteasome inhibitors potently sensitizes glioma cells to TRAIL-induced apoptosis. *FEBS J.* **275**(8), 1925–1936.

Kieran, D., Woods, I., et al. (2007). Deletion of the BH3-only protein puma protects motoneurons from ER stress-induced apoptosis and delays motoneuron loss in ALS mice. *Proc. Natl. Acad. Sci. USA* **104**(51), 20606–20611.

Lundholt, B. K., Scudder, K. M., et al. (2003). A simple technique for reducing edge effect in cell-based assays. *J. Biomol. Screen.* **8**(5), 566–570.

Masud, A., Mohapatra, A., et al. (2007). Endoplasmic reticulum stress-induced death of mouse embryonic fibroblasts requires the intrinsic pathway of apoptosis. *J. Biol. Chem.* **282**(19), 14132–14139.

Matsumoto, M., Minami, M., et al. (1996). Ectopic expression of CHOP (GADD153) induces apoptosis in M1 myeloblastic leukemia cells. *FEBS Lett.* **395**(2–3), 143–147.

Matsuzawa, A., Nishitoh, H., et al. (2002). Physiological roles of ASK1-mediated signal transduction in oxidative stress- and endoplasmic reticulum stress-induced apoptosis: Advanced findings from ASK1 knockout mice. *Antioxid. Redox Signal.* **4**(3), 415–425.

McCullough, K. D., Martindale, J. L., et al. (2001). Gadd153 sensitizes cells to endoplasmic reticulum stress by down-regulating Bcl2 and perturbing the cellular redox state. *Mol. Cell. Biol.* **21**(4), 1249–1259.

Morishima, N., Nakanishi, K., et al. (2002). An endoplasmic reticulum stress-specific caspase cascade in apoptosis. Cytochrome c-independent activation of caspase-9 by caspase-12. *J. Biol. Chem.* **277**(37), 34287–34294.

Morishima, N., Nakanishi, K., et al. (2004). Translocation of Bim to the endoplasmic reticulum (ER) mediates ER stress signaling for activation of caspase-12 during ER stress-induced apoptosis. *J. Biol. Chem.* **279**(48), 50375–50381.

Nagai, H., Noguchi, T., et al. (2007). Pathophysiological roles of ASK1-MAP kinase signaling pathways. *J. Biochem. Mol. Biol.* **40**(1), 1–6.

Nishitoh, H., Matsuzawa, A., et al. (2002). ASK1 is essential for endoplasmic reticulum stress-induced neuronal cell death triggered by expanded polyglutamine repeats. *Genes Dev.* **16**(11), 1345–1355.

Ohoka, N., Yoshii, S., et al. (2005). TRB3, a novel ER stress-inducible gene, is induced via ATF4-CHOP pathway and is involved in cell death. *EMBO J.* **24**(6), 1243–1255.

Ord, D., and Ord, T. (2005). Characterization of human NIPK (TRB3, SKIP3) gene activation in stressful conditions. *Biochem. Biophys. Res. Commun.* **330**(1), 210–218.

Ord, D., Meerits, K., et al. (2007). TRB3 protects cells against the growth inhibitory and cytotoxic effect of ATF4. *Exp. Cell Res.* **313**(16), 3556–3567.

Oyadomari, S., Koizumi, A., et al. (2002). Targeted disruption of the Chop gene delays endoplasmic reticulum stress-mediated diabetes. *J. Clin. Invest.* **109**(4), 525–532.

Park, J. W., Woo, K. J., et al. (2007). Resveratrol induces pro-apoptotic endoplasmic reticulum stress in human colon cancer cells. *Oncol. Rep.* **18**(5), 1269–1273.

Rao, R. V., Castro-Obregon, S., et al. (2002). Coupling endoplasmic reticulum stress to the cell death program. An Apaf-1-independent intrinsic pathway. *J. Biol. Chem.* **277**(24), 21836–21842.

Rincon, M., and Davis, R. J. (2009). Regulation of the immune response by stress-activated protein kinases. *Immunol. Rev.* **228**(1), 212–224.

Saleh, M., Vaillancourt, J. P., *et al.* (2004). Differential modulation of endotoxin responsiveness by human caspase-12 polymorphisms. *Nature* **429**(6987), 75–79.

Schwartzman, R. A., and Cidlowski, J. A. (1993). Apoptosis: The biochemistry and molecular biology of programmed cell death. *Endocr. Rev.* **14**(2), 133–151.

Shimazu, T., Degenhardt, K., *et al.* (2007). NBK/BIK antagonizes MCL-1 and BCL-XL and activates BAK-mediated apoptosis in response to protein synthesis inhibition. *Genes Dev.* **21**(8), 929–941.

Srivastava, R. K., Sollott, S. J., *et al.* (1999). Bcl-2 and Bcl-X(L) block thapsigargin-induced nitric oxide generation, c-Jun NH(2)-terminal kinase activity, and apoptosis. *Mol. Cell. Biol.* **19**(8), 5659–5674.

Urano, F., Wang, X., *et al.* (2000). Coupling of stress in the ER to activation of JNK protein kinases by transmembrane protein kinase IRE1. *Science* **287**(5453), 664–666.

Wang, X. Z., Kuroda, M., *et al.* (1998). Identification of novel stress-induced genes downstream of chop. *EMBO J.* **17**(13), 3619–3630.

Yamaguchi, H., and Wang, H. G. (2004). CHOP is involved in endoplasmic reticulum stress-induced apoptosis by enhancing DR5 expression in human carcinoma cells. *J. Biol. Chem.* **279**(44), 45495–45502.

Zinszner, H., Kuroda, M., *et al.* (1998). CHOP is implicated in programmed cell death in response to impaired function of the endoplasmic reticulum. *Genes Dev.* **12**(7), 982–995.

Zou, C. G., Cao, X. Z., *et al.* (2009). The molecular mechanism of endoplasmic reticulum stress-induced apoptosis in PC-12 neuronal cells: The protective effect of insulin-like growth factor I. *Endocrinology* **150**(1), 277–285.

CHAPTER FIVE

Quantitative Analysis of Amino Acid Oxidation Markers by Tandem Mass Spectrometry

Anuradha Vivekanandan-Giri, Jaeman Byun, *and* Subramaniam Pennathur

Contents

1. Introduction 75
2. Experimental Procedures 76
   2.1. Isotope dilution MS is a highly sensitive and specific method for detecting oxidized biomolecules *in vivo* 76
   2.2. Oxidized amino acid content correlates with degree of oxidative stress *in vivo* 77
   2.3. Sample preparation 78
   2.4. Preparation of authentic and isotope-labeled standards 80
   2.5. Separation and detection of analytes 81
3. Results 82
   3.1. LC/MS/MS detection of oxidized amino acids using authentic standards 82
   3.2. Quantitation of oxidized amino acids by MRM analysis 84
4. Conclusions 86
Acknowledgments 86
References 87

Abstract

Oxidative stress plays a central role in the pathogenesis of diverse chronic inflammatory disorders including diabetic complications, cardiovascular disease, aging, neurodegenerative disease, autoimmune disorders, and pulmonary fibrosis. Protein misfolding can lead to chronic endoplasmic reticulum (ER)

Division of Nephrology, Department of Internal Medicine, University of Michigan, Ann Arbor, Michigan, USA

stress which can exacerbate oxidative stress. This can trigger apoptotic cascades resulting in chronic inflammatory disorders. Despite intense interest in origins and magnitude of oxidative stress, ability to quantify oxidants has been limited because they are short lived. We have developed quantitative mass spectrometry (MS)-based analytical strategies to analyze stable end products of protein oxidation. These molecules provide quantitative and mechanistic assessment of degree of oxidative stress in cell cultures, tissues, and biofluids of animal models of disease and human samples. Our studies support the hypothesis that unique reactive intermediates generated in localized microenvironments of vulnerable tissues promote end-organ damage. The ability to quantify these changes and assess response to therapies will be pivotal in understanding disease mechanisms and monitoring efficacy of therapy.

Abbreviations

BHT	butylated hydroxytoulene
CI	chemical ionization
CID	collision–induced dissociation
DNA	deoxyribonucleic acid
DTPA	diethylenetriaminepentaacetic acid
EI	electron impact ionization
ESI	electrospray ionization
ER	endoplasmic reticulum
GC/MS	gas chromatography/mass spectrometry
HPLC	high performance liquid chromatography
LC/MS	liquid chromatography/mass spectrometry
MRM	multiple reaction monitoring
MS	mass spectrometry
MS/MS	tandem mass spectrometry
m/z	mass to charge ratio
RNS	reactive nitrogen species
ROS	reactive oxygen species
QQQ	triple quadrupole
HRP	horse radish peroxidase
TNM	tetranitromethane
TFA	trifluoroacetic acid
UPR	unfolded protein response

1. Introduction

Oxygen forms the basis of aerobic life but is only sparingly reactive by itself. However, it is well recognized that it can be activated by cellular metabolism to form reactive compounds termed as "reactive oxygen species" (ROS) which in turn can form a variety of intermediates including "reactive nitrogen species" (RNS). While physiological levels of such oxidants have a beneficial role in cellular signaling and host defense, excess oxidants can lead to pathological consequences.

Oxidative stress is due to the dysregulation in the synchronized balance between the oxidant-generating systems and the antioxidant defense in quenching the excessive free radicals. Proteins, lipids, and deoxyribonucleic acid (DNA) are the major targets of oxidative damage, eventually leading to the dysfunction of cellular physiology, as well as stimulating apoptosis. Alternatively, detrimental effects of ROS may be mediated by aberrant "redox signaling" in specific pathophysiological contexts. This latter effect may be attributed to disruption of the normal physiological roles of ROS in cell signaling (Thannickal, 2010).

During the past two decades, considerable evidence has implicated oxidative stress in several distinct human disorders, including aging (Beckman and Ames, 1998; Muller et al., 2007), atherosclerosis (Baynes and Thorpe, 2000; Pennathur and Heinecke, 2007), neurodegenerative diseases (Butterfield, 2006; Lin and Beal, 2006), diabetes (Baynes and Thorpe, 1999; Brownlee, 2001; Houstis et al., 2006; Jay et al., 2006; Vivekanadan-Giri et al., 2008), pulmonary fibrosis (Hecker et al., 2009), and end-stage renal disease (Cottone et al., 2008; Percy et al., 2005). A major focus of research has dwelt on the importance of endoplasmic reticulum (ER) stress and its link with oxidative stress. The unfolded protein response (UPR) is a signaling pathway that is triggered in response to protein misfolding in the ER. It has been noted that UPR is activated in response to oxidative stress as an adaptive mechanism (Malhotra and Kaufman, 2007) and antioxidants reduce ER stress (Back et al., 2009; Malhotra et al., 2008; Song et al., 2008). Persistent ER stress and oxidative stress in the susceptible individual can lead to end-organ damage (Fig. 5.1). Oxidative stress has also been implicated in many chronic fibrotic diseases as well (Thannickal, 2010). Indeed, fibrosis is a major cause of organ failure and death accounting for roughly 45% of mortality in the U.S. population (Wynn, 2004). Therefore, understanding pathways of oxidative damage and designing specific and sensitive methods of detecting oxidized biomolecules is of critical importance. In this chapter, we will discuss our work in developing tandem mass spectrometric (MS/MS) methods to quantify specific amino acid oxidation markers that provide mechanistic information on oxidative

Environmental, Genetic factors
↓
ER and Oxidative stress
↓
Production of reactive intermediates
↓
Oxidation of proteins, lipids and DNA
↓
End-Organ Damage and Chronic Disease

(Diabetic complications, Aging, Neurodegenerative disease, Pulmonary fibrosis, Autoimmune disease)

Figure 5.1 Roles of oxidative stress and endoplasmic reticulum (ER) stress in chronic disease. Interaction between environmental and genetic factors promotes ER and oxidative stress. Eventual damage to proteins, lipids, and DNA can lead to end-organ damage and chronic disease.

pathways operative *in vivo*. We will outline sample preparation from cell cultures, tissue, and other biospecimens, preparation of internal standards, liquid chromatography, and MS conditions and parameters for quantitative analysis of amino acid oxidation markers.

2. Experimental Procedures

2.1. Isotope dilution MS is a highly sensitive and specific method for detecting oxidized biomolecules *in vivo*

ROS and RNS intermediates are difficult to detect *in vivo* because they are extremely short lived due to their high reactivity with endogenous substrates; however, these oxidized substrates may serve as biomarkers for the activation of relevant oxidative stress pathways. Immunohistochemistry and dihydroethidium fluorescence have been extensively used to study oxidation-specific epitopes and oxidant production. These techniques are highly sensitive, and their ability to provide epitope-specific structural data can localize oxidative events to cell types or to subcellular locations. However, they are nonspecific as antibodies can bind to structurally similar compounds and, at best, only semiquantitative. The major drawback of high performance liquid chromatography (HPLC)-based methods is the appearance of coeluting structurally similar compounds (Shigenaga *et al.*, 1997). Spectrophotometric or flurometric detection of 3-nitrotyrosine can be confounded by contribution from tryptophan oxidation or other protein modifications (Guptasarma et al., 1992). In contrast, MS offers a highly sensitive and specific approach to quantify oxidative biomarkers. When combined with isotope dilution, in which a labeled internal standard which is identical to the target analyte except for the heavy isotope is added to the mixture, accurate quantitation can be achieved to low femtomole to attomole levels. Addition of the internal standard also allows normalization for loss of analytes during sample preparation.

2.2. Oxidized amino acid content correlates with degree of oxidative stress *in vivo*

To understand the molecular mechanisms that promote oxidative stress *in vivo*, we first identified the patterns of oxidation products that are formed by well-characterized oxidant-generating model systems. We then characterized patterns of products in tissue and plasma derived from animal models of disease and human samples. Since lipid peroxidation products readily undergo subsequent chain-propagating reactions and lose their initial oxidant imprint, we chose aromatic amino acids in proteins to study stable end products of oxidation. Using a combination of free radical generating systems *in vitro* and studying biospecimens from animal models of disease and humans, we and others defined patterns of these oxidative markers that accurately indicate pathways of oxidation that are activated (Back et al., 2009; Bergt et al., 2004; Brennan and Hazen, 2003; Hazen et al., 1997; Leeuwenburgh et al., 1997a,b; Malhotra et al., 2008; Okamura et al., 2009; Pennathur et al., 1999, 2001, 2005; Pop-Busui et al., 2009; Shishehbor et al., 2003; Shu et al., 2009; Song et al., 2008; Vincent et al., 2009; Wiggin et al., 2008; Zhang et al., 2008; Zheng et al., 2004). We have utilized this molecular fingerprinting strategy to identify tissue-specific pathways of oxidation *in vivo*. The phenylalanine residues in proteins when subjected to glycoxidation or hydroxyl radical damage forms *ortho*-tyrosine and *meta*-tyrosine. Oxidation reactions of tyrosine residues include dimerization (to form o,o'-dityrosine; mediated by tyrosine radicals, ROS and RNS), chlorination (to form 3-chlorotyrosine; catalyzed by myeloperoxidase), and nitration (to from 3-nitrotyrosine; mediated by RNS). Quantifying these unnatural amino acids characterizes the underlying oxidant pathway operative in target tissue.

The overall scheme for analysis for oxidative amino acids is outlined in Fig. 5.2. The analytical strategy can be divided into four parts: (1) sample preparation, (2) separation of analytes, (3) detection of analytes and data processing, (4) quantification of analytes.

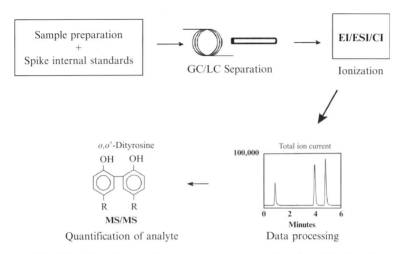

Figure 5.2 Workflow for tandem mass spectrometry-based oxidized amino acid analysis. Biological samples are processed and internal standards are added. Subsequently, the sample is subjected to either gas (GC) or liquid (LC) chromatography separation. The eluent is ionized by one of several modes of ionization such as EI, CI, or ESI. Following ionization, the resultant mass spectra are derived from the mass analyzers and the analytes of interest are identified. EI, electron impact ionization; CI, chemical ionization; ESI, electrospray ionization; MS/MS, tandem mass spectrometry.

2.3. Sample preparation

Oxidized amino acids can be measured from proteins derived from a variety of different samples including tissue, biofluids (plasma, urine, and cerebrospinal fluid), and cell culture. In general, plasma and urine measurements provide systemic levels of oxidant stress and tissue determinations provide local levels of oxidant stress. During procurement, it is important to ensure that samples are acquired in a sterile and uniform fashion. It is extremely important to prevent *ex vivo* oxidation as it will confound interpretation. Sample preparation typically entails isolation of proteins, hydrolysis to liberate individual amino acids, sample clean-up with solid-phase extraction, and processing prior to further analysis. Processing the samples is a double-edged sword because each step will result in some degree of analyte loss. It is imperative to add internal standards prior to processing to account for these losses.

2.3.1. Buffers and solutions required for sample preparation

Antioxidant buffer (Buffer A) consists of 100 μM diethylenetriaminepentaacetic acid (DTPA), 50 μM butylated hydroxy toluene (BHT) in 1% (v/v) ethanol, and 10 mM 3-amino-1,2,4-triazole in 50 mM sodium phosphate buffer, pH 7.4. DTPA is a metal chelator and prevents metal-catalyzed oxidation reactions; BHT is a lipid soluble antioxidant; aminotriazole is a peroxidase inhibitor. Fresh antioxidant buffer is prepared prior at the start of the experiment.

Delipidation buffer (Buffer B) consists of water, methanol, and water-saturated ether in 1:3:7 (v/v/v) ratio.

2.3.2. Tissue collection and protein isolation

The animals are perfused with antioxidant Buffer A prior to harvesting the tissue samples. This step ensures the removal of red blood cells and prevents *ex vivo* oxidation. The tissue samples are immediately stored in Buffer A in $-80\,°C$ until analysis. Prior to the start of the analytical procedure, the tissue samples are thawed, minced, and washed several times in aliquots of freshly prepared Buffer A. The samples are then homogenized with a hand tissue homogenizer. It is important to make sure that no residual tissue pieces are present. The tissue lysates are once again sonicated using stainless steel probe sonicator (Omni ruptor-250, Omni International, Marietta, GA). The tissue lysate is centrifuged at $1000 \times g$ for 10 min at $4\,°C$. The supernatant containing soluble proteins is aliquoted and stored for further processing.

2.3.3. Protein isolation from cell cultures

Prior to harvesting attached cells, the plates are washed several times with Buffer A. Following treatment with trypsin, the cells are diluted with Buffer A and spun briefly for 5 min at $1000 \times g$. The supernatant is discarded and the cell pellet is overlaid with fresh Buffer A and then resuspended and the process is repeated two more times. Subsequently, the cell suspension is subjected to sonication using stainless steel probe sonicator. The cell lysate is centrifuged at $1000 \times g$ for 10 min at $4\,°C$. The supernatant containing soluble proteins is aliquoted and stored for further processing.

2.3.4. Protein precipitation, delipidation, and hydrolysis of lysates derived from biological samples

All subsequent procedures are carried out at $4\,°C$. Tissue or cell lysates are taken in pyrolyzed hydrolysis vials and the volume is adjusted to 1 ml with 50 mM phosphate buffer, pH 7.4. For plasma, 5 µl is diluted with 50 mM phosphate buffer (pH 7.4) to a final volume of 1 ml in pyrolyzed hydrolysis vials. Subsequently the protein is precipitated by addition of trichloroacetic acid (10% final concentration). After centrifugation at $3000 \times g$ for 10 min at $4\,°C$, the protein precipitate is delipidated with Buffer B. This step ensures removal of lipids which can interfere with sample analysis. Following centrifugation, the lipid fraction is discarded. The protein pellet is dried in the fume hood under nitrogen to remove traces of the solvent. 4 N Methanesulfonic acid pretreated with benzoic acid (10 mg/ml) is added to the delipidated protein sample. Known concentration of isotope-labeled internal standards of *ortho*-tyrosine and *meta*-tyrosine, o,o'-dityrosine,

3-chlorotyrosine, 3-nitrotyrosine, $^{13}C_6$ phenylalanine, and $^{13}C_6$ tyrosine are spiked into the samples for absolute quantification of the modified amino acids and their precursors found in the biological samples. An alternate labeled precursor amino acid, [$^{13}C_9$,$^{15}N_1$]tyrosine, is added to monitor potential artifact formation which usually is a problem mainly for tyrosine halogenation, dimerization, and nitration products. The samples are hydrolyzed at 100 °C in sand bath for 24 h.

2.3.5. Solid-phase extraction of oxidized amino acids from biological samples

Chrom-P columns (ENVI-ChromP, 3 ml, Supelco, Bellefonte, PA) are activated with 100% methanol. Methanol is allowed to flow through gravity in the columns, to remove any trapped air bubbles. The column is preconditioned four times with 2 ml of 0.1% trifluoroacetic acid (TFA) in water at pH 4. The acid hydrolysate is subjected to solid-phase extraction and washed with 2 ml of 0.1% TFA of pH 4. The amino acids bound to the column are eluted with 50% methanol. The eluent is collected in clean glass vial and dried at 60 °C under vacuum. The dried samples are resolubilized in 50 μl of the 0.1% TFA.

2.4. Preparation of authentic and isotope-labeled standards

Isotope-labeled L-[$^{13}C_6$] phenylalanine and L-[$^{13}C_6$] tyrosine are obtained from Cambridge Isotope Laboratories (Andover, MA) and are used to synthesize all isotopically labeled standards. The overall synthetic strategy is outlined in Fig. 5.3. Isotopically labeled *ortho*-tyrosine and *meta*-tyrosine were synthesized using [$^{13}C_6$] phenylalanine, copper, and hydrogen peroxide (H_2O_2) as described previously (Huggins *et al.*, 1992). o,o'-[$^{13}C_{12}$] dityrosine was prepared with -[$^{13}C_6$] tyrosine, horse radish peroxidase (HRP), and H_2O_2 as described previously (Leeuwenburgh *et al.*, 1997b; Pennathur *et al.*, 1999; Pennathur *et al.*, 2005). L-3-Chloro[$^{13}C_6$] tyrosine was synthesized by reacting L-[$^{13}C_6$] tyrosine with an equimolar concentration of hypochlorous acid (HOCl) in the presence of 100 mM Cl$^-$ in 20 mM phosphate buffer (pH 4.0). Reactions were carried out for 60 min at 37 °C. They were initiated by adding HOCl and terminated with 0.1 mM methionine (Gaut *et al.*, 2002). Excess of tetranitromethane (TNM) was used to synthesize [$^{13}C_6$] 3-nitrotyrosine from 2 mM L-[$^{13}C_6$] tyrosine in 50 mM Tris buffer (pH 8.0). The mixture was vortexed for 1 h at room temperature (Gaut *et al.*, 2002). The resulting amino acids were isolated using solid-phase extraction with a 3-ml C18 column (Supelco, Inc., Bellefonte, PA). Concentrations of amino acids were determined by comparison with authentic standards using HPLC and monitoring of A_{276}. Further confirmation is obtained by verifying characteristic MS spectra by electrospray ionization tandem mass spectrometry (ESI/MS/MS)

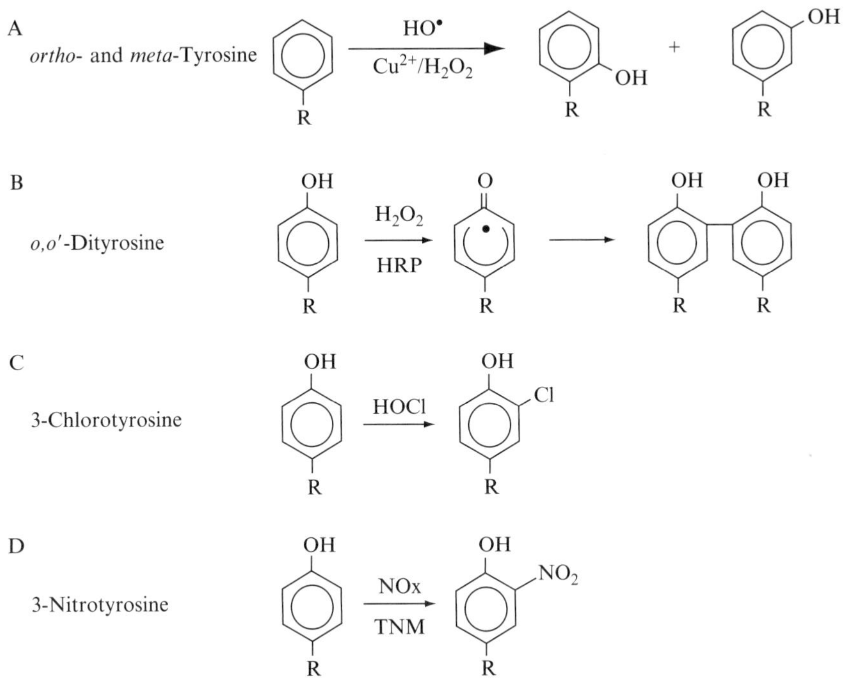

Figure 5.3 Structures and synthesis of oxidized amino acids. Oxidized amino acids are synthesized from the precursors as described in experimental procedures. H_2O_2, hydrogen peroxide; HOCl, hypochlorous acid; HRP, horse radish peroxidase; TNM, tetranitromethane.

2.5. Separation and detection of analytes

The two main separation platforms used are either gas chromatography (GC) or liquid chromatography (LC) coupled with MS.

2.5.1. Gas chromatography/Mass spectrometry (GC/MS)

GC/MS is a commonly used platform for measuring oxidized amino acids. As separation in GC occurs in an oven at high temperatures, analytes need to be volatile and thermally stable and it is therefore necessary to derivatize samples prior to analysis. The amino acids are converted to either heptafluorobutyryl derivatives or trimethylsilyl derivatives prior to analysis. The analytes entering the source in GC/MS are ionized by electron impact (EI) or chemical ionization (CI) and the mass analyzer is usually a single quadrupole. While necessary, it is important to keep in mind that derivatization is additional sample processing that can result in analyte loss and artifact formation and this is one of the major drawbacks of GC/MS. Additionally, GC/MS has limited mass range and therefore not suitable for larger molecular weight compounds such as *o,o′*-dityrosine. For these reasons, we no longer employ GC/MS for routine analysis of oxidized amino acids.

2.5.2. Liquid chromatography/mass spectrometry (LC/MS)

Biomedical mass spectrometry has recently been revolutionized by the advent of ionization methods that permit introduction of large, complex biomolecules into the mass spectrometer in liquid solutions. ESI is perhaps the most powerful of the new ionization methods for coupling LC with MS. In ESI/MS, large, complex, ionic molecules are introduced into the ion source as liquid solutions as a fine spray of charged droplets. This permits direct MS analysis of intact, simple, and complex analytes. Since high temperatures and need for volatility are not involved, sample derivatization is not needed.

The triple Quadrupole (QQQ) is a versatile mass analyzer. The basic subunit consists of four parallel metal rods—hence the term quadrupole. A radiofrequency voltage is applied to the rods and only ions within a particular mass to charge ratio (m/z) range will be able to traverse through the quadrupole to the detector. Ions outside the defined m/z range spiral out of control and are destroyed when they hit a rod. A QQQ combines three quadrupoles in series and allows MS/MS to be performed. In MS/MS, the first quadrupole is used to scan across a preset m/z range and select a user-defined parent ion of interest which is then fragmented in the second quadrupole by colliding with an inert gas such as nitrogen through the process of collision-induced dissociation (CID). The fragment daughter ions are sorted in the third quadrupole for analysis. Several functional groups (e.g., H_2O, COOH, etc.) characteristically fragment during CID, and thus analysis of the daughter ions can yield structural information about the parent ion. In addition, as compounds have unique fragmentation patterns, QQQ with multiple reaction monitoring (MRM) can be used to experimentally confirm potential analytes. QQQ is very sensitive and provides the most precise quantitation in the MRM mode and therefore ideally suited for analysis of oxidized amino acids.

3. RESULTS

3.1. LC/MS/MS detection of oxidized amino acids using authentic standards

Oxidized amino acids are quantified by LC/ESI/MS/MS in the MRM mode. MS experiments were performed using an Agilent 6410 Triple Quadrapole mass spectrometer coupled with an Agilent 1200 HPLC system (Agilent Technologies, New Castle, DE), equipped with a multimode source. C-18 column from Agilent with 1.8 μM particle size and 4.6×50 mm dimension is used for the HPLC separation. The amino acid hydrolysate is subjected to reverse-phase separation utilizing the following parameters. The mobile phase is 0.1% formic acid (Solvent A) and 0.1% formic acid in acetonitrile (Solvent B). The column was first equilibrated with 100% Solvent A. The gradient for the HPLC run is as follows: isocratic 2% acetonitrile for 0.5–3 min, and then a linear gradient of 10–95% acetonitrile, for 3–7 min with a flow rate of 0.6 ml/min. Under these chromatography conditions, authentic compounds and isotopically labeled standards exhibited retention times identical to those of analytes derived from tissue samples. The limit of detection (signal/noise >10) was < 100 fmol for all of the amino acids.

Positive LC/ESI/tandem MS was performed using following parameters: capillary spray voltage 4000 V, drying gas flow 11 l/min, drying gas temperature 350 °C, and nebulizer pressure 45 psi. Flow injection analysis was used to optimize the fragmentor voltage. Optimal fragmentor voltage for each amino acid in MS2 scan mode was obtained. Mass range between m/z 100 and m/z 400 was scanned to obtain full scan mass spectra.

Figure 5.4 depicts the MS/MS spectrum from *ortho*-tyrosine (Panel A), *o,o'*-dityrosine (Panel B), 3-chlorotyrosine (Panel C), and 3-nitrotyrosine (Panel D). *meta*-Tyrosine has an identical MS/MS spectrum as *ortho*-tyrosine

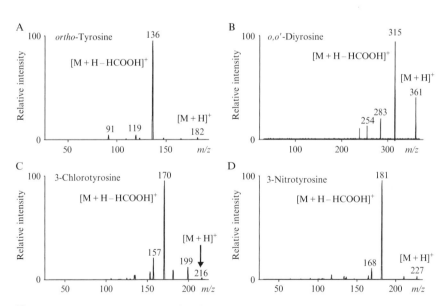

Figure 5.4 Liquid chromatography electrospray ionization tandem mass spectrometry of oxidized amino acids. The samples were separated by reverse-phase HPLC and subjected to ESI/MS as described in methods. MS/MS spectra reveal $[M + H]^+$ of *ortho*-tyrosine (Panel A), *o,o'*-dityrosine (Panel B), 3-chlorotyrosine (Panel C), and 3-nitrotyrosine (Panel D) are m/z 182, 361, 216, and 227. The neutral loss of m/z 46 is the most intense product ion for all the oxidized amino acids.

but can be distinguished due to differences in retention time (Table 5.1). The molecular ion $[M + H]^+$ of *ortho*-tyrosine, *o,o'*-dityrosine, 3-chlorotyrosine, and 3-nitrotyrosine are m/z 182, 361, 216, and 227 respectively. The neutral loss of m/z 46 is the most important product ion for all the oxidized amino acids. This yields an intense ion at m/z 136, 315, 170, and 181, respectively. These transitions form ideal candidates for MRM analysis.

3.2. Quantitation of oxidized amino acids by MRM analysis

The amino acids are quantified using isotope dilution LC/MS/MS in the MRM mode. The retention times and the MRM transitions utilized for quantification of the authentic compounds, internal standards, and artifact monitoring analytes are outlined in Table 5.1. Extracted ion chromatograms are derived from the MRM transitions and are shown in Fig. 5.5. The ion at m/z 136 and its respective isotopically labeled internal standard ion at m/z 142 were used to quantify *ortho*-tyrosine and *meta*-tyrosine. *o,o'*-Dityrosine is analyzed using the m/z 315 ion and its isotopically labeled internal standard ion at m/z 327. 3-Chlorotyrosine is quantified using ion at

Table 5.1 LC/ESI/MS/MS analysis of oxidized amino acids

Amino acid	Ion transition	Retention time
$^{13}C_6$ *ortho*-tyrosine	188 → 142	3.55
$^{12}C_6$ *ortho*-tyrosine	182 → 136	3.55
$^{13}C_6$ *meta*-tyrosine	188 → 142	2.95
$^{12}C_6$ *meta*-tyrosine	182 → 136	2.95
$^{13}C_6$ *o,o'*-dityrosine	373 → 327	2.25
$^{12}C_6$ *o,o'*-dityrosine	361 → 315	2.25
$^{13}C_6$ chlorotyrosine	222 → 176	4.22
$^{12}C_6$ chlorotyrosine	216 → 170	4.22
$^{13}C_6$ nitrotyrosine	233 → 187	4.88
$^{12}C_6$ nitrotyrosine	227 → 181	4.88
$^{13}C_9{}^{15}N_1$ *o,o'*-dityrosine (dimerization artifact)	381 → 334	2.25
$^{13}C_9{}^{15}N_1$ chlorotyrosine (chlorination artifact)	226 → 179	4.22
$^{13}C_9{}^{15}N_1$ nitrotyrosine (nitration artifact)	237 → 190	4.88
$^{13}C_6$ phenylalanine	172 → 126	3.75
$^{12}C_6$ phenylalanine	166 → 120	3.75
$^{13}C_6$ tyrosine	188 → 142	1.90
$^{12}C_6$ tyrosine	182 → 136	1.90
$^{13}C_9{}^{15}N_1$ tyrosine	192 → 145	1.90

The samples were separated by reverse-phase HPLC and subjected to ESI/MS as described in Section 3.1. The amino acid analytes, ion transitions utilized for MRM analysis, and the retention time are summarized.

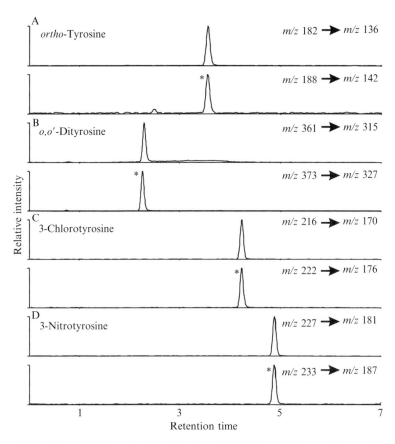

Figure 5.5 Quantification of oxidized amino acids in the multiple reaction monitoring mode utilizing extracted ion chromatograms. The samples were separated by reverse-phase HPLC and subjected to ESI/MS as described in methods. Extracted ion chromatograms were derived from the MRM transitions for *ortho*-tyrosine (Panel A), *o,o'*-dityrosine (Panel B), 3-chlorotyrosine (Panel C), and 3-nitrotyrosine (Panel D) as shown in the figure. The ratio of ion currents for each amino acid compared with the internal standard (depicted by *) was utilized to quantify the levels of the oxidized amino acids. Note coelution of authentic oxidized amino acid to the corresponding isotopically labeled internal standard.

170 m/z and its respective isotopically labeled internal standard ion at m/z 176. 3-Nitrotyrosine is quantified using the ion at m/z 181 and its respective isotopically labeled internal standard ion at m/z 187. To quantify phenylalanine and tyrosine (the precursor amino acids), the sample was diluted 1 in 1000 and analyzed in a separate injection as these precursor amino acids are typically present in 1000–10,000-fold excess to their oxidized counterparts. Phenylalanine, the precursor amino acid for *ortho*-tyrosine and *meta*-tyrosine, was quantified using the ion at m/z 126 and the isotopically labeled

internal standard ion at m/z 132. The ion at m/z 136 and the isotopically labeled internal standard ion at m/z 142 were used to quantify tyrosine, the precursor amino acid for o,o'-dityrosine, 3-chlorotyrosine, and 3-nitrotyrosine. We constructed an external calibration curve that used each amino acid as a standard and the corresponding isotopically labeled amino acid as an internal standard. The ratio of ion currents for each amino acid divided by that of the internal standard was a linear function of unlabeled amino acid for all ranges over which the amino acids were measured. The limit of detection (signal/noise >10) was <100 fmol for all the amino acids. Potential artifact during sample workup is usually an issue with o,o'-dityrosine, 3-chlorotyrosine, and 3-nitrotyrosine determination. Therefore, we routinely add L-[$^{13}C_9$, ^{15}N]tyrosine, a isotopically distinct tyrosine. We monitor artifact formation by analyzing appearance of m/z 334 (dimerization: o,o'-dityrosine artifact), m/z 179 (chlorination artifact), and m/z 190 (nitration artifact). The amount of artifact is then subtracted from the measured oxidized amino acid to correct for this issue. We do not routinely monitor for artifact formation for *ortho*-tyrosine and *meta*-tyrosine. If this is an issue, an isotopically distinct phenylalanine can be added to monitor this possibility. Finally, we normalize the oxidation markers to the precursor amino acids so that changes in tissue volume, amount of protein, or extraction efficiency are accounted for.

4. Conclusions

Many lines of evidence implicate oxidative stress in chronic diseases. However, accurate measurement of oxidant generation *in vivo* is challenging. The power of tandem mass spectrometry offers a unique opportunity to quantify low levels of oxidized biomolecules from complex bio specimens. We have utilized this approach to precisely assess oxidative stress in biological samples ranging from cell cultures, biofluids, and tissue from animal models of disease and humans. This analytical approach thus has a broad applicability to study a variety of disorders in which oxidative stress plays a causal role. More studies are needed to evaluate the utility of these markers to predict disease progression and to determine efficacy of potential interventions.

ACKNOWLEDGMENTS

This work is supported in part by grants from the National Institutes of Health (HL094230, DK089503, and DK082841), the Doris Duke Foundation Clinical Scientist Development Award, the American College of Rheumatology Within Our Reach Award, and by the Molecular Phenotyping Core of the Michigan Nutrition and Obesity Research Center (DK089503).

REFERENCES

Back, S. H., Scheuner, D., Han, J., Song, B., Ribick, M., Wang, J., Gildersleeve, R. D., Pennathur, S., and Kaufman, R. J. (2009). Translation attenuation through eIF2alpha phosphorylation prevents oxidative stress and maintains the differentiated state in beta cells. *Cell Metab.* **10,** 13–26.

Baynes, J. W., and Thorpe, S. R. (1999). Role of oxidative stress in diabetic complications: A new perspective on an old paradigm. *Diabetes* **48,** 1–9.

Baynes, J. W., and Thorpe, S. R. (2000). Glycoxidation and lipoxidation in atherogenesis. *Free Radic. Biol. Med.* **28,** 1708–1716.

Beckman, K. B., and Ames, B. N. (1998). The free radical theory of aging matures. *Physiol. Rev.* **78,** 547–581.

Bergt, C., Pennathur, S., Fu, X., Byun, J., O'Brien, K., McDonald, T. O., Singh, P., Anantharamaiah, G. M., Chait, A., Brunzell, J., Geary, R. L., Oram, J. F., et al. (2004). The myeloperoxidase product hypochlorous acid oxidizes HDL in the human artery wall and impairs ABCA1-dependent cholesterol transport. *Proc. Natl. Acad. Sci. USA* **101,** 13032–13037.

Brennan, M. L., and Hazen, S. L. (2003). Amino acid and protein oxidation in cardiovascular disease. *Amino Acids* **25,** 365–374.

Brownlee, M. (2001). Biochemistry and molecular cell biology of diabetic complications. *Nature* **414,** 813–820.

Butterfield, D. A. (2006). Oxidative stress in neurodegenerative disorders. *Antioxid. Redox Signal.* **8,** 1971–1973.

Cottone, S., Lorito, M. C., Riccobene, R., Nardi, E., Mule, G., Buscemi, S., Geraci, C., Guarneri, M., Arsena, R., and Cerasola, G. (2008). Oxidative stress, inflammation and cardiovascular disease in chronic renal failure. *J. Nephrol.* **21,** 175–179.

Gaut, J. P., Byun, J., Tran, H. D., and Heinecke, J. W. (2002). Artifact-free quantification of free 3-chlorotyrosine, 3-bromotyrosine, and 3-nitrotyrosine in human plasma by electron capture-negative chemical ionization gas chromatography mass spectrometry and liquid chromatography-electrospray ionization tandem mass spectrometry. *Anal. Biochem.* **300,** 252–259.

Guptasarma, P., Balasubramanian, D., Matsugo, S., and Saito, I. (1992). Hydroxyl radical mediated damage to proteins, with special reference to the crystallins. *Biochemistry* **31,** 4296–4303.

Hazen, S. L., Crowley, J. R., Mueller, D. M., and Heinecke, J. W. (1997). Mass spectrometric quantification of 3-chlorotyrosine in human tissues with attomole sensitivity: A sensitive and specific marker for myeloperoxidase-catalyzed chlorination at sites of inflammation. *Free Radic. Biol. Med.* **23,** 909–916.

Hecker, L., Vittal, R., Jones, T., Jagirdar, R., Luckhardt, T. R., Horowitz, J. C., Pennathur, S., Martinez, F. J., and Thannickal, V. J. (2009). NADPH oxidase-4 mediates myofibroblast activation and fibrogenic responses to lung injury. *Nat. Med.* **15,** 1077–1081.

Houstis, N., Rosen, E. D., and Lander, E. S. (2006). Reactive oxygen species have a causal role in multiple forms of insulin resistance. *Nature* **440,** 944–948.

Huggins, T. G., Staton, M. W., Dyer, D. G., Detorie, N. J., Walla, M. D., Baynes, J. W., and Thorpe, S. R. (1992). o-Tyrosine and dityrosine concentrations in oxidized proteins and lens proteins with age. *Ann. N. Y. Acad. Sci.* **663,** 436–437.

Jay, D., Hitomi, H., and Griendling, K. K. (2006). Oxidative stress and diabetic cardiovascular complications. *Free Radic. Biol. Med.* **40,** 183–192.

Leeuwenburgh, C., Hardy, M. M., Hazen, S. L., Wagner, P., Oh-ishi, S., Steinbrecher, U. P., and Heinecke, J. W. (1997a). Reactive nitrogen intermediates promote low density lipoprotein oxidation in human atherosclerotic intima. *J. Biol. Chem.* **272,** 1433–1436.

Leeuwenburgh, C., Rasmussen, J. E., Hsu, F. F., Mueller, D. M., Pennathur, S., and Heinecke, J. W. (1997b). Mass spectrometric quantification of markers for protein oxidation by tyrosyl radical, copper, and hydroxyl radical in low density lipoprotein isolated from human atherosclerotic plaques. *J. Biol. Chem.* **272,** 3520–3526.

Lin, M. T., and Beal, M. F. (2006). Mitochondrial dysfunction and oxidative stress in neurodegenerative diseases. *Nature* **443,** 787–795.

Malhotra, J. D., and Kaufman, R. J. (2007). Endoplasmic reticulum stress and oxidative stress: A vicious cycle or a double-edged sword? *Antioxid. Redox Signal.* **9,** 2277–2293.

Malhotra, J. D., Miao, H., Zhang, K., Wolfson, A., Pennathur, S., Pipe, S. W., and Kaufman, R. J. (2008). Antioxidants reduce endoplasmic reticulum stress and improve protein secretion. *Proc. Natl. Acad. Sci. USA* **105,** 18525–18530.

Muller, F. L., Lustgarten, M. S., Jang, Y., Richardson, A., and Van Remmen, H. (2007). Trends in oxidative aging theories. *Free Radic. Biol. Med.* **43,** 477–503.

Okamura, D. M., Pennathur, S., Pasichnyk, K., Lopez-Guisa, J. M., Collins, S., Febbraio, M., Heinecke, J., and Eddy, A. A. (2009). CD36 regulates oxidative stress and inflammation in hypercholesterolemic CKD. *J. Am. Soc. Nephrol.* **20,** 495–505.

Pennathur, S., and Heinecke, J. W. (2007). Oxidative stress and endothelial dysfunction in vascular disease. *Curr. Diab. Rep.* **7,** 257–264.

Pennathur, S., Jackson-Lewis, V., Przedborski, S., and Heinecke, J. W. (1999). Mass spectrometric quantification of 3-nitrotyrosine, ortho-tyrosine, and o, o'-dityrosine in brain tissue of 1-methyl-4-phenyl-1, 2, 3, 6-tetrahydropyridine-treated mice, a model of oxidative stress in Parkinson's disease. *J. Biol. Chem.* **274,** 34621–34628.

Pennathur, S., Wagner, J. D., Leeuwenburgh, C., Litwak, K. N., and Heinecke, J. W. (2001). A hydroxyl radical-like species oxidizes cynomolgus monkey artery wall proteins in early diabetic vascular disease. *J. Clin. Invest.* **107,** 853–860.

Pennathur, S., Ido, Y., Heller, J. I., Byun, J., Danda, R., Pergola, P., Williamson, J. R., and Heinecke, J. W. (2005). Reactive carbonyls and polyunsaturated fatty acids produce a hydroxyl radical-like species: A potential pathway for oxidative damage of retinal proteins in diabetes. *J. Biol. Chem.* **280,** 22706–22714.

Percy, C., Pat, B., Poronnik, P., and Gobe, G. (2005). Role of oxidative stress in age-associated chronic kidney pathologies. *Adv. Chronic Kidney Dis.* **12,** 78–83.

Pop-Busui, R., Oral, E., Raffel, D., Byun, J., Bajirovic, V., Vivekanandan-Giri, A., Kellogg, A., Pennathur, S., and Stevens, M. J. (2009). Impact of rosiglitazone and glyburide on nitrosative stress and myocardial blood flow regulation in type 2 diabetes mellitus. *Metabolism* **58,** 989–994.

Shigenaga, M. K., Lee, H. H., Blount, B. C., Christen, S., Shigeno, E. T., Yip, H., and Ames, B. N. (1997). Inflammation and NO(X)-induced nitration: Assay for 3-nitrotyrosine by HPLC with electrochemical detection. *Proc. Natl. Acad. Sci. USA* **94,** 3211–3216.

Shishehbor, M. H., Brennan, M. L., Aviles, R. J., Fu, X., Penn, M. S., Sprecher, D. L., and Hazen, S. L. (2003). Statins promote potent systemic antioxidant effects through specific inflammatory pathways. *Circulation* **108,** 426–431.

Shu, L., Park, J. L., Byun, J., Pennathur, S., Kollmeyer, J., and Shayman, J. A. (2009). Decreased nitric oxide bioavailability in a mouse model of Fabry disease. *J. Am. Soc. Nephrol.* **20,** 1975–1985.

Song, B., Scheuner, D., Ron, D., Pennathur, S., and Kaufman, R. J. (2008). Chop deletion reduces oxidative stress, improves beta cell function, and promotes cell survival in multiple mouse models of diabetes. *J. Clin. Invest.* **118,** 3378–3389.

Thannickal, V. J. (2010). Aging, antagonistic pleiotropy and fibrotic disease. *Int. J. Biochem. Cell Biol.* **42,** 1398–1400.

Vincent, A. M., Hayes, J. M., McLean, L. L., Vivekanandan-Giri, A., Pennathur, S., and Feldman, E. L. (2009). Dyslipidemia-induced neuropathy in mice: The role of oxLDL/LOX-1. *Diabetes* **58,** 2376–2385.

Vivekanadan-Giri, A., Wang, J. H., Byun, J., and Pennathur, S. (2008). Mass spectrometric quantification of amino acid oxidation products identifies oxidative mechanisms of diabetic end-organ damage. *Rev. Endocr. Metab. Disord.* **9,** 275–287.

Wiggin, T. D., Kretzler, M., Pennathur, S., Sullivan, K. A., Brosius, F. C., and Feldman, E. L. (2008). Rosiglitazone treatment reduces diabetic neuropathy in streptozotocin-treated DBA/2J mice. *Endocrinology* **149,** 4928–4937.

Wynn, T. A. (2004). Fibrotic disease and the T(H)1/T(H)2 paradigm. *Nat. Rev. Immunol.* **4,** 583–594.

Zhang, H., Saha, J., Byun, J., Schin, M., Lorenz, M., Kennedy, R. T., Kretzler, M., Feldman, E. L., Pennathur, S., and Brosius, F. C., 3rd. (2008). Rosiglitazone reduces renal and plasma markers of oxidative injury and reverses urinary metabolite abnormalities in the amelioration of diabetic nephropathy. *Am. J. Physiol. Renal Physiol.* **295,** F1071–F1081.

Zheng, L., Nukuna, B., Brennan, M. L., Sun, M., Goormastic, M., Settle, M., Schmitt, D., Fu, X., Thomson, L., Fox, P. L., Ischiropoulos, H., Smith, J. D., *et al.* (2004). Apolipoprotein A-I is a selective target for myeloperoxidase-catalyzed oxidation and functional impairment in subjects with cardiovascular disease. *J. Clin. Invest.* **114,** 529–541.

CHAPTER SIX

Animal Models in the Study of the Unfolded Protein Response

Hemamalini Bommiasamy *and* Brian Popko

Contents

1. Introduction		92
2. Activating Transcription Factor 6		93
3. IRE1/X-Box-Binding Protein-1		94
4. PKR-Like ER Kinase		97
5. eIF2α		98
6. ATF4		99
7. CHOP		100
8. GADD34		101
9. P58IPK		102
10. Transgenic Mouse Models for Monitoring ER Stress		102
11. UPR and Lipid Metabolism		103
12. UPR, Hypoxia, and Cancer		104
13. UPR and Inflammatory-Mediated Demyelination		105
14. Future Challenges		105
References		106

Abstract

The endoplasmic reticulum, a highly dynamic and complex organelle, is the site for synthesis, folding, and modification of transmembrane and secretory proteins. Any disruptions to the endoplasmic reticulum such as an accumulation of misfolded or unfolded proteins results in activation of the unfolded protein response (UPR). The UPR is comprised of three distinct signal transduction pathways that work to restore homeostasis to the endoplasmic reticulum. This

Department of Neurology, Center for Peripheral Neuropathy, The University of Chicago, Chicago, Illinois, USA review summarizes select mouse models available to study the UPR and the information learned from the analyses of these models.

1. INTRODUCTION

The unfolded protein response (UPR) is activated when the endoplasmic reticulum (ER) is exposed to stressful conditions, such as accumulation of unfolded or misfolded proteins. UPR signaling emanates from the ER and functions to maintain ER homeostasis during ER stress. Three mechanisms exist to respond to the accumulation of improperly or incompletely folded proteins. First, there is a decrease in the protein load through transient translational inhibition (Harding et al., 1999). Secondly, the ER increases in size and upregulates the expression of folding enzymes and chaperone proteins to accommodate the existing protein load (Kozutsumi et al., 1988; Shaffer et al., 2004; Sriburi et al., 2004). Thirdly, there is enhanced clearance of unfolded proteins through induction of ER-associated degradation (ERAD) components (Travers et al., 2000). Three ER transmembrane proteins, PKR-like ER kinase (PERK), activating transcription factor 6 (ATF6), and inositol-requiring enzyme 1(IRE1), mediate ER stress detection, serving as proximal transducers of the UPR. During normal conditions, BiP (immunoglobulin (Ig)-binding protein) associates with the luminal domains of PERK, IRE1, and ATF6; however, upon ER stress, BiP dissociates from these UPR regulators, resulting in their activation (Bertolotti et al., 2000). Together, these three pathways of the UPR alter the translational and transcriptional program of the cell, helping to alleviate ER stress and restore ER homeostasis. Nevertheless, severe or chronic ER stress leads to apoptosis.

This review focuses on select mouse models available to study the UPR (Table 6.1) and summarizes the invaluable information learned from these models. Knockout animals are generated through the inactivation of the endogenous gene by replacing it with a disrupted version via homologous recombination. One potential limitation of knockout technology is embryonic lethality, which can be overcome through the utilization of the Cre/ loxP DNA recombination system to generate tissue-specific knockout (conditional knockout) mice. This system requires two different genetically modified mouse lines. The first contains the target gene flanked by two loxP sites in the same orientation (floxed gene), and the second line bears a transgene expressing the Cre recombinase under the control of a cell or tissue-specific transcriptional control region. When these two mouse lines are crossed, the floxed gene is deleted in a cell- or tissue-specific manner through the Cre-mediated excision of the targeted gene segment between the loxP sites.

Table 6.1 UPR model systems in mouse

Gene name	Genetic models	References
Atf6	Atfα$^{-/-}$	Wu et al. (2007), Yamamoto et al. (2007)
	Atf6b$^{-/-}$	Wu et al. (2007)
Ire1	Ire1α$^{-/-}$	Zhang et al. (2005)
	Ire1b$^{-/-}$	Bertolotti et al. (2001)
Xbp1	Xbp1$^{-/-}$	Reimold et al. (2000)
	Xbp1$^{flox/flox}$	Hetz et al. (2008)
Perk	Perk$^{-/-}$	Harding et al. (2001)
	Perk$^{flox/flox}$	Zhang et al. (2002)
eIF2α	eIF2α$^{A/A}$	Scheuner et al. (2001)
Atf4	Atf4$^{-/-}$	Hettmann et al. (2000)
Chop	Chop$^{-/-}$	Zinszner et al. (1998)
Gadd34	Gadd34$^{-/-}$	Kojima et al. (2003)
	Gadd34$^{\Delta C/\Delta C}$	Novoa et al. (2003)
P58ipk	P58ipk$^{-/-}$	Ladiges et al. (2005)

2. Activating Transcription Factor 6

ATF6 exists in two isoforms, α and β, both of which are ubiquitously expressed and localized in the ER (Haze et al., 2001; Yoshida et al., 1998). ATF6 is a transmembrane protein whose activation and signaling is controlled by BiP. Upon ER stress, BiP dissociates from ATF6, revealing Golgi localization sequences (GLS; Shen et al., 2002). This results in translocation of ATF6 to the Golgi apparatus, where it undergoes regulated intramembrane proteolysis. Site 1 protease (S1P) cleaves ATF6 within the luminal domain and site 2 protease (S2P) cleaves within the phospholipid bilayer-spanning domain, releasing the transcriptionally active N-terminus cytosolic portion (Haze et al., 1999; Ye et al., 2000). Active ATF6 is a basic leucine zipper transcription factor that translocates to the nucleus, binds the ER stress response element (ERSE), and upregulates transcription of ER chaperone genes such as BiP (Yoshida et al., 1998, 2000, 2001b).

The function of ATF6 during ER stress was recently defined using ATF6-deficient mice. Single knockouts of *Atf6α* or *Atf6β* did not have any obvious defects, and the pups were born at the expected Mendelian frequency. *Atf6α/Atf6β* double knockout mice exhibited embryonic lethality, suggesting redundant functions for ATF6α and ATF6β during embryonic development (Yamamoto et al., 2007). *Atf6β*$^{-/-}$ mouse embryonic fibroblasts (MEFs) subjected to ER stress did not demonstrate a compromised ability to upregulate ER chaperone proteins. In contrast, *Atf6α*$^{-/-}$

MEFs are severely defective, upon ER stress, in the induction of ER chaperone proteins and folding enzymes such as BiP, GRP94, ERp72, and P5. Severe defects in the upregulation of ER-associated degradation components such as EDEM, HRD1, and Herp were also observed in $Atf6\alpha^{-/-}$ MEFs during ER stress (Wu et al., 2007; Yamamoto et al., 2007).

ER stress also resulted in decreased survival of $Atf6\alpha^{-/-}$ MEFs as compared to wild-type (WT) MEFs (Wu et al., 2007; Yamamoto et al., 2007). In vivo studies using intraperitoneal injection of the ER stress inducing agent tunicamycin (TM) yielded an 80% decrease in the survival of $Atf6\alpha^{-/-}$ mice as compared to WT mice, demonstrating that ATF6α is critical for protection against ER stress (Wu et al., 2007). Taken together, these studies demonstrate that ATF6α deficiency increases sensitivity to ER stress both in vitro and in vivo.

The generation of animals with a conditional (floxed) Atf6 allele to allow for the tissue-specific ablation of ATF6 will serve as a great resource to determine if this UPR-regulated transcription factor has cell-specific functions.

3. IRE1/X-Box-Binding Protein-1

IRE1 is an ER transmembrane protein found to have cytosolic kinase and endoribonuclease domains (Tirasophon et al., 1998). Two IRE1 isoforms are present in mammals, IRE1α and IRE1β. IRE1α is expressed ubiquitously, whereas IRE1β expression is limited to intestinal epithelial cells and the stomach (Tirasophon et al., 1998; Wang et al., 1998). Similar to ATF6, IRE1 is activated in response to BiP depletion following ER stress. BiP dissociation from IRE1 allows for IRE1 dimerization, which results in trans-autophosphorylation and activation of the endoribonuclease domain (Shamu and Walter, 1996; Welihinda and Kaufman, 1996). The site-specific endoribonuclease cleaves a 26-base pair intron from X-box binding protein 1 (Xbp1) mRNA, which encodes the transcription factor XBP1 (Calfon et al., 2002; Yoshida et al., 2001a). This excision and subsequent religation of the Xbp1 transcript alters the reading frame to increase the transcriptional activity of XBP1. Unspliced Xbp1 encodes a 33-kDa protein [XBP1(U)] and the spliced form of Xbp1 encodes a 54-kDa protein [XBP1(S)]. XBP1(U) and XBP1(S) share the same N-terminus but differ in their C-terminal transactivation domains. XBP1(S) is a stronger and more stable transcription factor than XBP1(U) (Calfon et al., 2002; Lee et al., 2002; Yoshida et al., 2001a).

$Ire1^{-/-}$ MEFS undergoing ER stress also display a defect in the ability to upregulate ERAD components, suggesting that both ATF6α and XBP1 are required for optimal expression of ERAD components, perhaps through heterodimerization (Lee et al., 2003). Immunoprecipitation studies revealed that active ATF6α and XBP1(S) heterodimerize during ER stress, suggesting that ATF6 and XBP1(S) work cooperatively during the UPR (Wu et al., 2007; Yamamoto et al., 2007).

Both $Ire1\alpha^{-/-}$ and $Xbp1^{-/-}$ mice exhibit embryonic lethality, demonstrating that both factors are necessary for embryonic development (Reimold et al., 2000; Zhang et al., 2005). XBP1 deficiency causes embryonic lethality due to liver hypoplasia and apoptosis (Reimold et al., 2000). Therefore, studying the role of the UPR transcriptional activator XBP1 in secretory cells has been difficult for the past several years. A liver-specific $Xbp1$ transgene was able to rescue $Xbp1^{-/-}$ mice ($Xbp1^{-/-}$; Liv^{Xbp1}) from dying *in utero*. While these mice were born at a normal frequency, they died a few days after birth. $Xbp1^{-/-}$; Liv^{Xbp1} mice display growth retardation and hypoglycemia that are suggestive of poor nutritional status. Additionally, these mice had undigested milk in their intestines and a poorly developed pancreas. Digestive enzymes, such as amylase and trypsin, were markedly decreased in $Xbp1^{-/-}$; Liv^{Xbp1} mice. Electron microscopy revealed a severely diminished amount of ER in $Xbp1^{-/-}$; Liv^{Xbp1} mice as compared to WT. There was also a decrease in the expression of genes encoding the ER-associated proteins SEC61α, EDEM, and PDI, as measured by real-time RT-PCR. These data correlated with an increased apoptosis of pancreatic acinar cells, which secrete digestive enzymes that facilitate breakdown of food in the small intestine, suggesting that XBP1 is necessary for the survival of these cells; however, the pancreatic islet cells, which produce the hormones insulin and glucagon, were unaffected. Salivary glands in the mutant mice were also underdeveloped, displayed poorly developed ER, and produced less amylase (Lee et al., 2005). These data reveal a critical role for XBP1 in the proper development and function of pancreatic acinar cells and salivary gland cells, suggesting that XBP1 is required for high-rate secretory cell function.

In addition to pancreatic studies, $Xbp1^{-/-}$ mouse models have been used to focus on high-rate secretory cells in the immune system. Because XBP1-deficient mice are embryonic lethal, the RAG2 (recombination-activating gene-2) complementation system was implemented to study the role of XBP1 in B cells. RAG2-deficient animals are unable to rearrange their antigen receptor genes; therefore, they lack B and T lymphocytes. $Xbp1^{-/-}$ embryonic stem (ES) cells were injected into RAG2-deficient blastocysts. The blastocysts were implanted into pseudopregnant females, generating chimeric mice. The lymphoid compartments in these chimeric mice could only be derived from XBP1-deficient ES cells, resulting in T and B cells that lacked XBP1. The B cell numbers, percentages, and phenotype were normal; however, baseline serum Ig levels were decreased in the chimeric mice (Reimold et al., 2001). There was also a marked decrease in Ig secretion upon *in vitro* stimulation of $Xbp1^{-/-}$ B cells with

LPS. Further studies revealed that the initial levels of IgM synthesis were similar in LPS-stimulated WT and $Xbp1^{-/-}$ B cells; however, over a 3-day time course there was a significant decrease in the level of IgM synthesis in $Xbp1^{-/-}$ B cells (Tirosh et al., 2005). These data suggest that XBP1 is not critical for initial onset of IgM synthesis upon stimulation, yet it is required for enhanced and sustained IgM synthesis and secretion. Interestingly, Ig light chain synthesis is normal in these cells, demonstrating that XBP1 is not required for light chain synthesis, but is critical for heavy chain synthesis (Tirosh et al., 2005). All other aspects of B cell functioning, aside from antibody secretion, were normal.

Consistent with the marked decrease in serum Ig, histological analysis of the jejunum revealed a 70-fold decrease in antibody-secreting plasma cells in the chimeric mice as compared to the control mice. Retroviral transduction with a construct expressing the spliced form of XBP1 [XBP1(S)] rescued the ability of $Xbp1^{-/-}$ B cells to secrete IgM. Transduction with this construct in mature B cells was sufficient to drive plasma cell differentiation (Reimold et al., 2001). These data demonstrate that XBP1 is required for terminal B cell differentiation and antibody production.

XBP1 is necessary for another type of immune cell, dendritic cells (DCs), which are critical in initiating the adaptive immune response. A specialized subset of DCs, plasmacytoid DCs (pDCs), secretes copious amounts of the cytokine interferon-α (IFN-α) upon activation by a virus or bacterial DNA (Liu, 2005). Similar to the terminal B cell differentiation process, pDCs expand the ER upon activation. $Rag2^{-/-}$ mice reconstituted with $Xbp1^{-/-}$ ES cells possessed severely reduced numbers of pDCs. $Xbp1^{-/-}$ pDCs also demonstrated decreased IFN-α production and poorly developed ER as compared to control chimeric mice (Iwakoshi et al., 2007). These data illustrate the need for XBP1 in the maintenance and proper development of pDCs.

Recently, floxed $Xbp1$ animals were generated, allowing for the production of tissue-specific knockout mice that have been utilized to delineate the function of XBP1 in various diseases, including Crohn's disease and amyotrophic lateral sclerosis (ALS; Hetz et al., 2008, 2009; Kaser et al., 2008). XBP1(S) and BiP levels were increased in biopsies from patients suffering from ulcerative colitis and Crohn's disease as compared to healthy individuals, implicating XBP1 in disease pathogenesis. Deletion of $Xbp1$ specifically in murine intraepithelial cells (IECs) led to spontaneous enteritis (inflammation of the small intestine; Kaser et al., 2008). Spontaneous enteritis in these animals correlates with an absence of Paneth cells due to apoptosis. Paneth cells are responsible for producing antimicrobial peptides and maintaining the gastrointestinal barrier. Moreover, chemically induced colitis using dextran sodium sulfate resulted in enhanced inflammation of the colon in mice lacking XBP1 in IECs as compared to WT littermates (Kaser et al., 2008). These data implicate the ER stress factor XBP1 in protection from inflammatory bowel disease by preservation of Paneth cells. Consistent with these data, mice deficient in *Ire1β*, which is exclusively expressed in the GI tract, develop more rapid and severe colitis as compared to WT mice in response to challenge with dextran sodium sulfate (Bertolotti *et al.*, 2001). This suggests that IRE1β plays a protective role against colitis. These data suggest that the ER stress response IRE1/XBP1 pathway is protective against inflammatory bowel disease, likely through preservation of Paneth cells.

A link also exists between neurodegenerative diseases characterized by misfolded proteins/protein aggregates and UPR activation. One such disease is ALS, also referred to as Lou Gehrig's disease. *Xbp1* was specifically deleted in the CNS of a mouse model for ALS. While no difference was found in male mice, female mice lacking XBP1 in the CNS exhibited a delay in onset of disease and experienced a prolonged life span. This correlated with enhanced autophagy and significantly reduced protein aggregates in the spinal cord of female mice lacking XBP1 in the CNS (Hetz *et al.*, 2009). It will be interesting to determine how gender, in combination with the absence of XBP1, affects the amelioration of disease symptoms, and to elucidate XBP1's role in other neurodegenerative diseases.

4. PKR-Like ER Kinase

The serine/threonine PERK mediates the reduction of the protein load in the ER in order to alleviate ER stress. Upon stress to the ER, BiP dissociates from PERK, resulting in PERK dimerization and transautophosphorylation. Activated PERK phosphorylates the alpha subunit of eukaryotic translation initiation factor 2 (eIF2α). Phosphorylated eIF2α binds to and inhibits the activity of the guanine nucleotide exchange factor eIF2B, thus preventing formation of the translation initiation complex (Harding *et al.*, 1999). This process results in the attenuation of global protein translation. Upon treatment with ER stress inducing agents, $Perk^{-/-}$ cells are unable to phosphorylate eIF2α and cannot slow the rate of protein translation, making them more sensitive to ER stress, as demonstrated by their decreased ability to survive (Harding *et al.*, 2000b). These data demonstrate that PERK is critical for cell survival and inhibiting translation during the UPR.

Phosphorylation of eIF2α results in a transient global decrease in translation initiation; however, it also facilitates selective translation of particular transcripts such as ATF4 (Harding *et al.*, 2000a).

Two known mechanisms exist to restore normal protein translation over the course of the UPR. GADD34, downstream of ATF4, associates with protein phosphatase 1 (PP1) and counters PERK activity by dephosphorylating eIF2α and restoring translation (Ma and Hendershot, 2003; Novoa et al., 2001, 2003). The second mechanism is thought to involve the protein P58IPK, which is upregulated by XBP1(S). P58IPK binds to and inhibits PERK phosphorylation and activation (Lee et al., 2003; van Huizen et al., 2003; Yan et al., 2002).

PERK has been found to be critical for pancreatic function. $Perk^{-/-}$ mice are hyperglycemic and develop diabetes mellitus (Harding et al., 2001). At $\sim$4 weeks of age, there is a decrease in the number of insulin-producing islet cells in $Perk^{-/-}$ mice. This correlates with diminished serum insulin levels and increased blood glucose levels. However, assessment of insulin production in prediabetic $Perk^{-/-}$ mice demonstrated increased insulin synthesis by islet cells. There was also increased apoptosis of pancreatic acinar cells. Electron microscopy analysis revealed disrupted ER morphology in $Perk^{-/-}$ islet cells and acinar cells as compared to WT cells (Harding et al., 2001). The pancreases of PERK-deficient mice appeared histologically normal at birth. However, after postnatal week 4, $Perk^{-/-}$ animals possessed very few detectable pancreatic β cells. These data reveal the importance of PERK for normal pancreatic function.

Another group also generated *Perk* null animals by creating floxed *Perk* animals and mating them to transgenic mice expressing Cre recombinase under the control of the EIIa promoter, resulting in the loss of PERK activity in all cells of the resulting animals (Zhang et al., 2002). These animals also demonstrated a loss of glucagon-secreting pancreatic α cells. $Perk^{-/-}$ animals display growth retardation, severe spinal curvature, splayed hind limbs, and reduced locomotor activity. At birth, $Perk^{-/-}$ mice exhibit osteoporosis and deficient mineralization throughout the entire skeletal system. Like the pancreatic cells, osteoblasts from the null animals also displayed abnormal ER morphology. PERK-deficient animals also had reduced levels of collagen type I α1 and α2, which are major components of the extracellular matrix of compact bones (Zhang et al., 2002). These data demonstrate an essential function for PERK in bone development. Floxed *Perk* mice have been generated and crossed to cell-specific Cre transgenic animals to gain further understanding of PERK functioning in the pancreas and osteoblasts (Iida et al., 2007; Wei et al., 2008).

5. EIF2α

The ability of PERK to inhibit global protein synthesis is dependent on the phosphorylation of serine 51 of murine eIF2α. Mutating serine 51 to alanine creates a nonphosphorylatable form of eIF2α. TM treatment of homozygous A/A eIF2α-mutant MEFs resulted in significant cell death in comparison to WT MEFs, demonstrating that the phosphorylation of eIF2α is required for cell survival in response to ER stress (Scheuner et al., 2001). Furthermore, MEFs with the nonphosphorylatable form of eIF2α were unable to produce ATF4, demonstrating that eIF2α phosphorylation is required for optimal ATF4 translation.

While homozygous A/A eIF2α-mutant mice were born at the expected Mendelian ratio and were phenotypically normal, these mutant neonates died within 18 h of birth (Scheuner et al., 2001). Blood glucose analysis showed severe hypoglycemia at 6–9 h after birth in homozygous A/A eIF2α mice. Hypoglycemia appeared to play a role in the lethality of the mutant neonates because injection with glucose for 2 days following birth rescued these animals for a short period. This hypoglycemia was linked to severely diminished insulin levels in the homozygous mutant embryonic and neonatal pancreas (Scheuner et al., 2001). Interestingly, the disease is much more severe in homozygous A/A eIF2α neonates than in $Perk^{-/-}$ animals, indicating that perhaps other eIF2α kinases, such as GCN2 or PKR, may compensate for the absence of PERK activity.

Heterozygous S/A eIF2α-mutant mice appear phenotypically normal despite demonstrating an ∼50% reduction in the phosphorylation of eIF2α; however, when these heterozygous mutant animals are fed a high fat diet, they become obese and develop diabetes (Scheuner et al., 2005). S/A eIF2α mice have decreased insulin secretion, which correlates with abnormal ER morphology and decreased quantity of insulin granules in β cells. Low levels of extracellular glucose inhibit translation in β cells, while high levels of extracellular glucose restore translation in β cells. Upon high glucose treatment, the rate of total protein synthesis did not fluctuate in WT animals, whether they were fed a low fat diet or a high fat diet. In contrast, there was an increased rate of translation in heterozygous mutant animals fed a high fat diet as compared to those fed a low fat diet. These data suggest that proper eIF2α phosphorylation is required to limit protein synthesis under increased glucose and high fat diet conditions. Interestingly, in mice receiving a high fat diet, there was enhanced binding of BiP to proinsulin in heterozygous S/A eIF2α mice as compared to WT (Scheuner et al., 2005). Since BiP is an ER chaperone known to bind unfolded proteins, this suggests that the protein folding capacity in heterozygous S/A eIF2α mice might be compromised.

6. ATF4

PERK activation and eIF2α phosphorylation result in the inhibition of global protein synthesis; however, phosphorylated eIF2α also leads to the preferential translation of *Atf4* mRNA. ATF4 is a transcription factor that upregulates genes involved in amino acid import, glutathione biosynthesis, and the antioxidative stress response (Harding et al., 2000a, 2003). ATF4 also upregulates CAAT/enhancer binding protein homologous protein (CHOP) (Ma and Hendershot, 2003).

Atf4 knockout mice demonstrate a high level of perinatal and postnatal mortality, with ~70% of animals dying between embryonic day 17.5 and postnatal day 14, which correlates with severe fetal anemia (Hettmann et al., 2000; Masuoka and Townes, 2002). The remaining 30% of mice lacking ATF4 survive until adulthood, but display growth retardation and exhibit severe micropthalmia, a developmental disorder of the eye (Hettmann et al., 2000). ER-stressed $Atf4^{-/-}$ MEFs were severely impaired in the expression of genes involved in amino acid import, glutathione biosynthesis, and resistance to oxidative stress in comparison to WT MEFs. ATF4-deficient MEFs were unable to survive and grow in culture unless supplemented with amino acids and reducing substances in the growth media. In the absence of supplementation, $Atf4^{-/-}$ cells displayed a buildup of intracellular peroxides followed by cell death (Harding et al., 2003). Taken together, these data suggest that the PERK/ATF4 pathway may be critical in protecting against oxidative damage during ER stress.

7. CHOP

CHOP is a bZIP transcription factor and its expression is strongly induced by ER stress (Wang et al., 1996). The PERK pathway is necessary for upregulating CHOP expression, although the other pathways of the UPR also contribute. Studies implicate CHOP in ER stress-induced apoptosis. Enforced expression of CHOP results in cell cycle arrest and/or apoptosis (Barone et al., 1994; Matsumoto et al., 1996; Maytin et al., 2001). MEFs lacking CHOP display enhanced resistance to ER stress-mediated apoptosis (Zinszner et al., 1998). CHOP-deficient mice exhibit no developmental defects and displayed decreased cell death in the renal tubular epithelium upon intraperitoneal injection of TM as compared to control animals (Zinszner et al., 1998). These data demonstrate that CHOP promotes apoptosis during ER stress. CHOP has been shown to repress expression of the antiapoptotic protein BCL-2, which could contribute to CHOP-mediated apoptosis (McCullough et al., 2001). Nevertheless, apoptosis still occurs in $Chop^{-/-}$ animals subjected to ER stress, suggesting that there are also CHOP-independent mechanisms of cell death that occur in response to ER stress. CHOP may also promote cell death through induction of the target genes *Gadd34* (discussed later) and *Ero1α* (Marciniak et al., 2004). $Ero1α^{-/-}$ MEFs experience reduced cell death in response to ER stress (Li et al., 2009). ERO1α creates a necessary oxidative environment in the ER, but in doing so produces reactive oxygen species that might promote apoptosis (Marciniak et al., 2004).

CHOP may play a central role in multiple mouse models of disease. In mouse models for type II diabetes, deletion of *Chop* resulted in decreased blood glucose levels and increased serum insulin levels. Additionally, improvement in pancreatic β cell function correlated with reduced cell death and decreased oxidative stress (Song et al., 2008). CHOP deficiency in two different atherosclerotic mouse models reduced the amount of necrosis and apoptosis in atherosclerotic lesions and decreased the overall lesion size, suggesting that CHOP activity may worsen atherosclerosis (Thorp et al., 2009). *Chop* ablation also reduced the severity of the neuropathy in a murine model of Charcot Marie Tooth disease, a neurological disorder affecting the peripheral nerves, as measured by enhanced motor capacity and increased nerve conduction velocity. Moreover, decreased neuropathy in $Chop^{-/-}$ mice correlates with reduced demyelination, suggesting a deleterious role for CHOP in demyelination during neuropathy (Pennuto et al., 2008).

While many studies demonstrate a harmful role for CHOP during disease, one study suggests a protective function for this protein. The neurodegenerative disorder Pelizaeus–Merzbacher disease is the result of mutations in the gene encoding the abundantly expressed integral myelin membrane protein proteolipid protein and is characterized by hypomyelination in the CNS resulting in varying degrees of delayed motor and intellectual function. A mouse model of Pelizaeus–Merzbacher disease, *rumpshaker* (*rsh*), exhibits tremor, ataxia, occasional seizures, and hypomyelination but has normal oligodendrocyte numbers and a normal life span. In contrast, *rsh* mice deficient for CHOP exhibit frequent seizures and only have an average life span of 10 weeks. Further studies showed that CHOP deficiency in *rsh* mice seems to result in oligodendrocyte apoptosis and enhanced hypomyelination, suggesting a protective role for CHOP (Southwood et al., 2002). Floxed *Chop* animals are currently unavailable but would be very useful to gain further understanding into the function of CHOP during disease pathogenesis.

8. GADD34

GADD34 interacts with the protein phosphatase PP1C to dephosphorylate eIF2α, resulting in a negative feedback loop. Dephosphorylation of eIF2α allows for translational recovery from ER stress. $Gadd34^{-/-}$ MEFs and $Gadd34^{\Delta C/\Delta C}$ MEFs, in which mutant GADD34 is unable to bind PP1C, displayed sustained levels of eIF2α phosphorylation and prolonged inhibition of protein synthesis during ER stress as compared to WT MEFs (Kojima *et al.*, 2003; Novoa *et al.*, 2003). *Gadd34*$^{-/-}$ and *Gadd34*$^{\Delta C/\Delta C}$ mice develop normally and were greatly protected from cell death in comparison to WT mice receiving an intraperitoneal TM injection, suggesting that GADD34 promotes cell death induced by ER stress (Kojima *et al.*, 2003; Novoa *et al.*, 2003). These results are similar to observations in *Chop*$^{-/-}$ animals and MEFs, further supporting that CHOP is required for optimal GADD34 expression upon ER stress. The generation of floxed *Gadd34* animals will be of great benefit to gain further insight into the contribution of GADD34 in a cell-specific manner during pathological conditions.

9. P58IPK

P58IPK binds to PERK and inhibits its kinase activity (Yan *et al.*, 2002). P58IPK is believed to be under the control of XBP1 since *Xbp1*$^{-/-}$ MEFs displayed reduced P58IPK expression upon UPR activation by TM treatment (Lee *et al.*, 2003). P58IPK-deficient mice are smaller and have significantly lower body weight in comparison to P58IPK heterozygous and WT animals. The decrease in body weight of *P58ipk*$^{-/-}$ mice correlates with decreased body fat. Interestingly, adult *P58ipk*$^{-/-}$ mice experience hypoglycemia, hypoinsulinemia, and increased apoptosis and loss of pancreatic β cells (Ladiges *et al.*, 2005). The phenotype of these animals is not as severe as that of *Perk*$^{-/-}$ or eIF2α-mutant animals, suggesting that P58IPK does not solely attenuate eIF2α phosphorylation.

P58IPK has been found to associate with SEC61, a component of the ER protein translocation channel (translocon). In *P58ipk*$^{-/-}$ mice, decreased levels of the chaperone HSP70 associated with translocon components in TM treated *P58ipk*$^{-/-}$ livers as compared to WT livers, suggesting that P58IPK functions in recruiting chaperones to the ER translocon. Studies also suggest that P58IPK plays a role in degrading proteins at the translocon during ER stress (Oyadomari *et al.*, 2006). These data suggest another function by which P58IPK contributes to ER homeostasis in cells undergoing ER stress.

10. TRANSGENIC MOUSE MODELS FOR MONITORING ER STRESS

Currently, two different transgenic mouse models exist to monitor ER stress *in vivo*. The first is the ERSE–LacZ transgenic mouse model. The transgene consists of a LacZ reporter gene driven by the rat *Grp78* promoter, which contains ERSEs (Mao et al., 2006). A control strain of mice carries a D300LacZ transgene that lacks the promoter region containing the ERSEs. By utilizing both the ERSE–LacZ and D300LacZ transgenic animals, one can identify tissues that have undergone ERSE-mediated responses to various ER stresses *in vivo* by detecting β-galactosidase activity (Mao et al., 2006).

A more recently generated model to detect ER stress *in vivo* is the "ER stress-activated indicator" (ERAI) model, which was generated by fusing Venus, a reporter gene that is a variant of green fluorescent protein, downstream of a partial sequence of human *XBP1* (Iwawaki et al., 2004). Under normal conditions, the mRNA of the fusion gene should not be spliced, which results in termination of translation at the stop codon upstream of the Venus sequence. In contrast, during ER stress, the 26 base pair intron of XBP1 is spliced out, leading to a frame shift that permits for translation of the Venus protein, which allows for monitoring of ER stress in various tissues by the detection of fluorescence. Intraperitoneal injection of TM into the ERAI transgenic mice resulted in strong fluorescence in the kidney, consistent with previous findings that the UPR is activated in the kidney after intraperitoneal injection of TM (Iwawaki et al., 2004). The ERAI mouse model might serve as a useful tool to monitor ER stress *in vivo* during development or pathophysiological conditions; however, limitations may include an undetectable signal from weak ER stress and a lack of ERAI transgene expression in certain cell types. Moreover, this model only monitors *XBP1* splicing and is not able to give insight into PERK and ATF6 activation. Nevertheless, the Venus system, in combination with *in vivo* imaging techniques, could prove to be a powerful tool for determining the timing and tissue-specific activation of the UPR in many disease models.

11. UPR AND LIPID METABOLISM

Studies from multiple labs demonstrate a link between the UPR and lipid metabolism and homeostasis. Intraperitoneal injection of the ER stress inducing agent TM in $Atf6\alpha^{-/-}$ animals resulted in hepatic steatosis (fatty liver; Rutkowski *et al.*, 2008; Yamamoto *et al.*, 2010). Ultrastructural analysis of the *Atf6α* null liver revealed increased intracellular triglyceride levels and accumulation of lipid droplets. Deletion of *Atf6α* resulted in the downregulation of key genes involved in fatty acid oxidation, gluconeogenesis, and lipogenesis (Rutkowski *et al.*, 2008; Yamamoto *et al.*, 2010). Moreover, genes involved in lipoprotein synthesis and transport were also suppressed. Hepatic steatosis is not unique to *Atf6* null animals, but is also found in mice with liver-specific deletion of *Ire1*, mice lacking phosphorylatable eIF2α in the liver, and *P58ipk*-deficient animals (Rutkowski *et al.*, 2008).

Intraperitoneal injection of TM led to hepatic steatosis in $Ire1\alpha$ null and mutant eIF2α livers, suggesting that the absence of any of the three pathways of the UPR can result in lipid disregulation in the liver. Interestingly, there was persistent CHOP expression in the $Atf6\alpha^{-/-}$ mice, $P58ipk^{-/-}$ mice, and mice harboring the liver-specific deletion of $Ire1$ after TM treatment, suggesting an inability to resolve ER stress (Rutkowski et al., 2008). It is possible that chronic ER stress might promote hepatic steatosis in part through CHOP. Other studies found that blocking eIF2α phosphorylation in the liver decreased hepatic steatosis in response to a high fat diet, suggesting that eIF2α phosphorylation leads to fatty liver disease during a high fat diet (Oyadomari et al., 2008). Lee et al. (2008) found that selective and inducible deletion of $Xbp1$ in the liver resulted in decreased lipid synthesis by the liver, leading to a marked decrease in cholesterol and triglyceride production, suggesting that XBP1 might be required for normal hepatic lipogenesis. Taken together, it appears that all three pathways of the UPR are important for regulating lipogenesis in response to chronic unresolved ER stress and physiological ER stress induced by diet.

12. UPR, Hypoxia, and Cancer

Cancer is an uncontrolled growth of abnormal cells, which invade and destroy healthy tissues. Cancerous cells are able to divide aggressively in a hypoxic environment, making them more likely to metastasize and more resistant to radiation. Transformed $Xbp1^{-/-}$ and WT MEFs were injected into the flanks of SCID mice, which lack B and T lymphocytes and are unable to reject tumors. Remarkably, transformed $Xbp1^{-/-}$ cells are unable to grow into tumors when implanted into these mice. Consistent with this finding, $Xbp1^{-/-}$ MEFs are compromised in their ability to survive hypoxic conditions as compared to WT cells (Romero-Ramirez et al., 2004). These data suggest that XBP1 may be required for tumor survival during hypoxic conditions. PERK has also been implicated in tumor growth. PERK-deficient cells exhibit decreased survival as compared to WT MEFs under hypoxic conditions (Koumenis et al., 2002). WT and $Perk^{-/-}$ MEFs immortalized with SV40 large/small T-antigen and transformed with the oncogene Ki-RasV12 were injected into the flanks of nude mice that have severely decreased T cells and are unable to reject tumors. $Perk^{-/-}$ tumors were markedly impaired in their ability to grow in vivo, suggesting that PERK is also essential for tumor growth (Bi et al., 2005; Blais et al., 2006). Active ATF6 is detected during hepatocellular carcinoma; therefore, it will be of great interest to determine if ATF6 also promotes tumorogenesis through utilization of ATF6-deficient MEFs (Shuda et al., 2003).

13. UPR AND INFLAMMATORY-MEDIATED DEMYELINATION

Multiple sclerosis (MS) is a disease that affects the central nervous system and is characterized by inflammation and the extensive death of specialized cells called oligodendrocytes. Oligodendrocytes function to produce myelin that forms a protective sheath around axons, and myelin is essential for the proper functioning of the nervous system. Loss of oligodendrocytes and damage to the myelin sheath cause serious neurological disorders such as MS.

Our laboratory believes that the high secretory demand on myelinating oligodendrocytes likely makes these cells sensitive to disruptions in the secretory pathway (Lin and Popko, 2009). We have found that CNS inflammation, through targeted expression of the proinflammatory cytokine IFN-γ in the CNS of mice, leads to tremor, ataxia, oligodendrocyte death, and hypomyelination. Consistent with our hypothesis, IFN-γ treatment *in vitro* and *in vivo* results in increased levels of phosphorylated eIF2α in oligodendrocytes (Lin *et al.*, 2005). $Perk^{+/-}$ mice with CNS inflammation exhibit reduced animal survival, increased loss of myelinating oligodendrocytes, and enhanced hypomyelination as compared to WT animals with CNS inflammation (Lin *et al.*, 2005, 2007). These data demonstrate that the PERK/eIF2α pathway is critical for dampening disease severity in response to enforced expression of IFN-γ in the CNS. Moreover, augmented activity of the PERK/eIF2α pathway through inactivation of GADD34 results in protection from the effects of IFN-γ. $Gadd34^{\Delta C/\Delta C}$; $Ifn-\gamma^{CNS+}$ mice display a diminished loss of myelinating oligodendrocytes and less hypomyelination as compared to WT mice expressing IFN-γ in the CNS (Lin *et al.*, 2008). Moreover, many of the $Gadd34^{\Delta C/\Delta C}$; $Ifn-\gamma^{CNS+}$ animals died by postnatal day 28, which correlated with enhanced medulloblastoma formation, suggesting that GADD34 might be involved in cancer regulation. Taken together, these data suggest a function for the UPR during inflammation-mediated demyelination and oligodendrocyte loss. Future studies might shed further light on the potential contribution of the other pathways of the UPR during inflammation-mediated demyelinating disorders such as MS.

14. FUTURE CHALLENGES

As discussed above, the UPR serves to protect cells undergoing ER stress by orchestrating transcriptional and translational changes within the cell. If the UPR is unsuccessful in restoring ER homeostasis, then apoptosis is triggered. Through the genetic manipulation of mice, much progress has been made in understanding the contribution of the UPR during physiological and disease conditions; however, much still remains to be elucidated. Creation of tissue-specific mutant or knockout animals, as well as utilization of available murine models to study the UPR, will be critical in furthering our knowledge of the UPR. Current and future studies have the potential to shed further light on the effects of manipulating UPR signals, which could lead to novel therapeutic approaches to ameliorate disease through selective targeting of the distinct branches of the UPR.

REFERENCES

Barone, M. V., et al. (1994). CHOP (GADD153) and its oncogenic variant, TLS-CHOP, have opposing effects on the induction of G1/S arrest. Genes Dev. **8**, 453–464.

Bertolotti, A., et al. (2000). Dynamic interaction of BiP and ER stress transducers in the unfolded-protein response. Nat. Cell Biol. **2**, 326–332.

Bertolotti, A., et al. (2001). Increased sensitivity to dextran sodium sulfate colitis in IRE1-beta-deficient mice. J. Clin. Invest. **107**, 585–593.

Bi, M., et al. (2005). ER stress-regulated translation increases tolerance to extreme hypoxia and promotes tumor growth. EMBO J. **24**, 3470–3481.

Blais, J. D., et al. (2006). Perk-dependent translational regulation promotes tumor cell adaptation and angiogenesis in response to hypoxic stress. Mol. Cell. Biol. **26**, 9517–9532.

Calfon, M., et al. (2002). IRE1 couples endoplasmic reticulum load to secretory capacity by processing the XBP-1 mRNA. Nature **415**, 92–96.

Harding, H. P., et al. (1999). Protein translation and folding are coupled by an endoplasmic-reticulum resident kinase. Nature **397**, 271–274.

Harding, H. P., et al. (2000a). Regulated translation initiation controls stress-induced gene expression in mammalian cells. Mol. Cell **6**, 1099–1108.

Harding, H. P., et al. (2000b). Perk is essential for translational regulation and cell survival during the unfolded protein response. Mol. Cell **5**, 897–904.

Harding, H. P., et al. (2001). Diabetes mellitus and exocrine pancreatic dysfunction in perk−/− mice reveals a role for translational control in secretory cell survival. Mol. Cell **7**, 1153–1163.

Harding, H. P., et al. (2003). An integrated stress response regulates amino acid metabolism and resistance to oxidative stress. Mol. Cell **11**, 619–633.

Haze, K., et al. (1999). Mammalian transcription factor ATF6 is synthesized as a transmembrane protein and activated by proteolysis in response to endoplasmic reticulum stress. Mol. Biol. Cell **10**, 3787–3799.

Haze, K., et al. (2001). Identification of the G13 (cAMP-response-element-binding protein-related protein) gene product related to activating transcription factor 6 as a transcriptional activator of the mammalian unfolded protein response. Biochem. J. **355**, 19–28.

Hettmann, T., et al. (2000). Microphthalmia due to p53-mediated apoptosis of anterior lens epithelial cells in mice lacking the CREB-2 transcription factor. Dev. Biol. **222**, 110–123.

Hetz, C., et al. (2008). Unfolded protein response transcription factor XBP-1 does not influence prion replication or pathogenesis. Proc. Natl. Acad. Sci. USA **105**, 757–762.

Hetz, C., et al. (2009). XBP-1 deficiency in the nervous system protects against amyotrophic lateral sclerosis by increasing autophagy. Genes Dev. **23**, 2294–2306.

Iida, K., et al. (2007). PERK eIF2 alpha kinase is required to regulate the viability of the exocrine pancreas in mice. *BMC Cell Biol.* **8,** 38.

Iwakoshi, N. N., et al. (2007). The transcription factor XBP-1 is essential for the development and survival of dendritic cells. *J. Exp. Med.* **204,** 2267–2275.

Iwawaki, T., et al. (2004). A transgenic mouse model for monitoring endoplasmic reticulum stress. *Nat. Med.* **10,** 98–102.

Kaser, A., et al. (2008). XBP1 links ER stress to intestinal inflammation and confers genetic risk for human inflammatory bowel disease. *Cell* **134,** 743–756.

Kojima, E., et al. (2003). The function of GADD34 is a recovery from a shutoff of protein synthesis induced by ER stress: Elucidation by GADD34-deficient mice. *FASEB J.* **17,** 1573–1575.

Koumenis, C., et al. (2002). Regulation of protein synthesis by hypoxia via activation of the endoplasmic reticulum kinase PERK and phosphorylation of the translation initiation factor eIF2alpha. *Mol. Cell. Biol.* **22,** 7405–7416.

Kozutsumi, Y., et al. (1988). The presence of malfolded proteins in the endoplasmic reticulum signals the induction of glucose-regulated proteins. *Nature* **332,** 462–464.

Ladiges, W. C., et al. (2005). Pancreatic beta-cell failure and diabetes in mice with a deletion mutation of the endoplasmic reticulum molecular chaperone gene P58IPK. *Diabetes* **54,** 1074–1081.

Lee, K., et al. (2002). IRE1-mediated unconventional mRNA splicing and S2P-mediated ATF6 cleavage merge to regulate XBP1 in signaling the unfolded protein response. *Genes Dev.* **16,** 452–466.

Lee, A. H., et al. (2003). XBP-1 regulates a subset of endoplasmic reticulum resident chaperone genes in the unfolded protein response. *Mol. Cell. Biol.* **23,** 7448–7459.

Lee, A. H., et al. (2005). XBP-1 is required for biogenesis of cellular secretory machinery of exocrine glands. *EMBO J.* **24,** 4368–4380.

Lee, A. H., et al. (2008). Regulation of hepatic lipogenesis by the transcription factor XBP1. *Science* **320,** 1492–1496.

Li, G., et al. (2009). Role of ERO1-alpha-mediated stimulation of inositol 1, 4, 5-triphosphate receptor activity in endoplasmic reticulum stress-induced apoptosis. *J. Cell Biol.* **186,** 783–792.

Lin, W., and Popko, B. (2009). Endoplasmic reticulum stress in disorders of myelinating cells. *Nat. Neurosci.* **12,** 379–385.

Lin, W., et al. (2005). Endoplasmic reticulum stress modulates the response of myelinating oligodendrocytes to the immune cytokine interferon-gamma. *J. Cell Biol.* **169,** 603–612.

Lin, W., et al. (2007). The integrated stress response prevents demyelination by protecting oligodendrocytes against immune-mediated damage. *J. Clin. Invest.* **117,** 448–456.

Lin, W., et al. (2008). Enhanced integrated stress response promotes myelinating oligodendrocyte survival in response to interferon-gamma. *Am. J. Pathol.* **173,** 1508–1517.

Liu, Y. J. (2005). IPC: Professional type 1 interferon-producing cells and plasmacytoid dendritic cell precursors. *Annu. Rev. Immunol.* **23,** 275–306.

Ma, Y., and Hendershot, L. M. (2003). Delineation of a negative feedback regulatory loop that controls protein translation during endoplasmic reticulum stress. *J. Biol. Chem.* **278,** 34864–34873.

Mao, C., et al. (2006). In vivo regulation of Grp78/BiP transcription in the embryonic heart: Role of the endoplasmic reticulum stress response element and GATA-4. *J. Biol. Chem.* **281,** 8877–8887.

Marciniak, S. J., et al. (2004). CHOP induces death by promoting protein synthesis and oxidation in the stressed endoplasmic reticulum. *Genes Dev.* **18,** 3066–3077.

Masuoka, H. C., and Townes, T. M. (2002). Targeted disruption of the activating transcription factor 4 gene results in severe fetal anemia in mice. *Blood* **99,** 736–745.

Matsumoto, M., et al. (1996). Ectopic expression of CHOP (GADD153) induces apoptosis in M1 myeloblastic leukemia cells. *FEBS Lett.* **395**, 143–147.

Maytin, E. V., et al. (2001). Stress-inducible transcription factor CHOP/gadd153 induces apoptosis in mammalian cells via p38 kinase-dependent and -independent mechanisms. *Exp. Cell Res.* **267**, 193–204.

McCullough, K. D., et al. (2001). Gadd153 sensitizes cells to endoplasmic reticulum stress by down-regulating Bcl2 and perturbing the cellular redox state. *Mol. Cell. Biol.* **21**, 1249–1259.

Novoa, I., et al. (2001). Feedback inhibition of the unfolded protein response by GADD34-mediated dephosphorylation of eIF2alpha. *J. Cell Biol.* **153**, 1011–1022.

Novoa, I., et al. (2003). Stress-induced gene expression requires programmed recovery from translational repression. *EMBO J.* **22**, 1180–1187.

Oyadomari, S., et al. (2006). Cotranslocational degradation protects the stressed endoplasmic reticulum from protein overload. *Cell* **126**, 727–739.

Oyadomari, S., et al. (2008). Dephosphorylation of translation initiation factor 2alpha enhances glucose tolerance and attenuates hepatosteatosis in mice. *Cell Metab.* **7**, 520–532.

Pennuto, M., et al. (2008). Ablation of the UPR-mediator CHOP restores motor function and reduces demyelination in Charcot-Marie-Tooth 1B mice. *Neuron* **57**, 393–405.

Reimold, A. M., et al. (2000). An essential role in liver development for transcription factor XBP-1. *Genes Dev.* **14**, 152–157.

Reimold, A. M., et al. (2001). Plasma cell differentiation requires the transcription factor XBP-1. *Nature* **412**, 300–307.

Romero-Ramirez, L., et al. (2004). XBP1 is essential for survival under hypoxic conditions and is required for tumor growth. *Cancer Res.* **64**, 5943–5947.

Rutkowski, D. T., et al. (2008). UPR pathways combine to prevent hepatic steatosis caused by ER stress-mediated suppression of transcriptional master regulators. *Dev. Cell* **15**, 829–840.

Scheuner, D., et al. (2001). Translational control is required for the unfolded protein response and in vivo glucose homeostasis. *Mol. Cell* **7**, 1165–1176.

Scheuner, D., et al. (2005). Control of mRNA translation preserves endoplasmic reticulum function in beta cells and maintains glucose homeostasis. *Nat. Med.* **11**, 757–764.

Shaffer, A. L., et al. (2004). XBP1, downstream of Blimp-1, expands the secretory apparatus and other organelles, and increases protein synthesis in plasma cell differentiation. *Immunity* **21**, 81–93.

Shamu, C. E., and Walter, P. (1996). Oligomerization and phosphorylation of the Ire1p kinase during intracellular signaling from the endoplasmic reticulum to the nucleus. *EMBO J.* **15**, 3028–3039.

Shen, J., et al. (2002). ER stress regulation of ATF6 localization by dissociation of BiP/GRP78 binding and unmasking of Golgi localization signals. *Dev. Cell* **3**, 99–111.

Shuda, M., et al. (2003). Activation of the ATF6, XBP1 and grp78 genes in human hepatocellular carcinoma: A possible involvement of the ER stress pathway in hepatocarcinogenesis. *J. Hepatol.* **38**, 605–614.

Song, B., et al. (2008). Chop deletion reduces oxidative stress, improves beta cell function, and promotes cell survival in multiple mouse models of diabetes. *J. Clin. Invest.* **118**, 3378–3389.

Southwood, C. M., et al. (2002). The unfolded protein response modulates disease severity in Pelizaeus-Merzbacher disease. *Neuron* **36**, 585–596.

Sriburi, R., et al. (2004). XBP1: A link between the unfolded protein response, lipid biosynthesis, and biogenesis of the endoplasmic reticulum. *J. Cell Biol.* **167**, 35–41.

Thorp, E., et al. (2009). Reduced apoptosis and plaque necrosis in advanced atherosclerotic lesions of Apoe−/− and Ldlr−/− mice lacking CHOP. *Cell Metab.* **9**, 474–481.

Tirasophon, W., et al. (1998). A stress response pathway from the endoplasmic reticulum to the nucleus requires a novel bifunctional protein kinase/endoribonuclease (Ire1p) in mammalian cells. *Genes Dev.* **12,** 1812–1824.

Tirosh, B., et al. (2005). XBP-1 specifically promotes IgM synthesis and secretion, but is dispensable for degradation of glycoproteins in primary B cells. *J. Exp. Med.* **202,** 505–516.

Travers, K. J., et al. (2000). Functional and genomic analyses reveal an essential coordination between the unfolded protein response and ER-associated degradation. *Cell* **101,** 249–258.

van Huizen, R., et al. (2003). P58IPK, a novel endoplasmic reticulum stress-inducible protein and potential negative regulator of eIF2alpha signaling. *J. Biol. Chem.* **278,** 15558–15564.

Wang, X. Z., et al. (1996). Signals from the stressed endoplasmic reticulum induce C/EBP-homologous protein (CHOP/GADD153). *Mol. Cell. Biol.* **16,** 4273–4280.

Wang, X. Z., et al. (1998). Cloning of mammalian Ire1 reveals diversity in the ER stress responses. *EMBO J.* **17,** 5708–5717.

Wei, J., et al. (2008). PERK is essential for neonatal skeletal development to regulate osteoblast proliferation and differentiation. *J. Cell. Physiol.* **217,** 693–707.

Welihinda, A. A., and Kaufman, R. J. (1996). The unfolded protein response pathway in Saccharomyces cerevisiae. Oligomerization and trans-phosphorylation of Ire1p (Ern1p) are required for kinase activation. *J. Biol. Chem.* **271,** 18181–18187.

Wu, J., et al. (2007). ATF6alpha optimizes long-term endoplasmic reticulum function to protect cells from chronic stress. *Dev. Cell* **13,** 351–364.

Yamamoto, K., et al. (2007). Transcriptional induction of mammalian ER quality control proteins is mediated by single or combined action of ATF6alpha and XBP1. *Dev. Cell* **13,** 365–376.

Yamamoto, K., et al. (2010). Induction of liver steatosis and lipid droplet formation in ATF6 {alpha}-knockout mice burdened with pharmacological endoplasmic reticulum stress. *Mol. Biol. Cell* **17,** 2975–2986.

Yan, W., et al. (2002). Control of PERK eIF2alpha kinase activity by the endoplasmic reticulum stress-induced molecular chaperone P58IPK. *Proc. Natl. Acad. Sci. USA* **99,** 15920–15925.

Ye, J., et al. (2000). ER stress induces cleavage of membrane-bound ATF6 by the same proteases that process SREBPs. *Mol. Cell* **6,** 1355–1364.

Yoshida, H., et al. (1998). Identification of the cis-acting endoplasmic reticulum stress response element responsible for transcriptional induction of mammalian glucose-regulated proteins. Involvement of basic leucine zipper transcription factors. *J. Biol. Chem.* **273,** 33741–33749.

Yoshida, H., et al. (2000). ATF6 activated by proteolysis binds in the presence of NF-Y (CBF) directly to the cis-acting element responsible for the mammalian unfolded protein response. *Mol. Cell. Biol.* **20,** 6755–6767.

Yoshida, H., et al. (2001a). XBP1 mRNA is induced by ATF6 and spliced by IRE1 in response to ER stress to produce a highly active transcription factor. *Cell* **107,** 881–891.

Yoshida, H., et al. (2001b). Endoplasmic reticulum stress-induced formation of transcription factor complex ERSF including NF-Y (CBF) and activating transcription factors 6alpha and 6beta that activates the mammalian unfolded protein response. *Mol. Cell. Biol.* **21,** 1239–1248.

Zhang, P., et al. (2002). The PERK eukaryotic initiation factor 2 alpha kinase is required for the development of the skeletal system, postnatal growth, and the function and viability of the pancreas. *Mol. Cell. Biol.* **22,** 3864–3874.

Zhang, K., et al. (2005). The unfolded protein response sensor IRE1alpha is required at 2 distinct steps in B cell lymphopoiesis. *J. Clin. Invest.* **115,** 268–281.

Zinszner, H., et al. (1998). CHOP is implicated in programmed cell death in response to impaired function of the endoplasmic reticulum. *Genes Dev.* **12,** 982–995.

CHAPTER SEVEN

MEASUREMENT OF FLUORIDE-INDUCED ENDOPLASMIC RETICULUM STRESS USING *GAUSSIA* LUCIFERASE

Ramaswamy Sharma,[*,†] Masahiro Tsuchiya,[‡] Bakhos A. Tannous,[‖,§] *and* John D. Bartlett[*,†]

Contents

1. Introduction	112
2. Properties of *Gaussia* Luciferase	114
3. Materials	115
3.1. Cells	115
3.2. Growth medium	116
3.3. Gluc lentivirus vector	116
3.4. Hexadimethrine bromide	117
3.5. Sodium fluoride	117
3.6. Anti-Gluc antibody	117
3.7. CTZ substrate	117
3.8. Luminometer	117
3.9. Microscope	118
3.10. Assay plates (96-well)	118
4. Procedure	118
4.1. Transduction (or transfection) with Gluc	118
4.2. Separation of Gluc-positive cells using FACS	120
4.3. Measurement of Gluc activity	120
4.4. Fluoride and Gluc activity	121
4.5. UPR activation is concurrent with decreased Gluc secretion	123
5. Conclusions	123
References	124

[*] Department of Cytokine Biology, Forsyth Institute, Cambridge Massachusetts, USA
[†] Department of Developmental Biology, Harvard School of Dental Medicine, Boston, Massachusetts, USA
[‡] Division of Aging and Geriatric Dentistry, Tohoku University, Japan
[‖] Departments of Neurology and Radiology, Massachusetts General Hospital, Harvard Medical School, Boston, Massachusetts, USA
[§] Program in Neuroscience, Harvard Medical School, Boston, Massachusetts, USA

Abstract

Endoplasmic reticulum (ER) stress and its consequent activation of the unfolded protein response (UPR) signaling pathway have been implicated in several pathophysiologic disorders as well as in drug resistance to treatment of tumors. Several techniques have been devised that qualitatively and quantitatively demonstrate the presence of ER stress and the activation of the UPR; however, most of these methods cannot be used to measure ER stress in real time. Here we describe the use of cells stably transduced with a secreted reporter, *Gaussia* luciferase (Gluc), to measure fluoride-induced ER stress. Factors that affect ER homeostasis, such as high-dose fluoride, will cause decreased Gluc secretion that can be measured as a decrease in Gluc activity in the culture medium supernatant. Gluc catalyzes the oxidative decarboxylation of coelenterazine (CTZ) to coeleneteramide, resulting in blue bioluminescence (λ_{max} 485 nm). Therefore, Gluc activity can be easily quantified by mixing a small aliquot of the medium supernatant with CTZ and measuring the resulting bioluminescence in a luminometer. Among the various reporters used so far, Gluc is regarded as the most sensitive indicator of ER stress. A second advantage for using Gluc is its ability to function in a wide pH range. This is especially useful for studying fluoride-mediated toxicity as fluoride-induced stress is enhanced under acidic conditions. Since Gluc can be measured in a noninvasive manner, it has been used in several *in vitro* and *in vivo* applications. In this chapter, we detail our methodology for using Gluc to monitor fluoride-induced ER stress.

1. INTRODUCTION

The endoplasmic reticulum (ER) plays multiple roles within the cell. Secreted or membrane-associated proteins are cotranslationally synthesized on ER-localized ribosomes and transit through the ER wherein they can be glycosylated or cross-linked via disulfide bonds. The ER is also a site for lipid and sterol biosynthesis and it functions as a storage center for calcium. Another important ER feature is its function as a quality control organelle that participates in chaperone-mediated protein folding. Only the proteins that fold into their proper conformation are allowed to exit the ER (Hammond and Helenius, 1995). Since several of these processes occur simultaneously with rapid kinetics, any interference in ER homeostasis will result in ER stress and consequent activation of an ER-to-nucleus signaling pathway, termed the unfolded protein response (UPR; reviewed by Schroder and Kaufman, 2005). While the UPR primarily serves to alleviate stress, it can also initiate apoptosis under conditions of chronic stress.

Several agents are now known to induce ER stress, including tunicamycin, which inhibits N-linked glycosylation; thapsigargin, which inhibits the Ca^{2+}-ATPase pump and causes ER calcium depletion; brefeldin A, which inhibits ER to Golgi transport; and dithiothreitol (DTT), which interferes with enzymes that mediate disulfide bonding. In addition, loss of ATP, inhibition of the proteasome and synthesis of mutant proteins also lead to ER stress. Recently, our laboratory has demonstrated that sodium fluoride (NaF) at high doses can also induce ER stress and activate the UPR *in vitro* and *in vivo* (Kubota *et al.*, 2005; Sharma *et al.*, 2008, 2010).

Perhaps, the gold standard for confirming the occurrence of ER stress is by demonstrating distended ER via transmission electron microscopy (TEM). Molecular biology techniques such as western or northern blotting or quantitative real-time PCR are also useful for identifying activation of the UPR components. These tests determine the phosphorylation status of proximal markers such as IRE1, PERK, and eIF2α; cleavage of ATF6 and upregulation of downstream target proteins such as BiP, CHOP, ATF4, ATF3; as well as splicing of the transcription factor, *Xbp-1*. Additional assays include measuring the induction of ER-associated stress elements (ERSE) by *in vitro* reporter systems such as chloramphenicol acetyltransferase (CAT), lacZ, luciferase, or GFP. We have used several of these methods to demonstrate that high-dose fluoride induces ER stress (Table 7.1). However, most of these conventional methods cannot be used for real-time analysis.

Kitamura and colleagues published a series of articles outlining the use of the secreted alkaline phosphatase reporter (SEAP) to quantitatively measure ER stress in real time (Hiramatsu *et al.*, 2006). These authors showed that treatment of SEAP-transfected cells with ER stress-inducing agents resulted in decreased SEAP secretion that could be easily monitored by assaying the cell culture medium supernatant for SEAP enzymatic activity. Importantly, the magnitude of decrease in SEAP secretion directly correlated to the extent of ER stress and to the increase in UPR target genes (Hiramatsu *et al.*, 2006). We have also successfully used SEAP to demonstrate ER stress induced by high-dose fluoride (Sharma *et al.*, 2008).

Fluoride (F^-) in drinking water at concentrations of 0.7–1.2 ppm is beneficial as an anti-cariogenic (CDC, 1995). However, chronic exposure

Table 7.1 UPR pathway components activated by fluoride

Method used	ER stress marker analyzed
Northern blots	BiP, Chop, XBP1
Reverse transcriptase PCR	Spliced XBP1, CA VI-type B
Western blots	BiP, p-PERK, p-eIF2α, Chop, XBP1, p-JNK, p-c-Jun
Immunocytochemistry	p-IRE1, p-PERK, Chop
Immunohistochemistry	p-IRE1, p-eIF2α
Reporter assays	SEAP secretion assays; Gluc secretion assays

Source: Kubota *et al.* (2005) and Sharma *et al.* (2008, 2010).

to high doses of fluoride, as occurs in several developing nations, can result in dental fluorosis (Dean and Elvove, 1936), skeletal fluorosis (Azar et al., 1961), as well as renal and thyroid toxicity (Ogilvie, 1953). The initial and most apparent effect of high-dose F$^-$ is on dental enamel, manifested as white spots of hypomineralized enamel (mild fluorosis) to pitted and stained enamel that chips easily (severe fluorosis). The ameloblast cells responsible for enamel formation are naturally exposed to pH values below 6.0. This occurs during the latter stages of enamel development (maturation stage) when large quantities of protons are released as the hydroxyapatite mineral precipitates (Smith et al., 1996). We have recently shown that fluoride-mediated cell stress in ameloblasts is enhanced at low pH, perhaps due to the formation of the weak acid, HF (Sharma et al., 2010). Therefore, for our experiments, we required a reporter that was stable at low pH values. SEAP activity assays require an alkaline pH since SEAP functions optimally at pH values above 8.0. Hence, we used a luciferase reporter from the marine copepod, *Gaussia princeps*, to assess ER stress under conditions of varying pH to mimic the *in vivo* environment of the maturation stage ameloblasts (Sharma et al., 2010).

2. Properties of *Gaussia* Luciferase

Among the known coelenterazine (CTZ)-catabolizing luciferases, such as *Renilla reniformis* (36 kD), *R. mullerei* (36.1 kD), Aequorin (22.3 kD), *Pleuromamma* (22.6 kD), and *Oplophorus* (21.53 kD), Gluc with 185 amino acids and a molecular weight of 19.9 kD is the smallest in size. Yet, its reaction product is ~750–2000-fold brighter than *R. reniformis* luciferase (Ballou et al., 2000). The humanized (codon-optimized for mammalian cell expression) version of Gluc produces over 1000-fold higher signal intensity as compared to the humanized versions of *Renilla* or firefly luciferases (Tannous et al., 2005). The advantages of *Gaussia* luciferase (Gluc) over other luciferase-based reporter systems such as firefly luciferase or *Renilla* luciferase are manifold. First, Gluc has a signal peptide that directs secretion in eukaryotes. More than 95% of Gluc protein is secreted from cells into the culture medium (Badr et al., 2007). Second, Gluc utilizes CTZ as a substrate and therefore does not require ATP for its enzymatic reaction. Third, Gluc is resistant to a wide range of conditions. It is halo-tolerant and tolerant to detergents such as cholate and deoxycholate. Gluc is also thermo-resistant up to 60 °C (Rathnayaka et al., 2010). Importantly, Gluc is active in a wide range of pH values (pH 3.0–11.0), although its activity peaks at pH 7.7 (Ballou et al., 2000). Finally, Gluc has been expressed in a variety of eukaryotic cell lines, including CHO, HEK-293, and LS8. Importantly, *in vivo* biological processes can be measured *ex vivo* by assaying a few microliters of blood for Gluc activity (Tannous, 2009; Wurdinger et al., 2008).

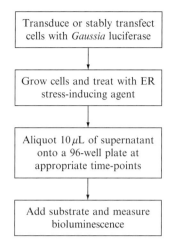

Figure 7.1 Step-wise schematic representation of the Gluc method to quantify ER stress. Cells of interest are transduced with a lentivirus expression vector containing the Gluc construct or stably transfected with a plasmid expressing Gluc. Positive clones are tested for ER stress. After treatment with the ER stress-inducing agent, 10 μL aliquots of the conditioned medium are transferred to a 96-well white plate at appropriate time points. CTZ, the Gluc substrate, is then injected into each well and the resulting bioluminescence is measured to quantify Gluc activity (secretion) using a luminometer.

Like SEAP, Gluc can be used to monitor the secretory pathway and detect ER stress in real time. However, Gluc activity assays to monitor the secretory pathway are quicker (seconds vs. hours) and can be 20,000-fold more sensitive than SEAP activity assays (Badr et al., 2007). Gluc transduction using lentivirus vector by itself does not induce ER stress; however, Gluc secretion decreases by exposure to chemicals that induce ER stress or that target the secretory pathway (Badr et al., 2007). Decreased Gluc secretion occurs in parallel with an increase in UPR markers such as splicing of the transcription factor XBP1 or an increase in the phosphorylation of the translation initiation factor eIF2α (Badr et al., 2007). These results have demonstrated that Gluc assay is a sensitive methodology for monitoring the secretory pathway in real time. The basic procedure for using the Gluc reporter is straightforward and is outlined in Fig. 7.1.

3. MATERIALS

3.1. Cells

Since excess fluoride primarily affects the ameloblasts that are responsible for enamel development, we use ameloblast-like LS8 cells (a kind gift from Dr. Malcolm Snead, University of Southern California, USA). We have also transduced HEK 293T cells with lentivirus vector to stably express Gluc under the control of CMV promoter. Cells are grown in an incubator at 37 °C in a 5% CO_2 atmosphere.

3.2. Growth medium

We grow LS8 cells in Minimal Essential Medium (α-MEM; Invitrogen, Carlsbad, CA, Cat. No. 32561), supplemented with 10% fetal bovine serum (FBS), sodium pyruvate (Invitrogen, Cat. No. 11360) and antibiotic–antimycotic (Invitrogen, Cat. No. 15240). HEK 293T cells are grown in Dulbecco's modified Eagle's medium (DMEM; Invitrogen, Cat. No. 11960), supplemented with Glutamax (Invitrogen, Cat. No. 35050), 10% FBS, sodium pyruvate, and antibiotic–antimycotic.

For generating media with a pH of 6.6 or 7.4 in a 5% CO_2 atmosphere, $NaHCO_3$ is added to DMEM base lacking pH buffer (Sigma, St. Louis, MO, Cat. No. D5523) at a final concentration of either 3 or 21 mM, respectively, as described previously (Gstraunthaler et al., 1992, 2000). Medium osmolarity is adjusted by adding NaCl.

3.3. Gluc lentivirus vector

A self-inactivating lentivirus vector, CSCW-IG, containing a cytomegalovirus (CMV) immediate early promoter and a GFP cDNA separated by an internal ribosomal entry site (IRES) element was used to generate two different Gluc constructs. In the first, a humanized (codon-optimized for mammalian cell expression) Gluc cDNA (Tannous et al., 2005) was cloned directly downstream of the CMV promoter and the GFP cDNA was replaced with cDNA encoding the cerulean fluorescent protein (CFP). Upon successful transduction and expression, this construct (CSCW-Gluc-IRES-CFP; Badr et al., 2007) generates two different proteins, Gluc and CFP. The presence of CFP is used to monitor transduction efficiency. This avoids potentially modifying the properties of Gluc that may occur upon fusing it as a tag adjacent to the Gluc coding sequence. A second construct was also generated that has yellow fluorescent protein (YFP) fused in-frame immediately downstream of the Gluc coding sequence (CSCW-Gluc-YFP). In this construct the Gluc sequence lacks a stop codon. The YFP part of CSCW-Gluc-YFP construct is used to visualize the exact location of Gluc within the cell using fluorescent microscopy. Both constructs were then packaged into lentivirus vectors described above (Badr et al., 2007). Cells were transduced at a multiplicity of infection (MOI) of 100 transducing units per cell, generating >90% transduction efficiency in cells used in this study.

A mammalian-expression plasmid or lentivirus vector encoding Gluc can now be obtained from Nanolight Technology, Prolume Ltd. (Pinetop, AZ) or Targeting Systems (El Cajon, CA), respectively.

3.4. Hexadimethrine bromide

(Also called polybrene; Sigma, St. Louis, MO, Cat. No. H9268.) A stock of 8 mg/mL polybrene is prepared in sterile water and stored in aliquots at −20 °C.

3.5. Sodium fluoride

(Fisher Scientific, Agawam, MA, Cat. No. S299-100.) Prepare a stock solution of 500 mM in water by dissolving 1.05 g sodium fluoride in 50 mL of autoclaved distilled water. Sterilize by filtration through a 0.22 μm filter.

3.6. Anti-Gluc antibody

We use a mouse monoclonal antibody to Gluc (Nanolight Technology, Prolume Ltd., Pinetop, AZ, Cat. No. 404A).

3.7. CTZ substrate

A stock of 5 mM CTZ (Nanolight Technology, Prolume Ltd., Pinetop, AZ, Cat. No. 303A) is prepared in acidified methanol, the latter prepared by adding 200 μL of 6 N HCl to 10 mL of 100% methanol. CTZ stocks can be aliquotted and stored at −80 °C. We typically prepare aliquots of 20 μL which is enough for one 96-well plate.

CTZ is unstable in aqueous solutions; hence fresh working stocks are required.

For preparing the working stock, we dilute the 5 mM stock to 20 μM in phosphate-buffered saline (PBS) containing 500 mM sodium chloride, pH 7.8 (1:250 dilution). Because CTZ will auto-oxidize, the working stock of CTZ requires incubation in dark for 30 min for stabilization prior to use. CTZ can also be diluted in DMEM. However, dilution in PBS results in higher light output and a more stable substrate.

3.8. Luminometer

We record our readings on a Dynex MLX luminometer (Dynex, Richfield, MN) that is equipped with automated injectors. It should be noted that luminometers from different companies vary in sensitivity. If the luminometer is not equipped with injectors, the commercially available GAR-2 reagent (Nanolight Technology, Prolume Ltd., Pinetop, AZ, Cat. No. 320) can be used. The GAR-2 reagent will stabilize the bioluminescent signal for a longer time period, enabling manual addition of the substrate to the sample prior to its measurement. However, the GAR-2 stabilizers decrease the signal intensity by an order of magnitude.

Prior to performing the actual experiment, it is important to determine the linear range of the luminometer for the assay. We usually use serial dilutions of the medium supernatant from Gluc-transduced cells for this purpose. Medium from untransduced control cells serves as a blank reading.

3.9. Microscope

We image our samples on a Zeiss LSM 510 confocal fluorescence microscope to monitor transduction efficiencies and to localize the presence of Gluc in the secretory pathway.

3.10. Assay plates (96-well)

We typically use 96-well Co-Star™ white plates without lid (Fisher Scientific, Agawam, MA, Cat. No. 07-200-589) for assaying Gluc activity.

4. PROCEDURE

4.1. Transduction (or transfection) with Gluc

For lentivirus transduction, plate cells of interest onto a 6-well plate at 40% confluency and incubate overnight at 37 °C in a 5% CO_2 atmosphere. If the cells are slow-growing, plate enough cells so as to obtain 80% confluency the next day. Aspirate the media and wash twice with PBS (pH 7.4). Add approximately 100 µL of the lentivirus vector to obtain a MOI of 50–100 in the presence of 8 µg/mL polybrene in 3 mL growth medium. Use the stock solution of 8 mg/mL polybrene to attain a final concentration of 8 µg/mL. As an option, you may spin the plate at 1000 $\times g$ for 90 min followed by incubation at 37 °C in a 5% CO_2 atmosphere for 5 h. This step will help in obtaining higher transduction efficiency in cells which are not easily transduced with lentivirus vectors. Wash the cells with PBS, pH 7.4 and add fresh growth medium. Incubate the cells for 48 h, and then trypsinize and transfer them. Transduction efficiencies can be calculated by observing the cells for CFP fluorescence using excitation/emission wavelengths of 426/466 nm, respectively (Fig. 7.2).

Transductions can also be performed in 96-well plates and are useful when the quantity of viral vector is low. Cells of interest are plated so that they become 70% confluent after incubation overnight (18–20 h) at 37 °C

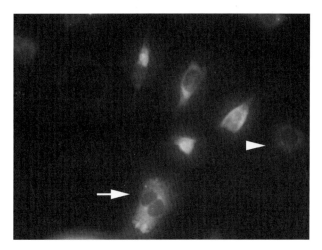

Figure 7.2 Identification of transduced cells. LS8 cells were transduced with a lentivirus construct expressing Gluc-YFP fusion. Fluorescence microscopy for YFP shows a successfully transduced cell (long arrow) and a nontransduced cell (short arrow). (See Color Insert.)

in a 5% CO_2 atmosphere. After washing the cells with PBS, pH 7.4, 110 μL of medium containing 8 μg/mL polybrene and 15 μL of the lentivirus vector are added per well which results in 50–100 transducing units per cell. The cells are incubated overnight at 37 °C in a 5% CO_2 atmosphere. Cells are then washed free of the virus and incubated in fresh medium for 48 h prior to transferring them into larger plates and monitoring transduction efficiencies.

Alternatively, mammalian cell-expression plasmids carrying the Gluc cDNA can be used instead of lentivirus vectors for transient transduction using transfection reagents such as Lipofectamine (Invitrogen, Cat. No. 18324). Disadvantages of using plasmid-based constructs include lower transfection efficiency, transient expression and potential stress to the cells. Lentivirus vectors integrate their DNA within the host genome, resulting in stable Gluc expression.

Lentivirus vectors or plasmids coexpressing Gluc with a fluorescent protein such as CFP or YFP are recommended as they avoid the need for immunostaining to determine transduction or transfection efficiencies. However, in their absence, positive cells can be identified using standard immunostaining procedures. Briefly, cells are grown on coverslips and fixed with 4% paraformaldehyde in PBS for 10 min at room temperature, washed with PBS and permeabilized with 0.1% NP-40 for 10 min. Cells are blocked with 10% goat serum (Vector Laboratories, Burlingame, CA) in PBS for 1 h and then incubated with a monoclonal mouse anti-Gluc antibody (1:100, Nanolight Technology, Prolume Ltd., Pinetop, AZ, Cat.

No. 404A) for 1 h at 37 °C. A donkey anti-mouse antibody conjugated to Cy3 (1:1000, Jackson Laboratories, West Grove, PA) is used to detect the presence of Gluc. Coverslips are mounted onto slides followed by addition of antifade (Vectastain, Vector Laboratories, Cat. No. H-1000) and visualized. If immunostaining is used to sort for viable Gluc-positive cells using fluorescence activated cell sorting (FACS) and the sorted cells are to be grown as clones, the fixation step is to be avoided. In this case, immunostaining procedures for FACS are to be followed.

4.2. Separation of Gluc-positive cells using FACS

If MOI or transduction efficiency are low, Gluc-expressing cells can be separated using FACS, after setting stringent gates based on untransfected cells (Fig. 7.3). Cells can either be separated based on CFP or YFP fluorescence or by immunostaining for Gluc using a suitable secondary antibody conjugated to a fluorochrome. Cells can be sorted directly onto a 96-well plate to separate clones. Clones can be measured for Gluc activity prior to performing the experiment.

4.3. Measurement of Gluc activity

To determine the amount of Gluc in the sample, aliquot 10 µL of cell-free conditioned medium onto a white (or black) 96-well plate. Transfer the plate to a luminometer containing an injector and inject 50 µL of

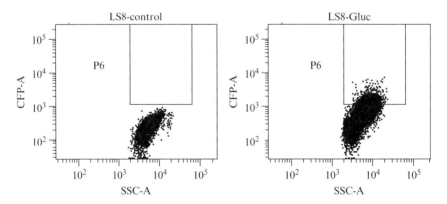

Figure 7.3 Fluorescence activated cell sorting (FACS). After lentivirus vector transduction, positive clones were identified and separated using FACS (BD Aria^{TM}) using appropriate gates based on results from the control nontransduced cells (LS8-control). Side scatter (X-axis) was plotted against fluorescence (CFP; Y-axis) to construct dot plots showing a subpopulation of positively transduced LS8 cells (LS8-Gluc cells in window). The gate was set to exclude background fluorescence, based on data from LS8-control cells.

preincubated CTZ (20 μM in PBS containing 500 mM sodium chloride, pH 7.8; see Section 3). Immediately, measure the bioluminescence generated over 10 s and integrate the signal for 2 s. A linear curve of Gluc activity is expected when plotted against cell number or against cell growth (Badr et al., 2007).

Since Gluc catalyzes flash-type bioluminescence reaction, it should be noted that maximum signal will be obtained within the first 10 s after injection of the CTZ substrate. The signal will drop by as much as 75% within 50 s of adding CTZ and will be just 10% of its initial intensity by 90 s (Tannous et al., 2005). Therefore, it is imperative that the bioluminescence be measured immediately. As outlined in Section 3.8, an alternative is to use commercially available reagents that can stabilize and prolong the bioluminescent signal. In these cases, please refer to the manufacturer's instructions on measurement of Gluc activity. Recently, mutated *Gaussia* luciferases have been generated that have a longer bioluminescence half-life while retaining their original intensity (Maguire et al., 2009; Welsh et al., 2009); these mutants could be used in the absence of a luminometer with a built-in injector.

In case the Gluc signal in the sample is beyond the linear range of the instrument, the sample can be diluted in culture medium prior to assay. Alternatively, the concentration of substrate can be reduced from 20 to 1 μM or, the signal integration time can be reduced.

Although greater than 95% of Gluc is secreted by the cell, intracellular Gluc can also be measured due to the high levels of bioluminescence generated (Tannous et al., 2005). This makes it possible to analyze cellular fractions for Gluc activity if required.

Finally, if samples cannot be analyzed immediately, they may be stored at 4 °C for several days without a loss in activity since Gluc half-life in conditioned medium is approximately 5 days (Wurdinger et al., 2008). Alternatively, they can be stored at −20 °C for a longer period.

4.4. Fluoride and Gluc activity

To determine the magnitude of ER stress induced by high-dose fluoride, we plated 0.25×10^5 LS8 cells expressing either Gluc and CFP or Gluc-YFP fusion in a 6-well plate in 2 mL growth medium. It should be noted that this cell number has to be optimized for different cell lines. Cells were allowed to settle overnight (∼18 h) at 37 °C in a 5% CO_2 atmosphere prior to treatment. After washing cells twice with PBS, pH 7.4 (to remove all traces of secreted Gluc), fresh medium (2 mL) containing fluoride (as NaF) was added to the wells and Gluc activity was measured at different time intervals. Figure 7.4 shows a significant decrease in Gluc activity after treatment with 0.5 mM NaF for 6 h ($p < 0.0001$) or after treatment with tunicamycin, a known ER stress-inducing agent that inhibits N-glycosylation of proteins. Incubating the cell-free medium supernatant with doses up to

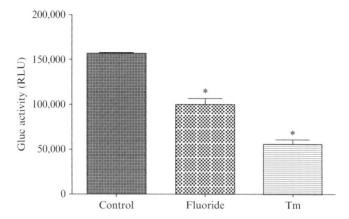

Figure 7.4 Decrease in Gluc secretion upon exposure to fluoride. LS8-Gluc cells were treated for 6 h with either 500 μM NaF or tunicamycin (Tm) which is a known ER stress-inducing agent. Cell-free conditioned medium were then assayed for Gluc activity. Data shown represent an average of two experimental sets performed in triplicate. A two-tailed unpaired Student's t-test indicated a significant difference in Gluc activity between the untreated controls and NaF-treated cells ($p < 0.0001$).

100 mM NaF showed no change in Gluc activity, showing that the observed fluoride-mediated decrease in Gluc secretion was indeed due to fluoride-mediated cell stress and not a direct effect of fluoride on Gluc enzymatic activity (Sharma et al., 2010). Bromide and iodide ions have been shown to stimulate Gluc activity. However, we showed that fluoride has no direct stimulatory effect on Gluc activity (Inouye and Sahara, 2008). Similarly, incubation with 2 mM NaCl had no effect on Gluc activity, indicating that it was the fluoride ion that was responsible for the decreased Gluc secretion. Also, we did not observe any significant changes in cell proliferation at 6 h, suggesting that the decrease in Gluc activity was not due to decreased cell numbers (Sharma et al., 2010).

Furthermore, fluorescent microscopy using LS8-Gluc-YFP fusion (Gluc is directly tagged to YFP in this case) indicated peri-nuclear accumulation of Gluc (Sharma et al., 2010). This result was similar to the observed colocalization of Gluc with the ER-marker PDI after treatment with the ER stress-inducing agent Brefeldin A (Badr et al., 2007). Both results indicate that the secretory pathway was affected. This intracellular accumulation of Gluc after fluoride treatment was further confirmed and quantified using western blot analysis (Sharma et al., 2010). Thus, the Gluc assay can be used to qualitatively demonstrate ER stress by identifying the presence of accumulated Gluc-YFP within the cell.

Finally, to determine the role of pH in fluoride-mediated cell stress, LS8-Gluc cells were treated with 0.5 or 1 mM NaF at pH 6.6 or 7.4 (see Section 3 for growth medium pH adjustment). We observed a significant

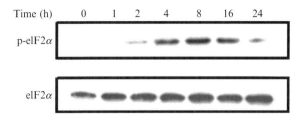

Figure 7.5 Induction of the UPR upon exposure to fluoride. LS8 cells treated with 2 mM NaF were analyzed for phosphorylation of the eukaryotic translation initiation factor 2α (eIF2α) at prespecified time points using western blotting procedures. Immunoblots show increased phosphorylation of eIF2α over time, correlating with the observed decrease in Gluc activity.

decrease in Gluc secretion within 2 h of treatment ($p < 0.05$) when cells were treated with 1 mM NaF at pH 6.6 as compared to treatment at pH 7.4 (Sharma et al., 2010). These data support our hypothesis that low pH enhanced the conversion of F^- to the weak acid HF which then can cross the cell membrane into the neutral environment of the cell cytosol. This causes ameloblast cell stress that compromises the cell's ability to efficiently form dental enamel, resulting in dental enamel fluorosis.

4.5. UPR activation is concurrent with decreased Gluc secretion

To determine if decreased Gluc secretion corresponds to increased ER stress and UPR activation, we have tested three UPR markers: spliced XBP1 transcripts, BiP protein levels, and phosphorylated eIF2α (Badr et al., 2007; Sharma et al., 2010). Treatment of LS8 cells with 2 mM NaF shows increased phosphorylation of eIF2α over time (Fig. 7.5), correlating with the observed decrease in Gluc activity.

Collectively, these results (Sections 4.4 and 4.5) indicate that the Gluc method can be used to quantitatively assess ER stress in general and fluoride-mediated cell stress in particular.

5. Conclusions

In this chapter, we have described a straightforward method for identifying compounds that induce ER stress. Because the Gluc cDNA is available packaged into lentivirus vectors, it can be used to transduce a wide range of cell lines that are relatively resistant to transfection procedures, thereby increasing the accessibility of this method to various experimental paradigms. Furthermore, the Gluc substrate, CTZ, can permeate the cell membrane making it possible to visualize the secretory pathway in live cells. For example, Gluc has been used to image protein trafficking to the cell surface and also protein exocytosis in real time (Suzuki *et al.*, 2007). Gluc can also be adapted for use in high throughput, robotic screens to determine the alleviating effect of specific chemicals on a faulty secretory pathway. Gluc can be introduced into cells that possess defects in the secretory pathway followed by treatment with the test compound. If the latter can reverse the defect, as shown by increased Gluc secretion, then that compound can be assessed for its therapeutic potential.

In summary, Gluc is a highly sensitive and a quantitative tool that can be used to analyze the secretory pathway and monitor ER stress in real time.

REFERENCES

Azar, H. A., Nucho, C. K., Bayyuk, S. I., and Bayyuk, W. B. (1961). Skeletal sclerosis due to chronic fluoride intoxication. Cases from an endemic area of fluorosis in the region of the Persian Gulf. *Ann. Intern. Med.* **55,** 193–200.

Badr, C. E., Hewett, J. W., Breakefield, X. O., and Tannous, B. A. (2007). A highly sensitive assay for monitoring the secretory pathway and ER stress. *PLoS One* **2,** e571.

Ballou, B., Szent-Gyorgyi, C., and Finley, G. (2000). Properties of a new luciferase from the copepod *Gaussia princeps*. 11th international symposium on bioluminescence and chemiluminescence, Asilomar, CA, USA. http://www.nanolight.com/pdf/nanolight.pdf.

CDC (1995). Engineering and administrative recommendations for water fluoridation, 1995. Centers for Disease Control and Prevention. *MMWR Recomm. Rep.* **44,** 1–40.

Dean, H. T., and Elvove, E. (1936). Some Epidemiological Aspects of Chronic Endemic Dental Fluorosis. *Am. J. Public Health Nations Health* **26,** 567–575.

Gstraunthaler, G., Landauer, F., and Pfaller, W. (1992). Ammoniagenesis in LLC-PK1 cultures: Role of transamination. *Am. J. Physiol.* **263,** C47–C54.

Gstraunthaler, G., Holcomb, T., Feifel, E., Liu, W., Spitaler, N., and Curthoys, N. P. (2000). Differential expression and acid-base regulation of glutaminase mRNAs in gluconeogenic LLC-PK(1)-FBPase(+) cells. *Am. J. Physiol. Renal Physiol.* **278,** F227–F237.

Hammond, C., and Helenius, A. (1995). Quality control in the secretory pathway. *Curr. Opin. Cell Biol.* **7,** 523–529.

Hiramatsu, N., Kasai, A., Hayakawa, K., Yao, J., and Kitamura, M. (2006). Real-time detection and continuous monitoring of ER stress *in vitro* and *in vivo* by ES-TRAP: evidence for systemic, transient ER stress during endotoxemia. *Nucleic Acids Res.* **34,** e93.

Inouye, S., and Sahara, Y. (2008). Identification of two catalytic domains in a luciferase secreted by the copepod Gaussia princeps. *Biochem. Biophys. Res. Commun.* **365,** 96–101.

Kubota, K., Lee, D. H., Tsuchiya, M., Young, C. S., Everett, E. T., Martinez-Mier, E. A., Snead, M. L., Nguyen, L., Urano, F., and Bartlett, J. D. (2005). Fluoride induces endoplasmic reticulum stress in ameloblasts responsible for dental enamel formation. *J. Biol. Chem.* **280,** 23194–23202.

Maguire, C. A., Deliolanis, N. C., Pike, L., Niers, J. M., Tjon-Kon-Fat, L. A., Sena-Esteves, M., and Tannous, B. A. (2009). Gaussia luciferase variant for high-throughput functional screening applications. *Anal. Chem.* **81,** 7102–7106.

Ogilvie, A. L. (1953). Histologic findings in the kidney, liver, pancreas, adrenal, and thyroid glands of the rat following sodium fluoride administration. *J. Dent. Res.* **32,** 386–397.

Rathnayaka, T., Tawa, M., Sohya, S., Yohda, M., and Kuroda, Y. (2010). Biophysical characterization of highly active recombinant Gaussia luciferase expressed in Escherichia coli. *Biochim. Biophys. Acta.* **1804,** 1902–1907.

Schroder, M., and Kaufman, R. J. (2005). The mammalian unfolded protein response. *Annu. Rev. Biochem.* **74,** 739–789.

Sharma, R., Tsuchiya, M., and Bartlett, J. D. (2008). Fluoride induces endoplasmic reticulum stress and inhibits protein synthesis and secretion. *Environ. Health Perspect.* **116,** 1142–1146.

Sharma, R., Tsuchiya, M., Skobe, Z., Tannous, B. A., and Bartlett, J. D. (2010). The acid test of fluoride: how pH modulates toxicity. *PLoS One* **5,** e10895.

Smith, C. E., Issid, M., Margolis, H. C., and Moreno, E. C. (1996). Developmental changes in the pH of enamel fluid and its effects on matrix-resident proteinases. *Adv. Dent. Res.* **10,** 159–169.

Suzuki, T., Usuda, S., Ichinose, H., and Inouye, S. (2007). Real-time bioluminescence imaging of a protein secretory pathway in living mammalian cells using Gaussia luciferase. *FEBS Lett.* **581,** 4551–4556.

Tannous, B. A. (2009). Gaussia luciferase reporter assay for monitoring biological processes in culture and in vivo. *Nat. Protoc.* **4,** 582–591.

Tannous, B. A., Kim, D. E., Fernandez, J. L., Weissleder, R., and Breakefield, X. O. (2005). Codon-optimized Gaussia luciferase cDNA for mammalian gene expression in culture and in vivo. *Mol. Ther.* **11,** 435–443.

Welsh, J. P., Patel, K. G., Manthiram, K., and Swartz, J. R. (2009). Multiply mutated Gaussia luciferases provide prolonged and intense bioluminescence. *Biochem. Biophys. Res. Commun.* **389,** 563–568.

Wurdinger, T., Badr, C., Pike, L., de Kleine, R., Weissleder, R., Breakefield, X. O., and Tannous, B. A. (2008). A secreted luciferase for ex vivo monitoring of in vivo processes. *Nat. Meth.* **5,** 171–173.

CHAPTER EIGHT

ANALYSIS OF NELFINAVIR-INDUCED ENDOPLASMIC RETICULUM STRESS

Ansgar Brüning

Contents

1. Introduction	128
2. Obtaining Nelfinavir	129
3. Using Nelfinavir	129
4. Analysis of Nelfinavir-Induced ER Stress	130
4.1. Phase contrast microscopy	130
4.2. Fluorescence microscopy	130
5. Immunoblot Analysis	134
6. RT-PCR Analysis	135
6.1. RNA preparation	138
6.2. cDNA synthesis	139
6.3. PCR and PCR conditions	139
7. Conclusions	141
Acknowledgment	142
References	142

Abstract

Nelfinavir (Viracept®) is an HIV protease inhibitor that has been shown to induce the endoplasmic reticulum (ER) stress reaction in human cancer cells. Although the presumed drug doses needed for an efficient ER stress reaction and ensuing apoptosis in cancer cells is somewhat higher than those prescribed for HIV-infected persons, nelfinavir represents one of the few clinically applicable ER stress-inducing agents, and is currently being tested in clinical studies on cancer patients. Therefore, this chapter describes how to obtain and use nelfinavir for *in vitro* and *in vivo* studies. In addition, methods are described that might facilitate the analysis and monitoring of the nelfinavir-induced ER stress response either in cancer cells in cell culture or in cancer tissue biopsies. These methods include various fluorescence-based ER staining techniques and the expression analysis of primary and secondary ER stress markers by immunoblotting and RT-PCR analysis. Among the several methods presented, the

Department of OB/GYN, Molecular Biology Laboratory, University Hospital Munich, Campus Innenstadt and Grosshadern, Munich, Germany analysis of an unconventional XBP1 splicing, caused by the ER stress sensor IRE1, is shown to present the most sensitive and most specific marker for nelfinavir-induced ER stress. Primers and PCR conditions suitable for XBP1 PCR and splicing analysis are presented. Such a PCR-based XBP1 splicing analysis might not only be suitable to monitor nelfinavir-induced ER stress, but could also be applied in drug screening programs to test for other ER stress-inducing agents with similar activities or synergistic activities with nelfinavir.

1. INTRODUCTION

Nelfinavir is a clinically applied HIV protease inhibitor that has been used to treat HIV-infected persons for more than a decade. Nelfinavir inhibits the retroviral aspartic protease that is necessary for the correct proteolytic cleavage of viral precursor proteins into their mature forms (Kaldor et al., 1997). In combination with reverse transcriptase inhibitors, this highly active antiretroviral combination therapy (HAART) has not only saved the lives of many individuals, but has also reduced the spread of HIV infections by suppressing the viral load to subinfectious levels. However, in current prescriptions, nelfinavir is widely replaced by newer generation HIV protease inhibitors, especially those that can be boosted with ritonavir.

Recently, nelfinavir has regained interest because it has been shown to be able to induce the endoplasmic reticulum (ER) stress reaction in human cancer cells (Brüning et al., 2009; Gills et al., 2007; Pyrko et al., 2007). Although primarily a cell protective mechanism (Lin et al., 2007), extensive ER stress can induce cell death by an apoptotic pathway different from those pathways activated by most established cancer treatments (Fels and Koumenis, 2006; Healy et al., 2009; Rao et al., 2004). Thus, nelfinavir could be of special interest as a medication against otherwise chemoresistant cancer cells that do not respond or have become resistant to standard chemotherapeutics such as the organoplatinum compounds and taxanes. Hence, several clinical studies on nelfinavir have been launched to investigate the potential use of nelfinavir as a second- or third-line treatment option for a variety of human cancer types (Brüning et al., 2010).

The molecular target of nelfinavir in humans is still unknown. Since the ER stress reaction reflects accumulation of misfolded proteins within the ER, it has been speculated that nelfinavir interacts with the proteolytic processing of newly synthesized proteins within the ER or the ER-associated degradation machinery (Brüning et al., 2009; Gupta et al., 2007).

Because of the potential interest of nelfinavir for cell biologists studying the mechanisms of the ER stress reaction, and for clinicians who depend on reliable assays to monitor the efficacy of nelfinavir on cancer tissues, this chapter describes methods that might be useful for the analysis of nelfinavir-induced ER stress in cell culture as well as in cancer tissue biopsies.

2. Obtaining Nelfinavir

Nelfinavir is an HIV drug under prescription and, until recently, was not available from any commercial supplier of laboratory reagents. Originally developed by Agouron Pharmaceuticals, Inc. (USA), nelfinavir is distributed and marketed as Viracept® either by Pfizer Inc. (Groton, CA, USA) or by the Roche AG (Switzerland), depending on the geographical distribution. Viracept® pills usually contain 250 mg of nelfinavir in its mesylate form, which is sufficient to test its efficacy in preliminary experiments even in a large number of cell biology studies. After removal of the outer tablet film with a scalpel, a small weighed amount of the inner tablet mass can be taken for the elution of nelfinavir with dimethylsulfoxide or ethanol, as previously performed by us and others (Brüning et al., 2009; Pyrko et al., 2007). For in-depth experiments, we recommend the use of chemically pure nelfinavir mesylate, which, upon request, can be obtained directly from Pfizer after signing a conventional material transfer agreement. Recently, nelfinavir became additionally available by Sigma-Aldrich (www.sigmaaldrich.com). Experiments demonstrated in this book chapter are based on the use of chemically pure nelfinavir.

3. Using Nelfinavir

In HIV-infected persons taking the prescribed oral doses of nelfinavir, mean plasma concentrations of 2.2 ± 1.25 µg/ml nelfinavir are achieved (Brüning et al., 2010). These low and well-tolerated concentrations of nelfinavir are sufficient for the effective and long-term inhibition of the retroviral proteinase, but not efficient for the induction of the ER stress reaction in cancer cells. To induce an ER stress reaction in epithelial cancer cells that is strong enough to result in apoptosis, at least 8–15 µg/ml nelfinavir have to be applied. For cell culture experiments, nelfinavir can directly be applied from the stock solution (10–50 mg/ml in DMSO or ethanol) to the cell culture medium. Otherwise, prepare a working solution of nelfinavir in a blue cap and add this solution to the cells. For the use of nelfinavir in animal studies, several methods have been adapted. Pyrko et al. (2007), studying subcutaneous glioblastoma xenografts in mice, administered 120 mg of nelfinavir per kg in a solution containing 25% ethanol directly to the stomach with a feeding needle. Gills et al. (2007), using a lung cancer xenograft model, relied on the intraperitoneal route and applied

100 mg of nelfinavir per kg in a solution of 4% DMSO, 5% polyethylene glycol, and 5% Tween 80 in saline.

Since nelfinavir is produced as a ready-to-use drug for oral application in humans, there is no difficulty in applying it to patients in clinical studies. Although the number of pills (around 10 tablets of 250 mg daily) that must be ingested by HIV patients is cumbersome, the availability of a 625 mg bolus in some countries has slightly improved this situation. As clinical studies with nelfinavir are currently in progress to determine not only its efficacy in cancer patients, but likewise the maximum tolerable dose and associated adverse effects, no recommendations on the optimal amount of nelfinavir for cancer patients can currently be given.

4. Analysis of Nelfinavir-Induced ER Stress

4.1. Phase contrast microscopy

Application of nelfinavir to cells in cell culture induces an extensive and remarkable intracellular vacuolation process that can readily be observed and documented by phase contrast microscopy (Fig. 8.1). Vacuolation appears to be most prominent and visible in cancer cells with an epithelial morphology, such as HeLa (ATCC CCL-2) cervical cancer cells and OVCAR3 (ATCC HTB-161) ovarian cancer cells (Fig. 8.1). Other widely available human cancer cell lines with epithelial morphology such as the breast cancer cell line MCF7 (ATCC HTB-22) and the lung cancer cell line A549 (ATCC CCL-185) are similarly suited to visualize the dramatic morphological changes caused by nelfinavir. However, the formation of these intracellular vacuoles is a late process of nelfinavir-induced ER stress and can take up to 24 h to become visible. Longer incubation times result in cell death and detachment of the cells from the substrate. However, since intracellular vacuolation is not an exclusive feature of ER stress and can be caused by other factors too, the use of specific organelle markers is recommended for an accurate and unambiguous identification of these vacuoles as ER-derived compartments.

4.2. Fluorescence microscopy

4.2.1. Fluorescent antibodies

Several methods for the specific labeling and identification of the ER by fluorescence microscopy have been developed that can be applied for the characterization of nelfinavir-induced ER vacuolation. Antibodies against specific and abundant ER-resident proteins, such as calreticulin, calnexin, and PDI (protein disulfide isomerase) are obviously the first choice. The ER-resident protein BiP (binding protein, GRP78) is especially suited for

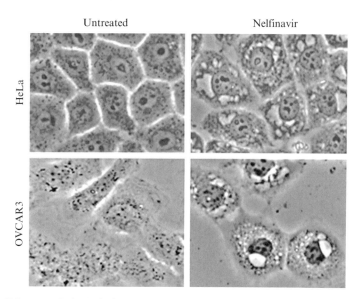

Figure 8.1 Morphological characteristics of nelfinavir-induced ER stress. Nelfinavir-induced swelling of the ER in epithelial cancer cells can be easily viewed and documented by phase contrast microscopy. Most often a large, paranuclear vacuole can be seen. Vacuolation is a late stage of nelfinavir-induced ER stress and takes up to 24 h to become visible, and might reflect autophagy or reticulophagy as well. Samples: OVCAR3 cells treated for 24 h with 8 μg/ml nelfinavir, and HeLa cells treated for 24 h with 15 μg/ml nelfinavir.

nelfinavir-induced ER stress analysis, since this protein, expressed at low levels in unstimulated cells, is strongly induced by nelfinavir and can thus serve as a morphological as well as a physiological marker for the ER stress reaction in immunofluorescence studies.

The following protocol can be used for BiP staining in nelfinavir-treated cancer cells:

- Plate cells at least 1 day before the application of nelfinavir on glass cover slips or chamber slides to allow cells to become adherent. The use of chamber slides is recommended, since it facilitates the handling of nelfinavir solutions, fixation procedures, and antibody incubation steps. When 8-well chamber slides are used (e.g., Lab-Tek®II Chamber Slides™, order number 154534, Nunc International, www.nuncbrand.com), an initial inoculation of 5×10^3 cells per well is recommended.
- Incubate the adherently growing cells for 24 h with nelfinavir at 8–15 μg/ml, depending on the cell line used.
- Wash cells once with phosphate-buffered saline (PBS) and fix and permeabilize cells for 5 min with $-20\ °C$ cold methanol.
- Replace the methanol by PBS and allow cells to rehydrate for 5 min.

- Incubate the fixed cells for 2 h with a 1:200 dilution of anti-BiP antibody (see Table 8.1) in PBS/5% horse serum.
- Wash cells three times with PBS for 5 min each.
- Incubate cells for 1 h with an appropriate secondary antibody suitable for immunofluorescence analysis (Fig. 8.2A: Cy3-conjugated goat anti-rabbit antibody, 1:500 dilution in PBS/5% horse serum).
- Wash cells three times with PBS for 5 min each, rinse once with distilled water, mount the cells with an appropriate mounting medium, and then subject the cells to immunofluorescence analysis.

Figure 8.2A shows the accumulation of BiP immunostaining in the vacuoles of nelfinavir-treated OVCAR3 cells (Fig. 8.2A, upper panel). Notably, at late stages of the vacuolation process, staining for BiP expression appears to be restricted to the ER membranes, and the content of the larger vacuoles appears to be void of BiP immunostaining (Fig. 8.2B, lower panel), indicating the formation of autophagosomes with degraded contents.

4.2.1. Fluorescent proteins

Gills et al. (2007) used a yellow fluorescent protein tagged with an ER-retaining KDEL (Lys-Asp-Glu-Leu) sequence to identify the nature of nelfinavir-induced vacuoles and the fate of ER-resident proteins after

Table 8.1 Antibodies for the analysis of nelfinavir-induced ER stress

Antigen	Size (kDa)	Antibody	Order number (clone)	Supplier
BiP[a]	78	Rabbit mAb	#3177 (C50B12)	Cell Signaling Technology
BiP	78	Rabbit	sc-13968	Santa Cruz Biotechnology
ATF3	28	Rabbit	sc-188	Santa Cruz Biotechnology
Phospho-eIF2α[a]	38	Rabbit	#9721	Cell Signaling Technology
eIF2α[a]	38	Rabbit	#9722	Cell Signaling Technology
IRE1α[a]	130	Rabbit mAb	#3294 (14C10)	Cell Signaling Technology
CHOP[a]	27	Mouse mAb	#2895 (L63F7)	Cell Signaling Technology

[a] Included in an ER stress antibody sampler kit (#9956; Cell Signaling Technology). mAb, monoclonal antibody. Cell Signaling Technology, www.cellsignal.com; Santa Cruz Biotechnology, www.scbt.com.

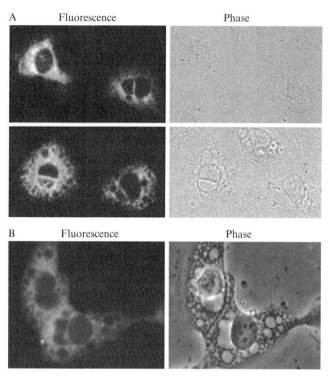

Figure 8.2 Fluorescence analysis of nelfinavir-treated cancer cells. A specific ER organelle staining in nelfinavir-treated cancer cells can either be achieved by BiP immunostaining (A), or with fluorescent glibenclamide to stain for ER membranes (B; previously published in Brüning et al. 2009). Samples: OVCAR3 cells treated for 24 h with 8 μg/ml nelfinavir. (See Color Insert.)

nelfinavir treatment. Since the use of direct fluorescent proteins does not necessarily need fixation of the cells prior to fluorescence analysis, this method is applicable even to viable cells, which potentially allows time-lapse studies on the generation and fate of nelfinavir-derived vesicles. However, this technique is based on an additional transfection step and is only applicable on cell lines that reveal a sufficient high transfection rate. Plasmids for cloning and the expression of ER-targeted proteins fused to various fluorescent proteins are available from various suppliers of molecular biological reagents.

4.2.2. Fluorescent glibenclamide

The oral antidiabetic glibenclamide (5-chloro-N-(4-[N-(cyclohexylcarbamoyl)sulfamoyl]phenethyl)-2-methoxybenzamide) is a drug that stimulates insulin release from pancreatic β-cells by binding to the sulfonylurea receptors of ATP-dependent potassium channels. As these receptors are highly enriched on the membranes of the ER not only in pancreatic cells, coupling of fluorescent dyes to this synthetic drug allows vital staining of ER membranes by the binding of fluorescent glibenclamide to its receptors. We have previously used BODIPY-coupled glibenclamide (ER-Tracker[TM] Green, Molecular Probes/Invitrogen, www.invitrogen.com) to label ER membranes in nelfinavir-treated cancer cells (Brüning et al., 2009). Figure 8.2B shows ER-Tracker Green-stained OVCAR3 cells after treatment for 24 h with 8 µg/ml nelfinavir. The following protocol is recommended for staining the ER membranes of viable cancer cells treated with nelfinavir:

- Plate cells at least 1 day before the application of nelfinavir on glass cover slips or chamber slides and allow cells to become adherent.
- Incubate for 24 h with nelfinavir at 8–15 µg/ml.
- Dissolve 100 µg of lyophilized ER-Tracker dye in 128 µl DMSO to generate a 1 mM stock solution.
- Wash cells once with Hank's balanced salt solution (HBSS) with calcium and magnesium, and treat the viable cells for 15 min with a working solution of 1 µM ER-Tracker dye in HBSS/Ca/Mg under cell culture conditions.
- Replace staining solution by normal cell culture medium.

The cells, grown on glass cover slips, can then directly be subjected to fluorescence analysis using filter systems adapted for FITC (fluorescein isothiocyanate) staining. Alternatively, ER Tracker[TM] Red (Molecular Probes/Invitrogen) can be used for detection by the rhodamine filter system. Since the methods constitute vital staining, care has to be taken that the cover slips do not dry out, which can be achieved by steadily replenishing the medium or by sealing the rims of the glass cover slips with paraffin or similar agents.

5. IMMUNOBLOT ANALYSIS

Since the ER stress response causes several characteristic changes in the expression level and phosphorylation status of cellular proteins (diagrammed in Fig. 8.3), some of them are expected to be detectable by Western blot analysis. The ER stress reaction is primarily sensed and transduced by three ER transmembrane proteins: ATF6 (activating transcription factor 6), IRE1 (inositol requiring protein kinase and endonuclease 1), and PERK (double-stranded RNA-activated protein kinase-like ER kinase) (Fig. 8.3). Proteolytic cleavage of ATF6 and phosphorylation of IRE1 and PERK then mediate both cytoplasmic signaling cascades and nuclear transcription

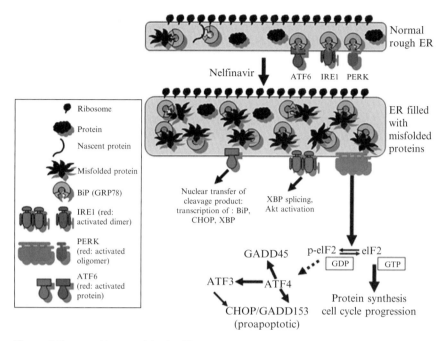

Figure 8.3 Working model of nelfinavir-induced ER stress. Current working model of the mechanisms of nelfinavir-induced ER membrane swelling and ER stress signaling (previously published in Brüning et al., 2009).

activities (Fels et al., 2006; Lin et al., 2007). Unfortunately, we, as well as others (Samali et al., 2010), were not convinced by the quality of several of the commercially available antibodies against the so-called proximal ER stress sensors, including ATF6 and PERK. Apparently, no commercial phospho-specific IRE1 antibody is available, although an IRE1 antibody exists that gives quite reasonable results (Table 8.1; Brüning et al., 2009). Therefore, several researchers have relied on the use of antibodies against secondary targets of the ER stress reaction that include BiP, phospho-eIF2 α (eukaryotic initiation factor-2α), and ATF3. Examples of immunoblots with selected antibodies to monitor nelfinavir-induced ER stress in the HeLa cell line are shown in Fig. 8.4. A summary of antibodies useful for monitoring nelfinavir-induced ER stress is given in Table 8.1.

6. RT-PCR Analysis

As nelfinavir is currently being tested in clinical trials for its potential use as a new anticancer drug, it would be helpful to have a reliable and cost-effective assay to assess the efficacy of nelfinavir on cancer tissues in humans

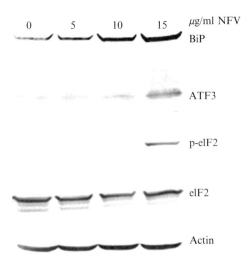

Figure 8.4 Immunoblot analysis for nelfinavir-induced ER stress markers. Immunoblots of nelfinavir-treated HeLa cells (0–15 μg/ml nelfinavir) were performed to demonstrate the applicability of antibodies that have reproducibly been shown to detect ER stress-related proteins. Antibody suppliers and clonal designations are listed in Table 8.1.

or in animal studies. Cancer biopsies might be investigated by immunohistochemical or immunoblot analysis for the expression of ER stress-related proteins such as ATF3, BiP, or CHOP (CCAAT/enhancer-binding protein homologous protein) but, as can be seen in Figs. 8.4 and 8.5A, most of these proteins exhibit a basic expression level and might not be useful as unique markers of nelfinavir-induced cell stress. Further, as mentioned above, some of these proteins represent only secondary factors of the ER stress reaction, or can be induced by other signaling pathways as well. In contrast, the splicing of the X box-binding protein 1 (XBP1) by IRE1 is a unique and early hallmark of ER stress (Yoshida et al., 2001). Under normal cell physiological conditions, XBP1 is transcribed from a 273 bp open reading frame to give rise to a 29 kDa protein that is rapidly degraded by proteasomal activity (Yoshida et al., 2001). Under ER stress conditions, the integral ER membrane protein IRE1 unfolds its endonuclease activity and excises 26 bp from an exon of the XBP1 mRNA (Fig. 8.6). The ensuing frame shift then gives rise to a different open reading frame and a markedly longer form of XBP1. This unconventionally spliced form of XBP1 (XBP1-S, 50 kDa) displays a significantly longer half-time and a higher transcriptional activity than its unspliced form. By using primers flanking the XBP1 splicing region, this mRNA modification can be easily detected by RT-PCR analysis by the appearance of a faster migrating band as demonstrated in Fig. 8.5A.

Nelfinavir and ER Stress

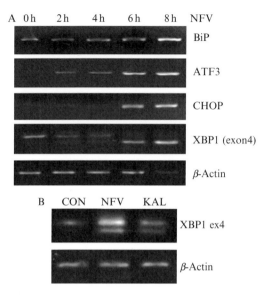

Figure 8.5 RT-PCR analysis for nelfinavir-induced ER stress markers. (A) A conventional RT-PCR analysis was performed to demonstrate the applicability of specific primers for ER stress markers in nelfinavir-treated HeLa cells (incubated with 15 μg/ml nelfinavir for different time points). Primer pairs used are listed in Table 8.2. (B) HeLa cells were treated for 6 h with 10 μg/ml of either nelfinavir (NFV) or Kaletra® (KAL), and subjected to XBP1 RT-PCR analysis.

→
TGCT<u>GAAGAGGAGGCGGAAGC</u>CAAGGGGAATGAAGTGAGG
CCAGTGGCCGGGTCT<u>GCTGAGTCCG</u>:CAGCACTCAGACTA
CGTGCACCTCTG:<u>CAGCAGGTGC</u>AGGCCCAGTTGTCACCC
CTCCAGAACATCTCCCCATGGATTCTGGCGGTATTGACTC
TTCAGATTCAGAGT<u>CTGATATCCTGTTGGGCATT</u>CTGGAC
←

Figure 8.6 XBP1 splicing site and primer orientation. The 26 nucleotides that are spliced within an exon of XBP1 (exon 4, marked in blue) are indicated in red. Nucleotides of the adjacent exons are printed in black. Binding sites of PCR primers are underlined and their orientation (5′–3′) is indicated by an arrow. Primers were placed before the exon–exon boundaries to avoid amplification of exon 4 from genomic DNA. A suggested sequence for a TaqMan Probe (e.g., 6-FAM-GCTGAGTCCG-CAGCAGGTG-TAMRA) that specifically recognizes the spliced form of XBP1 and would allow quantitative real-time PCR analysis is underlined in blue. (See Color Insert.)

Furthermore, splicing of XBP1 occurs within hours of nelfinavir treatment, indicating XBP1 splicing as an early marker of the action of nelfinavir. In addition, this assay could be used to screen for ER stress-inducing effects of other HIV protease inhibitors. For example, as can be seen in Fig. 8.5B, splicing of XBP1 also occurs in HeLa cells treated with Kaletra® (lopinavir plus baby dose ritonavir), another widely applied HIV protease inhibitor. Thus, analysis of XBP1 splicing might be a suitable screening assay to identify new drugs for their ability to induce or to interfere with the ER stress reaction. Furthermore, this splicing analysis might be especially suited for testing the efficacy of nelfinavir in cancer tissue biopsies by conventional RT-PCR analysis or by real-time PCR analysis. Therefore, an RNA extraction protocol and PCR conditions are described that can be applied to analyze XBP1 splicing in samples generated *in vitro* and *in vivo*.

6.1. RNA preparation

(a) Cell culture cells:
- Seed 2.5×10^5 cells in 6-well plates and allow cells to grow for 24 h.
- Add 8–15 μg/ml of nelfinavir to the cells. Provide the control cells with an equal amount of either DMSO or ethanol.
- Incubate cells for 6–8 h with nelfinavir. Longer incubations times are not necessary and recommended for Western blotting only.
- Prepare RNA by dissolving cells directly on the cell culture plates with the appropriate denaturation buffer. Both the phenol/chloroform extraction method, simplified by the TRIzol (Invitrogen, www.invitrogen.com) or TRI reagent (Sigma-Aldrich, www.sigma-aldrich.com) formulation, as well as spin columns (Macherey-Nagel, www.mn-net.com; Qiagen, www.qiagen.com) will generate a sufficient amount of total RNA. However, when using tissue samples, we strongly recommend the use of TRIzol or TRI reagent as described below for the TRIzol method.

(b) Cancer tissues:
- Cut and mince parts of the cancer tissue into small pieces with a scalpel. Avoid using larger tissue clumps for RNA extraction. Since cancer tissues are heterogeneous, taking samples from different regions of the tumor is recommended.
- Transfer small pieces of the cancer tissue samples (10–100 mg) into a 1.5 ml sterile reaction vial and immediately freeze the sample in liquid nitrogen.
- Take a single-use 1.5 ml plastic pestle (#P7339-901, Argos Technologies, www.argos-tech.com; also distributed by VWR International, www.vwr.com) that fits into the 1.5 ml reaction vial and grind the frozen tissue sample by hand. Please be sure to wear safety goggles and gloves. Grinding is performed on the frozen tissue samples after remnants of the liquid nitrogen have evaporated. Do not allow the tissue sample to thaw. When the tissue gets a gum-like consistency during grinding, place the reaction vial with the pestle in liquid

nitrogen again for a short time. Repeat the procedure until the tissue becomes pulverized.
- Add 1 ml TRIzol reagent to the sample and dissolve the ground tissue by pipetting up and down several times. Allow the TRIzol to react with the samples for 5 min at room temperature. At this stage, samples can be collected and kept frozen for up to 1 month or longer when stored at $-80\ °C$.
- Add 200 μl of chloroform and shake vigorously by hand (no vortex). Incubate for 2 min at room temperature and centrifuge the homogenate at $12,000 \times g$ for 15 min at $4\ °C$.
- Transfer the clear upper phase, containing the RNA, to a fresh microcentrifuge tube and mix with 500 μl isopropyl alcohol. After incubation for 10 min at room temperature, centrifuge at $12,000 \times g$ for 10 min at $4\ °C$.
- Remove the supernatant and wash the RNA pellet once with 75% ethanol by vortexing and subsequent centrifugation for 5 min at $7500 \times g$ at $4\ °C$. Allow the pellet to dry for 5 min in the same fume hood that has been used for TRIzol and chloroform handling. Dissolve the pellet in 16.5–50 μl RNase-free water.

6.2. cDNA synthesis

- Add 0.75 μl of a 100 μM solution of oligo(dT)$_{15}$ to 16.5 μl RNA as obtained above (approximately 2 μg RNA).
- Incubate for 5 min at 70 °C, then cool immediately on ice and add:
  - 5 μl M-MLV reverse transcriptase 5× reaction buffer (#M531A, Promega, www.promega.com).
  - 1.25 μl PCR nucleotide mix, 10 mM each (# C114H, Promega).
  - 0.5 μl (20 U) recombinant RNasin® ribonuclease inhibitor (#N2511, Promega).
  - 1 μl (200 U) M-MLV (Moloney Murine Leukemia Virus) reverse transcriptase (#M170B, Promega).
- Incubate for 60 min at 42 °C, and then cool immediately on ice.

6.3. PCR and PCR conditions

Mix in a sterile PCR tube the following components:

- 1 μl of cDNA as obtained above.
- 1 μl of primer mix (5 μM of each forward and reverse primer; for sequences, see Table 8.2).
- 12.5 μl 2× PCR master mix (containing Taq polymerase and dNTP; #M750C, Promega).

Table 8.2 Primer pairs for the analysis of nelfinavir-induced ER stress

Gene product	Forward primer (5′–3′)	Reverse primer (5′–3′)	PCR product (bp)
BiP(GRP78)	GCTGTAGCGTATGGTGCTGC	ATCAGTGTCTACAAC TCATC	794
ATF3	AGCAAAATGATGCTTCAACAC	GCTCTGCAATGTTCCTTCTTTC	553
CHOP	CTGCAGAGATGGCAGCTGAGTC	TGCTTGGTGCAGATTCACCATTC	515
XBP1(exon 4)	GAAGAGGAGGCGGAAGCCAAG	GAATGCCCAACAGGATATCAG	189/163
β-Actin	GGAGAAGCTGTGCTACGTCG	CGCTCAGGAGGAGCAATGAT	366

All primers were designed and tested by the author. Primers for ATF3 and CHOP were generated to amplify the full-length gene product.

- 10.5 μl sterile water.

Place samples in a thermocycler, and apply the following PCR conditions:

XBP1

- 5 min 95 °C, followed by 32 cycles of:
- 0.5 min 95 °C.
- 0.5 min 63 °C.
- 0.5 min 72 °C.

ATF3, CHOP, BiP

- 5 min 95 °C, followed by 25 cycles of:
- 0.5 min 95 °C.
- 0.5 min 57 °C.
- 1.0 min 72 °C.

Apply samples with loading buffer (final concentrations: 5% glycerol, 10 mM Tris, pH 7.5, 0.05% bromophenol blue) to a 2–4% agarose gel stained with ethidium bromide or SYBR® Safe (Invitrogen). To separate fragments differing in length by 26 bp, longer runs are recommended. Other cDNA synthesis and PCR systems offered by the various suppliers of molecular biological reagents will do as well. Primers to be used for the described analyses, including oligo $d(T)_{15}$, can be synthesized by a company specialized in oligonucleotide synthesis.

7. Conclusions

As long as the primary cellular targets of nelfinavir in human cancer cells are unknown, the analysis of the effect of nelfinavir on human cancer cells has to rely on a monitoring of the ER stress reaction or apoptosis. The detection of nelfinavir-induced ER stress can be performed by several of the morphological and biochemical analyses presented here that could likewise be applied, or have been applied, to detect the effect of other ER stress-inducing agents. While the fluorescence-based techniques are applicable for *in vitro* experiments only, Western blot and RT-PCR analyses could be applied to cell culture cells as well as to extracts of cancer tissue biopsies. Unfortunately, not all of the primary and secondary ER stress sensors are easily detectable due to a lack of reliable commercially available antibodies, and some of these factors even display basic expression levels or activities. Since ER stress-induced splicing of XBP1 is a highly specific marker for the ER stress reaction and appears to be absent in unstressed cells, we recommend a PCR-based analysis of XBP1 splicing as a reliable and early marker for nelfinavir-induced ER stress.

ACKNOWLEDGMENT

We are indebted to the Publisher Landes Bioscience for kindly permitting the reproduction of Figs. 8.2B and 8.3, previously published in Brüning *et al.* (2009).

REFERENCES

Brüning, A., Burger, P., Vogel, M., Rahmeh, M., Gingelmaier, A., Friese, K., Lenhard, M., and Burges, A. (2009). Nelfinavir induces the unfolded protein response in ovarian cancer cells, resulting in ER vacuolization, cell cycle retardation and apoptosis. *Cancer Biol. Ther.* **8,** 226–232.

Brüning, A., Gingelmaier, A., Friese, K., and Mylonas, I. (2010). New prospects for nelfinavir in non-HIV-related diseases. *Curr. Mol. Pharmacol.* **3,** 91–97.

Fels, D. R., and Koumenis, C. (2006). The PERK/eIF2alpha/ATF4 module of the UPR in hypoxia resistance and tumor growth. *Cancer Biol. Ther.* **5,** 723–738.

Gills, J. J., Lopiccolo, J., Tsurutani, J., Shoemaker, R. H., Best, C. J., Abu-Asab, M. S., Borojerdi, J., Warfel, N. A., Gardner, E. R., Danish, M., Hollander, M. C., Kawabata, S., *et al.* (2007). Nelfinavir, A lead HIV protease inhibitor, is a broad-spectrum, anticancer agent that induces endoplasmic reticulum stress, autophagy, and apoptosis in vitro and in vivo. *Clin. Cancer Res.* **13,** 5183–5194.

Gupta, A. K., Li, B., Cerniglia, G. J., Ahmed, M. S., Hahn, S. M., and Maity, A. (2007). The HIV protease inhibitor nelfinavir downregulates Akt phosphorylation by inhibiting proteasomal activity and inducing the unfolded protein response. *Neoplasia* **9,** 271–278.

Healy, S. J., Gorman, A. M., Mousavi-Shafaei, P., Gupta, S., and Samali, A. (2009). Targeting the endoplasmic reticulum-stress response as an anticancer strategy. *Eur. J. Pharmacol.* **625,** 234–246.

Kaldor, S. W., Kalish, V. J., Davies, J. F. 2nd, Shetty, B. V., Fritz, J. E., Appelt, K., Burgess, J. A., Campanale, K. M., Chirgadze, N. Y., Clawson, D. K., Dressman, B. A., Hatch, S. D., *et al.* (1997). Viracept (nelfinavir mesylate, AG1343): A potent, orally bioavailable inhibitor of HIV-1 protease. *J. Med. Chem.* **40,** 3979–3985.

Lin, J. H., Li, H., Yasumura, D., Cohen, H. R., Zhang, C., Panning, B., Shokat, K. M., Lavail, M. M., and Walter, P. (2007). IRE1 signaling affects cell fate during the unfolded protein response. *Science* **9,** 944–949.

Pyrko, P., Kardosh, A., Wang, W., Xiong, W., Schönthal, A. H., and Chen, T. C. (2007). HIV-1 protease inhibitors nelfinavir and atazanavir induce malignant glioma death by triggering endoplasmic reticulum stress. *Cancer Res.* **67,** 10920–10928.

Rao, R. V., Ellerby, H. M., and Bredesen, D. E. (2004). Coupling endoplasmic reticulum stress to the cell death program. *Cell Death Differ.* **11,** 372–380.

Samali, A., Fitzgerald, U., Deegan, S., and Gupta, S. (2010). Methods for monitoring endoplasmic reticulum stress and the unfolded protein response. *Int. J. Cell Biol.* **2010,** Article ID 830307, 11 pages. Doi: 10.1155/2010/830307.

Yoshida, H., Matsui, T., Yamamoto, A., Okada, T., and Mori, K. (2001). XBP1 mRNA is induced by ATF6 and spliced by IRE1 in response to ER stress to produce a highly active transcription factor. *Cell* **107,** 881–891.

CHAPTER NINE

Using Temporal Genetic Switches to Synchronize the Unfolded Protein Response in Cell Populations *In Vivo*

Alexander Gow

Contents

1. Introduction 144
2. Variable Functions of CHOP in the PERK Signaling Pathway 146
3. Mutations in the *PLP1* Gene: A Model UPR Disease 147
4. Problems with Current Paradigms Used to Characterize the UPR 149
5. A Novel *In Vivo* Genetic Switch Solves Many of These Problems 150
   - 5.1. Part 1: Knockin construct to enable temporal activation of the mutant *Plp1* gene and UPR induction in oligodendrocytes 150
   - 5.2. Part 2: A transgene construct to induce Cre recombinase expression in oligodendrocytes and trigger the genetic switch 153
   - 5.3. Detection of oligodendrocytes in which the genetic switch has been triggered 154
   - 5.4. Alternative knockin constructs for simultaneous UPR induction and labeling of oligodendrocytes 155
   - 5.5. Expected phenotypes in genetic switch technology (GST) mice 156
   - 5.6. Applications of GST to the study of UPR diseases and neurodegeneration 156
6. Generalizing GST to Study Other UPR Diseases 157
Acknowledgments 157
References 158

Abstract

In recent years, recognition of the importance of protein aggregation in human diseases has increasingly come to the fore and it is clear that many degenerative disorders involve activation of a metabolic signaling cascade known as the unfolded protein response (UPR). The UPR encompasses conserved mechanisms in cells to monitor and react to changes in metabolic flux through the secretory pathway. Such changes reflect an imbalance in cell homeostasis and

Center for Molecular Medicine and Genetics, Carman and Ann Adams Department of Pediatrics, Department of Neurology, Wayne State University School of Medicine, Detroit, Michigan, USA the UPR integrates several signaling cascades to restore homeostasis. As such, the UPR is simply interpreted as a protection mechanism for cells as they perform their normal functions. A number of groups have suggested that the UPR also can eliminate cells in which homeostasis is lost, for example, during disease. This notion has kindled the rather paradoxical concept that inhibiting the UPR will ameliorate degenerative disease. However, several *in vivo* studies in the nervous system indicate that curtailing UPR function either exacerbates disease or may reduce severity through unexpected or unidentified pathways. Perhaps the notion that the UPR protects cells or eliminates them stems from widespread use of suboptimal paradigms to characterize the UPR; thus, too little is currently known about this homeostatic pathway. Herein, I describe the development of genetic switch technology (GST) to generate a novel model for studying UPR diseases. The model is geared toward obtaining high resolution *in vivo* detail for oligodendrocytes of the central nervous system, but it can be adapted to study other cell types and other UPR diseases.

1. INTRODUCTION

The unfolded protein response (UPR) is a near-ubiquitous metabolic stress signaling cascade in higher animals emanating from the endoplasmic reticulum (ER) to the nucleus and the cytoplasm that has been studied extensively over the last two decades in a multitude of experimental paradigms and cell types *in vitro* and *in vivo*. A common dictum that has emerged from the plethora of reviews that summarize the essential features of this work is that the UPR protects cells from the toxic effects of unfolded/misfolded proteins in the secretory pathway but, perhaps paradoxically, triggers apoptosis. Several models have been proposed to define circumstances under which positive or negative outcomes (i.e., cell survival vs. cell death) arise during metabolic stress so as to account for this apparent dichotomy in UPR function (Gow and Wrabetz, 2009; Lin *et al.*, 2009; Rutkowski and Kaufman, 2007).

Whichever the model, the dichotomy has galvanized perceptions of the UPR and popularized the search for components that switch the cascade between protective and destructive outputs, particularly for signaling that involves the PRKR-like ER kinase (PERK/eIF2ak3). This PERK pathway comprises a negative-feedback, temporal delay circuit (Fig. 9.1) that inhibits ribosome assembly through a phosphorylation–dephosphorylation cycle of the alpha subunit of eukaryotic initiation factor 2 (eIF2α) and promotes expression of several transcription factors and a regulatory subunit of protein phosphatase 1 (reviewed in Wek and Cavener, 2007).

Herein, I briefly comment on some of the key variable findings reported for the function of the PERK pathway, which have revolved around the

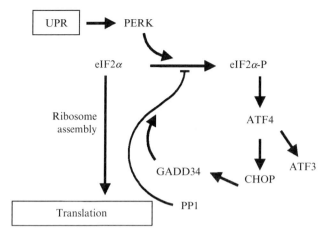

Figure 9.1 Essential features of the PERK signaling temporal delay circuit. UPR-induced dimerization and transautophosphorylation of PERK in the ER results in phosphorylation of eIF2α, which inhibits global translation in the cell by suppressing ribosome assembly, induces the specific translation of the ATF4 transcription factor. ATF4 activates expression of other transcription factors including ATF3 and CHOP. CHOP induces expression of GADD34, which is a regulatory subunit of the protein phosphatase, PP1, and targets this complex to effect the dephosphorylation of phospho-eIF2α. The time taken to express this set of proteins constitutes the time delay, during which the cell reduces protein accumulation in the ER by dislocation of the misfolded nascent chains and proteolysis through the ubiquitin–proteasome pathway.

role played by the transcription factor, C/EBP-homologous protein (CHOP/GADD153/Ddit3). CHOP was the first protein reported to trigger the switch from cell survival to cell death in the PERK pathway (Zinszner et al., 1998). However, subsequent studies do not show a strong relationship between CHOP expression and cell death, at least in the nervous system (Lin et al., 2005; Pennuto et al., 2008; Southwood et al., 2002). The ambiguities in CHOP function that are revealed by these studies highlight the need for new ideas and new tools to characterize the function of the PERK pathway. Accordingly, I describe a novel set of genetic tools that may help to achieve this goal and increase the effectiveness with which in vivo studies can be used to examine the pathophysiology of UPR diseases. To illustrate the application of these tools, I focus on animal models for a well-characterized neurodegenerative disease, Pelizaeus-Merzbacher disease (PMD), that is caused by mutations in the *PROTEOLIPID PROTEIN 1 (PLP1)* gene and serves as a model UPR disease (Southwood et al., 2002). Importantly, the genetic tools detailed below are generalizable to other UPR diseases and may benefit the investigation of these diseases in animal models in vivo.

2. Variable Functions of CHOP in the PERK Signaling Pathway

In early studies, Ron and colleagues (Zinszner et al., 1998) proposed CHOP as a proapoptotic transcription factor and reported that ablation of the *Chop* gene abolishes UPR-induced cell death in kidney. More recently, ablation of CHOP expression in *Myelin P_0 (Mpz)* gene mutant mice was shown to attenuate demyelination in the peripheral nervous system, although not as a consequence of reducing Schwann cell apoptosis but rather through an unknown mechanism (Pennuto et al., 2008).

In contrast, studies from my group have indicated that CHOP is a prosurvival protein in the central nervous system (CNS). We have demonstrated that ablation of the *Chop* gene in *Plp1* mutant mice significantly increases apoptosis in oligodendrocytes undergoing a UPR (Southwood et al., 2002). Importantly, these data are consistent with findings from other studies that have used gene ablation to inactivate components at several different levels of the PERK pathway (Lin et al., 2005, 2007, 2008). Two examples illustrate this point.

The first example involves data from *Perk*-null heterozygous and homozygous mice. If CHOP is a proapoptotic protein, then ablating the *Perk* gene should suppress *Chop* gene expression and reduce cell death under UPR conditions; however, the reverse appears to be the case. Ablating *Perk* in embryonic stem (ES) cells increases their susceptibility to UPR-induced cell death *in vitro* (Harding et al., 2000). Inactivating *Perk* in mice causes pancreatic β-cell death and diabetes, presumably because the PERK pathway (and CHOP expression) is critical for normal development and the regulation of metabolic activity in the secretory pathway (Harding et al., 2001). Furthermore, immune-mediated demyelination and oligodendrocyte apoptosis are increased during allergic experimental encephalomyelitis in mice that are hypomorphic for the *Perk* gene (Lin et al., 2007) most likely because the PERK pathway serves to protect oligodendrocytes.

In the second example, *Gadd34*-null homozygotes have been examined. If CHOP represents a proapoptotic signal, then inactivation of the *Gadd34* gene to suppress the dephosphorylation of phospho-eIF2α should sustain UPR signaling, prolong CHOP expression, and exacerbate cell death. Again, the reverse is the case. Oligodendrocyte apoptosis is reduced in response to IFNγ–mediated induction of the UPR in *Gadd34*–null mice (Lin et al., 2008). Together, these examples indicate that inactivation of the *Perk*, *Chop*, or *Gadd34* genes in oligodendrocytes during a UPR yields the consistent result that PERK pathway signaling, which invariably involves CHOP expression and its nuclear localization, promotes oligodendrocyte survival, and diminishes the negative impact of UPR disease.

The inconsistencies that these and other studies reveal for the function of PERK signaling, and the role played by CHOP, necessitate a reevaluation and an expansion of experimental methods used to define this pathway. Because of the difficulty of teasing out mechanistic details *in vivo*, most studies have been performed *in vitro*. However, such experiments appear prone to cell culture artifacts (Sharma and Gow, 2007; Sharma *et al.*, 2007); thus, the development of novel paradigms should be focused on tools to characterize the UPR at high resolution *in vivo*.

3. Mutations in the *PLP1* Gene: A Model UPR Disease

Pelizaeus-Merzbacher disease is a rare, allelic X-linked recessive leukodystrophy caused by deletion, duplication or mutation of the *PROTEOLIPID PROTEIN 1* (*PLP1*) gene (Gencic *et al.*, 1989; Hudson *et al.*, 1989; Raskind *et al.*, 1991). By and large, males are affected (but see Hurst *et al.*, 2006; Nance *et al.*, 1993) with pleiotrophic clinical symptoms manifested in the first year of life and perinatally in the most severely affected individuals (reviewed in Garbern, 2007; Inoue, 2005). The primary defect originates in oligodendrocytes, which express the *PLP1* gene and synthesize myelin principally within the white matter of the CNS.

Proteins encoded by the *PLP1* gene (PLP1 and its smaller splice-isoform DM20) are the major polytopic transmembrane proteins targeted to CNS myelin sheaths through the secretory pathway (Fig. 9.2A). These sheaths are large lipid-rich membrane sheets that are synthesized as lamellipodial expansions of oligodendrocyte processes and spirally wrap and electrically insulate axons to enable saltatory conduction (recently reviewed by Baron and Hoekstra, 2010). During myelinogenesis, as much as 10% of the mRNA in oligodendrocytes encodes *PLP1* gene products, rendering it one of the most highly expressed genes known (Milner *et al.*, 1985). Accordingly, genetic lesions that significantly increase expression level or alter splicing of the primary transcript, or mutations in the coding region that destabilize the three-dimensional structure of the proteins, might be expected to perturb the secretory pathway and induce metabolic stress leading to activation of the UPR.

To date, missense and nonsense mutations in *PLP1* are the only genetic lesions in PMD that have been found to induce the UPR (Gow, 2004; Gow *et al.*, 1994, 1998; Southwood *et al.*, 2002). Thus, mutant PLP1, and in some cases DM20 (Gow and Lazzarini, 1996), accumulates in the ER causing metabolic stress and activation of the UPR (Fig. 9.2B). Resolution of the stress, by degrading the accumulated nascent protein chains negatively regulates the UPR and allows resumption of global protein synthesis.

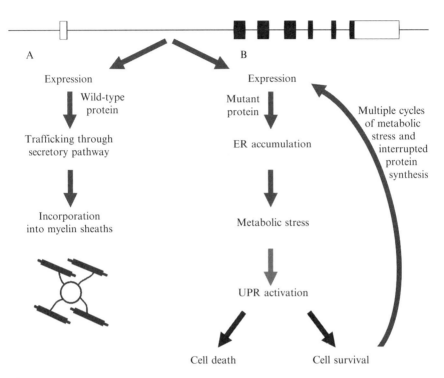

Figure 9.2 Summary of wild-type or mutant PLP1 expression and intracellular trafficking in oligodendrocytes. (A) Wild-type PLP1 and DM20 are synthesized on the rough ER and transported in vesicles along microtubules for incorporation into myelin sheaths. Oligodendrocytes (cell with four processes to myelin sheaths) simultaneously synthesize myelin sheaths around multiple axons and generate several hundred-fold more membrane within a few days than the surface area of the cell body. (B) Missense and nonsense mutations in the *PLP1* gene destabilize the primary structure of PLP1 and DM20, causing ER accumulation, metabolic stress and activation of the UPR. Global protein translation is temporarily arrested via PERK pathway signaling. ER-associated protein degradation clears the accumulated protein to shut down the UPR and protein translation resumes. Mutant PLP1 and DM20 again accumulate in the ER leading to additional cycles of UPR activation. In each cycle, the cell can resolve the stress but there is an increased risk of undergoing apoptosis when protein expression resumes (Gow and Wrabetz, 2009).

However, resumption of the expression of mutant *Plp1* gene products again causes accumulation in the secretory pathway and triggers another cycle of metabolic stress. Such iterative cycles of UPR activation increase the risk of apoptosis for individual cells (reviewed in Gow and Wrabetz, 2009). Overexpression of PLP1, which accounts for approximately 50% of *PLP1* mutations, is caused by duplication or triplication of the gene. The pathophysiology appears to involve cholesterol accumulation in lysosomes or altered lipid metabolism (Karim *et al.*, 2010; Simons *et al.*, 2002). Alternatively, the absence of *PLP1* gene expression leads to a subtle late-onset phenotype characterized by length-dependent axonal degeneration (Garbern *et al.*, 2002; Griffiths *et al.*, 1998; Rosenbluth *et al.*, 2006).

Excellent naturally occurring rodent models of PMD have been available for decades (Billings-Gagliardi *et al.*, 1995; Dautigny *et al.*, 1986; Gencic and Hudson, 1990; Nadon and Duncan, 1995; Schneider *et al.*, 1992) and have been characterized in great detail (reviewed in Werner *et al.*, 1998). Moreover, the specific missense mutations found in two of the mouse mutants, *msd* (A243V) and *rsh* (I187T), are also found in patients (Kobayashi *et al.*, 1994; Yamamoto *et al.*, 1998) and confer analogous symptoms in terms of relative disease severity, degree of pathology, extent of oligodendrocyte apoptosis and relative decreases in life span.

4. Problems with Current Paradigms Used to Characterize the UPR

In diseases for which the UPR is thought to be the primary trigger of pathology, the target cells are often postmitotic, differentiated and fully functional prior to the metabolic stressor. In contrast, the majority of experiments focused on defining UPR pathways and the functions of its signaling cascades are performed *in vitro* with undifferentiated cells, such as transformed cell lines or mouse embryonic fibroblasts that are perpetually proliferative or merely contact inhibited at high density. Accordingly, it may be of little surprise that components of the UPR found to be critical for function *in vitro*, turn out to be irrelevant *in vivo* (e.g., Di Sano *et al.*, 2006; Lamkanfi *et al.*, 2004; Obeng and Boise, 2005; Sharma and Gow, 2007; Sharma *et al.*, 2007). Perhaps this disconnect between *in vitro* and *in vivo* studies arises from the paradigms used, as well as nuances in UPR-activated pathways in different cell types and/or under conditions of proliferation versus differentiation.

Prime advantages of using proliferative cells in culture include their cellular homogeneity and the ease with which they can be expanded, treated and analyzed to generate time-course data of UPR-related changes. Thus, tunicamycin or a comparable UPR-inducing agent is added to the medium, *which defines time zero for UPR induction*, and essentially all of the cells progress in synchrony down a common pathway culminating in metabolic stress. In contrast, activation of the UPR *in vivo* is typically a dyssynchronous event that plays out stochastically at the level of individual cells. *In other words, there is no time zero in tissue samples.* Furthermore, tissues are typically heterogeneous and the UPR may be induced in only one cell type. In such instances, all phases of UPR pathogenesis are simultaneously represented in the target cell population. In addition, any changes that can be measured in population-based assays from the harvested tissue, such as altered gene expression by northern blotting or qPCR, appear muted by the background of cells in which the UPR is not active. These problems severely constrain the sensitivity and types of experiments that can be performed *in vivo* and often increase variability in the results.

5. A Novel *In Vivo* Genetic Switch Solves Many of These Problems

To overcome many disadvantages of *in vivo* paradigms in which to characterize the UPR, my laboratory is developing novel genetic switch technology (GST) in transgenic mice. UPR induction in these animals will be cell type specific, synchronous and will activate a fluorescent tag to enable the identification, characterization and harvesting of stressed cells at a known time interval after induction of the UPR. The description of components of GST is divided into two parts.

Part 1 describes the generation of a knockin mutant *msd* allele of the *Plp1* gene that is constitutively inactivated (i.e., transcriptionally silent) during mouse development. This allele is activated by inducing loxP-Cre-mediated recombination in oligodendrocyte nuclei, which induces mutant *Plp1* gene expression and the UPR. Subsequently, oligodendrocytes undergo apoptosis *en masse*.

Part 2 describes a transgene for oligodendrocyte-specific expression of the Cre recombinase, CreER^{T2}. This recombinase is a fusion protein with a variant of the human estrogen receptor (Leone *et al.*, 2003) and is expressed by mature myelinating oligodendrocytes as an inactive protein bound to hsp90 in the cytoplasm. Upon treatment of mice with tamoxifen, CreER^{T2} translocates to the nucleus and recombines genomic DNA at loxP sites, thereby engaging the genetic switch described in Part 1.

5.1. Part 1: Knockin construct to enable temporal activation of the mutant *Plp1* gene and UPR induction in oligodendrocytes

To induce the UPR in oligodendrocytes, a targeting vector (designated Neo^{msd}) was generated for homologous recombination in ES cells. This genomic construct introduces into the *Plp1* gene the *msd* mutant allele and several other elements (Fig. 9.3A).

5.1.1. Required materials

Sources of DNA for the targeting constructs

- Source of genomic DNA for the gene of interest from the library of choice. This library should be derived from the ES cells that will be used

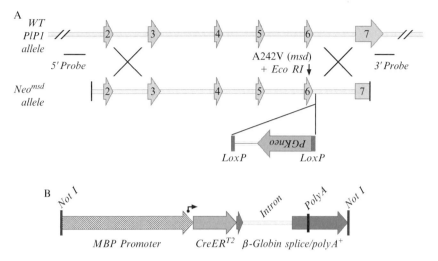

Figure 9.3 Components of the GST used to generate an *in vivo* model of a UPR disease in the CNS. (A) A mouse *Plp1* gene targeting construct for homologous recombination in ES cells. A portion of the wild-type *Plp1* gene comprising exons 2–7 is shown for comparison with a targeting construct of this region. A missense mutation, which specifies the *msd* allele, as well as a neutral polymorphism to generate an *Eco*RI site (for genotyping purposes) is inserted into exon 6. An antisense *PGKneo* selectable marker gene is cloned into the 5' end of intron 6 to allow neomycin resistance selection of homologous recombinant ES cells as well as to confer antisense-mediated suppression of *Plp1* gene expression in mice during development. The *PGKneo* cassette is flanked by loxP sites to allow Cre recombinase-mediated recombination. Excision of the *PGKneo* cassette activates expression of the mutant *msd* allele of *Plp1*, which induces the UPR. Locations are shown for the external 5' and 3' genomic probes used to verify homologous recombination in ES cells. (B) A transgene, designated $MCreER^{T2}G$, containing 1.9 kb of the mouse MBP promoter and the human β-globin splice and polyadenylation signals, is used to drive expression of a tamoxifen-activated Cre recombinase ($CreER^{T2}$) in mature myelinating oligodendrocytes. In the presence of tamoxifen, $CreER^{T2}$ translocates to the nucleus and excises the *PGKneo* cassette to activate mutant *Plp1* gene expression.

for gene targeting. Alternatively, long-range PCR using TaKaRa DNA polymerase (Clontech, Mountain View, CA) can be used to clone the desired region from the gene of interest.
- Plasmid vectors for subcloning genomic fragments and the selectable neomycin resistance marker gene (e.g., pKJ-1, Rudnicki *et al.*, 1992), designated the *PGKneo* cassette. These reagents can usually be obtained from a transgenic Core facility or a company that will perform the gene targeting work in the ES cells.
- The *PGKneo* cassette needs to be flanked by loxP sites (i.e., a floxed *PGKneo* cassette) to facilitate its removal when activating expression of the mutant *Plp1* gene. In the event that a floxed cassette is not available

from the transgenic core, it can be generated by cloning two loxP sites into a bacterial plasmid and inserting the *PGKneo* cassette between them. The DNA sequence of a loxP site is 34 bases long: 5′–ATA ACT TCG TAT AGC ATA CAT TAT ACG AAG TTA T–3′.
- The CreER^{T2} coding region can be obtained through a materials transfer agreement with Dr P. Chambon at the Institut de Génétique et de Biologie Moléculaire et Cellulaire (IGBMC) in France.

Kits

- Quickchange kit (Agilent Technologies, Santa Clara, CA) for mutagenesis of genomic DNA to introduce coding region mutations or restriction endonuclease sites as well as to insert the floxed *PGKneo* cassette.

Other reagents

- DNA restriction endonucleases for cloning (New England Biolabs, Ipswich, MA).
- Corn oil (Cat. # C8267, Sigma-Aldrich, St Louis, MO).
- Tamoxifen (Cat. # T5648, Sigma-Aldrich) for intraperitoneal injection or gavaging of mice. Heat 5 ml of corn oil in an 18 ml glass vial at 42 °C for 30 min. Add 100 mg of tamoxifen and vortex intermittently over several hours until dissolved. Store up to 1 month at 4 °C protected from light. For single doses, up to 0.15 mg/g body weight can be used. For multiple doses use no more than 0.075 mg/g/day.

Genetically engineered mice

The *ROSA26* allele is a defined locus on mouse chromosome 6 that has been used extensively for the targeted insertion of a wide variety of transgenes (Nyabi *et al.*, 2009) including fluorescent protein reporters under transcriptional control of ubiquitously expressed housekeeping genes. We use a *ROSA26-EGFP* strain designated *Tomato* mice (Jackson Laboratories, Gt(ROSA)26Sor$^{tm4(ACTB-tdTomato,-EGFP)Luo}$/J, Stock # 007576) in which EGFP is tethered to the plasma membrane of expressing cells (Muzumdar *et al.*, 2007).

5.1.2. Construct design

Thus, an *Eco*RI genomic fragment containing exons 2–7 of the *Plp1* gene was cloned from a *129SvEv* genomic library (Stecca *et al.*, 2000). The *msd* mutation, which substitutes valine for alanine in PLP1 (Gencic and Hudson, 1990), and an additional neutral polymorphism to create an *Eco*RI restriction endonuclease site, were inserted into exon 6 of the construct using site-directed mutagenesis (Quickchange kit, Agilent Technologies, Santa Clara, CA). A second round of mutagenesis was used to create an *Avr*II restriction site located 70 bp downstream of the 5′ splice site in intron 6. A floxed *PGKneo* cassette was subcloned into the *Avr*II site in the antisense orientation with respect to transcription of the *Plp1* gene. The *PGKneo* positive selectable marker has two functions. Initially it is used for selection of ES cell clones in which the targeting vector has undergone homologous or nonhomologous integration into the genome. Subsequently, the cassette is used to regulate expression of the mutant *Plp1* allele.

In knockin mice, the targeted insertion of the Neo^{msd} allele into the *Plp1* gene constitutively silences the locus during development, presumably by antisense-mediated degradation. Thus, strong transcription from the ubiquitously expressed *PGKneo* cassette yields antisense *Plp1* transcripts that hybridize to sense *Plp1* transcripts, suppress their splicing and reduce their stability (Boison and Stoffel, 1994; Stecca *et al.*, 2000). Male Neo^{msd} mice, which harbor a single X-chromosome, are functionally null for *Plp1* expression and females with one Neo^{msd} allele are carriers.

Targeting vectors frequently include negative selectable markers, such as thymidine kinase or diphtheria toxin under transcriptional control of a ubiquitous promoter. We have found in previous studies that negative selection reduces the number of ES cell colonies obtained after electroporation, but they do not significantly increase the likelihood of obtaining homologous recombinants (Stecca *et al.*, 2000). Accordingly, we do not use negative selection.

5.2. Part 2: A transgene construct to induce Cre recombinase expression in oligodendrocytes and trigger the genetic switch

Previous studies in transgenic mice have demonstrated that the proximal 1.9 kb of the canonical mouse myelin basic protein (MBP) promoter/enhancer region drives high level oligodendrocyte-specific expression for the entire life of the mouse, with no expression outside of the CNS (Gow *et al.*, 1992). This contrasts with a number of other MBP promoter cassettes that are leaky in other cell types or are suppressed within a few months after birth (Foran and Peterson, 1992; Turnley *et al.*, 1991). As illustrated in Fig. 9.3B, 1.9 kb of the MBP promoter is used to express the $CreER^{T2}$ fusion protein and a 1.6 kb genomic *Bam*HI–*Pst*I fragment from the human β-globin gene comprising part of exon 2 through the end of exon 3 is used for splicing and polyadenylation signals.

The net effect of $MCreER^{T2}G$ transgene expression is accumulation of the recombinase in the cytoplasm of mature, myelinating oligodendrocytes. Immature premyelinating cells do not express the transgene. A single dose of tamoxifen (≤ 0.15 mg/g) dissociates $CreER^{T2}$ from the hsp90 complex in a proportion of oligodendrocytes, depending on the dose, and enables nuclear translocation. The dose–response for nuclear translocation does not appear to be linear; more than a 30-fold decrease in the number of CreER^{T2}–positive oligodendrocyte nuclei is observed for a 16-fold decrease in tamoxifen (unpublished data). Multiple tamoxifen doses on successive days substantially increase oligodendrocyte labeling.

Recombination of the Neo^{msd} allele at loxP sites (Fig. 9.3A), which constitutes the genetic switch, removes the *PGKneo* cassette and spliced mutant *Plp1* mRNA accumulates in the cytoplasm. Translation of this mRNA results in accumulation of mutant *Plp1* gene products in the ER, which activates the UPR. Because *Plp1* gene expression is triggered by tamoxifen treatment, a large number of oligodendrocytes throughout the CNS will undergo synchronized induction of the UPR and will subsequently undergo apoptosis. The half-life of tamoxifen in the mice is less than eight hours (Borgna and Rochefort, 1981) so the wave of *Plp1* expression, UPR induction and oligodendrocyte death is quite sharply defined.

5.3. Detection of oligodendrocytes in which the genetic switch has been triggered

In several of the applications of the GST, it is not of great importance to identify the cells that induce the UPR, only the regions where pathology is emerging. For example, repeated single doses of tamoxifen every 1–2 weeks for several months will induce successive waves of oligodendrocyte death in the CNS. Each wave will be accompanied by focal demyelinating lesions which will subsequently be remyelinated by the expansion, migration and differentiation of oligodendrocyte progenitors from local stem cell pools. The purpose of this experiment would be to examine the demyelination or remyelination processes after the oligodendrocytes had undergone apoptosis.

For applications of the genetic switch that are focused on characterizing the different aspects of the UPR signaling cascade, the *in vivo* identification of individual stressed cells will be critical. To accomplish this task, oligodendrocytes in which the UPR has been activated will also require a detectable tag, such as green fluorescent protein. For this purpose, a transgene driving expression of a loxP-activated membrane-tethered EGFP reporter is ideal (e.g., *Tomato* mice). Thus, a time-course experiment would begin with a single dose of tamoxifen given to a cohort of mice at time zero, and the harvesting of CNS from several animals every 12–24 h thereafter. The tissues would be processed by cryostat sectioning and laser capture to isolate fluorescent oligodendrocytes. Alternatively, the tissue would be homogenized for cell purification using fluorescence-activated cell sorting. The samples would then be processed for microarray analysis.

5.4. Alternative knockin constructs for simultaneous UPR induction and labeling of oligodendrocytes

A limitation of the experimental design in Section 5.3 is the number of alleles that need to come together in individual mice to carry out the experiment: the Neo^{msd} allele, the $MCreER^{T2}G$ transgene and the *Tomato* allele. In addition, the mice must be male to ensure mutant *Plp1* gene (on the X-chromosome) expression in oligodendrocytes. From such a breeding colony, the frequency of generating these mice is 1-in-16. The colony can be simplified somewhat by breeding the *Tomato* allele to homozygosity, which will increase the frequency to 1-in-8. Control mice for this experiment, which do not harbor the Neo^{msd} allele, will also be generated with a frequency of 1-in-8.

The frequency of the required genotypes for experiment can be further improved to 1-in-4 by combining the Neo^{msd} and an EGFP reporter in a single locus, as illustrated in Fig. 9.4. Thus, the spliced *Plp1* transcript that is generated by activating the $Neo^{msd:egfp}$ allele will be bicistronic. Ribosomes will be recruited to the 5′ cap to synthesize mutant PLP1 and DM20, and also to the internal ribosome entry site (IRES) located in the 3′ untranslated region of the *Plp1* gene to synthesize the EGFP reporter. Some studies report that the efficiency of IRES-dependent translation in different cell types varies according to the translation initiation factors available. In addition, the efficiency of IRES- and 5′ cap-dependent translation can shift under conditions of metabolic stress (reviewed by Fitzgerald and Semler, 2009). Accordingly, an alternative approach is to fuse the EGFP reporter to the carboxyl-terminus of the *Plp1* open reading frame in the targeting construct.

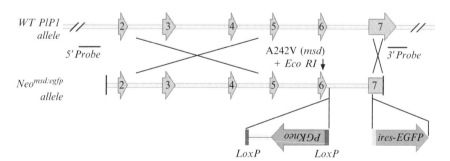

Figure 9.4 Variation on the targeting construct to enable fluorescence-based detection of cells expressing the *msd* mutant *Plp1* gene. In addition to insertion of the *msd* mutation in exon 6 and the *PGKneo* cassette in intron 6, an IRES-EGFP cassette has been inserted into exon 7 in the 3′ untranslated region. Expression of this allele after Cre-mediated recombination yields a bicistronic mRNA that encodes both mutant PLP1 and DM20 as well as EGFP. Locations are shown for the external 5′ and 3′ genomic probes used to verify homologous recombination in ES cells.

5.5. Expected phenotypes in genetic switch technology (GST) mice

Analogous to humans, the phenotype of male mice lacking a functional *Plp1* gene is very mild. Indeed, male *Plp1*-null mice develop normally, synthesize myelin of appropriate thickness and remain to be phenotypically normal until 1–1.5 years of age (Klugmann et al., 1997; Rosenbluth et al., 1996; Stecca et al., 2000). Abnormal neuronal metabolism at the level of vesicular trafficking in axons is increasingly apparent from approximately 2 months of age (Griffiths et al., 1998).

In contrast to the *Plp1*-null mice, the phenotype of male *msd* mice is very severe. Clinical signs comprising intention tremors and whole body shudders are apparent from 11 to 12 days of age and these symptoms intensify over the following 2 weeks. By 20–25 days, these mice develop stimulus-induced seizures, which intensify in severity and frequency until the death of the animals at 25–30 days. Histology of the CNS at the time of death reveals the virtual absence of myelin membrane and oligodendrocyte lineage cells.

Depending on the size and frequency of the tamoxifen dose, the phenotype of the Neo^{msd} mice will range from near wild type to approximately *msd*. Some variability in the phenotype might be expected because the precise number and neuroanatomical locations of the demyelinating lesions will vary in different mice. Thus, if a significant demyelinating lesion develops in the brainstem after tamoxifen treatment, it will be more likely to cause significant motor disability or kill the mouse than would a similar size lesion in the optic nerve.

5.6. Applications of GST to the study of UPR diseases and neurodegeneration

Perhaps the simplest application of the GST in Neo^{msd} mice is to generate a timeline for UPR signaling in oligodendrocytes *in vivo*. Following a single dose of tamoxifen, oligodendrocytes will begin to express mutant PLP1 and DM20, which will accumulate in the ER and trigger the UPR. This pathology will evolve gradually over a few days or so, which will provide us with a number of time points to purify RNA from the cells and determine their expression profile (e.g., every 12 h). Changes in the expression profiles of known UPR genes will be used to rank order the genes in the UPR pathway. In addition, we use these profiles to identify novel candidate genes that are coordinately regulated.

A second application for these mice is to examine the demyelination/remyelination cycle in adults. Thus, a single dose of tamoxifen, or a tightly clustered series of doses over several days, would trigger demyelination throughout the CNS. The evolution and dissolution of these lesions would be examined with respect to cell death, the infiltration of oligodendrocyte progenitors, the involvement of the immune system. A number of experimental demyelinating models are available in mice, but most are relatively complex to initiate. For example, some models require stereotactic injections of toxins into the parenchyma or extensive treatment with toxins in the diet, while others require specific genetic background strains for susceptibility to autoimmune demyelination (Blakemore and Franklin, 2008; Gao et al., 2000; Reynolds et al., 1996).

An extension to the single-dose strategy to induce demyelination would be repetitive tamoxifen doses (every 1–2 weeks) to drive repeated demyelination/ remyelination cycles in the mice. This model would be useful for studying the long term affects of oligodendrocyte cell death on the integrity and function of large white matter tracts such as corpus callosum or brainstem. A great deal of anecdotal evidence in the clinical literature suggests that patients with multiple sclerosis eventually develop behavioral disorders, presumably stemming from the cumulative effects of remodeling white matter tracts following multiple demyelinating lesions (reviewed in Fields, 2008; Stewart and Davis, 2004).

6. Generalizing GST to Study Other UPR Diseases

The current chapter draws upon the research program in my laboratory to characterize the UPR in oligodendrocytes expressing a mutant allele of the *Plp1* gene. However, GST is easily generalized to generate mouse models of other genetic forms of UPR disease. In this regard, it is important to consider that UPR diseases principally arise from gain-of-function mechanisms and that one or two alleles of the targeted gene will induce a UPR in activated cells (Gow and Sharma, 2003). Accordingly, the design of GST targeting constructs described here will be applicable to most genes and cell types and the uses for these animals will be similar to Neo^{msd} mice. For targeted mutations of autosomal genes, there may be significant advantage to comparing the phenotype caused by activation of one or two mutant alleles (i.e., heterozygotes or homozygotes for the mutant allele) in terms of the strength of UPR induction. In the case of the *Plp1* gene, only one allele is expressed in any cell as a result of X-inactivation in females or the presence of the Y-chromosome in males.

ACKNOWLEDGMENTS

Thanks to Drs. Birdal Bilir and Hao Zhang, CMMG, Wayne State University, for proofreading this chapter. This work is supported by grants from the National Institutes of Health, National Institute of Neurological Disorders and Stroke (R01NS43783) and the National Multiple Sclerosis Society (RG2891).

REFERENCES

Baron, W., and Hoekstra, D. (2010). On the biogenesis of myelin membranes: Sorting, trafficking and cell polarity. *FEBS Lett.* **584,** 1760–1770.
Billings-Gagliardi, S., Kirschner, D. A., Nadon, N. L., DiBenedetto, L. M., Karthigasan, J., Lane, P., Pearsall, G. B., and Wolf, M. K. (1995). Jimpy 4J: A new X-linked mouse mutation producing severe CNS hypomyelination. *Dev. Neurosci.* **17,** 300–310.
Blakemore, W. F., and Franklin, R. J. (2008). Remyelination in experimental models of toxin-induced demyelination. *Curr. Top. Microbiol. Immunol.* **318,** 193–212.
Boison, D., and Stoffel, W. (1994). Disruption of the compacted myelin sheath of axons of the central nervous system in proteolipid protein-deficient mice. *Proc. Natl. Acad. Sci. USA* **91,** 11709–11713.
Borgna, J. L., and Rochefort, H. (1981). Hydroxylated metabolites of tamoxifen are formed in vivo and bound to estrogen receptor in target tissues. *J. Biol. Chem.* **256,** 859–868.
Dautigny, A., Mattei, M. G., Morello, D., Alliel, P. M., Pham-Dinh, D., Amar, L., Arnaud, D., Simon, D., Mattei, J. F., Guenet, J. L., et al. (1986). The structural gene coding for myelin-associated proteolipid protein is mutated in jimpy mice. *Nature* **321,** 867–869.
Di Sano, F., Ferraro, E., Tufi, R., Achsel, T., Piacentini, M., and Cecconi, F. (2006). Endoplasmic reticulum stress induces apoptosis by an apoptosome-dependent but caspase 12-independent mechanism. *J. Biol. Chem.* **281,** 2693–2700.
Fields, R. D. (2008). White matter in learning, cognition and psychiatric disorders. *Trends Neurosci.* **31,** 361–370.
Fitzgerald, K. D., and Semler, B. L. (2009). Bridging IRES elements in mRNAs to the eukaryotic translation apparatus. *Biochim. Biophys. Acta* **1789,** 518–528.
Foran, D. R., and Peterson, A. C. (1992). Myelin acquisition in the central nervous system of the mouse revealed by an MBP-Lac Z transgene. *J. Neurosci.* **12,** 4890–4897.
Gao, X., Gillig, T. A., Ye, P., D'Ercole, A. J., Matsushima, G. K., and Popko, B. (2000). Interferon-gamma protects against cuprizone-induced demyelination. *Mol. Cell. Neurosci.* **16,** 338–349.
Garbern, J. Y. (2007). Pelizaeus-Merzbacher disease: Genetic and cellular pathogenesis. *Cell. Mol. Life Sci.* **64,** 50–65.
Garbern, J. Y., Yool, D. A., Moore, G. J., Wilds, I. B., Faulk, M. W., Klugmann, M., Nave, K. A., Sistermans, E. A., van der Knaap, M. S., Bird, T. D., Shy, M. E., Kamholz, J. A., et al. (2002). Patients lacking the major CNS myelin protein, proteolipid protein 1, develop length-dependent axonal degeneration in the absence of demyelination and inflammation. *Brain* **125,** 551–561.
Gencic, S., and Hudson, L. D. (1990). Conservative amino acid substitution in the myelin proteolipid protein of jimpy[msd] mice. *J. Neurosci.* **10,** 117–124.
Gencic, S., Abuelo, D., Ambler, M., and Hudson, L. D. (1989). Pelizaeus-Merzbacher disease: An X-linked neurologic disorder of myelin metabolism with a novel mutation in the gene encoding proteolipid protein. *Am. J. Hum. Genet.* **45,** 435–442.
Gow, A. (2004). Protein misfolding as a disease determinant. In "Myelin Biology and Disorders," (R. A. Lazzarini, ed.). Vol. 1, pp. 877–885. Elsevier, Amsterdam.
Gow, A., and Lazzarini, R. A. (1996). A cellular mechanism governing the severity of Pelizaeus-Merzbacher disease. *Nat. Genet.* **13,** 422–428.
Gow, A., and Sharma, R. (2003). The unfolded protein response in protein aggregating diseases. *Neuromolecular Med.* **4,** 73–94.
Gow, A., and Wrabetz, L. (2009). CHOP and the endoplasmic reticulum stress response in myelinating glia. *Curr. Opin. Neurobiol.* **19,** 1–6. PMC2787654.

Gow, A., Friedrich, V. L., and Lazzarini, R. A. (1992). Myelin basic protein gene contains separate enhancers for oligodendrocytes and Schwann cell expression. *J. Cell Biol.* **119**, 605–616.

Gow, A., Friedrich, V. L., and Lazzarini, R. A. (1994). Many naturally occurring mutations of myelin proteolipid protein impair its intracellular transport. *J. Neurosci. Res.* **37**, 574–583.

Gow, A., Southwood, C. M., and Lazzarini, R. A. (1998). Disrupted proteolipid protein trafficking results in oligodendrocyte apoptosis in an animal model of Pelizaeus-Merzbacher disease. *J. Cell Biol.* **140**, 925–934. PMC2141744.

Griffiths, I., Klugmann, M., Anderson, T., Yool, D., Thomson, C., Schwab, M. H., Schneider, A., Zimmermann, F., McCulloch, M., Nadon, N., and Nave, K.-A. (1998). Axonal swellings and degeneration in mice lacking the major proteolipid of myelin. *Science* **280**, 1610–1613.

Harding, H. P., Zhang, Y., Bertolotti, A., Zeng, H., and Ron, D. (2000). Perk is essential for translational regulation and cell survival during the unfolded protein response. *Mol. Cell* **5**, 897–904.

Harding, H. P., Zeng, H., Zhang, Y., Jungries, R., Chung, P., Plesken, H., Sabatini, D. D., and Ron, D. (2001). Diabetes mellitus and exocrine pancreatic dysfunction in perk-/- mice reveals a role for translational control in secretory cell survival. *Mol. Cell* **7**, 1153–1163.

Hudson, L. D., Puckett, C., Berndt, J., Chan, J., and Gencic, S. (1989). Mutation of the proteolipid protein gene PLP in a human X chromosome-linked myelin disorder. *Proc. Natl. Acad. Sci. USA* **86**, 8128–8131.

Hurst, S., Garbern, J., Trepanier, A., and Gow, A. (2006). Quantifying the carrier female phenotype in Pelizaeus-Merzbacher disease. *Genet. Med.* **8**, 371–378.

Inoue, K. (2005). PLP1-related inherited dysmyelinating disorders: Pelizaeus-Merzbacher disease and spastic paraplegia type 2. *Neurogenetics* **6**, 1–16.

Karim, S. A., Barrie, J. A., McCulloch, M. C., Montague, P., Edgar, J. M., Iden, D. L., Anderson, T. J., Nave, K. A., Griffiths, I. R., and McLaughlin, M. (2010). PLP/DM20 Expression and turnover in a transgenic mouse model of pelizaeus-merzbacher disease. *Glia* **58**, 1727–1738.

Klugmann, M., Schwab, M. H., Puhlhofer, A., Schneider, A., Zimmermann, F., Griffiths, I. R., and Nave, K.-A. (1997). Assembly of CNS myelin in the absence of proteolipid protein. *Neuron* **18**, 59–70.

Kobayashi, H., Hoffman, E. P., and Marks, H. G. (1994). The *rumpshaker* mutation in spastic paraplegia. *Nat. Genet.* **7**, 351–352.

Lamkanfi, M., Kalai, M., and Vandenabeele, P. (2004). Caspase-12: An overview. *Cell Death Differ.* **11**, 365–368.

Leone, D. P., Genoud, S., Atanasoski, S., Grausenburger, R., Berger, P., Metzger, D., Macklin, W. B., Chambon, P., and Suter, U. (2003). Tamoxifen-inducible glia-specific Cre mice for somatic mutagenesis in oligodendrocytes and Schwann cells. *Mol. Cell. Neurosci.* **22**, 430–440.

Lin, W., Harding, H. P., Ron, D., and Popko, B. (2005). Endoplasmic reticulum stress modulates the response of myelinating oligodendrocytes to the immune cytokine interferon-gamma. *J. Cell Biol.* **169**, 603–612.

Lin, W., Bailey, S. L., Ho, H., Harding, H. P., Ron, D., Miller, S. D., and Popko, B. (2007). The integrated stress response prevents demyelination by protecting oligodendrocytes against immune-mediated damage. *J. Clin. Invest.* **117**, 448–456.

Lin, W., Kunkler, P. E., Harding, H. P., Ron, D., Kraig, R. P., and Popko, B. (2008). Enhanced integrated stress response promotes myelinating oligodendrocyte survival in response to interferon-gamma. *Am. J. Pathol.* **173**, 1508–1517.

Lin, J. H., Li, H., Zhang, Y., Ron, D., and Walter, P. (2009). Divergent effects of PERK and IRE1 signaling on cell viability. *PLoS ONE* **4,** e4170.

Milner, R. J., Lai, C., Nave, K. A., Lenoir, D., Ogata, J., and Sutcliffe, G. (1985). Nucleotide sequences of two mRNAs for rat brain myelin proteolipid protein. *Cell* **42,** 931–939.

Muzumdar, M. D., Tasic, B., Miyamichi, K., Li, L., Luo, L. (2007). A global double-fluorescent Cre reporter mouse. *Genesis* **45**(9), 593–605.

Nadon, N. L., and Duncan, I. D. (1995). Gene expression and oligodendrocyte development in the myelin deficient rat. *J. Neurosci. Res.* **41,** 96–104.

Nance, M. A., Pratt, V. M., Boyadjiev, S., Taylor, S., Hodes, M. E., and Dlouhy, S. R. (1993). Adult-onset neurological disorder in a Pelizaeus-Merzbacher disease carrier mother. *Am. J. Hum. Genet.* **53,** 1745.

Nyabi, O., Naessens, M., Haigh, K., Gembarska, A., Goossens, S., Maetens, M., De Clercq, S., Drogat, B., Haenebalcke, L., Bartunkova, S., De Vos, I., De Craene, B., et al. (2009). Efficient mouse transgenesis using Gateway-compatible ROSA26 locus targeting vectors and F1 hybrid ES cells. *Nucleic Acids Res.* **37,** e55.

Obeng, E. A., and Boise, L. H. (2005). Caspase-12 and caspase-4 are not required for caspase-dependent endoplasmic reticulum stress-induced apoptosis. *J. Biol. Chem.* **280,** 29578–29587.

Pennuto, M., Tinelli, E., Malaguti, M., Del Carro, U., D'Antonio, M., Ron, D., Quattrini, A., Feltri, M. L., and Wrabetz, L. (2008). Ablation of the UPR-mediator CHOP restores motor function and reduces demyelination in Charcot-Marie-Tooth 1B mice. *Neuron* **57,** 393–405.

Raskind, W. H., Williams, C. A., Hudson, L. D., and Bird, T. D. (1991). Complete deletion of the proteolipid protein gene (PLP) in a family with X-linked Pelizaeus-Merzbacher disease. *Am. J. Hum. Genet.* **49,** 1355–1360.

Reynolds, R., di Bello, I. C., Meeson, A., and Piddlesden, S. (1996). Comparison of a chemically mediated and an immunologically mediated demyelinating Lesion model. *Methods* **10,** 440–452.

Rosenbluth, J., Stoffel, W., and Schiff, R. (1996). Myelin structure in proteolipid protein (PLP)-null mouse spinal cord. *J. Comp. Neurol.* **371,** 336–344.

Rosenbluth, J., Nave, K. A., Mierzwa, A., and Schiff, R. (2006). Subtle myelin defects in PLP-null mice. *Glia* **54,** 172–182.

Rudnicki, M. A., Braun, T., Hinuma, S., and Jaenisch, R. (1992). Inactivation of MyoD in mice leads to up-regulation of the myogenic HLH gene Myf-5 and results in apparently normal muscle development. *Cell* **71,** 383–390.

Rutkowski, D. T., and Kaufman, R. J. (2007). That which does not kill me makes me stronger: Adapting to chronic ER stress. *Trends Biochem. Sci.* **32,** 469–476.

Schneider, A., Montague, P., Griffiths, I., Fanarraga, M., Kennedy, P., Brophy, P., and Nave, K.-A. (1992). Uncoupling of hypomyelination and glial cell death by a mutation in the proteolipid protein gene. *Nature* **358,** 758–761.

Sharma, R., and Gow, A. (2007). Minimal role for caspase 12 in the unfolded protein response in oligodendrocytes in vivo. *J. Neurochem.* **101,** 889–897.

Sharma, R., Jiang, H., Zhong, L., Tseng, J., and Gow, A. (2007). Minimal role for activating transcription factor 3 in the oligodendrocyte unfolded protein response in vivo. *J. Neurochem.* **102,** 1702–1713.

Simons, M., Kramer, E. M., Macchi, P., Rathke-Hartlieb, S., Trotter, J., Nave, K. A., and Schulz, J. B. (2002). Overexpression of the myelin proteolipid protein leads to accumulation of cholesterol and proteolipid protein in endosomes/lysosomes: Implications for Pelizaeus-Merzbacher disease. *J. Cell Biol.* **157,** 327–336.

Southwood, C. M., Garbern, J., Jiang, W., and Gow, A. (2002). The unfolded protein response modulates disease severity in Pelizaeus-Merzbacher disease. *Neuron* **36,** 585–596.

Stecca, B., Southwood, C. M., Gragerov, A., Kelley, K. A., Friedrich, V. L. J., and Gow, A. (2000). The evolution of lipophilin genes from invertebrates to tetrapods: DM-20 cannot replace PLP in CNS myelin. *J. Neurosci.* **20,** 4002–4010.

Stewart, D. G., and Davis, K. L. (2004). Possible contributions of myelin and oligodendrocyte dysfunction to schizophrenia. *Int. Rev. Neurobiol.* **59,** 381–424.

Turnley, A. M., Morahan, G., Okano, H., Bernard, O., Mikoshiba, K., Allison, J., Bartlett, P. F., and Miller, J. A. F. P. (1991). Dysmyelination in transgenic mice resulting from expression of class I histocompatibility molecules in oligodendrocytes. *Nature* **353,** 566–569.

Wek, R. C., and Cavener, D. R. (2007). Translational control and the unfolded protein response. *Antioxid. Redox Signal.* **9,** 2357–2371.

Werner, H., Jung, M., Klugmann, M., Sereda, M., Griffiths, I., and Nave, K. A. (1998). Mouse models of myelin diseases. *Brain Pathol.* **8,** 771–793.

Yamamoto, T., Nanba, E., Zhang, H., Sasaki, M., Komaki, H., and Takeshita, K. (1998). Jimpy(msd) mouse mutation and connatal Pelizaeus-Merzbacher disease. *Am. J. Med. Genet.* **75,** 439–440.

Zinszner, H., Kuroda, M., Wang, X., Batchvarova, N., Lightfoot, R. T., Remotti, H., Stevens, J. L., and Ron, D. (1998). CHOP is implicated in programmed cell death in response to impaired function of the endoplasmic reticulum. *Genes Dev.* **12,** 982–995.

CHAPTER TEN

GLYCOPROTEIN MATURATION AND THE UPR

Andreas J. Hülsmeier, Michael Welti, *and* Thierry Hennet

Contents

1. Introduction	163
2. N-glycosylation	165
2.1. Dolichol phosphate analysis	165
2.2. Dolichol-linked oligosaccharide analysis	169
2.3. N-glycosylation site occupancy	172
3. O-glycosylation	175
3.1. Release of *O*-glycans by the β-elimination reaction	176
3.2. Release of *O*-glycans by hydrazinolysis	179
Acknowledgment	181
References	181

Abstract

Glycosylation is a complex form of protein modification occurring in the secretory pathway. The addition of *N*- and *O*-glycans affects intracellular processes like the folding and trafficking of most glycoproteins. To better understand the impact of glycosylation in protein folding and maturation, parameters like glycosylation site occupancy and oligosaccharide structure must be measured quantitatively. In this chapter, we describe current methods enabling the determination of N-glycosylation by assessment of cellular dolichol phosphate levels, dolichol-linked oligosaccharides, and the occupancy of N-glycosylation sites. We also provide detailed methods for the analysis of O-glycosylation, whose role in intracellular protein maturation is often overlooked.

1. INTRODUCTION

Glycosylation is a widespread and complex form of modification, which adds signals and specific functions to glycoproteins. N-linked and O-linked glycosylation, which are the main forms of protein glycosylation, are structurally and functionally distinct. N-linked glycosylation is initiated in the endoplasmic reticulum (ER), where the oligosaccharide

Institute of Physiology, University of Zürich, Winterthurerstrasse, Zürich, Switzerland

$GlcNAc_2Man_9Glc_3$ is transferred cotranslationally to selected asparagine residues of nascent glycoproteins. While still in the ER, oligosaccharides on glycoproteins are trimmed by glucosidase and mannosidase enzymes to $GlcNAc_2Man_8$. Later in the Golgi apparatus, N-linked oligosaccharides undergo further trimming and elongation steps, which contribute to the structural diversity of N-linked glycosylation. In contrast to the cotranslational beginning of N-glycosylation, O-linked glycosylation takes place on folded proteins and is initiated by the transfer of monosaccharides to serine and threonine residues. Some forms of O-glycosylation like O-fucosylation (Harris and Spellman, 1993) and O-mannosylation (Lommel and Strahl, 2009) begin in the ER, whereas mucin-type O-glycosylation (Tian and Ten Hagen, 2009) and the biosynthesis of glycosaminoglycan chains (Bishop et al., 2007) begin in the Golgi apparatus.

N-glycosylation is important for the folding and trafficking of many glycoproteins (Helenius and Aebi, 2004). The inhibition of N-linked glycosylation, for example, by tunicamycin, results in the accumulation of misfolded proteins in the ER, which is a strong stimulus of the UPR (Shamu et al., 1994). After trimming by glucosidase-I in the ER, N-linked oligosaccharides are bound by the chaperone proteins calnexin and calreticulin, which contribute to the folding of glycoproteins (Ellgaard et al., 1999). Furthermore, N-linked oligosaccharides are also recognized by the EDEM protein, which redirects misfolded proteins from the ER lumen to the cytosol for proteasome-mediated degradation. Another important signal carried by N-linked oligosaccharides is the mannose-6-phosphate epitope, which mediates the recognition and transfer of lysosomal proteins to their target organelle (Dahms et al., 1989).

The most common form of O-glycosylation, that is, mucin-type glycosylation, does not contribute to glycoprotein folding, since it occurs in the Golgi apparatus. However, the trafficking and secretion of some glycoproteins require proper mucin-type glycosylation, as demonstrated in the case of the hormone FGF23. Deficient initiation of mucin-type O-glycosylation alters the susceptibility of FGF23 to Golgi-localized proteases, thereby inhibiting the maturation of active FGF23. Mucin-type glycosylation of FGF23 is initiated by the polypeptide N-acetylgalactosaminyltransferase-3 enzyme. Loss of this enzymatic activity leads to the disease tumoral calcinosis, which is also caused by FGF23 deficiency (Chefetz and Sprecher, 2009).

Another example of glycosyltransferase-assisted trafficking is given by the OFUT1 O-fucosyltransferase enzyme. This ER-localized glycosyltransferase adds fucose to the EGF-like repeats of proteins such as Notch and its ligands. Independent of its glycosyltransferase activity, OFUT1 also acts as a chaperone, which is essential for the transfer of Notch from the ER to the Golgi apparatus (Okajima et al., 2005). The contribution of O-linked glycosylation to glycoprotein folding and secretion is much less documented than in the case of N-linked glycosylation. This limited knowledge is certainly related to the technically difficult investigation of O-linked glycan structures. In fact, O-linked glycans are often densely clustered and these glycans cannot be released without significantly altering the polypeptide backbone. The O-glycanase enzyme from *Streptococcus pneumoniae* has a specificity limited to the O-linked disaccharide Gal(β1-3)GalNAc-O. By contrast, N-linked glycans can be conveniently cleaved at the first GlcNAc unit by the N-glycosidase F enzyme.

2. N-GLYCOSYLATION

Proper N-linked glycosylation is required for folding and intracellular trafficking of glycoproteins. Deficiency in the biosynthesis of the precursor of N-glycans or in the processing of N-glycans in the ER can alter glycoprotein folding and induce UPR. The following methods can be applied to identify defects of oligosaccharide assembly and of N-glycosylation site occupancy in target cells.

2.1. Dolichol phosphate analysis

The oligosaccharide core consisting of $GlcNAc_2Man_9Glc_3$ is first assembled on the polyisoprenoid carrier dolichol (Dol). Moreover, dolichol phosphate (Dol-P) is also part of the glycosylation substrates Dol-P-mannose and Dol-P-glucose, which account for the final extension of the oligosaccharide core in the ER. Considering the low levels of Dol-P-based structures in cells, a sensitive Dol-P detection can be achieved by labeling with the fluorescent compound 9-anthryldiazomethane (ADAM). The procedure was originally described for tissue sample extraction (Elmberger *et al.*, 1989) and we have adapted the method for cultured cells (Haeuptle *et al.*, 2009). Detection of unlabeled Dol-P can also be achieved by negative ion electrospray mass spectrometry (ESI-MS), which, in combination to precursor ion fragmentation, provides detailed information on Dol-P composition and on the saturation state of the α-isoprene unit.

2.1.1. Required materials

Devices and materials

- LaChrom D-7000 HPLC system (Merck), equipped with an Inertsil ODS-3 column (5 μm, 4.6×250 mm; GL Sciences, Inc., Japan) with a precolumn and a LaChrom 7485 fluorescence detector.
- Nano-flow ESI-ion trap MS (Eksigent nano-LC, 3200 QTRAP-MS, AB/SCIEX)
- C_{18} Sep-Pak and Silica Sep-Pak cartridges (Waters)

Other reagents

- ADAM (Sigma-Aldrich)
- C_{80}-polyprenol and C_{95}-Dol-P standards (Larodan Fine Chemicals, Sweden)
- Diazald (Sigma-Aldrich)
- Carbitol (Sigma-Aldrich)

2.1.2. Extraction and purification of Dol and Dol-P

Adherent cells (about 3×10^8 cells) are washed in phosphate buffered saline (PBS) after trypsinization and resuspended in 6 ml of water and then fixed by addition of 6 ml of methanol. The resulting suspension is transferred to a round bottom glass flask and 3 ml of 15 M potassium hydroxide are added for alkaline hydrolysis. An aliquot of the C_{80} polyprenyl phosphate (e.g., 15 µg) can be added as internal standard for subsequent quantification. The round bottom flask is attached to a reflux cooler and incubated in a preheated oil bath at 100 °C for 1 h. This alkaline hydrolysis removes carbohydrates from the Dol-P carrier. The resulting lysate is transferred to a glass tube and methanol is added to a total volume of 15 ml. The suspension is mixed with 30 ml of methanol/dichloromethane (1:4, v:v) and incubated at 40 °C for 1 h. Discard the upper phase (methanol) and transfer the lower phase (dichloromethane) containing the extracted lipids to a glass tube suitable for centrifugation. Wash the organic phase with 10 ml of dichloromethane/methanol/water (3:48:47, v:v:v). For phase separation, centrifuge at $6800 \times g$ at room temperature for 10 min and discard the upper phase. Repeat this washing step four times, which ensures the removal of free sugars and salt from the sample. The organic phase is then dried under nitrogen gas.

Dol and Dol-P are purified in two steps using C_{18}- and silica-based chromatography resins. First, equilibrate the C_{18} Sep-Pak cartridge with 5 ml of methanol, followed by 5 ml of chloroform/methanol (2:1, v:v) and 10 ml of methanol/water (98:2, v:v), 20 mM phosphoric acid. The dried lipids from the extraction step are dissolved in 400 µl of chloroform/methanol (2:1, v:v) and diluted by adding 10 ml of methanol/water (98:2, v:v), 20 mM phosphoric acid. Load the solution onto the C_{18} Sep-Pak cartridge and wash with 20 ml of methanol/water (98:2, v:v), 20 mM phosphoric acid. To remove phosphoric acid from the cartridge, wash again with 10 ml of methanol/water (98:2, v:v) devoid of phosphoric acid. Dol and Dol-P can be eluted with 20 ml of chloroform/methanol (2:1, v:v) and collected in a new glass tube. The eluate should be neutralized by adding 100 µl of 25% ammonium hydroxide.

The second purification step through Silica Sep-Pak cartridges allows the separation of Dol from Dol-P. First, equilibrate the cartridge with 10 ml of chloroform/methanol/water (10:10:3, v:v:v) and 40 ml of chloroform/methanol (2:1, v:v, in 0.5% ammonium hydroxide). Place a new glass tube to collect the dolichol fraction, which will be in the flowthrough of the chromatography. Apply the neutralized eluate from the previous step onto the Silica Sep-Pak cartridge. Rinse the glass tube of the eluate with 10 ml chloroform/methanol (2:1, v:v, in 0.5% ammonium hydroxide) and load it onto the Silica cartridge. Repeat the process with 10 ml chloroform/methanol (2:1, v:v, in 0.5% ammonium hydroxide). Elute Dol-P into another glass tube by applying 30 ml chloroform/methanol/water (10:10:3, v:v:v) onto the Silica cartridge. Dry the Dol and Dol-P fractions under nitrogen gas at 50 °C.

2.1.3. Anthryldiazomethane labeling of Dol-P

Dol-P labeling is made in two steps. First, Dol-P is dimethylated and then selectively demethylated to produce monomethylated Dol-P (Dol-P-Me). The procedure yields an enhanced fluorescence signal when labeling Dol-P-Me with 9-anthryl derivatives (Yamada *et al.*, 1986). The second step is the derivatization of Dol-P-Me with ADAM. We use the diazomethane generator system (Sigma-Aldrich) for the dimethylation of Dol-P.[1] It consists of two glass tubes: a small internal tube placed inside a larger outer tube. The dried Dol-P fraction is first dissolved in 1 ml of diethylether and transferred into the large outer glass tube of the diazomethane generator. The glass that contained the Dol-P fraction is washed twice with 1 ml of diethylether, which is added to the outer tube of the diazomethane generator. To produce diazomethane, add 0.367 g Diazald (explosive!) to the inner tube of the diazomethane generator system. Dissolve Diazald by addition of 1 ml of Carbitol. Assemble the two parts of the diazomethane generator system and immerse it into an ice bath. Make sure that the generator system is sealed. Using a syringe, carefully add 1.5 ml of 37% potassium hydroxide dropwise to the inner tube of the diazomethane generator system. Add few drops first and wait until the reaction starts, as seen by formation of gaseous diazomethane in the Diazald solution. In this way, the content of the inner tube is prevented from spilling into the outer tube. Then, carefully add the rest of the potassium hydroxide. Gently mix the reactants and incubate the reaction for 1 h in the ice bath. The diethylether phase containing Dol-P should turn yellow by then, indicating an excess of diazomethane. After disassembly of the diazomethane generator, leftover reactants can be neutralized by addition of 0.15 g of silicic acid to the inner tube.

---

[1] Diazomethane is very toxic in contact with eye, skin, or by inhalation. Diazomethane may explode in contact with sharp edges and when heated beyond 100 °C.

The outer tube should be sealed with parafilm and left for 2 h at room temperature. The Dol-P-Me$_2$ solution is transferred to a glass extraction tube and dried under nitrogen gas. To selectively demethylate Dol-P-Me$_2$, dissolve it in 4 ml of *tert*-butylamine and incubate at 70 °C for 14 h. Make sure to seal the extraction tube tightly to prevent the loss of *tert*-butylamine. Then, evaporate *tert*-butylamine under nitrogen gas and dissolve Dol-P-Me in 1.5 ml of 0.1 M hydrochloric acid adjusted to pH 3 with sodium hydroxide. Extract Dol-P-Me twice with 3 ml of diethylether and dry the combined extracts under nitrogen gas. Labeling is performed in 1 ml of diethylether saturated with ADAM. Prepare the ADAM solution in a separate glass tube by gradually adding the reddish ADAM powder to 0.7 ml of diethylether until saturation. Take 600 μl of the yellowish ADAM solution and add it to the dried Dol-P-Me fraction. Incubate the labeling reaction at 4 °C in the dark for 8 h. After drying under nitrogen gas, dissolve the reaction products in 2.4 ml of n-hexane. Wash the hexane phase three times with 1.8 ml of acetonitrile. Discard each time the acetonitrile lower phase. Dry the n-hexane phase under nitrogen gas. The ADAM-labeled Dol-P-Me fraction can be dissolved in acetonitrile/dichloromethane (3:2, v:v) for HPLC analysis.

2.1.4. HPLC analysis of fluorescently labeled Dol-P
ADAM-derivatized Dol-P is separated by HPLC using an Intertsil ODS-3 column and applying isocratic elution according to Yamada *et al.* (1986). The mobile phase consists of acetonitrile/dichloromethane (3:2, v:v) supplemented with 0.01% diethylamine. ADAM fluorescence is detected at 412 nm using an excitatory wavelength of 365 nm.

2.1.5. Negative ion electrospray mass spectrometry of Dol-P
Purified Dol-P can also be directly analyzed by ESI-MS (Haeuptle *et al.*, 2009) without derivatization. The dried Dol-P sample is first dissolved in 90% acetonitrile, 10% n-hexane, containing 0.01% diethylamine, and then vortexed and centrifuged to eliminate any particulate matter. The nano-flow system consists of an injector fitted with a 10 μl fused silica capillary loop, coupled in-line to a nano-flow pump and emitter tip of a nano-electrospray ion source via 20 μm ID fused silica capillary tubing. An aliquot of the sample is loaded into the injector loop and switched in-line to the nano-flow system, introducing the sample directly into the ion source with the dissolution solvent at a flow rate of 400 nl/min. Using a 3200 AB/Sciex QTRAP-MS, all mass spectra are acquired manually in the Enhanced MS scan mode (ion trapping mode), scanning from m/z 1000 to 1700 at a scan rate of 1000 amu/s and a step size of 0.06 amu. The curtain gas flow is maintained at 10 psi, collision gas pressure at 4×10^{-5} Torr (high), interface

heater temperature at 150 °C, declustering potential at −200 V, entrance potential at −10 V, collision energy at −10 V, and the ion spray voltage is varied between −2400 and −4500 V, depending on the condition of the emitter needle. Fragment ion spectra of selected Dol-P precursor masses are acquired in the Enhanced Product Ion mode (EPI mode, fragmentation in ion trapping mode) with the collision energy set to −100 V and linear scanning starting at m/z 100. Extensive rinsing of the nano-flow system in between sample injections is necessary to avoid cross contamination.

2.2. Dolichol-linked oligosaccharide analysis

Defects of dolichol-linked oligosaccharide (DLO) assembly decrease the availability of the mature Dol-PP-GlcNAc$_2$Man$_9$Glc$_3$ core for glycosylation of acceptor Asn sites on nascent glycoproteins (Hülsmeier et al., 2007). Such a DLO shortage leads to the nonoccupancy of N-glycosylation sites, which impacts on protein folding and trafficking (Helenius and Aebi, 2004). DLOs are conveniently analyzed either by metabolic labeling using [^{3}H]-mannose, or by derivatization with the fluorochrome 2-aminobenzamide (2AB). The extraction, hydrolysis, and purification procedure of DLOs is common to both methods.

2.2.1. Required materials

Chemicals

In addition to the solvents mentioned in Section 2.1.1. the reagents listed below are required.
- ^{3}H-labeled D-mannose, 5 mCi (Hartmann Analytics, Germany)
- 2-Propanol (Sigma-Aldrich)
- Trizma base (Sigma-Aldrich)
- 2AB (Sigma-Aldrich)
- Sodium cyanoborohydride NaBH$_3$CN, (Sigma-Aldrich)

Devices and other materials

- Dowex AG 1×4, Dowex AG 50×8 (Bio-Rad)
- ENVI-C$_{18}$ and ENVI-Carb resin (Supelco)
- Ultrafree-MC centrifugal filter devices (Millipore, Catalog number: UFC30LH25)
- Supelcosil LC-NH2 column, 5 μm particle size, 250×4.6 mm (Supelco)
- Scintillation flow monitor FLO-ONE A-525 (Packard)

2.2.2. [^{3}H]-labeling of DLOs

Approximately 2×10^8 cells are washed three times with PBS to remove glucose from the medium. Cells are then incubated in DMEM without serum and glucose at 37 °C for 45 min. After this starving step, [^{3}H]-mannose (125 µCi/10^8 cells) is added to the cells, which are incubated further at 37 °C for 1 h.

2.2.3. Extraction of DLOs

The labeling medium is aspirated and cells are washed quickly with 13 ml of ice-cold PBS. Cells are then fixed by adding 11 ml of ice-cold methanol/ 0.1 M Tris–HCl (8:3, v:v) and collected by scraping when dealing with adherent cells. The cell suspension is transferred to a 50-ml conical tube and 12 ml of chloroform are added. After vigorous mixing using a vortex, the suspension is centrifuged at $6000\times g$ for 10 min. The upper methanol phase and the lower chloroform phase are discarded, while the interface precipitate is kept for further extraction of DLOs. To this end, add 3 ml of chloroform/methanol/water (10:10:3, v:v:v), vortex for 2 min, and spin at $6000\times g$ for 10 min. Transfer the supernatant to a glass extraction tube. Repeat the extraction three times using 2 ml of chloroform/methanol/water (10:10:3, v:v:v). Then dry the extracted DLOs under nitrogen gas.

2.2.4. Hydrolysis and purification of oligosaccharides

Acid hydrolysis is used to release the oligosaccharide from the lipid carrier. For this, incubate the extracted DLOs in 2 ml of 0.1 M hydrochloric acid (in 50% 2-propanol) at 50 °C for 1 h and then neutralize by adding 1 ml of 0.2 M NaOH. To purify oligosaccharides, prepare a disposable chromatography column with 1 ml of Dowex AG 1×4 resin, followed by 0.6 ml of Dowex AG 50×8 resin. Equilibrate the column with 20 ml of 30% 2-propanol and apply the 3 ml hydrolyzate, then wash with 3 ml of 30% 2-propanol. Collect the flowthrough (6 ml total) into a new glass tube and evaporate under nitrogen gas. A second column system is used to further purify the DLOs. Prepare a column with 0.2 ml of ENVI-C_{18} resin and 1 ml of ENVI-Carb 120/400 resin. Place a C_{18} Sep-Pak cartridge on top of the prepared column. Wash the system with 5 ml of 100% methanol, 5 ml of 100% acetonitrile, and 5 ml of water/acetonitrile/0.1 M ammonium acetate. Finally, equilibrate with 9 ml of 2% acetonitrile/0.1 M ammonium acetate. Resuspend the dried oligosaccharide sample in 1 ml of acetonitrile/ 0.1 M ammonium acetate and apply it onto the column. Wash the sample vial with 1 ml of acetonitrile/0.1 M ammonium acetate, which is then loaded onto the column. Make sure that the lower ENVI resin column does not run dry during the loading process. Wash the column system with 9 ml of acetonitrile/0.1 M ammonium acetate. To elute DLOs, add 6 ml of water/acetonitrile (3:1, v:v) and collect the eluate in a new glass tube. Water and acetonitrile are then removed from the pure oligosaccharide sample by evaporation under nitrogen gas.

2.2.5. HPLC analysis of [^{3}H]-labeled oligosaccharides

A Supelco LC-NH2 normal phase column (including a LC-NH2 guard column) and a two solvent system are used for HPLC separation (Zufferey et al., 1995). Prepare acetonitrile/water (7:3, v:v) and acetonitrile/water (1:1) as mobile phases. Dissolve the purified oligosaccharides in 50–100 μl water for injection. The separation is carried out by running a gradient from acetonitrile/water (7:3, v:v) to acetonitrile/water (1:1) over 75 min at a flow rate of 1 ml/min. The [^{3}H]-labeled oligosaccharides are detected using a flow scintillation detector. When the level of [^{3}H]-incorporation is too low for detection by flow scintillation, fractions of the HPLC run can be collected and measured separately in a scintillation counter (Müller et al., 2005). While more time consuming, the latter detection mode yields a superior signal to noise ratio.

2.2.6. 2AB labeling of DLOs

The labeling of oligosaccharides with 2AB is performed according to Bigge et al. (1995). For 2AB labeling, DLOs are extracted as described under Section 2.2.3. and oligosaccharides are purified as outlined under Section 2.2.4. Dried oligosaccharides are resuspended in 200 μl water/acetonitrile (3:1, v:v) and transferred to a 1.5 ml, screw cap Eppendorf tube, where they are dried again. The 2AB labeling reagent is prepared by dissolving 24 mg 2AB in 500 μl acetic acid/DMSO (3:7, v:v) and adding 31 mg sodium cyanoborohydride, thereby achieving a 0.35 M 2AB and 1 M sodium cyanoborohydride solution. Add 20 μl of the labeling reagent to the dried oligosaccharides, mix by vortexing and incubate at 65 °C for 2 h. After cooling, add 380 μl acetonitrile to the sample. Excess labeling reagent is removed using the paper disk clean up procedure (Müller et al., 2005). To this end, punch paper disks from a Whatman 3MM paper using an office hole puncher. Place two paper disks in a centrifugal filter device and wash with 450 μl of water by spinning 30 s at 2000×g. Wash the filter twice with 450 μl 95% acetonitrile and load the labeled samples. Let the sample pass through the paper disks without spinning the filter device. Wash six times with 450 μl 95% acetonitrile followed by a quick spin at 2000×g. Elute the labeled oligosaccharides by applying three times 50 μl of water and spinning briefly at 2000×g.

2.2.7. HPLC analysis of 2AB-labeled oligosaccharides

To separate 2AB-labeled oligosaccharides, we have adapted the procedure of Royle et al. (2002) using a three-buffer solvent system (Grubenmann et al., 2004). We use a GlykoSep-N column maintained at 30 °C. Solvent A consists of 80% acetonitrile with 10 mM formic acid (pH 4.4) and solvent B of 40% acetonitrile with 30 mM formic acid (pH 4.4). The 50 mM formic acid stock solution is adjusted to pH 4.4 with ammonium hydroxide. Solvent C is 0.5% formic acid (pH is not adjusted). The gradient is run from 100% solvent A to 100% solvent B over 160 min at a flow rate of 0.4 ml/min. A transition from 100% solvent B to 100% solvent C follows within 2 min at a flow rate of 1 ml/min.

2.3. N-glycosylation site occupancy

Most defects of N-glycosylation result in partial occupancy of N-glycosylation sites on proteins. The quantitative analysis of glycosylation site occupancy is not trivial because glycosylated glycosylation sites usually ionize less efficiently than native peptides and it is not feasible to synthesize adequate standard glycopeptides for quantification purposes. Due to the structural microheterogeneity of glycopeptides, the relatively low sensitivity in detection is further compromised. Therefore, we developed a strategy to simplify the analysis and quantification of glycosylation site occupancy by deglycosylating the N-linked glycopeptides with N-glycosidase F. As an example, we describe here the analysis of human serum transferrin, which carries two N-glycans at Asn_{413} and Asn_{611}. We showed previously that Asn_{611} glycosylation is most sensitive to cellular stress imposed by congenital disorders of glycosylation or alcohol abuse (Hülsmeier et al., 2007). Serum transferrin is immune-affinity purified from 5 μl serum, proteolytically digested, and N-glycans are removed by digestion with N-glycosidase F, which converts the corresponding glycosylated Asn to Asp. N-glycosidase F digestion in buffer constituted with isotopic water ($H_2^{18}O$) leading to the incorporation of ^{18}O into the respective Asp, which allows to monitor for potential spontaneous deamination of unoccupied Asn prior to digestion with N-glycosidase F. The inclusion of isotopic labeled standard peptide enables sensitive and accurate quantitation of deglycosylated versus unoccupied glycosylation sites. The generated glycosylation sequon peptides are then analyzed by liquid chromatography multiple reaction monitoring mass spectrometry (LC-MRM-MS).[2] We are using a nano-flow LC (Eksigent) coupled to a hybrid type QTRAP-MS (ABSciex 3200 QTRAP-MS) equipped with a nano-electrospray ionization source. With this instrumentation, the triple quadrupole mode MRM-MS can be combined sequentially with product ion scanning in ion trapping mode, or in a MRM signal-induced product ion scanning mode. MRM has the advantage that selected fragment transitions are measured resulting in high specificity, sensitivity, and reproducible signals for quantitation over a wide range of signal intensities. In our setup, the product ion scanning serves as an additional scanning mode to confirm the sequence of the peptide ion selected in the first quadrupole ion filter.

---

[2] MRM is also referred to as single reaction monitoring mass spectrometry (SRM MS) and can be considered synonymous to MRM.

2.3.1. Required devices and materials

- Multiple Affinity Removal LC Column, 4.6×50 mm, with proprietary buffer system (MARS LC column "Human 6" or custom made "Human 3" for IgG, transferrin, and antitrypsin, Agilent Technologies)
- Iodoacetamide (Sigma Ultra grade)
- Endoproteinase Asp-N (Roche Diagnostics GmbH)
- Trypsin (Roche Diagnostics GmbH, proteomics grade)
- N-glycosidase F (Roche Diagnostics GmbH)
- Aprotinin (Roche Diagnostics GmbH)
- 20 mM ammonium bicarbonate, dissolved in 97 atom% $H_2^{18}O$ (Cambridge Isotope Laboratories, Inc.)
- Custom synthesized, isotope labeled peptides ($\geq$90% purity, HeavyPeptide Basic kit, Thermo Electron Corp.)
- Multiple reaction monitoring- (MRM-) capable nano-LC-MS system
- C_{18} PepMap 100 trap column (300-μm inner diameter×5 mm, Dionex)
- C_{18} PepMap100 column (75-μm inner diameter×15 cm, 3 μm, 100 Å, Dionex)

2.3.2. Purification of transferrin

A sample aliquot containing transferrin corresponding to 5-μl human serum is diluted 10 times with Agilent buffer A[3] and loaded onto the MARS column according to the manufacturer instructions. Transferrin will be eluted from the column in one step with 100% Agilent buffer B. The chromatography can be monitored by UV absorption at 280 nm and the transferrin containing peak fraction is collected. The eluate is desalted and concentrated by TCA precipitation. The samples are cooled on ice and adjusted to 12% TCA by adding 0.136 volumes of 100% TCA. The samples are kept on ice for 30 min or ideally at $-20\ °C$ overnight and centrifuged for 30 min at 14,000×g at 4 °C. The supernatants are discarded and the pellet is washed twice with 300 μl ice-cold ($-20\ °C$) acetone. Centrifuge for 15 min at 4 °C and 14,000×g each time after adding the acetone.

2.3.3. Proteolytic digestion and deglycosylation of glycopeptides

The TCA precipitated protein pellet is dissolved by sonication in 250 μl of 0.57 M Tris–HCl, pH 8.5, 50 mM DTT (1 M DTT diluted 20-fold with 0.6 M Tris–HCl, pH 8.5). Proteins are incubated for 5 min at 80 °C, cooled to room temperature, and cysteine residues are alkylated by adding 63 μl of 1 M iodoacetamide (fivefold excess over DTT, freshly prepared).

---

[3] Buffers A and B are supplied by Agilent together with the MARS column and their composition is proprietary. Therefore, we name these buffers "Agilent buffer A" and "Agilent buffer B," respectively.

The samples are incubated at room temperature in the dark for 40 min and subsequently desalted by TCA precipitation (see Section 2.3.2). The alkylated proteins are redissolved in 50 μl of 20 mM ammonium bicarbonate, containing 10% acetonitrile. Then 0.5 μg of endoproteinase Asp-N and 0.5 μg of trypsin are added and the sample is incubated at 37 °C for 16 h. Proteolysis is minimized by heating the sample to 80 °C for 5 min, adding 0.5 μmol EDTA and 100 μg aprotinin. Then 8 pmol of each standard peptide is added and the sample is lyophilized overnight. The sample is redissolved in 50 μl of 20 mM ammonium bicarbonate in 95 atom% $H_2^{18}O$, 1.5 units of N-glycosidase F (dissolved in 95 atom% $H_2^{18}O$) is added and the sample is digested at 37 °C for 8 h. The duration of the N-glycosidase F digestion was empirically optimized and it is recommended to keep these conditions to avoid sample degradation or incomplete N-glycosidase F digestion. The reaction is stopped by heat inactivation at 80 °C for 5 min. One tenth of the sample is adjusted to 20 μl of 0.1% formic acid in 2% acetonitrile and filtered through a 0.22 μm centrifugal filter device.

2.3.4. Liquid chromatography multiple reaction monitoring mass spectrometry

The sample is loaded on the trap column and the column is washed with 60 μl of 0.1% formic acid, 2% acetonitrile. After washing, the trap column is switched in-line to the PepMap column and a binary gradient at 250 nl/min is applied with buffer A 0.1% formic acid, and buffer B 80% ACN containing 0.1% formic acid. Buffer B is held at 14% for 3 min, increased to 54% over 102 min and to 100% over 8 min, held at 100% for 12 min, and decreased to 14% over 12 min, and the column is re-equilibrated at 14% buffer B for 16 min. The Asn413 peptides elute between 8 and 14 min and the Asn611 peptides elute between 22 and 34 min retention time. The parameters used for the detection of the transferrin peptides were determined empirically with synthetic peptides and are listed in Table 10.1.

2.3.5. Calculation of the N-glycosylation site occupancies

The average counts per second (cps) for each MRM transition are calculated by integrating the counts over the time of peak elution with the instrument acquisition software. The molar relative response factor (MRRF) for each peptide sequence is calculated by dividing the cps of the target peptide to the cps of the corresponding standard peptide.

$$MRRF = \frac{cps_{Target\ peptide}}{cps_{Standard\ peptide}}$$

Table 10.1 Peptides used for transferrin glycosylation sites Asn_{413} and Asn_{611}

		m/z value		
Designation	Sequence	Standard peptide	Target peptide	Q3 ion
Asn413-NK	CGLVP**V**LAENYNK	749.4	742.4	y_9
Asn413-NKS	CGLVP**V**LAENYNKS	792.9	785.9	y_{10}
Asn413-DK	CGLVP**V**LAENYDK	755.9	744.9	y_9
Asn413-DKS	CGLVP**V**LAENYDKS	799.4	788.4	y_{10}
Asn611-NVT	QQQHL**F**GSNVT	640.3	633.3	b_5-NH_3
Asn611-DVT	QQQHL**F**GSDVT	645.3	635.3	b_9

Designation	Dwell time (ms)	DP (V)	EP (V)	CEP (V)	CE (V)	CXP (V)
Asn413-NK	100	40	9.0	40	41	45
Asn413-NKS	100	41	9.5	43	44	48
Asn413-DK	100	41	9.0	40	41	45
Asn413-DKS	100	41	9.5	43	44	48
Asn611-NVT	150	40	9.0	35	37	43
Asn611-DVT	150	40	9.0	35	37	44

Amino acids containing the isotope label are marked bold: F, +10 Da; L, +7 Da; A, +4 Da. Due to a misscleavage at Lys_{414}, two sets of peptides are required. Two Asn_{413} sequon variants are generated at approximately equal abundance in the reaction mixture. The fragmentation conditions for each peptide were optimized empirically for maximum signal intensities. Dwell time, time per transition; DP, declustering potential; EP, entrance potential; CEP, collision cell entrance potential; CE, collision energy; CXP, collision cell exit potential.

The percentage of site occupancies is calculated as follows:

$$\% \text{ at } Asn_{611} = \frac{MRRF_{DVT}}{MRRF_{DVT} + MRRF_{NVT}} \times 100$$

$$\% \text{ at } Asn_{413} = \frac{MRRF_{DK} + MRRF_{DKS}}{MRRF_{DK} + MRRF_{DKS} + MRRF_{NK} + MRRF_{NKS}} \times 100$$

3. O-GLYCOSYLATION

Mucin-type glycosylation is initiated in the Golgi apparatus and is therefore not directly involved in the UPR. However, mucin-type O-glycosylation is involved in the intracellular processing of glycoproteins and contributes to their stability at the cell surface.

3.1. Release of O-glycans by the β-elimination reaction

O-glycans linked to the hydroxy amino acids Ser or Thr can be released by mild alkali treatment. This mild alkali treatment induces the β-elimination reaction, releasing O-linked carbohydrates from the β-carbon of Ser or Thr and leads to the formation of 2-aminopropenoic acid or 2-amino-2-butenocic acid, respectively (Fig. 10.1). A decrease of Ser or Thr after mild alkali treatment can be monitored by amino acid analysis, providing further information about the presence of O-glycosylation and the hydroxy amino acid involved in the O-glycosyl linkage (Hülsmeier et al., 2002, 2010). The released O-glycans are unstable under alkaline conditions and undergo stepwise degradation reactions termed "peeling reaction." A second β-elimination reaction occurs at the reducing end GalNAc residue, resulting in the formation of a furanosyl compound, the Morgan Elson chromogen (Fig. 10.1). The "peeling reaction" can be minimized by including sodium borohydride ($NaBH_4$) to the reaction mixture, thereby reducing the aldehydic group of GalNAc to the corresponding primary alcohol N-acetylgalactosaminitol (GalNAc-ol). GalNAc-ol is stable under alkaline conditions. Further GalNAc can be isotopically labeled by using either sodium borodeuterite ($NaBD_4$) or sodium borotritiate as reducing agents. A 2:8 (mol:mol) mixture of $NaBH_4$ and $NaBD_4$ can be used, which introduces a "fingerprint" isotope distribution into the O-glycans, facilitating the identification of O-glycans in MALDI-MS and alleviating the assignments of fragmentation ion spectra by marking the reducing end C1 carbohydroxy group (Hülsmeier et al., 2002).

3.1.1. Required devices and materials

- Sodium borohydride, sodium borodeuterite (Sigma-Aldrich)
- Sodium hydroxide, purest available grade (Sigma-Aldrich)
- Clean glass rod and glass tube
- Methyl iodide (ReagentPlus, 99.5%, Sigma-Aldrich)
- Sodium thiosulphate (Sigma-Aldrich)
- 2,5-Dihydroxybenzoic acid (Fluka)

3.1.2. Reductive β-elimination

The glycoprotein sample (1 mg or less) is lyophilized in a 1.5 ml screw cap polypropylene vial and dissolved in 200 µl freshly prepared 0.1 M sodium hydroxide, containing 0.2 M $NaBH_4$ and 0.8 M $NaBD_4$ (prepared by mixing 1 M $NaBH_4$ and 1 M $NaBD_4$ solutions 2:8 by volume). The sample is incubated at 37 °C for 24 h. Then further 100 µl of 0.1 M sodium hydroxide, containing 0.2 M $NaBH_4$ and 0.8 M $NaBD_4$ are added and incubation continues for further 48 h at 37 °C. The reaction is stopped by carefully adding acetic acid until gas development ceases. The sample will be acidified to approximately pH 4.

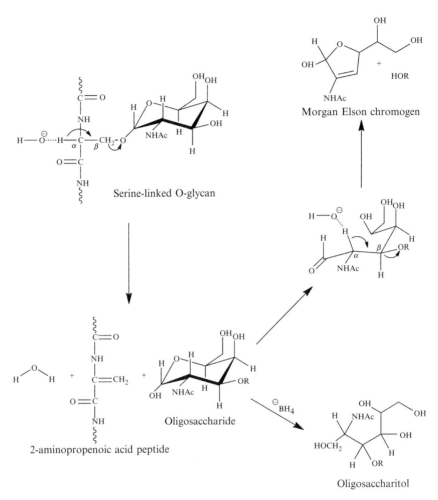

Figure 10.1 The proposed reaction scheme for the β-elimination reaction is exemplified with a serine-linked mucin-type O-glycan. Under alkaline conditions, hydroxide ions attack the α-carbon of the amino acid serine, inducing the liberation of the oligosaccharide from the β-carbon. Initially, water and a free reducing end oligosaccharide are generated from the polypeptide and as a result 2-aminopropenoic acid is formed. Inclusion of borohydride to the reaction mixture reduces the oligosaccharide to its corresponding primary alcohol (oligosaccharitol). Without reduction of the oligosaccharide, the glycans would undergo a second β-elimination sequence and the reducing end GalNac would form the Morgan Elson chromogen. The second β-elimination is accelerated, if the β-carbon is substituted, that is, participating in a glycosydic linkage.

3.1.3. Purification of the β-eliminated O-glycans

A C18 Sep-Pak cartridge is conditioned with 5 ml methanol and 5 ml propanol and equilibrated with two times 5 ml 1% acetic acid. A 0.6 ml Dowex AG50W-X12 column is conditioned with three times 5 ml 4 M HCl and washed with water until the flowthrough becomes neutral. The Sep-Pak cartridge is mounted on top of the Dowex column and both columns are equilibrated with three times 5 ml of 1% acetic acid. The β-elimination products are passed through the combined columns and the columns are washed twice with 2 ml of 1% acetic acid. The flowthrough is collected into a glass vial and dried under vacuum evaporation. Residual boric acid is removed by coevaporation twice with 250 µl 1% acetic acid in methanol and twice with 250 µl methanol. Residual acetic acid is removed by two evaporations with 50 µl of toluene.

3.1.4. Permethylation

The permethylation derivatization is carried out according to the NaOH method (Ciucanu and Kerel, 1984). The purified O-glycans are dried from 100 µl of 10 mM triethylamine to improve the solubility of negatively charged glycans as triethylamine salts in dimethyl sulfoxide (DMSO). Then, 50 µl DMSO is added and the reaction vial is agitated for 20 min. A 120 mg/ml slurry of NaOH in DMSO is prepared by crushing NaOH pellets with a glass rod in a glass tube. The NaOH will not dissolve completely and precipitate. Mixing of the slurry is required prior application to the derivatization reaction. A 50 µl aliquot of the NaOH/DMSO slurry is added to the reaction vial containing the O-glycans and the vial is shaken for additional 20 min. Then, 10 µl of methyl iodide are added twice, and the vial is agitated for 10 min after each addition, followed by a final addition of 20 µl of methyl iodide and agitation of the vial for further 20 min. Permethylated glycans are extracted by adding 250 µl of chloroform and 500 µl of 1 M sodium thiosulphate. The vial is agitated thoroughly and the liquid phases are separated by centrifugation. The aqueous phase is discarded and the chloroform phase, containing the permethylated O-glycans is washed 10 times with 1 ml of water. The chloroform phase is vacuum dried and the permethylated O-glycans are redissolved in 50% acetonitrile for application to MALDI-mass spectrometry (MALDI-MS).

3.1.5. MALDI-MS

The MALDI matrix is prepared by dissolving 10 mg 2,5-dihydroxybenzoic acid (DHB) in 1 ml of 50% acetonitrile, containing 1 mM sodium chloride (Hülsmeier et al., 2010). Aliquots of the permethylated O-glycans are mixed on the MALDI plate with DHB matrix 1:1 (v:v) and allowed to dry at room temperature. The dried spots are recrystallized by applying less than 0.1 µl of ethanol. A 10-µl pipette tip is dipped into ethanol and some solvent will be taken up by capillary force. Ethanol outside the tip is evaporated by waving the tip in the air for a few seconds. Then the tip is placed over the dried DHB spot so that the ethanol solvent just touches the MALDI plate surface. The DHB crystals will dissolve quickly and are left to air dry. As a result, an even layer of DHB crystals will be formed, permitting sensitive detection of the analytes. The recrystallization can be repeated, if necessary. Additional DHB matrix can be added, if difficult samples are to be analyzed. MALDI-mass spectra are recorded in positive ion mode and glycans are mainly detected as their sodium ion adducts, due to the presence of 1 mM sodium chloride in the DHB matrix.

3.2. Release of *O*-glycans by hydrazinolysis

Releasing *O*-glycans by hydrazinolysis has the advantage, that the glycans are liberated with a free reducing end saccharide. This allows subsequent labeling with fluorophores like 2AB or anthranilic acid, facilitating high-resolution chromatographic separation combined with unparalleled sensitive fluorescence detection (see Sections 2.2.6 and 2.2.7, Fig. 10.2). It is important for this reaction to occur under anhydrous conditions and that the sample is free of salt, metal ions, detergent, dyes, and stains. The reaction mechanism during hydrazinolysis has not been elucidated so far. However, an initial formation of a hydrazone derivative with concomitant release of water seems to be likely. After re-N-acetylation of the released glycans, acetohydrazone derivatives are formed and *O*-glycans in unreduced form can be recovered after passage through cation exchange resin and the addition of copper(II) acetate in mild acid (Patel *et al.*, 1993).

3.2.1. Required devices and materials

- Water and methanol rinsed, oven dried 250 µl glass syringe
- Glass reaction vials
- Lyophilizer
- Pure anhydrous hydrazine (Ludger Ltd., UK)
- Ice-cold, saturated sodium bicarbonate solution
- Acetic acid anhydride (Sigma-Aldrich)
- Dowex AG50 [$^+$H-form] resin (Bio-Rad)
- Copper(II) acetate (Sigma-Aldrich)
- Supelclean^{TM} ENVI-18 resin (Sigma-Aldrich)

3.2.2. Hydrazinolysis

The glycoprotein sample is dialyzed against 0.1% TFA and transferred to a clean glass reaction vial. The sample is lyophilized extensively for 1–3 days, depending on the amount of protein (up to 5 mg protein). Then, anhydrous hydrazine is added immediately using a glass syringe, prerinsed with hydrazine. Hydrazine is added in excess to the sample to give a less than 5 mg protein per milliliter of hydrazine solution. The reaction vessel is capped

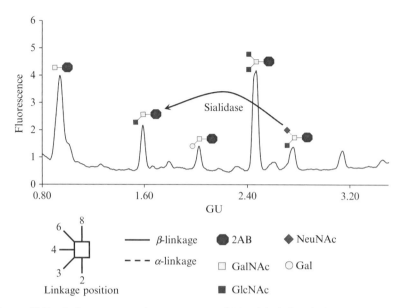

Figure 10.2 A fluorescence chromatogram of 2AB-labeled O-linked glycans from bovine submaxillary mucin. The O-glycans were released by hydrazinolysis, derivatized with 2-AB, and subjected to normal phase HPLC. The retention times are converted into glucose units (GU) via calibration with a 2AB-labeled dextran hydrosylate in consecutive runs. The calculation of GU facilitates the assignment of glycan structures by comparison with published GU values. These assignments can further be corroborated by digestion with exoglycosidase enzymes. For example, digestion of the sample with sialidase would lead to a reduction of the GlcNAcβ1-3[NeuNAcα2-6]GalNAc-2AB peak and a corresponding increase of the GlcNAcβ1-3GalNAc-2AB peak. The linkage positions are indicated by the orientation of the substituent and are decoded in the displayed pictogram. Anomericity is encoded as a solid line for β-linkage and dotted line for α-linkage, respectively. The glycan constituents are displayed in different shaped symbols according to the figure legend.

securely and mixed gently to bring the majority of sample into solution. The sample is transferred in a heating block at 60 °C and incubated for 5 h. The sample is cooled to room temperature and vacuum dried or lyophilized. Residual hydrazine can be removed from the sample by three times redrying from 100 μl methanol and finally evaporation from 50 μl toluene. The vial is placed on ice and 100 μl of ice-cold saturated sodium bicarbonate solution is added, followed by the addition of twice 10 μl acetic acid anhydride.[4] The sample is mixed and incubated at room temperature for

---

[4] From our experience, re-N-acetylation with sodium bicarbonate can lead to partial O-acetylation reactions. O-acetylation is not evident in HPLC analyses, but can be detected in MALDI-MS. Here it can serve as an indicator for the presence of saccharide in precursor ion scanning experiments. Saccharides would be detected in MALDI with a characteristic satellite peak increment of 42 Da.

10 min. Then, a second aliquot of acetic acid anhydride is added and the incubation proceeds for further 20 min. The sample is passed through a 3-ml Dowex AG50 [$^+$H-form] column, followed by 4 ml of water. The eluate is collected, dried, and redissolved in 2 ml of 1 mM copper(II) acetate in 1 mM acetic acid. The sample is incubated at room temperature for 1 h and the O-glycans are purified by passage through a column of 2 ml ENVI-18 resin over 1 ml Dowex AG50 [$^+$H-form] resin, eluted with water. The O-glycans are now ready for 2AB-derivatization and HPLC analysis (see Sections 2.2.6 and 2.2.7). 2AB-labeled glycan can also be subjected to permethylation for analysis by MALDI-MS. In our hands, permethylation of 2AB-glycans leads to significant higher signal intensities in MALDI-MS compared to permethylation of alkaline borohydride reduced glycans.

ACKNOWLEDGMENT

This work was supported by the Swiss National Foundation grant 31003A-116039 to TH.

REFERENCES

Bigge, J. C., Patel, T. P., Bruce, J. A., Goulding, P. N., Charles, S. M., and Parekh, R. B. (1995). Nonselective and efficient fluorescent labeling of glycans using 2-amino benzamide and anthranilic acid. *Anal. Biochem.* **230,** 229–238.

Bishop, J. R., Schuksz, M., and Esko, J. D. (2007). Heparan sulphate proteoglycans fine-tune mammalian physiology. *Nature* **446,** 1030–1037.

Chefetz, I., and Sprecher, E. (2009). Familial tumoral calcinosis and the role of O-glycosylation in the maintenance of phosphate homeostasis. *Biochim. Biophys. Acta* **1792,** 847–852.

Ciucanu, I., and Kerel, F. (1984). A simple and rapid method for the permethylation of carbohydrates. *Carbohydr. Res.* **131,** 209–217.

Dahms, N. M., Lobel, P., and Kornfeld, S. (1989). Mannose 6-phosphate receptors and lysosomal enzyme targeting. *J. Biol. Chem.* **264,** 12115–12118.

Ellgaard, L., Molinari, M., and Helenius, A. (1999). Setting the standards: Quality control in the secretory pathway. *Science* **286,** 1882–1888.

Elmberger, P. G., Eggens, I., and Dallner, G. (1989). Conditions for quantitation of dolichyl phosphate, dolichol, ubiquinone and cholesterol by HPLC. *Biomed. Chromatogr.* **3,** 20–28.

Grubenmann, C. E., Frank, C. G., Hülsmeier, A. J., Schollen, E., Matthijs, G., Mayatepek, E., Berger, E. G., Aebi, M., and Hennet, T. (2004). Deficiency of the first mannosylation step in the N-glycosylation pathway causes congenital disorder of glycosylation type Ik. *Hum. Mol. Genet.* **13,** 535–542.

Haeuptle, M. A., Hülsmeier, A. J., and Hennet, T. (2010). HPLC and mass spectrometry analysis of dolichol-phosphates at the cell culture scale. *Anal Biochem.* **396,** 133–138.

Harris, R. J., and Spellman, M. W. (1993). O-linked fucose and other post-translational modifications unique to EGF modules. *Glycobiology* **3,** 219–224.

Helenius, A., and Aebi, M. (2004). Roles of N-linked glycans in the endoplasmic reticulum. *Annu. Rev. Biochem.* **73,** 1019–1049.

Hülsmeier, A. J., Gehrig, P. M., Geyer, R., Sack, R., Gottstein, B., Deplazes, P., and Kohler, P. (2002). A major Echinococcus multilocularis antigen is a mucin-type glycoprotein. *J. Biol. Chem.* **277,** 5742–5748.

Hülsmeier, A. J., Paesold-Burda, P., and Hennet, T. (2007). N-glycosylation site occupancy in serum glycoproteins using multiple reaction monitoring liquid chromatography-mass spectrometry. *Mol. Cell. Proteomics* **6,** 2132–2138.

Hülsmeier, A. J., Deplazes, P., Naem, S., Nonaka, N., Hennet, T., and Köhler, P. (2010). An *Echinococcus multilocularis* coproantigen is a surface glycoprotein with unique O-glycosylation. *Glycobiology* **20,** 127–135.

Lommel, M., and Strahl, S. (2009). Protein O-mannosylation: Conserved from bacteria to humans. *Glycobiology* **19,** 816–828.

Müller, R., Hülsmeier, A. J., Altmann, F., Ten Hagen, K., Tiemeyer, M., and Hennet, T. (2005). Characterization of mucin-type core-1 beta1-3 galactosyltransferase homologous enzymes in *Drosophila melanogaster. FEBS J.* **272,** 4295–4305.

Okajima, T., Xu, A., Lei, L., and Irvine, K. D. (2005). Chaperone activity of protein O-fucosyltransferase 1 promotes notch receptor folding. *Science* **307,** 1599–1603.

Patel, T., Bruce, J., Merry, A., Bigge, C., Wormald, M., Jaques, A., and Parekh, R. (1993). Use of hydrazine to release in intact and unreduced form both N- and O-linked oligosaccharides from glycoproteins. *Biochemistry* **32,** 679–693.

Royle, L., Mattu, T. S., Hart, E., Langridge, J. I., Merry, A. H., Murphy, N., Harvey, D. J., Dwek, R. A., and Rudd, P. M. (2002). An analytical and structural database provides a strategy for sequencing O-glycans from microgram quantities of glycoproteins. *Anal. Biochem.* **304,** 70–90.

Shamu, C. E., Cox, J. S., and Walter, P. (1994). The unfolded-protein-response pathway in yeast. *Trends Cell Biol.* **4,** 56–60.

Tian, E., and Ten Hagen, K. G. (2009). Recent insights into the biological roles of mucin-type O-glycosylation. *Glycoconj. J.* **26,** 325–334.

Yamada, K., Abe, S., Suzuki, T., Katayama, K., and Sato, T. (1986). A high-performance liquid chromatographic method for the determination of dolichyl phosphates in tissues. *Anal. Biochem.* **156,** 380–385.

Zufferey, R., Knauer, R., Burda, P., Stagljar, I., te Heesen, S., Lehle, L., and Aebi, M. (1995). STT3, a highly conserved protein required for yeast oligosaccharyl transferase activity *in vivo. EMBO J.* **14,** 4949–4960.

CHAPTER ELEVEN

Monitoring and Manipulating Mammalian Unfolded Protein Response

Nobuhiko Hiramatsu,*,[1] Victory T. Joseph,*,†,[1] and Jonathan H. Lin*

Contents

1. Introduction	184
2. Monitoring Mammalian UPR	185
2.1. Monitoring IRE1 activity	185
2.2. Monitoring PERK activity	187
2.3. Monitoring ATF6 activity	188
2.4. Alternative approaches to UPR evaluation	188
3. Chemical–Genetic Manipulation of Mammalian UPR	189
3.1. Chemical–genetic manipulation of IRE1	190
3.2. Chemical–genetic manipulation of PERK	193
3.3. Chemical–genetic activation of ATF6	193
4. Concluding Remarks	194
Acknowledgments	195
References	195

Abstract

The unfolded protein response (UPR) is a conserved, intracellular signaling pathway activated by endoplasmic reticulum (ER) stress. In mammalian cells, the UPR is controlled by three ER-resident transmembrane proteins: inositol-requiring enyzme-1 (IRE1), PKR-like ER kinase (PERK), and activating transcription factor-6 (ATF6), by which cytoprotective mechanisms are initiated to restore ER functions. However, if cellular homeostasis is not restored by the UPR's initial events, UPR signaling triggers apoptotic cell death, which correlates with the pathogenesis of a wide range of human diseases. The intrinsic function of the UPR in regulating cell survival and death suggests its importance as a

---

* Department of Pathology, School of Medicine, University of California, San Diego, La Jolla, California, USA
† Department of Neuroscience, School of Medicine, University of California, San Diego, La Jolla, California, USA
[1] These authors equally contributed to this work

mechanistic link between ER stress and disease pathogenesis. Understanding UPR regulatory molecules or signaling pathways involved in disease pathogenesis is critical to establishing therapeutic strategies. For this purpose, several experimental tools have been developed to evaluate individual UPR components. In this chapter, we present methods to monitor and quantify activation of individual UPR signaling pathways in mammalian cells and tissues, and we review strategies to artificially and selectively activate individual UPR signaling pathways using chemical–genetic approaches.

1. INTRODUCTION

The endoplasmic reticulum (ER) is the entrance site for newly synthesized polypeptides of membrane and secreted proteins. In the ER, residential chaperones and enzymes fold and assemble polypeptides before export to the rest of the secretory pathway. The ER's role in protein quality control is pivotal to maintaining cellular homeostasis. Pathologic and physiologic processes can disrupt ER protein folding and assembly, and under these circumstances, misfolded proteins accumulate in the ER lumen, leading to ER stress. ER stress is correlated with and causally linked to the pathogenesis of a wide range of human diseases, including neurodegenerative disease, metabolic and inflammatory disease, infection, and subtypes of cancer (He, 2006; Hotamisligil, 2010; Lin et al., 2008; Moenner et al., 2007). Why and how ER stress causes diseases and what molecular and cellular changes lead to diseases are questions that remain largely unanswered and require further investigation.

The unfolded protein response (UPR) is a conserved, intracellular signaling process activated by ER stress (Ron and Walter, 2007). In mammalian cells, the UPR comprises at least three parallel intracellular signaling pathways controlled by ER-resident transmembrane proteins: inositol-requiring enzyme-1 (IRE1), PKR-like ER kinase (PERK), and activating transcription factor-6 (ATF6). IRE1, PERK, and ATF6 detect disrupted ER homeostasis and then transmit this information to the rest of the cell by activating downstream signal transduction effectors. In experimentally induced conditions of acute ER stress, all three UPR branches are simultaneously activated and initiate cytoprotective events that suppress global protein translation to reduce the protein folding burden of the ER, upregulate the transcription of molecular chaperones and protein folding enzymes to increase folding capacity of the ER, and enhance the degradation of unfolded proteins via proteasomal machinery. However, if cellular homeostasis is not restored despite these actions, the UPR then triggers apoptosis.

The intrinsic function of the UPR in regulating cell survival and death in response to ER stress suggests its importance as a mechanistic link between

ER stress and disease pathogenesis. However, the concomitant activation of multiple UPR signaling pathways by ER stress in mammalian cells has hindered precise assessments of the contribution of each specific UPR signaling pathway to disease pathogenesis. Identifying specific UPR regulatory molecules or signaling pathways involved in disease pathogenesis is critical to developing targeted therapies that prevent the development or progression of disease. Here, we review methods to monitor and quantify activation of individual UPR signaling pathways in mammalian cells and tissues, and we review strategies to artificially and selectively activate individual UPR signaling pathways using chemical–genetic approaches.

2. Monitoring Mammalian UPR

Here we present conventional protocols to detect and quantify the activity of IRE1, PERK, or ATF6 signaling in mammalian cell lines and tissues. Additionally, we review reporter systems developed for monitoring mammalian UPR activity.

2.1. Monitoring IRE1 activity

Upon induction of ER stress, IRE1α endonuclease (RNase) is activated and subsequently cleaves its substrate, X-box binding protein-1 (*Xbp-1*) messenger RNA (mRNA) (Calfon *et al.*, 2002; Yoshida *et al.*, 2001). *Xbp-1* mRNA has two conserved, overlapping open reading frames (ORFs). Activated IRE1α removes a 26-nucleotide (nt) intron from unspliced *Xbp-1* mRNA, leading to a translational frame shift to produce a protein encoded from the two ORFs. The product from spliced *Xbp-1* mRNA, XBP-1s, is an active transcription factor that upregulates the expression of ER associated degradation (ERAD) components and ER chaperones. Specificity of *Xbp-1* splicing by activated IRE1α is evidenced by mutation or knockout of IRE1α (Lee *et al.*, 2002). Therefore, monitoring *Xbp-1* splicing status is a specific molecular readout for IRE1α activity.

2.1.1. *Xbp-1* splicing assay for mammalian cells and tissues

Determination of *Xbp-1* splicing is performed by reverse transcriptase PCR (RT-PCR). Total RNA is extracted from cells or tissue lysate, and cDNA is synthesized by RT reaction with oligo-dT primer and Superscript III (Invitrogen, Carlsbad, CA). Sample cDNA is used as a template for PCR amplification with specific primers for *Xbp-1* (Table 11.1). PCR conditions are as follows:

Step 1: 95 °C for 5 min;
Step 2: 95 °C for 1 min;

Table 11.1 PCR primer sets

Gene	Species	Sequence (5′–3′) Forward	Reverse	Product (bp)
RPL19	Human	ATGTATCACAGCCTGTACCTG	TTCTTGGTCTCTTCCTCCTTG	233
	Mouse	ATGCCAACTCCCGTCAGCAG	TCATCCTTCTCATCCAGGTCACC	198
	Rat	TGGACCCCAATGAAACCAAC	TACCCTTCCTCTTCCCTATGCC	180
	Hamster	ATGCCAACTCCCGTCAGCAG	TCATCCTTCTCATCCAGGTCACC	198
BiP/Grp78	Human	CGGGCAAAGATGTCAGGAAAG	TTCTGGACGGGCTTCATAGTAGAC	211
	Mouse	CCTGCGTCGGTGTGTTCAAG	AAGGGTCATTCCAAGTGCG	201
	Rat	CCTGCGTCGGTGTATTCAAG	AAGGGTCATTCCAAGTGCG	201
	Hamster	TGGACCTGTTCCGATCTACC	GACAGCAGCACCGTATGCTA	206
Chop	Human	ACCAAGGGAGAACCAGGAAACG	TCACCATTCGGTCAATCAGAGC	201
	Mouse	ACGGAAACAGAGTGGTCAGTGC	CAGGAGGTGATGCCCACTGTTC	219
	Rat	ACGGACCACGAGTGGTCAGTGC	CAGGAGGTGATGCCAACAGTTC	219
	Hamster	AACGAGAGGAAAGTGGCTCA	TGTTCTCCCTACCCAGCATC	222
Xbp-1[a]	Human	TTACGAGAGAAAACTCATGGC	GGGTCCAAGTTGTCCAGAATGC	s-257, u-283
	Mouse	GAACCAGGAGTTAAGAACACG	AGGCAACAGTGTCAGAGTCC	s-179, u-205
	Rat	GAGTGGAGTAAGGCTGGTGG	TGGGTAGACCTCTGGGAGTTCC	s-223, u-249
	Hamster	TTGAGAGAGAAAACTCATGGC	GGGTCCAACTTGTCCAGAATGC	s-263, u-289

[a] Spliced form of xbp-1 (s); unspliced form of xbp-1 (u)

Step 3: 58 °C for 30 s;
Step 4: 72 °C for 30 s;
Repeat steps 2–4 for 35 cycles;
Step 5: 72 °C for 5 min.

Differentiation of unspliced and spliced *Xbp-1* can be resolved on a 2.0–3.0% agarose/1×TAE gel. A minor hybrid amplicon species consisting of unspliced *Xbp-1* annealed to spliced *Xbp-1* is also typically observed and appears above the unspliced amplicon (Back *et al.*, 2005; Lin *et al.*, 2007).

2.2. Monitoring PERK activity

In response to ER stress, PERK forms dimers and undergoes autophosphorylation (Harding *et al.*, 1999). Consequently, PERK's kinase activity phosphorylates the alpha subunit of eukaryotic translation initiation factor-2 (eIF2α). Phosphorylation of eIF2α impairs ribosomal assembly on mRNA, thereby attenuating global protein translation (Harding *et al.*, 2000b). However, translation of activating transcription factor 4 (ATF4) is increased when eIF2α is phosphorylated because of the presence of multiple upstream ORFs in its cognate *Atf4* mRNA (Harding *et al.*, 2000a; Lu *et al.*, 2004a; Vattem and Wek, 2004). ATF4 subsequently induces expression of its target genes, including transcription factor C/EBP-homologous protein (CHOP). CHOP, ATF4, and phosphorylated eIF2α are therefore useful molecular markers for monitoring activity of PERK.

2.2.1. Quantitative PCR (qPCR) analysis of CHOP

Relative *Chop* mRNA expression is evaluated by qPCR. Total RNA extraction and RT reaction is performed to generate cDNA. *Rpl19*, a gene for ribosomal protein L19, whose transcription is not affected by ER stress, can be used as an internal control for normalization (Hollien and Weissman, 2006). Primer information for *Chop* and *Rpl19* is shown in Table 11.1 and qPCR conditions are as follows:

Step 1: 95 °C for 5 min;
Step 2: 95 °C for 30 s;
Step 3: 58 °C for 30 s;
Step 4: 72 °C for 30 s;
Repeat steps 2–4 for 40 cycles.

2.2.2. Western blotting of eIF2α and ATF4

Cells are homogenized by sonication in 1% NP-40 or 6 M urea, 20 mM HEPES pH = 8.0, containing protease and phosphatase inhibitors (Pierce, Rockford, IL). Twenty to thirty micrograms of total cell lysate are loaded onto SDS-PAGE mini gels and analyzed by Western blot. The following antibody dilutions are used: anti-eIF2α at 1:2000 (Cell Signaling, Natick, MA); anti-phospho-eIF2α at 1:500 (Cell Signaling); anti-ATF4 at 1:2000 (Santa Cruz Biotechnologies, Santa Cruz, CA). After overnight incubation with primary antibody, membranes are washed in PBS (or TBS) containing 0.1% Tween-20 and incubated in horseradish peroxidase (HRP)-coupled secondary antibody (Amersham, Piscataway, NJ) diluted 1:5000 in wash buffer. Immunoreactivity is detected using the enhanced chemiluminescence assay (Pierce). These conditions have been optimized for human protein samples and may require modification when applied to other mammalian cell types and tissues.

2.3. Monitoring ATF6 activity

The third stress sensor for ER stress in mammalian cells is ATF6 (Haze et al., 1999). Upon induction of ER stress, ATF6 translocates from the ER to the Golgi apparatus where Site-1 Protease (S1P) and Site-2 Protease (S2P) cleave it (Ye et al., 2000). After cleavage by S1P/S2P, the amino-terminus of ATF6, containing a basic leucine zipper (bZIP) transactivating domain, translocates to the nucleus as an active transcription factor. ATF6 increases ER protein folding capacity and assembly by transcriptional upregulation of molecular chaperones and/or enzymes. In particular, immunoglobulin heavy chain binding protein (*BiP/Grp78*) is robustly induced in response to ER stress and is a direct transcriptional target of ATF6 (Haze et al., 1999). A simple and effective way to evaluate ATF6 activity is to measure *BiP/Grp78* mRNA or protein levels. In addition, Ron Prywes and colleagues established a method to monitor proteolytic cleavage using a FLAG-tagged ATF6 construct (Shen and Prywes, 2005). In this section, we briefly describe these methods of evaluating ATF6 activation.

2.3.1. qPCR analysis of BiP
Conditions for analysis of *BiP/Grp78* by qPCR are identical to *Chop*. Primer sequences for *BiP/Grp78* are shown in Table 11.1.

2.3.2. Western blotting of FLAG-ATF6
Cells are transfected with a p3×FLAG-ATF6 plasmid and protein lysate is prepared with lysis buffer as described above (Shen and Prywes, 2005). Detection is performed by Western blotting as described above using a 1:5000 dilution of primary antibody to FLAG (Sigma, St. Louis, MO).

2.4. Alternative approaches to UPR evaluation

Several reporter systems have been developed to monitor mammalian UPR activity. ER stress response elements (ERSE) (Yoshida et al., 1998) and UPR elements (UPRE) (Yoshida et al., 2000) have been identified within the promoter regions of UPR target genes, and UPR-dependent transcription factors bind to these regions to start their transcription. Constructs with conventional reporter proteins, such as luciferase and β-galactosidase, fused to UPRE and ERSE sequences can report ER stress, as evidenced by increased luciferase and β-galactosidase activities, in both cell culture and transgenic mice (Mao et al., 2004; Wang et al., 2000; Yoshida et al., 1998). However, in these systems, the contribution of individual UPR pathways (IRE1 vs. PERK vs. ATF6) cannot be resolved.

Additionally, Iwawaki et al. (2004) established in vitro and in vivo green fluorescent protein (GFP) reporter systems to monitor the IRE1 pathway. They constructed a reporter plasmid containing a FLAG-tagged, partial sequence (410–633 nt) of hXBP-1 and a GFP variant (Venus) sequence under the control of the chicken β-actin promoter. This partial sequence of hXBP-1 includes a 26-nt intron that is spliced out by activated IRE1. Under normal condition, expression of Venus-fused h*Xbp-1* mRNA contains the 26-nt intron, thereby halting translation via the stop codon located between the coding regions of hXBP-1 and Venus. Under ER stress conditions, the 26-nt intron is spliced out, inducing a frame shift and resulting in production of the hXBP-1-Venus fusion protein. Using this construct observation or quantification of Venus fluorescence allows monitoring of IRE1 activation. However, interpreting the on/off kinetics of UPR signaling might be difficult with this GFP-based approaches because the intrinsic stability of GFP may result in a prolonged fluorescence signal that remains elevated even after cessation of ER stress or the activation of a particular UPR pathway being monitored.

3. Chemical–Genetic Manipulation of Mammalian UPR

Common physiological and pharmacological inducers of ER stress (e.g., hypoxia, tunicamycin, and thapsigargin) concomitantly activate all UPR signaling pathways. To precisely identify the downstream effects of each signaling pathway, multiple chemical–genetic tools have been developed to selectively activate one UPR pathway at a time. Here we review chemical–genetic tools that enable selective activation of individual UPR signaling pathways in mammalian cells independent of ER stress.

The techniques described within take advantage of UPR initiation events by creating cell lines that mimic the activation of IRE1, PERK, or ATF6 proteins upon addition of cell-permeable, small molecules. Unlike simple gene deletion or constitutive overexpression, drug-induced activation acts quickly and reversibly. Furthermore, it avoids activating compensatory mechanisms in the cell arising from long-term UPR signaling. The establishment of stable, isogenic cell lines to study each pathway is recommended to (1) minimize ER stress arising from transfection or transduction protocols; (2) minimize epigenetic influences resulting from variable transgenic insertion; (3) permit longer analysis since the transgene is stably expressed; and (4) prevent overexpression bias by selecting stable clones with no elevation in basal ER stress compared to wild-type cells.

3.1. Chemical–genetic manipulation of IRE1

3.1.1. Activation of wild-type IRE1

IRE1α is a type I membrane protein consisting of an ER stress-sensor domain protruding into the ER lumen coupled to cytosolic kinase and RNase domains (Cox et al., 1993). In response to ER stress, IRE1α oligomerizes, which induces trans-autophosphorylation of its cytosolic domains and activation of its RNase activity (Cox and Walter, 1996; Shamu and Walter, 1996; Sidrauski and Walter, 1997) (Fig. 11.1A). The activated RNase cleaves Xbp-1 mRNA in the cytosol to generate spliced Xbp-1 mRNA which encodes a potent bZIP-containing transcription factor that upregulates ERAD components and ER chaperones (Lee et al., 2003b). In addition to catalyzing Xbp-1 mRNA splicing, mammalian IRE1α also binds to a growing number of ER membrane and cytosolic proteins including TRAF2 (tumor necrosis factor receptor-association factor 2), BAX (Bcl-2-associated X protein), BAK (Bcl-2 homologous antagonist killer), BI-1 (BAX inhibitor-1), HSP90 (heat shock protein 90), and RACK1 (receptor for activated C-kinase 1) (Hetz et al., 2006; Hetz and Glimcher, 2009; Lisbona et al., 2009; Marcu et al., 2002; Urano et al., 2000). The IRE1–TRAF2 interaction leads to the activation of the c-Jun N-terminal kinase (JNK) signaling pathway (Urano et al., 2000). The physiologic functions of other IRE1 protein–protein interactions are less well understood.

3.1.2. Activation of IRE1[I642G]

Several chemical–genetic approaches can selectively recapitulate IRE1's RNase activity in mammalian cells. ATP binding to IRE1's kinase domain is believed to activate IRE1's RNase function (Korennykh et al., 2009). A generalized chemical–genetic strategy for sensitizing protein kinases to cell-permeable molecules, while not affecting wild-type kinases, was first applied in yeast (Bishop et al., 2000; Papa et al., 2003) and has since been adapted and demonstrated to be effective in regulating mammalian IRE1α (Han et al., 2008; Hollien et al., 2009; Lin et al., 2007). A human IRE1α mutant bearing an isoleucine-to-glycine missense mutation at the gatekeeper position in the kinase domain, IRE1[I642G], enables

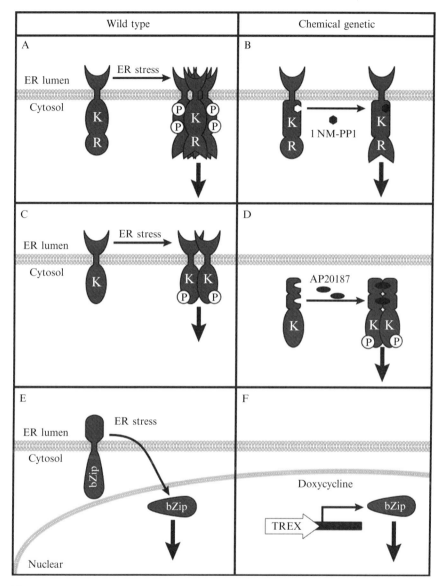

Figure 11.1 Comparison of UPR activation by ER stress or chemical–genetics. (A) In response to ER stress, wild-type IRE1 oligomerizes, *trans*-autophosphorylates (P) its kinase domain (K), and activates its endoribonuclease (RNase) activity (R). (B) In response to 1NM-PP1, IRE1[I642G] activates its RNase activity. (C) In response to ER stress, wild-type PERK dimerizes, autophosphorylates, and activates its kinase. (D) In response to AP20187, Fv2E–PERK conditionally dimerizes, autophosphorylates, and activates its kinase function. (E) In response to ER stress, full-length of ATF6 is cleaved and its N-terminus containing a bZip DNA binding domain translocates to the nucleus to activate transcription of target genes. (F) In response to doxycycline, the cleaved form of ATF6 under the control of T-REx promoter is expressed and activates transcription of its target genes.

ATP-competitive ligands to selectively bind its kinase domain thereby activating the RNase activity of this IRE1α allele (Fig. 11.1B).

1NM-PP1 (1-(tert-butyl)-3-(naphthalen-1-ylmethyl)-1H-pyrazolo[3,4-d]pyrimidin-4-amine) is the most common ligand used to regulate IRE1 [I642G] and is commercially available (EMD Chemicals, Gibbstown, NJ). Addition of 1NM-PP1 to cells expressing IRE1[I642G] results in rapid *Xbp-1* mRNA splicing in multiple mammalian cell types including human embryonic kidney (HEK293), mouse embryonic fibroblast (MEF), and human prostate cancer (PC-3) (Han *et al.*, 2008; Hollien *et al.*, 2009; Lin *et al.*, 2007; Thorpe and Schwarze, 2010). This strategy robustly reproduces IRE1α signaling through *Xbp-1* mRNA splicing without perturbing other UPR pathways.

When using the IRE1[I642G] mutant, however, it is important to note its differences from WT IRE1α and what consequences it might have in evaluating the IRE1α signaling pathway as a whole. It is unclear whether the mutant IRE1α forms dimers and/or oligomers or whether this step is necessary for activation in this system whereas *trans*-autophosphorylation is absent and unnecessary for RNase activity in the mutant. Differences in activation mechanism may be critical, especially in light of any cross talk between UPR pathways. While typically thought of as cytoprotective, recently, Hollien *et al.* (2009) and Han *et al.* (2009) discovered that mammalian IRE1α promotes the cleavage of numerous additional mRNAs and that IRE1 may also promote apoptosis depending on the specific mRNAs degraded. These authors demonstrated that IRE1α with active phosphotransfer was able to reduce levels of ER-localized mRNA which preceded apoptosis—this activity was absent in kinase catalysis-compromised mutants such as IRE1[I642G]. This suggests that WT IRE1α can send proapoptotic signals independent of *Xbp-1* splicing and that IRE1[I642G] uncouples these endonucleolytic outputs. This uncoupling may be therapeutically advantageous but may confound understanding of the endogenous role of IRE1 and the UPR.

Back *et al.* (2006) employed a different strategy to activate IRE1α via induced dimerization. By fusing the kinase and RNase domains of IRE1α to a modified FK506-binding domain (Fv1E), addition of a small organic molecule, AP20187, can serve as a dimerizer and recapitulate IRE1α's RNase activity. This strategy has also been adapted for inducible PERK activation and is described in more detail below. Additionally, others have successfully developed inducible spliced XBP-1 (XBP-1s) using the Tet-Off system (Lee *et al.*, 2003a); however, solely evaluating XBP-1s expression ignores additional targets of IRE1α RNase activity that may have physiological significance (Han *et al.*, 2009; Hollien *et al.*, 2009). Notably, neither system is suitable to study the physiologic functions of other IRE1 protein–protein interactions, as described above (Han *et al.*, 2009; Hollien *et al.*, 2009).

3.2. Chemical–genetic manipulation of PERK

3.2.1. Activation of wild-type PERK

PERK is an ER stress sensor protein that has an ER-luminal domain, bearing sequence and functional similarities to IRE1's luminal domain, and a cytosolic domain encoding a kinase. In response to ER stress, PERK oligomerizes, activating its kinase function (Fig. 11.1C). PERK kinase activity phosphorylates eIF2α which in turn attenuates protein translation by inhibiting ribosome assembly on mRNAs.

3.2.2. Activation of Fv2E–PERK

Chemical–genetic approaches have been developed that selectively recapitulate PERK's kinase activity in mammalian cells. Dimerization of PERK is a proximal step in its activation following ER stress. Fv2E–PERK comprises PERK's kinase domain fused to tandem-modified FK506-binding domains, Fv2E. Fv2E–PERK dimerization can be artificially induced by application of AP20187 (Fig. 11.1D). AP20187 is a cell permeable synthetic mimic of FK1012 (dimer of FK506) that can simultaneously bind two Fv2E domains and physically allow dimerization of Fv2E-fused proteins (Pollock and Rivera, 1999). Fv2E–PERK dimerization by AP20187 causes autophosphorylation of Fv2E–PERK, *trans*-phosphorylation of eIF2α, and downstream PERK signaling (Lu *et al.*, 2004a,b). This strategy robustly reproduces PERK signaling without necessitating ER stress or activating other UPR signaling pathways. AP20187-based activation of the PERK pathway has been demonstrated in Chinese hamster ovary (CHO), HEK293, MEF, and human mammary epithelial (MCF10A) cells, as well as in mice (Lin *et al.*, 2009; Lu *et al.*, 2004a,b; Oyadomari *et al.*, 2008; Sequeira *et al.*, 2007).

Fv2E–PERK fusion proteins can be created using standard cloning techniques and the ARGENT Regulated Homodimerization Kit (ARIAD Pharmaceuticals, Cambridge, MA). Fv2E is a modified FKBP domain with high affinity to AP20187, therefore commercially available anti-FKBP12 antibodies (Affinity Bioreagents, Rockford, IL) can be used to detect expression of the chimeric PERK protein by Western blot. AP20187 is commercially available from ARIAD Pharmaceuticals, can be diluted in DMF (diamethylformamide), and can be added directly to cell culture preparations.

3.3. Chemical–genetic activation of ATF6

3.3.1. Activation of wild-type ATF6

In contrast to IRE1 and PERK, ATF6 is a type II transmembrane protein with a carboxyl-terminus sensor domain facing the ER lumen coupled to a bZIP-containing transcription factor domain in its cytosolic amino terminus. In response to ER stress, ATF6 exits the ER and translocates to the Golgi (Shen *et al.*, 2002). ATF6 is sequentially cleaved into its active form by two proteases, S1P and S2P (Ye *et al.*, 2000) in the Golgi. Cleavage of ATF6 liberates the cytosolic fragment of ATF6 containing the bZIP transcription factor. The cleaved ATF6 fragment then translocates to the nucleus and upregulates transcription of target genes including ER protein-folding enzymes and ERAD proteins (Fig. 11.1E) (Haze *et al.*, 1999; Wu *et al.*, 2007; Yamamoto *et al.*, 2007).

3.3.2. Activation of TET-ATF6

Chemical–genetic approaches have been developed that selectively recapitulate ATF6's transcriptional factor activity in mammalian cells. Tetracycline-inducible induction of the cleaved ATF6 fragment encoding the cytosolic bZIP transactivational domain reproduces native ATF6 signaling in human cervical cancer cells (HeLa) and HEK293 cells (Okada *et al.*, 2002). In this system, the ATF6 fragment is placed under the control of a strong promoter containing multiple regulatory tet operator elements. In the absence of tetracycline or its derivative, doxycycline, tet repressors bind to the tet operators and inhibit transcription of ATF6. However, in the presence of doxycycline, the tet repressors bind the drug and undergo a conformational change that dissociates it from the tet operator. As a result, transcription of ATF6 is strongly induced and downstream transcriptional targets of ATF6 are upregulated (Fig. 11.1F). Other UPR signaling pathways are unaffected under these conditions.

There are several ATF6 isoforms; however, amino acid residues 1–373 of human ATF6α correspond to the active product of S1P/S2P-mediated proteolysis. This truncated ATF6α constitutively exists in the nucleus (Haze *et al.*, 1999). We recommend establishing stable Tet-On cell lines expressing this truncated form of ATF6α. For the establishment of a Tet-On system, there are several commercially available vectors and cell lines such as the T-REx system (Invitrogen, Carlsbad, CA) that makes creating of a stable ATF6α(373) cell line convenient and efficient.

Conditional regulation of ATF6 activity can also be achieved with the ERT2 (mutated estrogen receptor) system (Thuerauf *et al.*, 2007). Using this system, the functional domain of cleaved ATF6 is fused to ERT2. In the absence of ERT2 ligands (e.g., tamoxifen), ERT2 forms a complex with various cellular proteins, such as Hsp90, which mask ATF6. The addition of tamoxifen, however, blocks Hsp90 binding to ERT2, allowing ERT2–ATF6 to induce transcriptional activity of ATF6.

4. Concluding Remarks

UPR signaling pathways are likely to contribute to the pathogenesis and progression of numerous human diseases arising from protein misfolding or ER stress. Strategies to monitor and manipulate individual UPR signaling pathways can offer valuable tools to investigate the precise role of the UPR in disease pathogenesis and may offer new ways to combat diseases arising from ER stress.

ACKNOWLEDGMENTS

We thank C. Hetz, H. Li, C. Zhang, and members of the Lin laboratory for comments and suggestions.

REFERENCES

Back, S. H., Schroder, M., Lee, K., Zhang, K., and Kaufman, R. J. (2005). ER stress signaling by regulated splicing: IRE1/HAC1/XBP1. *Methods* **35,** 395–416.

Back, S. H., Lee, K., Vink, E., and Kaufman, R. J. (2006). Cytoplasmic IRE1alpha-mediated XBP1 mRNA splicing in the absence of nuclear processing and endoplasmic reticulum stress. *J. Biol. Chem.* **281,** 18691–18706.

Bishop, A. C., Ubersax, J. A., Petsch, D. T., Matheos, D. P., Gray, N. S., Blethrow, J., Shimizu, E., Tsien, J. Z., Schultz, P. G., Rose, M. D., Wood, J. L., Morgan, D. O., et al. (2000). A chemical switch for inhibitor-sensitive alleles of any protein kinase. *Nature* **407,** 395–401.

Calfon, M., Zeng, H., Urano, F., Till, J. H., Hubbard, S. R., Harding, H. P., Clark, S. G., and Ron, D. (2002). IRE1 couples endoplasmic reticulum load to secretory capacity by processing the XBP-1 mRNA. *Nature* **415,** 92–96.

Cox, J. S., and Walter, P. (1996). A novel mechanism for regulating activity of a transcription factor that controls the unfolded protein response. *Cell* **87,** 391–404.

Cox, J. S., Shamu, C. E., and Walter, P. (1993). Transcriptional induction of genes encoding endoplasmic reticulum resident proteins requires a transmembrane protein kinase. *Cell* **73,** 1197–1206.

Han, D., Upton, J. P., Hagen, A., Callahan, J., Oakes, S. A., and Papa, F. R. (2008). A kinase inhibitor activates the IRE1alpha RNase to confer cytoprotection against ER stress. *Biochem. Biophys. Res. Commun.* **365,** 777–783.

Han, D., Lerner, A. G., Vande Walle, L., Upton, J. P., Xu, W., Hagen, A., Backes, B. J., Oakes, S. A., and Papa, F. R. (2009). IRE1alpha kinase activation modes control alternate endoribonuclease outputs to determine divergent cell fates. *Cell* **138,** 562–575.

Harding, H. P., Zhang, Y., and Ron, D. (1999). Protein translation and folding are coupled by an endoplasmic-reticulum-resident kinase. *Nature* **397,** 271–274.

Harding, H. P., Novoa, I., Zhang, Y., Zeng, H., Wek, R., Schapira, M., and Ron, D. (2000a). Regulated translation initiation controls stress-induced gene expression in mammalian cells. *Mol. Cell* **6,** 1099–1108.

Harding, H. P., Zhang, Y., Bertolotti, A., Zeng, H., and Ron, D. (2000b). Perk is essential for translational regulation and cell survival during the unfolded protein response. *Mol. Cell* **5,** 897–904.

Haze, K., Yoshida, H., Yanagi, H., Yura, T., and Mori, K. (1999). Mammalian transcription factor ATF6 is synthesized as a transmembrane protein and activated by proteolysis in response to endoplasmic reticulum stress. *Mol. Biol. Cell* **10,** 3787–3799.

He, B. (2006). Viruses, endoplasmic reticulum stress, and interferon responses. *Cell Death Differ.* **13,** 393–403.

Hetz, C., and Glimcher, L. H. (2009). Fine-tuning of the unfolded protein response: Assembling the IRE1alpha interactome. *Mol. Cell* **35,** 551–561.

Hetz, C., Bernasconi, P., Fisher, J., Lee, A. H., Bassik, M. C., Antonsson, B., Brandt, G. S., Iwakoshi, N. N., Schinzel, A., Glimcher, L. H., and Korsmeyer, S. J. (2006). Proapoptotic BAX and BAK modulate the unfolded protein response by a direct interaction with IRE1alpha. *Science* **312,** 572–576.

Hollien, J., and Weissman, J. S. (2006). Decay of endoplasmic reticulum-localized mRNAs during the unfolded protein response. *Science* **313,** 104–107.

Hollien, J., Lin, J. H., Li, H., Stevens, N., Walter, P., and Weissman, J. S. (2009). Regulated Ire1-dependent decay of messenger RNAs in mammalian cells. *J. Cell Biol.* **186,** 323–331.

Hotamisligil, G. S. (2010). Endoplasmic reticulum stress and the inflammatory basis of metabolic disease. *Cell* **140,** 900–917.

Iwawaki, T., Akai, R., Kohno, K., and Miura, M. (2004). A transgenic mouse model for monitoring endoplasmic reticulum stress. *Nat. Med.* **10,** 98–102.

Korennykh, A. V., Egea, P. F., Korostelev, A. A., Finer-Moore, J., Zhang, C., Shokat, K. M., Stroud, R. M., and Walter, P. (2009). The unfolded protein response signals through high-order assembly of Ire1. *Nature* **457,** 687–693.

Lee, K., Tirasophon, W., Shen, X., Michalak, M., Prywes, R., Okada, T., Yoshida, H., Mori, K., and Kaufman, R. J. (2002). IRE1-mediated unconventional mRNA splicing and S2P-mediated ATF6 cleavage merge to regulate XBP1 in signaling the unfolded protein response. *Genes Dev.* **16,** 452–466.

Lee, A. H., Iwakoshi, N. N., Anderson, K. C., and Glimcher, L. H. (2003a). Proteasome inhibitors disrupt the unfolded protein response in myeloma cells. *Proc. Natl. Acad. Sci. USA* **100,** 9946–9951.

Lee, A. H., Iwakoshi, N. N., and Glimcher, L. H. (2003b). XBP-1 regulates a subset of endoplasmic reticulum resident chaperone genes in the unfolded protein response. *Mol. Cell. Biol.* **23,** 7448–7459.

Lin, J. H., Li, H., Yasumura, D., Cohen, H. R., Zhang, C., Panning, B., Shokat, K. M., Lavail, M. M., and Walter, P. (2007). IRE1 signaling affects cell fate during the unfolded protein response. *Science* **318,** 944–949.

Lin, J. H., Walter, P., and Yen, T. S. (2008). Endoplasmic reticulum stress in disease pathogenesis. *Annu. Rev. Pathol.* **3,** 399–425.

Lin, J. H., Li, H., Zhang, Y., Ron, D., and Walter, P. (2009). Divergent effects of PERK and IRE1 signaling on cell viability. *PLoS ONE* **4,** e4170.

Lisbona, F., Rojas-Rivera, D., Thielen, P., Zamorano, S., Todd, D., Martinon, F., Glavic, A., Kress, C., Lin, J. H., Walter, P., Reed, J. C., Glimcher, L. H., *et al.* (2009). BAX inhibitor-1 is a negative regulator of the ER stress sensor IRE1alpha. *Mol. Cell* **33,** 679–691.

Lu, P. D., Harding, H. P., and Ron, D. (2004a). Translation reinitiation at alternative open reading frames regulates gene expression in an integrated stress response. *J. Cell Biol.* **167,** 27–33.

Lu, P. D., Jousse, C., Marciniak, S. J., Zhang, Y., Novoa, I., Scheuner, D., Kaufman, R. J., Ron, D., and Harding, H. P. (2004b). Cytoprotection by pre-emptive conditional phosphorylation of translation initiation factor 2. *EMBO J.* **23,** 169–179.

Mao, C., Dong, D., Little, E., Luo, S., and Lee, A. S. (2004). Transgenic mouse model for monitoring endoplasmic reticulum stress in vivo. *Nat. Med.* **10,** 1013–1014, author reply 1014.

Marcu, M. G., Doyle, M., Bertolotti, A., Ron, D., Hendershot, L., and Neckers, L. (2002). Heat shock protein 90 modulates the unfolded protein response by stabilizing IRE1alpha. *Mol. Cell. Biol.* **22,** 8506–8513.

Moenner, M., Pluquet, O., Bouchecareilh, M., and Chevet, E. (2007). Integrated endoplasmic reticulum stress responses in cancer. *Cancer Res.* **67,** 10631–10634.

Okada, T., Yoshida, H., Akazawa, R., Negishi, M., and Mori, K. (2002). Distinct roles of activating transcription factor 6 (ATF6) and double-stranded RNA-activated protein kinase-like endoplasmic reticulum kinase (PERK) in transcription during the mammalian unfolded protein response. *Biochem. J.* **366,** 585–594.

Oyadomari, S., Harding, H. P., Zhang, Y., Oyadomari, M., and Ron, D. (2008). Dephosphorylation of translation initiation factor 2alpha enhances glucose tolerance and attenuates hepatosteatosis in mice. *Cell Metab.* **7,** 520–532.

Papa, F. R., Zhang, C., Shokat, K., and Walter, P. (2003). Bypassing a kinase activity with an ATP-competitive drug. *Science* **302,** 1533–1537.

Pollock, R., and Rivera, V. M. (1999). Regulation of gene expression with synthetic dimerizers. *Meth. Enzymol.* **306,** 263–281.

Ron, D., and Walter, P. (2007). Signal integration in the endoplasmic reticulum unfolded protein response. *Nat. Rev. Mol. Cell Biol.* **8,** 519–529.

Sequeira, S. J., Ranganathan, A. C., Adam, A. P., Iglesias, B. V., Farias, E. F., and Aguirre-Ghiso, J. A. (2007). Inhibition of proliferation by PERK regulates mammary acinar morphogenesis and tumor formation. *PLoS ONE* **2,** e615.

Shamu, C. E., and Walter, P. (1996). Oligomerization and phosphorylation of the Ire1p kinase during intracellular signaling from the endoplasmic reticulum to the nucleus. *EMBO J.* **15,** 3028–3039.

Shen, J., and Prywes, R. (2005). ER stress signaling by regulated proteolysis of ATF6. *Methods* **35,** 382–389.

Shen, J., Chen, X., Hendershot, L., and Prywes, R. (2002). ER stress regulation of ATF6 localization by dissociation of BiP/GRP78 binding and unmasking of Golgi localization signals. *Dev. Cell* **3,** 99–111.

Sidrauski, C., and Walter, P. (1997). The transmembrane kinase Ire1p is a site-specific endonuclease that initiates mRNA splicing in the unfolded protein response. *Cell* **90,** 1031–1039.

Thorpe, J. A., and Schwarze, S. R. (2010). IRE1alpha controls cyclin A1 expression and promotes cell proliferation through XBP-1. *Cell Stress Chaperones* **15,** 497–508.

Thuerauf, D. J., Marcinko, M., Belmont, P. J., and Glembotski, C. C. (2007). Effects of the isoform-specific characteristics of ATF6 alpha and ATF6 beta on endoplasmic reticulum stress response gene expression and cell viability. *J. Biol. Chem.* **282,** 22865–22878.

Urano, F., Wang, X., Bertolotti, A., Zhang, Y., Chung, P., Harding, H. P., and Ron, D. (2000). Coupling of stress in the ER to activation of JNK protein kinases by transmembrane protein kinase IRE1. *Science* **287,** 664–666.

Vattem, K. M., and Wek, R. C. (2004). Reinitiation involving upstream ORFs regulates ATF4 mRNA translation in mammalian cells. *Proc. Natl. Acad. Sci. USA* **101,** 11269–11274.

Wang, Y., Shen, J., Arenzana, N., Tirasophon, W., Kaufman, R. J., and Prywes, R. (2000). Activation of ATF6 and an ATF6 DNA binding site by the endoplasmic reticulum stress response. *J. Biol. Chem.* **275,** 27013–27020.

Wu, J., Rutkowski, D. T., Dubois, M., Swathirajan, J., Saunders, T., Wang, J., Song, B., Yau, G. D., and Kaufman, R. J. (2007). ATF6alpha optimizes long-term endoplasmic reticulum function to protect cells from chronic stress. *Dev. Cell* **13,** 351–364.

Yamamoto, K., Sato, T., Matsui, T., Sato, M., Okada, T., Yoshida, H., Harada, A., and Mori, K. (2007). Transcriptional induction of mammalian ER quality control proteins is mediated by single or combined action of ATF6alpha and XBP1. *Dev. Cell* **13,** 365–376.

Ye, J., Rawson, R. B., Komuro, R., Chen, X., Dave, U. P., Prywes, R., Brown, M. S., and Goldstein, J. L. (2000). ER stress induces cleavage of membrane-bound ATF6 by the same proteases that process SREBPs. *Mol. Cell* **6,** 1355–1364.

Yoshida, H., Haze, K., Yanagi, H., Yura, T., and Mori, K. (1998). Identification of the cis-acting endoplasmic reticulum stress response element responsible for transcriptional induction of mammalian glucose-regulated proteins. Involvement basic leucine zipper transcription factors. *J. Biol. Chem.* **273,** 33741–33749.

Yoshida, H., Okada, T., Haze, K., Yanagi, H., Yura, T., Negishi, M., and Mori, K. (2000). ATF6 activated by proteolysis binds in the presence of NF-Y (CBF) directly to the cis-acting element responsible for the mammalian unfolded protein response. *Mol. Cell. Biol.* **20,** 6755–6767.

Yoshida, H., Matsui, T., Yamamoto, A., Okada, T., and Mori, K. (2001). XBP1 mRNA is induced by ATF6 and spliced by IRE1 in response to ER stress to produce a highly active transcription factor. *Cell* **107,** 881–891.

CHAPTER TWELVE

A Screen for Mutants Requiring Activation of the Unfolded Protein Response for Viability

Guillaume Thibault[*] and Davis T. W. Ng[*,†]

Contents

1. Introduction	200
2. Screening for Mutants That Require UPR Activation for Viability	202
2.1. Primary screen	202
2.2. Secondary screens of per mutants	206
2.3. Protocol for pulse-chase immunoprecipitation assay	206
3. Cloning and Sequencing per Genes	209
3.1. Cloning by complementation	209
3.2. Library transformation and plasmid isolation	210
3.3. Identification of mutant genes	211
4. Monitoring the UPR Activity	212
4.1. The β-galactosidase assay	212
4.2. Alternative procedures to monitor the UPR activity	213
5. SGA Database	213
6. Media Recipes	214
7. Closing Remarks	214
Acknowledgment	215
References	215

Abstract

The unfolded protein response (UPR) is an intracellular signal transduction pathway that monitors endoplasmic reticulum (ER) homeostasis. Activation of the UPR is required to alleviate the effects of ER stress. However, our understanding of what physiologically constitutes ER stress or disequilibrium is incomplete. The current view suggests that stress manifests as the functional capacity of the ER becomes limiting. To uncover the range of functions under the purview of the UPR, we previously devised a method to isolate mutants that (1) activate the UPR and (2) require UPR activation for viability. These mutants that

[*] Temasek Life Sciences Laboratory, National University of Singapore, Singapore
[†] Department of Biological Sciences, National University of Singapore, Singapore

represent functions, when compromised, cause specific forms of disequilibrium perceived by the UPR. Making UPR activation essential to these mutants ensures a stringent physiological link and avoids stimuli causing nonproductive UPR activation. Thus far, the screen has revealed that the range of functions monitored is surprisingly diverse. Beyond the importance of the screen to understand UPR physiology, it has proven to be useful in discovering new genes in many aspects of protein biosynthesis and quality control.

1. Introduction

The unfolded protein response (UPR) is an intracellular signal transduction pathway that regulates the expression of target genes to maintain cell homeostasis. Although metazoans have three UPR outputs, yeast cells rely exclusively on the Ire1p pathway, which is conserved in all eukarya (Cox et al., 1997; Mori et al., 1993; Ron and Walter, 2007). Upon endoplasmic reticulum (ER) dysfunction, the ER-localized sensor Ire1p self-associates and activates the pathway by splicing an inhibitory intron from a mRNA encoding a transcription factor called Hac1p (Cox and Walter, 1996; Sidrauski and Walter, 1997). Newly synthesized Hac1p then induces the transcription of UPR target genes. Whole genome transcriptional profiling in *Saccharomyces cerevisiae* has revealed nearly 400 genes specifically regulated by the UPR pathway (Travers et al., 2000). Their activation facilitates the refolding of proteins and enhances the recognition and degradation of misfolded proteins under stress conditions. The prominent UPR target genes include ER molecular chaperones, ER-associated degradation (ERAD) proteins, genes that encompass the secretory pathway, and lipid biosynthesis proteins. However under extreme stress condition, the UPR may fail to fix the acute ER stress which leads to cell death (Bernales et al., 2006; Nakagawa et al., 2000). In recent years, diseases linked to ER stress include Alzheimer, Parkinson, diabetes, and hepatic steatosis (for extended review see Yoshida, 2007).

To better understand the role of the UPR, several approaches have been developed to identify and characterize linked genes. Budding yeast has been useful for this purpose because its dependence on a single UPR pathway. The first major breakthrough was achieved by monitoring the transcriptional scope of the UPR using DNA microarrays (Travers et al., 2000). Among the 400 genes upregulated by the UPR, half the targets of known function are genes that cover the secretory pathway in yeast. Using the genetic-based *per* (protein processing in the ER) screen, several genes were also identified to activate the UPR when deficient (Ng et al., 2000). This screening is discussed in great details below. More recently, a UPR reporter-based approach was used to identify genes with roles in protein maturation (Jonikas *et al.*, 2009). However, regardless the value of the different approaches, the discovery of new genes was limited due to the complexity of the UPR. Additionally, the UPR only regulates a subset of genes for any given function.

To identify new genes physiologically linked to the UPR, a genetic screen was used to isolate mutants that are dependent on the UPR activation for viability (Ng *et al.*, 2000). The screen was based on the fact that the UPR pathway is not necessary in budding yeast under normal growth condition but become essential under conditions of ER stress. Mutants that display synthetic lethality with an *ire1* null mutation were searched. The *ire1* mutation blocks UPR activation to prevent amelioration of ER stress (Cox *et al.*, 1993). Synthetic lethality, which is defined as a combination of two otherwise nonlethal mutations that results in cell death, has been used to identify genes in critical biological process in different model organisms (Fay *et al.*, 2002; Giaever *et al.*, 2002; Lucchesi, 1968; St Onge *et al.*, 2007). An epistatic miniarray profile (E-MAPs) of genes involved in the early secretory pathway was used to identify genetic interactions between logically connected subsets of genes (Schuldiner *et al.*, 2005). Genes involved in the yeast early secretory pathway were chosen because they are spatially restricted with central functions in addition of being highly conserved and crucial to many metazoan processes. An extensive study on the whole genome of *S. cerevisiae* named synthetic genetic array (SGA) was recently published (Costanzo *et al.*, 2010). Although it provides an invaluable amount of information on gene functions, this approach is mostly limited to nonessential genes. However, a classical synthetic lethality screen based on the *ire1* null mutant also includes the identification of essential genes.

The UPR synthetic lethality screen generated mutants covering important players of the secretory protein biosynthesis and protein quality control. Strains were identified to be defective in *N*- and *O*-linked glycosylation, glycosylphosphatidylinositol (GPI)-anchored protein biosynthesis, protein folding, ER quality control and ERAD, protein trafficking, and ER luminal ion homeostasis (Ng *et al.*, 2000 and unpublished data). For example, the essential genes *RFT1* and *GPI10* were identified in the screen to be physiologically linked to the UPR.

The initial analysis of *per* mutants (*per*: protein processing in the *ER*) demonstrated the potential and specificity of the screen. All 20 *per* mutants isolated were shown to be defective in some aspects of the ER function indicating the specificity of the screen. Thus, classical and systematic synthetic lethality approaches, when combined, provide rich sources of information that will help to better understand the complex cell responses to ER stress. Here, we describe the method used in the *PER* screen in addition to a short overview of the SGA in the UPR context.

2. Screening for Mutants That Require UPR Activation for Viability

2.1. Primary screen

To perform a classical synthetic lethality screen for interactors of the UPR, we applied a colony-color sectoring assay. This approach exploits the phenotype of the *ade2* adenine biosynthetic mutant, which accumulates the intermediate aminoimidazole ribotide as a red pigment when adenine is limiting (Zimmermann, 1975). An *ade2ade3* double mutant, however, does not accumulate this pigment so it grows as white colonies on agar plates (Barbour and Xiao, 2006). Accordingly, transformed with a plasmid bearing a functional copy of the *ADE3* gene, the *ade2ade3* strain gives rise to red colonies under plasmid selection. Without selective pressure, the same strain gives rise to red/white sectoring colonies because centromeric plasmids missegregate mitotically approximately once in every fifty cell divisions. Using this principle, a Δ*ire1,ade2ade3* strain was constructed in the W303 background (DNY419) and transformed with pDN336, a yeast *CEN/ARS* vector with a *URA3* selectable marker that contains the wild-type *ADE3* and *IRE1* genes (Table 12.1). The resulting strain, DNY421, gives rise to sectoring colonies on YPD and low adenine containing SC media due to the low frequency plasmid loss expected for otherwise wild-type strains (Fig. 12.1). Mutants displaying synthetic lethal or synthetic growth defects grow as red, nonsectoring colonies (Fig. 12.1B, *per16-1*). Mutants of this class are designated *per* for protein processing in the ER because most isolates display defects in the biogenesis or quality control of secretory pathway proteins (Ng *et al.*, 2000).

Although it is possible to screen for spontaneous *per* mutants to avoid off-target mutations, it is more practical to mutagenize cells to increase the frequency. We prefer to use short wave ultraviolet (UV) light as a mutagen but chemical mutagens can also be used to increase the range of alleles (see Guthrie and Fink, 1991, for detailed protocols of chemical mutagenesis). For UV mutagenesis, DNY421 cells are streaked onto a YPD plate and a single sectoring colony is used to inoculate a 100 mL culture in SC Ura-dropout media to select for pDN336. The flask is incubated at 30 °C overnight with aeration. In the morning, the culture should be at mid to late log phase. 50 A_{600} OD units are harvested and pelleted in a 50 mL conical centrifuge tube using a clinical centrifuge at $1000 \times g$ for 5 min. The pellet is washed once in 50 mL ice-cold sterile water and resuspended in 25 mL 0.9% KCl. Using a support stand, a hand-held UV lamp (254 nm, 4 W) is mounted 15 cm above the surface of a plate stirrer. It is essential to set this up in a darkroom equipped with a low intensity safety light. Normal ambient light can activate DNA repair mechanisms so it is to be avoided. Transfer the cell suspension to a disposable 100 mm Petri dish containing a

Table 12.1 Strains used in the *per* screening

Strain	Genotype	Source
W303a	*MAT**a**, leu2-3-112, his3-11, trp1-1, ura3-1, can1-100, ade2-1*	Cox et al. (1993)
DNY420	*MAT**α**, ire1::TRP1, ura3-1, can1-100, ade2-1, ade3, leu2-3-112::LEU2-UPRE LacZ, his3-11::HIS3-UPRE LacZ* (pDN336)	Ng et al. (2000)
DNY421	*MAT**a**, ire1::TRP1, ura3-1, can1-100, ade2-1, ade3, leu2-3-112, his3-11::HIS3-UPRE LacZ* (pDN336)	Ng et al. (2000)
DNY486	*MAT**a**, per1-1,* DNY421 background	Ng et al. (2000)
DNY499	*MAT**a**, per1-2,* DNY421 background	Ng et al. (2000)
DNY489	*MAT**a**, per2-1,* DNY421 background	Ng et al. (2000)
DNY491	*MAT**a**, per3-1,* DNY421 background	Ng et al. (2000)
DNY493	*MAT**a**, per4-1,* DNY421 background	Ng et al. (2000)
DNY478	*MAT**a**, per4-2,* DNY421 background	Ng et al. (2000)
DNY495	*MAT**a**, per5-1,* DNY421 background	Ng et al. (2000)
DNY497	*MAT**a**, per6-1,* DNY421 background	Ng et al. (2000)
DNY503	*MAT**a**, per6-2,* DNY421 background	Ng et al. (2000)
DNY501	*MAT**a**, per7-1,* DNY421 background	Ng et al. (2000)
DNY505	*MAT**a**, per8-1,* DNY421 background	Ng et al. (2000)
DNY507	*MAT**a**, per9-1,* DNY421 background	Ng et al. (2000)
DNY509	*MAT**a**, per10-1,* DNY421 background	Ng et al. (2000)
DNY484	*MAT**a**, per10-2,* DNY421 background	Ng et al. (2000)
DNY488	*MAT**a**, per11-1,* DNY421 background	Ng et al. (2000)
DNY470	*MAT**a**, per12-1,* DNY421 background	Ng et al. (2000)
DNY472	*MAT**a**, per13-1,* DNY421 background	Ng et al. (2000)
DNY475	*MAT**a**, per14-1,* DNY421 background	Ng et al. (2000)
DNY479	*MAT**a**, per15-1,* DNY421 background	Ng et al. (2000)
DNY481	*MAT**a**, per16-1,* DNY421 background	Ng et al. (2000)

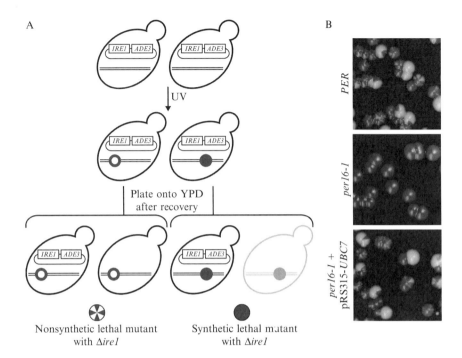

Figure 12.1 Primary genetic screening based on colony color phenotype. (A) Schematic representation of genetic screening based on color colony phenotypes. Nonsynthetic and synthetic lethal mutants with Δ*ire1*, generated by UV light mutagenesis, are represented with open and closed blue circles, respectively. (B) Color colony phenotypes of *PER*, *per16-1*, and *per16-1* (pRS315-*UBC7*) strains grown on YPD plate at 30 °C for 3 days. Colors were contrasted to illustrate the sectoring phenotype. (For interpretation of the references to color in this figure legend, the reader is referred to the Web version of this chapter.)

sterile stir bar and place it on top of the stirrer. Collect and set aside a 1 mL aliquot. Replace the plastic lid to decrease contamination. Next, the cells are exposed to sequential 30-s pulses of UV light by turning on the UV lamp and starting the timer when the lid is removed. Replace the lid to terminate the pulse and remove a 1 mL aliquot. We find that repeating the expose-and-collect procedure a total of six times is sufficient to provide a good mutagenic dose range for strains tested. To determine the kill rate at each mutagenic dose, remove 100 μL from each 1 mL portion and set up a series of 10-fold dilutions (10^{-1}–10^{-6}). From the 10^{-4}–10^{-6} dilutions, spread 100 μL of each onto individually labeled 100 mm YPD Petri dishes. These steps should also be performed in the darkroom with illumination provided only from the safety light. The YPD plates are wrapped in aluminum foil and placed into a 30 °C incubator for 48 h. To determine the kill rate for each dose, compare colony numbers to the nonmutagenized control. For this screen, we found that a 60–70% kill rate as ideal. Higher kill rates increase the frequency of nonsectoring isolates but also increase the number of undesirable secondary phenotypes including poor growth and sterility. The remaining undiluted mutagenized cell aliquots are transferred to separate, prelabeled culture tubes containing 10 mL YPD media and wrapped in aluminum foil. Place the culture tubes in a rotating roller drum for aeration at 30 °C for 18 h to stabilize mutations. Place each culture on ice to terminate growth (exposure to normal ambient light is fine at this point) and perform a dilution series. To determine the cell titre, spread 100 μL from each of the 10^{-4}–10^{-6} dilutions onto YPD plates and incubate at 30 °C for 2 days. In the meantime, store the mutagenized cells at 4 °C. No significant loss of viability was detected under these conditions. Once the cell titre has been determined, select the culture with the desirable kill rate and spread 300 colony-forming units per plate (50–100 plates). Either low adenine SC media or YPD media can be used as growth media. For YPD plates, we first prepare an autoclaved 5X YP stock (50 g peptone and 25 g yeast extract in 500 mL water). This is mixed with dextrose and agar and autoclaved again. The additional autoclave step of the yeast extract/peptone decreases the adenine concentration to enhance the red coloration of the sectoring phenotype. Extended autoclaving of the complete YPD media should be avoided because it causes media caramelization and deterioration of agar. The cells are incubated at 30 °C for 3 days to allow colonies to fully develop the red pigmentation. Growth at lower temperatures can also be performed if temperature sensitive mutants are desired. Using sterile flat toothpicks, nonsectoring colonies are picked and streaked on plates containing the same media for single colonies. False-positives are revealed at this step as sectoring colonies (approx. 50% of initial isolates) and the relative health of the remaining nonsectoring mutants can be assessed (colony sizes relative to DNY421 control) and noted.

To identify recessive mutants, mutant isolates are crossed with DNY420, the *MATα* counterpart strain to DNY421. After 5 h at 30 °C, diploid zygotes (trilobed cells) are picked using a yeast tetrad dissection microscope and grown on YPD plates. Diploid cells are grown on YPD plates for 2–3 days at 30 °C. Colony phenotypes are inspected by eye. Heterozygous diploids from recessive mutants give rise to sectoring or white colonies. Those arising from dominant mutants will form red, nonsectoring colonies. Complementation tests to group *per* mutants are not feasible because many display unlinked noncomplementation phenotypes when relying on the sectoring assay. Linkage analysis can be used instead but with the advent of extensive genome resources and rapid sequencing technologies, it can be more efficient to directly clone each mutant gene (see Section 3.3) and sequence mutant loci. In addition, genetic linkage analysis performed for the first 20 *per* mutant isolates demonstrated that the extent of the screen represented a large number of genes (Ng *et al.*, 2000).

Nevertheless, tetrad analysis should be performed to confirm that the mutant phenotype is caused by a single genetic locus (2:2 segregation, mutant:wild type).

2.2. Secondary screens of *per* mutants

Because the large number of pathways represented by *per* genes, it can be advantageous to perform functional profiling to categorize mutants. We previously devised a simple approach that provides a surprising amount of information. Because most *per* mutants are expected to disrupt protein biosynthesis and quality control in the ER, the assays focus on these functions. The first is a simple metabolic pulse-chase immunoprecipitation assay. Key to the experiment is the choice of proteins for analysis. We selected the vacuolar protease carboxypeptidase Y (CPY) and the GPI-linked plasma membrane protein Gas1p. Both CPY and Gas1p are glycoproteins, so the integrity of glycosylation functions (N-linked for CPY; N- and O-linked for Gas1p) can be assessed by increased gel mobility of glycosylation intermediates (Fig. 12.2A). Because glycosylation occurs in the ER, defects in translocation can be detected by the appearance of an aglycosylated form without intermediates (Hann and Walter, 1991). Golgi enzymes extend the glycans of Gas1p and CPY so ER-to-Golgi transport defects are also easily detected by gel mobility (Fig. 12.2A, Golgi Gas1p and p2 CPY). However, transport blocks could be the consequence of either general membrane trafficking defects or defects in ER protein maturation. GPI-anchorage defects can be deduced in those mutants having transport defects in Gas1p but not CPY. Of course, the simple pulse-chase experiment provides profiles that are consistent with particular defects. More direct, specialized experiments are needed to confirm the defects. Analysis of CPY and Gas1p does not address defects in protein quality control. For that, quantitative pulse-chase experiments are performed using HA-epitope tagged substrates that use different ERAD pathways. The model substrates CPY*, Sec61-2p, and Ste6-166p use the ERAD-L (luminal), ERAD-M (membrane), and ERAD-C (cytosol) pathways, respectively (Carvalho et al., 2006). Mutants defective in ERAD would display stabilization of one or more of these substrates.

2.3. Protocol for pulse-chase immunoprecipitation assay

This protocol can be used to monitor protein biogenesis such as Gas1p and CPY and the degradation of ERAD substrates such as CPY*. The procedure is slightly different where the pulse and chase times are longer to monitor ERAD substrates than protein biogenesis.

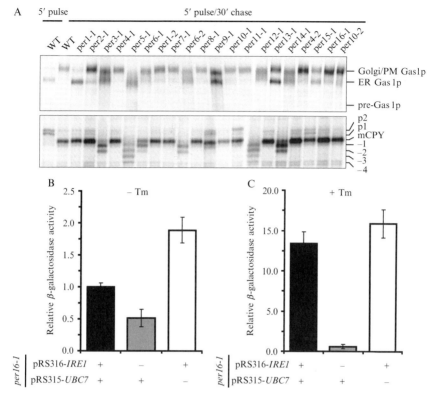

Figure 12.2 Secondary screening of synthetic lethal mutants with $\Delta ire1$. (A) Processing of Gas1p and CPY in *per* mutant strains by pulse-chase analysis. The various processed forms of the proteins are indicated as follow for Gas1p: PM, plasma membrane; pre-Gas1p; position of cytosolic nontranslocated Gas1p and for CPY: P1, ER pro-CPY; P2 Golgi pro-CPY; mCPY, vacuolar mature CPY; -1, -2, -3, and -4 represent underglycosylated CPY. Normalized β-galactosidase activity, in the absence (B) and the presence (C) of 2.5 μg/mL tunicamycin (Tm), are shown for the *per16-1* strain which contains a genomic copy of the *UPRE-LacZ* reporter gene. Values in (B) and (C) are normalized to *per16-1* (pRS315-*UBC7*, pRS316-*IRE1*) in the absence of tunicamycin. (A) Adapted from Figure 2, Ng *et al.* (2000). Error bars indicate SD from three measurements.

2.3.1. Pulse-chase

Mutant and control *per* strains are grown in SC media lacking uracil (to select for pDN366), methionine, and cysteine at 30 °C to early log phase (OD_{600} 0.35–0.5). Three OD_{600} cell equivalents per time point are harvested at $1000 \times g$ for 5 min in 50-mL conical centrifuge tubes. The cells are resuspended in a volume of 0.9 mL per time point of the same media (i.e., 2.7 mL for 3 time points) and preincubated at 30 °C for 30 min to allow recovery from the centrifugation. Pulse-labeling begins with the addition of

80 µCi of L-[^{35}S]-methionine/cysteine mix per time point. The cells are labeled for 5 or 10 min if monitoring protein biogenesis or ERAD, respectively. The chase is initiated by the addition of 100×methionine/cysteine chase solution to a final concentration of 2 mM methionine/cysteine. The mixture is rapidly and gently mixed by vortexing. A 0.9 mL aliquot is immediately transferred to a 2 mL screw cap microcentrifuge tube containing 100 µL of 100% trichloroacetic acid (TCA) that is kept on ice. Subsequent time points maybe aliquoted at 15 and 30 min chase for the biogenesis of Gas1p and CPY. However, aliquots are typically taken at 30 and 60 min to monitor the degradation of CPY★. After the chase, about 0.8 mL of 0.5 mm zirconia/silica beads (Biospec Products 11079105z) is added to each aliquoted sample. The cells are homogenized using an eight-position "Beadbeater" (Biospec Products, Bartlesville, OK) with two times 30 s agitation cycles separated by 5 min incubation on ice. Cell lysate is transferred to a new 1.5 mL screw cap tube. Beads are washed with 0.4 mL 10% TCA and the washing solution is transferred to the cell lysate. The suspension is centrifuged at 16,000×g for 10 min, 4 °C. The pellet is resuspended in 150 µL denaturing buffer (100 mM Tris base, 3% SDS, 1 mM PMSF) per time point and heated for 10 min at 100 °C. Insoluble debris is pelleted by centrifuging 10 min at 16,000×g, 4 °C. The supernatant is transferred to a new 1.5 mL screw cap tube. At this step, samples may be stored at −20 °C before proceeding to the immunoprecipitation assay.

2.3.2. Immunoprecipitation assay

The samples must be boiled and centrifuged if they were stored at −20 °C. To ensure equal amount of labeled proteins across all time points of the same strain, the radioactivity is counted by adding 5 µL of the detergent lysate to 4 mL liquid scintillation cocktail (GE Healthcare NBCS204) and vortexed. After 30 min incubation at room temperature, the radioactivity is measured in a multipurpose scintillation counter (Beckman Coulter LS 6500). The volume of detergent lysate used for immunoprecipitation assay is set at 50 µL for the time point giving the lowest count per minute (cpm) for the same strain. The remaining samples are normalized accordingly. The normalization step is only necessary to monitor the degradation of ERAD substrates. A volume of 50 µL of detergent lysate is used for samples monitoring protein biogenesis. The calculated or fixed volume of detergent lysate is added to 700 µL IPS II (13.3 mM Tris–HCl, pH 7.5, 150 mM NaCl, 1% Triton X-100, 0.02% NaN$_3$, 1 mM PMSF, 0.15% protease inhibitor cocktail (Sigma P8215)), and the appropriate antisera. Samples are gently mix by inversion and incubated on a nutator plate for 1 h at 4 °C. Samples are centrifuged for 20 min at 16,000×g, 4 °C, and the supernatant is transferred to a fresh tube containing 30 µL of 50% protein A-Sepharose® (from *Staphylococcus aureus*) beads suspension (Sigma P3391). Samples are gently mix by inversion and incubated on a nutator plate for 2 h at 4 °C.

The beads are washed three times with 1 mL IPS I (20 mM Tris–HCl, pH 7.5, 150 mM NaCl, 0.2% SDS, 1% Triton X-100, 0.02% NaN$_3$) and once with 1 mL of phosphate-buffered saline (PBS). Immunoprecipitated proteins are eluted from beads with 10 μL of protein loading buffer and boiled for 10 min. Samples are separated on 8% SDS-PAGE gel to resolved CPY, Gas1p, or CPY★. The resulting gel is coomassie blue stained (optional) and dried. Proteins of interest are visualized by autoradiography. A typical profile of *per* mutants to monitor biogenesis of Gas1p and CPY is shown in Fig. 12.2A. To assess the degradation rate of CPY★, the corresponding bands must be quantified and normalized to the zero time point for each strain using ImageQuant TL. To be conclusive, the degradation kinetic of ERAD substrates should be obtained from the average of at least three independent experiments including standard deviation.

3. Cloning and Sequencing *per* Genes

3.1. Cloning by complementation

Isolated *per* mutants can be screened for conditional phenotypes including growth sensitivity at 37 °C (ts: temperature sensitive) and 16 °C (cs: cold sensitive). Once linkage of conditional phenotypes to synthetic lethality has been confirmed by tetrad analysis, mutant genes can be cloned by complementation of ts or cs phenotypes, which is highly favorable because of its simplicity. However, many *per* mutants do not display conditional phenotypes. Their genes can be cloned also, but with somewhat greater effort. For nonconditional *per* mutants, we use one of the two methods. The first takes advantage of the *IRE1/ADE3* pDN336 plasmid bearing the *URA3* marker. As defined, *per* mutants cannot survive without the plasmid because of their dependence on an intact UPR. Reversing synthetic lethality by receiving a complementing clone from a yeast genomic plasmid library eliminates the requirement for pDN336. Such clones can be isolated by plating transformants on media containing 1 mg/L 5-fluoroorotic acid (5-FOA). Although 5-FOA is not inherently toxic to yeast, its metabolism by Ura3p produces a toxic compound, thus all *URA3*-containing cells perish. Thus, only cells receiving the complementing clone (or *IRE1*) will survive on 5-FOA plates. Alternatively, complementing clones could be identified by their reversal of the nonsectoring phenotype. Here, transformants are plated on low adenine SC dropout (to select for library plasmids) media at 300 colonies per plate. A total of 10,000–20,000 transformants should be screened for good coverage. Plates are incubated at 30 °C for 3 days and scored for sectoring colonies. For a plasmid library marked with *URA3*, pDN336 can be replaced with pDN388, which is similar except that it is marked with *LEU2*. For this option, 5-FOA counterselection cannot be used.

Although centromeric plasmid libraries are generally favored to identify mutant genes, multi-copy libraries are often useful for identifying dosage suppressors of mutant strains. Here, the method is the same except additional steps are required (see Section 3.3) to differentiate dosage suppressors from complementing clones.

3.2. Library transformation and plasmid isolation

A plasmid-based genomic library is transformed into the recipient *per* mutant at high efficiency. A culture of 50 mL is grown at the permissive temperature to an OD_{600} of 0.4 and cells are harvested at $1000 \times g$ for 10 min. The cells are washed with 20 mL sterile LiOAc mix (10 mM Tris–HCl, pH 7.4, 100 mM lithium acetate, 1 mM EDTA) and resuspended in 500 μL LiOAc mix. A volume of 5 μg DNA library with 100 μg denatured sonicated salmon sperm DNA are added to five tubes containing 700 μL sterile PEG mix (10 mM Tris–HCl, pH 7.4, 100 mM lithium acetate, 1 mM EDTA, 40% polyethylene glycol 3350 MW (Sigma P4338)). Then, 100 μL resuspended cells are added to each tubes and mix by vortexing. After 45 min incubation at the permissive temperature, 100 μL DMSO is added and the mixture is incubated at 42 °C for 22 min (temperature sensitive strains are heat shocked at 30 °C). Cells are harvested at $16,000 \times g$ for 15 s and resuspended in 1 mL of selective media. To evaluate the transformation efficiency, 1% of the transformed cells is plated on selective media and incubated for 2 days at the permissive temperature. In all, 15,000–25,000 transformants are generally sufficient for excess coverage of most genomic libraries. After a recovery period of 12 h at the permissive temperature, the remaining cells are plated onto selective media. The colonies are picked and streaked on YPD or low adenine SC plates to confirm the restoration of the sectoring phenotype. After dropping out the *IRE1/ADE3* centromeric plasmid (pDN336 or pDN388), the isolated white colonies are grown overnight in liquid media selecting only for library plasmids at the permissive temperature. The plasmids originated from the library plasmid are isolated from yeast and transformed in *Escherichia coli* cells (Hoffman, 2001).

To isolate plasmids originated from the library, cells are pelleted by centrifugation at $1000 \times g$ for 10 min and resuspended in 200 μL breaking buffer (10 mM Tris–HCl, pH 8.0, 100 mM NaCl, 1 mM EDTA, 2% Triton X-100, 1% SDS). Another 200 μL phenol:chloroform:isoamyl alcohol (25:24:1) and 200 μL 0.5 mm zirconia/silica beads (Biospec Products 11079105z) are added. The cells are homogenized using an eight-position "Beadbeater" (Biospec Products, Bartlesville, OK) with two times 30 s agitation cycles. The mixture is centrifuged at $16,000 \times g$ for 5 min. From the aqueous layer, 2 μL is transformed to electrocompetent *E. coli* cells. Then, plasmids are isolated from the transformants.

3.3. Identification of mutant genes

Following amplification and preparation of a putative complementing clone in *E. coli*, it is retransformed into its respective *per* mutant for confirmation. True complementing or dosage suppressor clones will recapitulate restoration of the sectoring phenotype. Next, the genomic DNA insert is determined. For this, only flanking and peripheral sequences need to be sequenced. Primers are designed approximately 100 bp upstream on both sides of the insert. For plasmid libraries based on YCp50 (*URA3, CEN/ARS*) and YEp13 (*LEU2, 2μ* replicon), genomic sequences are inserted in the unique *BamHI* site and using the primer sequences 5'-CGCTACTTG-GAGCCACTATCGAC-3'and 5'-ATCGGTGATGTCGGCGATAT-3'. Once insert peripheral sequences are determined (50 bp from each side are sufficient), they are used to query the yeast genome database to obtain their genome coordinates (www.yeastgenome.org). With this information, the sequence of the entire insert can be obtained.

Often, clones obtained from genomic libraries contain multiple genes. To identify the complementing gene, sometimes the isolation of multiple plasmid clones can narrow down the candidates. If all the clones are from the same region of DNA, sequences common to all clones is likely to contain the complementing gene. With some luck, a single gene will emerge. If multiple candidates remain, a number of approaches can be employed. The classical approach is to use conventional molecular biology techniques to produce an insert deletion series. These plasmids are transformed in the corresponding *per* mutant so that only those constructs carrying the complementing gene will restore the sectoring phenotype. Although rather tedious, this approach is recommended for multi-copy libraries to maintain copy number for dosage suppressors. For centromeric libraries, we found it effective to amplify each candidate gene from the corresponding *per* mutant genomic DNA using the polymerase chain reaction (PCR). Each PCR product is sequenced directly. Almost invariably, only a single gene will emerge as mutated. We favor this method because it is rapid and the sequence of the mutant allele is immediately obtained. This approach can also be used for multi-copy libraries. All wild-type sequences indicate that the clone is a dosage suppressor and a single mutated gene suggests the plasmid carried that as the complementing clone. Ultimate confirmation is determined by converting the mutant locus to wild type through plasmid integration/excision (Guthrie and Fink, 1991).

4. Monitoring the UPR Activity

4.1. The β-galactosidase assay

The intracellular signal transduction pathway UPR monitors and maintains ER homeostasis. Upon ER stress, Ire1p activates the pathway by splicing an inhibitory intron from a mRNA encoding a transcription factor called Hac1p. Newly synthesized Hac1p then induced the transcription of UPR target genes. A β-galactosidase activity based assay can be used to monitor the constitutive UPR activation in cells. In the *per* strain, a copy of a *lacZ* reporter gene fused to a minimal UPR reporter was integrated in the genome to ease monitoring the UPR activity (Cox *et al.*, 1993). Alternatively, a centromeric plasmid expressing a copy of a *lacZ* reporter gene fused to a minimal UPR reporter can be used.

Mutants in the presence and absence of the complement clones in addition of the *per* strain DNY420 as control are grown in 5 mL of selective SC dropout media at 30 °C for about 8 h. The cultures are diluted to an OD_{600} of 0.00025 in fresh 50 mL of SC dropout media and incubated overnight at 30 °C. The cells must never grow beyond absorbance of 0.5 at 600 nm. Cells must be diluted down to OD_{600} of 0.01 and grow again for 7–16 h if the OD_{600}/mL exceeds 0.5. This is an important point to keep in mind since the UPR will be activated in cells under glucose deprivation conditions. Each culture is separated equally into two flasks 1 h before harvest and a final concentration of 2.5 μg/mL tunicamycin (Sigma T7765) is added to one of the two flasks. Tunicamycin strongly induces the UPR by blocking protein *N*-glycosylation (Cox and Walter, 1996). Incubation at 30 °C is resumed for 1 h. The cells are harvested at $1000 \times g$ for 10 min and resuspended in 1 mL LacZ buffer (125 mM sodium phosphate, pH 7, 10 mM KCl, 1 mM MgSO$_4$, 50 mM β-mercaptoethanol). The cells are transferred to a 1.5 mL microfuge tube and centrifuged at $1000 \times g$ for 5 min. The cell pellet is resuspended in 75 μL LacZ buffer. An aliquot of 25 μL cells resuspension is transferred into 975 μL distilled water for absorbance measurement at 600 nm. Monitoring cell density from this step is essential to normalize the β-galactosidase activity. To the remaining 50 μL cells, 50 μL CHCl$_3$, and 20 μL 0.1% SDS are added and vortexed vigorously for 20 s to break the cells. The tubes are equilibrated at 30 °C for 5 min. To start the reaction, 700 μL of 2 mg/mL ONPG (2-nitrophenyl β-D-galactopyranoside; Sigma N1127) in LacZ buffer preincubated at 30 °C is added. Then, the reaction is quenched with 500 μL of 1 M Na$_2$CO$_3$ when sufficient yellow color has developed without exceeding a 10 min reaction. The exact incubation time after adding ONPG should be noted to calculate the β-galactosidase activity. The mixture is centrifuged at $13,000 \times g$ for

1 min to remove cell debris and chloroform. The absorbance is measured at 420 and 550 nm. The β-galactosidase activity is calculated using Eq. (12.1):

$$\text{Miller units} = 1000 \times (\text{OD}_{420} - 1.75 \times \text{OD}_{550})/(t \times (V_A/V_R) \times \text{OD}_{600}) \quad (12.1)$$

where OD_{420} is the absorbance of *o*-nitrophenol which is the yellow colored product hydrolyzed by β-galactosidase from ONPG, OD_{550} is the scatter from cell debris which is multiplied by 1.75 to correct the scatter observed at 420 nm, *t* is the reaction time in minutes, V_A is the volume of cells of the assay (50 μL), V_R is the volume of cells used to read the absorbance at 600 nm (25 μL), and OD_{600} indicates cell density. The values are then normalized to "wild-type" strain. As shown in Fig. 12.2B, the UPR is roughly induced by twofold in the *per16-1* mutant (DNY481) in the absence of *UBC7* compared to its presence (pDN444). The addition of tunicamycin to *per16-1* strain −/+ UBC7 both strongly induce the UPR (Fig. 12.2C). As a negative control, the UPR is not induced in *per16-1* strain lacking *IRE1* (Fig. 12.2B and C).

4.2. Alternative procedures to monitor the UPR activity

The splicing of *HAC1* mRNA, by activated Ire1p, can be visualized to monitor the UPR activity. The classical technique northern blot allows the detection of unspliced and spliced *HAC1* mRNA. The DNA probe should be a complementary sequence corresponding to the *HAC1* mRNA region that is not spliced. The *HAC1* mRNA generate 993 and 741 bp bands for the unprocessed and spliced forms, respectively. Alternatively, the splicing of *HAC1* mRNA may be monitored by reverse transcriptase PCR (RT-PRC) or quantitative real-time PCR (qPCR). Finally, it is possible to visualize the UPR activity in live cells under fluorescence microscopy by expressing a GFP reporter gene controlled by a minimal UPR reporter (Jonikas *et al.*, 2009).

5. SGA Database

Genetic interaction data provide invaluable information on gene functions. SGA was recently carried out on the whole genome of *S. cerevisiae* by Costanzo and colleagues where they inspected 5.4 million gene–gene pairs for synthetic genetic interactions (Costanzo *et al.*, 2010). Interestingly, about 75% of all genes in yeast generated quantitative genetic interaction. Among the genes examined, the transcription factor *HAC1*, which acts downstream of the ER stress sensor *IRE1*, was screened for synthetic lethal interactions. From the stringent cutoff dataset of *HAC1*, many genes were found to have synthetic genetic interactions falling in diverse functions for instance protein translocation and folding, glycosylation and mannosylation, protein degradation, protein processing, protein trafficking, protein retrieval and retention, lipid biosynthesis, and other functions. The great advantage of this approach, in comparison to the *per* screen, is that the synthetic lethal genes interaction with *HAC1* are known. However, SGA is mostly limited to nonessential genes although conditional or hypomorphic alleles of essential genes may be screened if they have been generated prior to SGA. Overall, SGA gives the opportunity to explore the function and role of previously unidentified genes that take part in ER stress.

6. Media Recipes

YPD: 20 g peptone, 10 g yeast extract, and 20 g dextrose in 1 L water; autoclave.

Synthetic complete (SC): For 1 L mix, 100 mL of 10 × amino acid mix (1 L stock: 0.6 g adenine, 0.3 g uracil, 0.8 g L-tryptophan, 0.6 g L-histidine, 0.2 g L-arginine, 0.2 g L-methionine, 0.3 g L-tyrosine, 0.8 g L-leucine, 0.3 g L-isoleucine, 0.3 g L-lysine, 0.5 g L-phenylalanine, 1.0 g L-glutamic acid, 1.0 g L-aspartic acid, 1.5 g valine, 2.0 g threonine, and 2.0 g serine; sterilize by filtration), 100 mL of 10×yeast nitrogen base without amino acids (1 L stock: 70 g in water; autoclave), 40 mL 50% sterile dextrose, and 760 mL sterile water. For selective media, leave out the appropriate component in the 10× amino acid mix. Ten times low adenine stock is prepared by reducing adenine to 60 mg/L. For solid media, a final concentration of 2% agar is added to the media.

Liquid sporulation media: 3 g potassium acetate, 0.2 g raffinose in 1 L water; sterilize by filtration.

7. Closing Remarks

The UPR is a critical regulatory circuit involved in general cellular homeostasis, developmental regimes, acute stress, and disease pathology. Taken together, the multitude of studies suggests the UPR as a key pathway that underlies the robustness of living systems. How the output of the UPR performs these roles remains incompletely understood. This is widely acknowledged as nonphysiological, partly due to earlier studies relying on chemical inducers of ER stress, mainly tunicamycin and DTT. Thus, recently developed approaches using more physiological conditions to evoke ER disequilibrium will provide richer biological insight into this fascinating system. However, such advances would still merely scratch the surface. In budding yeast, fewer than half the genes with changed expression under ER stress can be attributed to the UPR. Thus, how other regulatory systems synergize with the UPR to adapt to internal and environmental changes remains an unanswered challenge for years to come.

ACKNOWLEDGMENT

This work was supported by funds from the Temasek Trust and by a grant from the Singapore Millennium Foundation to G.T. (postdoctoral fellowship).

REFERENCES

Barbour, L., and Xiao, W. (2006). Synthetic lethal screen. *Methods Mol. Biol.* **313,** 161–169.
Bernales, S., McDonald, K. L., and Walter, P. (2006). Autophagy counterbalances endoplasmic reticulum expansion during the unfolded protein response. *PLoS Biol.* **4,** e423.
Carvalho, P., Goder, V., and Rapoport, T. A. (2006). Distinct ubiquitin–ligase complexes define convergent pathways for the degradation of ER proteins. *Cell* **126,** 361–373.
Costanzo, M., Baryshnikova, A., Bellay, J., Kim, Y., Spear, E. D., Sevier, C. S., Ding, H., Koh, J. L., Toufighi, K., Mostafavi, S., Prinz, J., St Onge, R. P., *et al.* (2010). The genetic landscape of a cell. *Science* **327,** 425–431.
Cox, J. S., and Walter, P. (1996). A novel mechanism for regulating activity of a transcription factor that controls the unfolded protein response. *Cell* **87,** 391–404.
Cox, J. S., Shamu, C. E., and Walter, P. (1993). Transcriptional induction of genes encoding endoplasmic reticulum resident proteins requires a transmembrane protein kinase. *Cell* **73,** 1197–1206.
Cox, J. S., Chapman, R. E., and Walter, P. (1997). The unfolded protein response coordinates the production of endoplasmic reticulum protein and endoplasmic reticulum membrane. *Mol. Biol. Cell* **8,** 1805–1814.
Fay, D. S., Keenan, S., and Han, M. (2002). fzr-1 and lin-35/Rb function redundantly to control cell proliferation in *C. elegans* as revealed by a nonbiased synthetic screen. *Genes Dev.* **16,** 503–517.
Giaever, G., Chu, A. M., Ni, L., Connelly, C., Riles, L., Veronneau, S., Dow, S., Lucau-Danila, A., Anderson, K., Andre, B., Arkin, A. P., Astromoff, A., *et al.* (2002). Functional profiling of the *Saccharomyces cerevisiae* genome. *Nature* **418,** 387–391.
Guthrie, C., and Fink, G. R. (eds.), (1991). Guide to Yeast Genetics and Molecular Biology, Academic press, San Diego.
Hann, B. C., and Walter, P. (1991). The signal recognition particle in *S. cerevisiae*. *Cell* **67,** 131–144.
Hoffman, C. S. (2001). Preparation of Yeast DNA. *Current Protocols in Molecular Biology* 13.11.1-13.11.4.
Jonikas, M. C., Collins, S. R., Denic, V., Oh, E., Quan, E. M., Schmid, V., Weibezahn, J., Schwappach, B., Walter, P., Weissman, J. S., and Schuldiner, M. (2009). Comprehensive characterization of genes required for protein folding in the endoplasmic reticulum. *Science* **323,** 1693–1697.
Lucchesi, J. C. (1968). Synthetic lethality and semi-lethality among functionally related mutants of *Drosophila melanfgaster*. *Genetics* **59,** 37–44.

Mori, K., Ma, W., Gething, M. J., and Sambrook, J. (1993). A transmembrane protein with a cdc2+/CDC28-related kinase activity is required for signaling from the ER to the nucleus. *Cell* **74,** 743–756.

Nakagawa, T., Zhu, H., Morishima, N., Li, E., Xu, J., Yankner, B. A., and Yuan, J. (2000). Caspase-12 mediates endoplasmic-reticulum-specific apoptosis and cytotoxicity by amyloid-beta. *Nature* **403,** 98–103.

Ng, D. T., Spear, E. D., and Walter, P. (2000). The unfolded protein response regulates multiple aspects of secretory and membrane protein biogenesis and endoplasmic reticulum quality control. *J. Cell. Biol.* **150,** 77–88.

Ron, D., and Walter, P. (2007). Signal integration in the endoplasmic reticulum unfolded protein response. *Nat. Rev. Mol. Cell Biol.* **8,** 519–529.

Schuldiner, M., Collins, S. R., Thompson, N. J., Denic, V., Bhamidipati, A., Punna, T., Ihmels, J., Andrews, B., Boone, C., Greenblatt, J. F., Weissman, J. S., and Krogan, N. J. (2005). Exploration of the function and organization of the yeast early secretory pathway through an epistatic miniarray profile. *Cell* **123,** 507–519.

Sidrauski, C., and Walter, P. (1997). The transmembrane kinase Ire1p is a site-specific endonuclease that initiates mRNA splicing in the unfolded protein response. *Cell* **90,** 1031–1039.

St Onge, R. P., Mani, R., Oh, J., Proctor, M., Fung, E., Davis, R. W., Nislow, C., Roth, F. P., and Giaever, G. (2007). Systematic pathway analysis using high-resolution fitness profiling of combinatorial gene deletions. *Nat. Genet.* **39,** 199–206.

Travers, K. J., Patil, C. K., Wodicka, L., Lockhart, D. J., Weissman, J. S., and Walter, P. (2000). Functional and genomic analyses reveal an essential coordination between the unfolded protein response and ER-associated degradation. *Cell* **101,** 249–258.

Yoshida, H. (2007). ER stress and diseases. *FEBS J.* **274,** 630–658.

Zimmermann, F. K. (1975). Procedures used in the induction of mitotic recombination and mutation in the yeast *Saccharomyces cerevisiae*. *Mutat. Res.* **31,** 71–86.

CHAPTER THIRTEEN

Signaling Pathways of Proteostasis Network Unrevealed by Proteomic Approaches on the Understanding of Misfolded Protein Rescue

Patrícia Gomes-Alves, Sofia Neves, *and* Deborah Penque

Contents

1. Introduction	218
2. 2DE-Based Proteomics Approach Analysis to Study Protein Expression Profile	219
3. Experimental Design	220
3.1. Cell lines	220
3.2. Low-temperature treatment and *in vivo* labeling	220
4. 2D Gel Electrophoresis	221
4.1. Sample preparation	221
4.2. 2D-PAGE	221
4.3. Coomassie and fluorography	222
4.4. 2D maps image analysis	223
4.5. Statistical analysis	223
5. Protein Identification by MS	223
5.1. Protein gel cut and digestion	224
5.2. MALDI analysis—Equipment and specifications	224
6. Biochemical Validation	225
7. Data Mining and Publication	226
7.1. Bioinformatics analysis	226
7.2. Clustering analysis	228
7.3. MIAPE: The Minimum Information About a Proteomics Experiment	231
Acknowledgments	231
References	231

Laboratório de Proteómica, Departamento de Genética, Instituto Nacional de Saúde Dr Ricardo Jorge (INSA, I.P.), Av. Padre Cruz, Lisboa, Portugal

Abstract

Attempts to promote normal processing and function of F508del-CFTR, the most common mutant in cystic fibrosis (CF), have been described as potential therapeutic strategies in the management of this disease.

Here we described the proteomic approaches, namely two-dimensional electrophoresis (2DE), mass spectrometry (MS), and bioinformatics tools used in our recent studies to gain insight into the proteins potentially involved in low-temperature or mutagenic treatment-induced rescue process of F508del-CFTR. The proteins identified are part of the proteostasis network, such as the unfolded protein response (UPR) signaling pathways that may regulate the processing of CF transmembrane conductance regulator (CFTR) through the folding and trafficking progression. The complete characterization of these signaling pathways and their regulators in CF will certainly contribute to the development of novel therapeutic strategies against CF.

1. Introduction

The unfolded protein response (UPR) is one of the signaling pathways composing the proteostasis network, which regulates and maintains the protein folding and function in the face of many cellular challenges during cell lifetime (Powers *et al.*, 2009). The UPR is a complex molecular cascade activated in response to cellular stressors that induce the endoplasmic reticulum (ER) overloading with unfolded proteins. The UPR is mainly characterized by upregulation of ER chaperones to improve the folding capacity of the ER and reestablish ER homeostasis, by facilitating proteins correct folding or assembly and preventing protein aggregation in the ER lumen and consequently improving cell survival (Citterio *et al.*, 2008; Kelsen *et al.*, 2008). Activation of ER-associated degradation (ERAD) pathways, another signaling pathways composing proteostasis network responsible for the degradation of the excess of misfolded proteins in the ER (van Anken and Braakman, 2005), is also a charactcristic feature of UPR.

Dysregulation of cellular proteostasis by a constant flux of misfolded and mistrafficking proteins contributes to a broad range of human diseases such as cystic fibrosis (CF).

CF is mostly caused by the inherited F508del mutation that leads to CF transmembrane conductance regulator (CFTR) misfolding and excessive degradation, ultimately resulting in reduced availability of functional protein.

Under F508del-CFTR overexpression, cells induce ER stress and/or the UPR signaling pathways as a response to a loss of proteostatic control (Bartoszewski *et al.*, 2008; Gomes-Alves and Penque, 2010; Gomes-Alves *et al.*, 2009, 2010a; Kerbiriou *et al.*, 2007). We have shown by proteomics that treatments aimed at rescuing F508del-CFTR, such as low-temperature incubation or "RXR" mutagenic repair (by inactivation of four RXR motifs of F508del-CFTR-4RK mutant), modulate/readjust the proteome components of pathways composing proteostasis network, including the UPR and ERAD (Gomes-Alves and Penque, 2010; Gomes-Alves *et al.*, 2009, 2010a).

Most therapies designed to rescue the trafficking defect of F508del-CFTR cause the UPR/ER stress induction (Gomes-Alves *et al.*, 2010b; Loo *et al.*, 2005; Singh *et al.*, 2008) involving, however, a larger number or higher expression of proteins than those induced by the misfolded protein itself (Gomes-Alves and Penque, 2010; Gomes-Alves *et al.*, 2010b; Loo *et al.*, 2005; Singh *et al.*, 2008). This treatment-induced UPR/ER stress stimulus seems also to readjust some proteostasis imbalance associated with F508del-CFTR expression, which probably endows the cell with the ability to rescue at least partially the mutant CFTR.

We showed that overexpression in BHK cell lines of "repaired" form of F508del-CFTR, F508del-4RK-CFTR, or low-temperature treatment on BHK cell lines overexpressing the mutant F508del-CFTR induce upregulation of GRP78/BiP (the UPR hallmarker), several ER stress (e.g., GRP94 and GRP75), and several foldases (e.g., PDI) and readjust (to wt type level) some proteome imbalance associated with the proteosome degradation (e.g., Psme2 and Cops5) and folding and transport (e.g., RACK1) of CFTR (Gomes-Alves and Penque, 2010; Gomes-Alves *et al.*, 2009, 2010a).

We are convinced that the complete characterization of the proteostasis signaling pathways and their regulators in CF by high-throughput proteomics will confidently culminate in the development of novel therapeutic strategies against CF.

2. 2DE-BASED PROTEOMICS APPROACH ANALYSIS TO STUDY PROTEIN EXPRESSION PROFILE

Proteome profiling of cells/tissues under different conditions (e.g., health vs. disease, treated vs. nontreated) by two-dimensional gel electrophoresis (2DE) and mass spectrometry (MS)-based proteomic approach has contributed to the elucidation of the basic cellular mechanisms of disease by discovering candidate biomarkers and disease targets for new drug development (Penque, 2009).

The main advantage of 2DE technology is its capacity to provide a global view of a sample proteome at a given time by resolving hundreds to thousands of proteins simultaneously on a single gel. In spite of the inability of 2DE to resolve all proteins present in a sample (e.g., highly hydrophobic proteins, low abundance proteins, extremely basic or acidic pI values, molecular size out of gel limits) or the growing developments in gel-free MS-based approaches, 2DE in combination with MS still remains the most popular technology in proteomics (Penque, 2009).

In the next sections we provide detailed information on the optimal conditions and procedures for 2DE/MS-based proteomics analysis used in our previous publications (Gomes-Alves and Penque, 2010; Gomes-Alves et al., 2009), as an approach able to uncover cellular mechanisms involved on treatments-induced mistrafficking protein rescue.

3. Experimental Design

3.1. Cell lines

The importance of using recombinant CFTR-expressing cell lines as heterologous model systems is recognized from previous studies (Chang et al., 1999; Mendes et al., 2003; Roxo-Rosa et al., 2006), since endogenous CFTR-expressing models are limited and the expression level is usually too low to study CFTR biogenesis, trafficking and function.

Adherent cell lines, such as BHK cells, not expressing (BHK-L) or stably expressing wild type (wt)-, F508del-CFTR (mutant), or F508del/4RK-CFTR ("repaired") are cultivated in DMEM/F12 media containing 5% fetal calf serum (Gibco, Invitrogen) and 500 μM methotrexate (MTX—Teva Pharma) in culture flasks (Nunc Prand Products) until 80% of confluence (Roxo-Rosa et al., 2006). Cells are then trypsinized and washed once in PBS (Gibco, Invitrogen) and pellets with 5×10^6 cells should be quickly frozen on dry ice and stored at $-80\ °C$ until analysis. All BHK cell lines were provided by Roxo-Rosa et al. (2006).

3.2. Low-temperature treatment and *in vivo* labeling

Low temperature (26–30 °C) apparently stabilizes the mutant protein during folding in the ER, allowing some CFTR molecules to escape ER quality control and traffic through the Golgi apparatus to the plasma membrane (Denning et al., 1992; Mendes et al., 2003; Sharma et al., 2001). Alternatively, Rennolds et al. (2008) have suggested that at low-temperature, F508del-CFTR may use a nonconventional trafficking pathway and bypasses the Golgi apparatus where CFTR becomes Endo-H resistant.

To investigate the effect of low temperature in the proteostasis environment involved in F508del-CFTR rescue, 6×10^5 BHK cells stably expressing wild type (wt)- or F508del-CFTR should be seeded per 35 mm culture dishes (Nunc Prand Products), grown to about 80% of confluence and then incubated at 37 °C (control) or 26 °C for 24 or 48 h in the same media without MTX. Cells are incubated 30 min in methionine-free a-minimal essential medium (MEM, Gibco, Invitrogen) prior to the metabolic labeling in the same medium with 140 mCi/ml of [^{35}S]-methionine (MP Biomedicals), in the last 3 h of incubation at 37 or 26 °C, respectively.

The 3-h cell metabolic labeling allows the visualization of proteins that are synthesized and further processed during this time of incubation with the [^{35}S]-methionine (*de novo* synthesis—visualization of treatment effects on protein synthesis), while Coomassie staining allows the visualization of all proteins at steady state present in the sample.

4. 2D Gel Electrophoresis

4.1. Sample preparation

Pellets of 5×10^6 (or 0.8×10^6 for radiolabeled cells) cells are lysed directly in 450 ml of lysis buffer [7 M urea (Sigma-Aldrich), 2 M thiourea (Sigma-Aldrich), 4% (w/v) CHAPS (Sigma-Aldrich), 60 mM DTE (Sigma-Aldrich), 0.75% (v/v) ampholine 3.5–10.0 (GE Healthcare), 0.25% (v/v) ampholine 4.0–6.0 (GE Healthcare)], and 0.12 μl of protease inhibitor mixture (leupectin 1 mg/ml; aprotinin 2 mg/ml; pefabloc 50 mg/ml; E64 3.5 mg/ml; benzamidin 50 mg/ml). After 30 min at 37 °C rocking, samples are sonicated three times 15 s in a waterbath sonifier (Selecta) and then clarified by centrifugation at 14,000×g for 30 min. Supernatants are collected for a new microtube.

4.2. 2D-PAGE

In the first dimension of 2DE (isoelectric focusing, IEF), proteins are separated on the basis of their isoelectric point in an immobilized pH gradient gel (IPG) and in the second-dimension (SDS-PAGE) proteins undergo an additional separation according to their molecular weight under denaturing conditions. There are different 2DE methodologies or possibilities and all require the use of biological and technical replicates to produce meaningful results (Penque, 2009). A guidelines called MIAPE (Minimum Information About Proteomics Experiments) gel electrophoresis that act as checklists of the essential information that should be reported about particular 2DE experimental techniques has been also recently created (http://psidev.sourceforge.net) (see below).

In our 2DE procedure, total extracted protein samples (supernatants) are loaded onto 18 cm Immobiline DryStrips (GE Healthcare) with a nonlinear wide range pH gradient (pH 3–10) with a top overlay of 3 ml of mineral oil

(Amershan Bioscience) to prevent evaporation and urea crystallization during next steps of IEF on an IPGphor IEF system (Amershan Bioscience). For IEF, an active rehydration at 50 V for at least 12 h at 20 °C is performed before the focusing program of a total of about 60 kVh as follows: 200 V for 1 h, 500 V for 1 h, 1000 V for 2 h, 2000 V for 2 h, 3000 V for 2 h, increased gradient voltage to 8000 V for 1 h, and 8000 V for 5 h. After IEF is run, IPG strips were equilibrated in IEF equilibration solution (50 mM Tris–HCl, pH 8.8 (Merck), 6 M urea (Sigma-Aldrich), 30% (v/v) glycerol (Sigma-Aldrich), 2% (w/v) sodium dodecyl sulfate (SDS) (GE Healthcare), and traces of bromophenol blue (Merck) with 2% (w/v) DTE (Sigma-Aldrich) for 15 min and with 2.8% (w/v) iodoacetamide (Sigma-Aldrich) for another 15 min, to, respectively, reduce and alkylate proteins before application onto the top of a SDS-PAGE for molecular weight separation (second dimension), constituted by 8–16% (w/v) gradient polyacrylamide and run overnight at 3 W/gel (Ettan DALTwelve System, GE Healthcare). The strip gels are "glued" onto the top of a SDS-PAGE with boiled 0.5% (w/v) agarose in running buffer. If the second dimension is not carried out immediately after IEF, strips can be stored for prolonged times at -80 °C.

4.3. Coomassie and fluorography

Proteins once displayed on the 2D gel must be visualized in a valuable form to enable their accurate qualitative/quantitative analysis by dedicated software. The critical factors of staining methods are their sensitivity, reproducibility and the linear range of detection. The most used standard procedures of protein staining include Coomassie brilliant blue (CBB), and variants such as colloidal CBB (C-CBB) or blue silver (C-CBB modified), and MS-compatible silver stain (Gorg et al., 2004). Other choices include DIGE (difference gel electrophoresis), SYPRO® Ruby (fluorescent dyes), or radiolabeling (e.g., ^{35}S-methionine or ^{32}P-phosphorous) (Minden, 2007; Roxo-Rosa et al., 2004).

As referred before in Section 3.2, the proteins labeled *in vivo* with [^{35}S]-methionine are visualized by autoradiography as follows: the gel is fixed in 30% (v/v) methanol and 10% (v/v) acetic acid for 30 min, washed four times 15 min in water, and treated for 1 h with 1 M Na salicylate, pH 7. The gels are then dried in a vacuum system for 2 h at 80 °C and film exposed (Fuji Film) with intensifying screen. "Cold" proteins (without radiolabeling) are visualized by Coomassie "blue silver" staining exactly as described (Candiano et al., 2004).

Metabolic labeling is much more sensitive [<0.1–0.01 ng (Penque, 2009)] than the "blue silver" Coomassie (detection limit 1–100 ng/spot) (Candiano et al., 2004). Moreover, [^{35}S]-methionine labeling detects the presence of a number of proteins that are undetectable by silver staining or Western blotting (Chen and Casadevall, 1999). The digitalized images from both Coomassie stained and fluorographic 2DE-maps are obtained using a proper scanner (obtained image quality attributes must be compatible with the requests of the analysis software), as for example ImageScanner (GE Healthcare).

4.4. 2D maps image analysis

Subsequent analysis of the spot pattern on a gel by specialized software enables multiple-gel comparison, whereby quantitative and qualitative changes are detected by comparison with control gels. The critical parameter in computer-based image analysis is the quality of the images. This requires high quality and reproducible 2D gels and also high quality acquisition of the image with image capturing devices, as mentioned above. Image analysis software is designed for automated spot detection, matching, normalization and quantification. However corrections by user intervention are usually necessary.

In our studies the dedicated software used for gel image analysis was Progenesis PG200v2006-Nonlinear. An average gel image is created for each group of gels and a maximum of absences allowed must be defined (according to the analysis objectives and depending on the gels replica number). In order to standardize the intensities of the labeling among all spots present in the several 2DE-maps, analysis is carried out by taking into account the normalized volume of spots (or %vol., i.e., the volume of each spot over the volume of all spots in the gel).

4.5. Statistical analysis

The difference in expression levels between groups (control, disease, and treatment) for a given protein is statistically assessed by using the ANOVA test, for n observations, where n is the number conditions (groups) analyzed. Differences over 1.5-fold are considered statistically significant when $p < 0.05$. Proteins of interest are then selected and excised from gels for subsequently MS analysis (see below).

5. PROTEIN IDENTIFICATION BY MS

Two specific types of MS analysis can be used for protein identification: the peptide-mass fingerprint (PMF), which involves determination of the masses of the peptides in the digest and is normally performed in a MALDI-TOF instrument, and the amino acid sequencing of the peptides (MS/MS), normally obtained by ESI tandem MS or MALDI-TOF/TOF (Penque, 2009).

MS or MS/MS spectra data are normally used to search a predicted mass map or protein sequence within a database to identify the protein of interest using software such as Sequest (Thermo) or Mascot (MatrixScience) (Aebersold and Mann, 2003; Steen and Mann, 2004). Using a database search, candidate proteins are ranked from a list of most closely matched candidates using various scoring algorithms.

5.1. Protein gel cut and digestion

After spot excision from gels, protein spots are distained as follows: wash spots with type I (Millipore) water for 20 min and then incubate with 50% (v/v) acetonitrile (ACN) for 30 min, at 37 °C with shaking. After distaining, spots are dehydrated by incubation with ACN first for 5 min and further 30 min with fresh ACN, at RT, with shaking. The gel pieces are lyophilized by a Speedvac evaporator (Savant, Farmingdale, USA) for 30–60 min, followed by trypsin digestion. For trypsin digestion spots are rehydrated with digestion buffer containing 6.7 ng/l of trypsin (Promega, Madison, WI, USA) in 50 mM NH_4HCO_3 and incubated for 30 min at RT. Digestion buffer is removed and 50 mM NH_4HCO_3 is added to the spots, which are incubated overnight (12–16 h) at 37 °C.

After overnight digestion with trypsin, supernatants are collected for new tubes and extraction of tryptic peptides is performed by successive incubation of the gel piece with 1% (v/v) TFA/50% (v/v) ACN and with 100% (v/v) ACN. For each step samples are sonified for 10 min in a waterbath sonifier (Selecta). The respective supernatants are pooled and dehydrated with a speed vacuum rotor (Savant). The pellet is resuspended in 5 μl of 0.1% (v/v) TFA/50% (v/v) ACN and finally sonified for 15 min in the waterbath sonifier. The sample (0.5 μl) is spotted on the MALDI plate and 0.5 μl of α-CHCA are added to the sample, the mixture is allowed to dry at RT (Bensalem et al., 2007).

5.2. MALDI analysis—Equipment and specifications

Peptides are analyzed on an Applied Biosystems 4700 ProteomicsAnalyzer with TOF/TOF ion optics. Data are acquired in positive MS reflector mode with six spots of protein standards (Calibration Mixture 2, Applied Biosystem) used for calibration (4000 Series Explorer Software v3.0 RC1). Three or five S/N best precursors of each spectrum can be selected for MS/MS analysis. The interpretation of the MS+MS/MS data is carried out by the GPS Explorer software (Version 3.5, Applied Biosystems) and a local copy of the MASCOT search engine (Version 2.0). Monoisotopic peptide-mass values are considered, a MS mass tolerance is set at 50 ppm and a MS/MS fragment tolerance set at 0.25 Da. Identification restrictions can be applied. Trypsin is normally the digestion enzyme used and one missed cleavage site is usually allowed. Cys carbamidomethylation and Met oxidation are set as fixed and variable amino acid modifications, respectively. A taxonomic restriction to specific species protein sequences may be included. For MS all peaks with S/N greater than 5 and for MS/MS all peaks with S/N greater than 3 are searched against the database (e.g., Swissprot, NCBInr, etc.). The criteria to accept the identification are significant homology scores achieved in Mascot ($p < 0.05$), but depending on the goals of the study or on the standards of the journal where the author wishes to publish, more stringent criteria may be used (e.g., minimum of matched peptides, minimum of sequence coverage, etc.).

6. BIOCHEMICAL VALIDATION

The generated data from a quantitative proteomics analysis has required validation/verification that can be performed by traditional biochemical assays (e.g., Western blotting, immunocyto/hystochemistry, or other) or more recently by MS-based approaches (e.g., selected/multiple reaction monitoring (SRM/MRM)-based workflows) (Kitteringham et al., 2009; Picotti et al., 2010)

In our study, Western blot assay using antibodies specific to relevant proteins is used to validate/verify the proteomic data obtained. Briefly, protein extracts from cell lines are quantified (Bradford; Bio-Rad protein assay) and 15 μg of each cell sample (triplicates) are separated on NuPAGE® Novex 4-12% (w/v) Bis–Tris gel 1.0 mm, 15 well (Invitrogen), transferred to nitrocellulose membranes (Schleicher & Schuell). After transference, the membranes are blocked with PBS plus 5% (w/v) fat free milk for 2 h and probed in the same solution for 2 h at RT with: 1:300 rabbit anti-GRP78/BiP Ab (Sigma-Aldrich), 1:500 mouse anti-PDI (Santa Cruz Biotechnology), 1:2000 rabbit anti-GRP94 (Abcam), 1:500 mouse anti-GRP75 (Abcam), 1:1000 rabbit anti-S100A6 (Abcam), 1:1000 mouse anti-HSP72 (Stressgen); 1:1000 goat anti-Psme2 (Abcam), 1:10000 mouse anti-COPS5 (Abcam) or 1:1000 rabbit anti-RACK1 (Sigma-Aldrich). After three washes of 15 min each with PBS–0.5% (v/v) Tween 20, membranes were probed with appropriated secondary antibody, 1:5000 anti-mouse IgG HRP (from sheep) or 1:5000 anti-rabbit IgG HRP (from donkey) (both from GE Healthcare). The immunoreactions are developed on washed membranes as before using enhanced chemiluminescence-ECL (GE Healthcare).

For Western blots normalization, all membranes are washed with the stripping buffer [1.5% (w/v) glycine, 0.1% (w/v) SDS; 1% (v/v) Tween 20; pH 2.2] five times for 10 min, washed with PBS and one last time with PBS-T, and reprobed as above with 1:1000 mouse anti-α tubulin (Sigma-Aldrich) and developed using ECL. The abundance of proteins is calculated from densitometry of immunoblots ($n = 3$ replicates) using the Progenesis PG200v2006. Figures 13.1 and 13.2 are illustrated examples of 2DE-maps and validation data obtained for PSME and GRP78/BiP, respectively, under the different conditions studied.

7. Data Mining and Publication

7.1. Bioinformatics analysis

Bioinformatics tools that deal with high-throughput data and allow its efficient and accurate analysis, interpretation and correlation, have accompanied the important progresses in the proteomics field as they are fundamental for the continuous development of proteomics technology.

Gene Ontology (GO) (www.geneonthology.org) is used to search for biological processes, cellular components and molecular functions associated to the proteins identified. In order to collect all the available

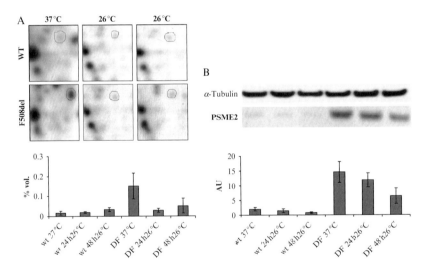

Figure 13.1 2DE analysis (A) and Western blot validation (B) of Psme2 protein expression in BHK-wt 37 °C, BHK-wt 26 °C/24 h, BHK-wt 26 °C/48 h, BHK-F508del 37 °C, BHK-F508del 26 °C/24 h and BHK-26 °C/48 h cells. Close-up views of representative 2DE areas of Psme (A), spots are shown. The 2DE graphic shows the normalized % volume, that is, the volume of each spot over the volume of all spots in the gel (ANOVA test, $p < 0.05$; $n = 4/5$ gels/group, total $n = 26$ gels). The Western blot graphic shows the relative normalized abundance of Psme2 (B), in those BHK cell lines calculated from densitometry of immunoblots using α-tubulin as internal standard ($n = 3$ independent replicates/each Ab reaction). Progenesis PG200v2006 software (Nonlinear) was used for both Western blot and 2DE densitometry analyses.

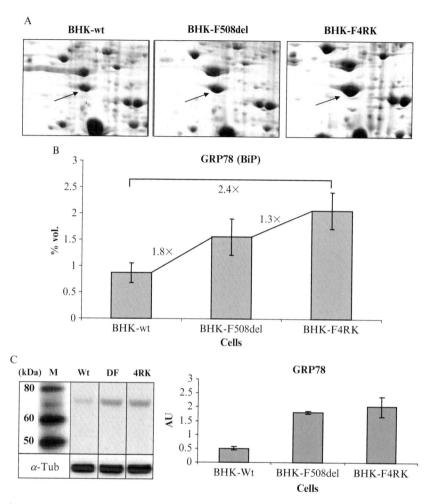

Figure 13.2 GRP78 (BiP) gel spot representation in BHK-wt, -F508del, and -F4RK 2D-maps (A). %Volume of GRP78 in each cell line is graphically presented. The fold change between the three groups is also given (B). GRP78 Western blot analysis (C). Fifteen micrograms of total protein lysates from BHK-wt (lane 1), BHK-F508del (DF) (lane 2), and BHK-F4RK (lane 3) cells was electrophoretically separated on precasting 4–12% (w/v) gradient polyacrylamide mini gel and immunoblotted for GRP78 using rabbit anti-GRP78 Ab (1:300) and for α-tubulin using mouse anti-α-tubulin (1:10000) as normalizer standard of the reaction. The antigen–antibody complex was detected by ECL system. The graphic shows the relative normalized abundance GRP78 in those BHK cell lines calculated from densitometry of immunoblots ($n = 3$ replicates/each Ab reaction) using the Progenesis PG200v2006 software (Nonlinear).

information about each protein, PIKE (Protein Information and Knowledge Extractor) software can also be used, this bioinformatics tool retrieves a set of biological information like: gene name, location, tissue specificity, function, GO terms and keywords (http://proteo.cnb.uam.es:8080/pike/) (see Fig. 13.3A). Protein interactions networks can also be accessed through web-free software, such as Cytoscape, using available protein–protein interaction databases (http://www.cytoscape.org/). Other commercial available tools such as Ingenuity Pathway Analysis (Ingenuity Systems Inc.) are able to retrieve functional biological networks where proteins of interest are involved.

7.2. Clustering analysis

By hierarchical clustering, a powerful data mining approach for a first exploration of proteomic data, proteins can be group blindly according to their expression profiles facilitating their interpretation. The clustering results depend on parameters such as data preprocessing, between-profile similarity measurement, and the dendrogram construction procedure—methodology reviewed elsewhere (Meunier et al., 2007).

In the "low-temperature" study and after the differential protein expression analysis amongst the six experimental conditions, each spot identified showing statistical significant difference relative to the BHK-wt 37 °C condition are included in the relationship analysis amongst each other (columns) and the six phenotypes (rows), using the normalized volume quantity data.

To avoid distortion in clustering analysis on the account of a small subset of abundant protein spots, standardization of data is performed using the z-score calculation Eq. (13.1); where $\bar{x}$ is the arithmetic mean and s is the standard deviation, calculated from the volume quantity data for a given protein spot in the six phenotypes, and x is the individual volume quantity to be normalized. Special care is taken relating to the missing values in the six experimental conditions for an identified spot, which are regarded as a missing spot coloration in 2D gel and as so, interpreted as zero volume quantity. For a given spot, the null information are first substituted by the $\bar{x}$ data in the analyzed phenotypes (column), and then submitted to the standardization process.

$$z = (x - \bar{x})\frac{1}{s} \qquad (13.1)$$

In order to classify the individual protein spots (columns), and the phenotypes(rows), according to the global expression profile, a two way hierarchical clustering analysis can be executed using geWorkbench, version 1.5.1 (http://www.geworkbench.org/), a free open source genomic analysis platform developed at Columbia University with funding from the NIH Roadmap Initiative (1U54CA121852-01A1) or the National Cancer Institute. The normalization process is done directly in the platform choosing the option "Mean-variance normalizer," which was followed by "Fast

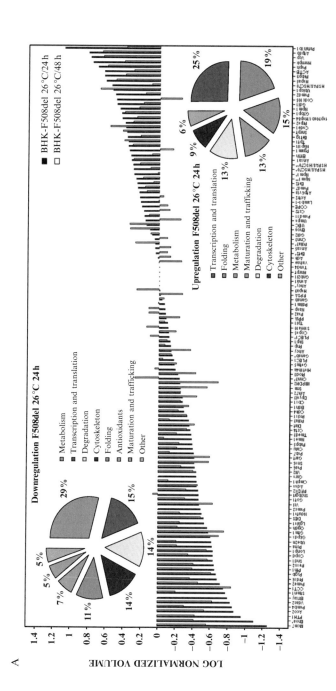

Figure 13.3 (Continued)

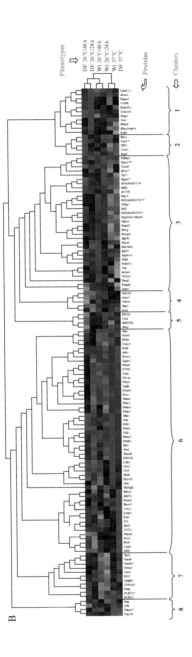

Figure 13.3 Graphical representation of the log normalized volume of proteins in BHK-F508del (A). Expression levels of those proteins in BHK-F508del at 37 °C were considered as reference to determine the up- or downregulation of the correspondent protein in the other phenotype. Proteins up- and downregulated in F508del cells were also classified according to their molecular function. Unsupervised hierarchical clustering (Euclidean distances and average linkage) of the expression profiles of the 139 protein spots differentially expressed in the six phenotypes studied (B). Each horizontal row in the heat map represents the expression level for one protein in the different phenotypes (columns). The relative abundance is displayed by color intensity (light red—more abundant; light green—less abundant). (See Color Insert.)

Hierarchical Clustering Analysis." The euclidean distance metrics Eq. (13.2) was chosen to measure the dissimilarity between the data and the best agglomerative classification for our data is obtained from "Average Linkage."

$$d(x, y) = \left[\Sigma_i (x_i - y_i)^2 \right]^{1/2} \qquad (13.2)$$

Figure 13.3B shows the output of the geWorkbench, where the two generated dendograms (protein spots and phenotypes) are linked by a heat map graph of the standardized data, allowing a better interpretation of the clustering results.

7.3. MIAPE: The Minimum Information About a Proteomics Experiment

MIAPE (http://www.psidev.info/index.php?q=node/91) is a project of the HUPO proteomics standards initiative that establishes the minimum information that must be reported on a proteomics experiment (where samples came from, how analyses of them were performed, etc.) so other authors can reproduce the results of the original work. The requirements of the various journals also differ, so important details may be lacking in some cases, or presented in an inappropriate form.

As so, when authors are interested in publish results obtained from proteomics approaches it is practical and useful to consult the MIAPE guidelines already published, most information is available at the webpage http://www.psidev.info/index.php?q=node/91. Before submitting a paper for publication it is also essential to consult the requisites of the journal of interest about the data that authors must present for the type of approach they have been using in their investigation.

ACKNOWLEDGMENTS

The research relevant to this paper was supported by Fundação da Ciência e a Tecnologia (FCT) e Fundo Europeu de Desenvolvimento Regional (FEDER) research grants POCTI/MGI/40878/2001 and POCI/SAU-MMO/56163/2004, FCT-Poly-Annual Funding Program and FEDER-Saúde XXI Program, Portugal. PGA was a recipient of FCT-PhD fellowship SFRH/BD/17744/2004 and SN is a recipient of Ricardo Jorge-research fellowship.

REFERENCES

Aebersold, R., and Mann, M. (2003). Mass spectrometry-based proteomics. *Nature* **422**, 198–207.

Bartoszewski, R., et al. (2008). Activation of the unfolded protein response by DeltaF508 CFTR. *Am. J. Respir. Cell. Mol. Biol.* **39**, 448–457.

Bensalem, N., et al. (2007). High sensitivity identification of membrane proteins by MALDI TOF-MASS spectrometry using polystyrene beads. *J. Proteome Res.* **6**, 1595–1602.

Candiano, G., et al. (2004). Blue silver: A very sensitive colloidal Coomassie G-250 staining for proteome analysis. *Electrophoresis* **25**, 1327–1333.

Chang, X. B., et al. (1999). Removal of multiple arginine-framed trafficking signals overcomes misprocessing of delta F508 CFTR present in most patients with cystic fibrosis. *Mol. Cell* **4**, 137–142.

Chen, L. C., and Casadevall, A. (1999). Labeling of proteins with [^{35}S]methionine and/or [^{35}S]cysteine in the absence of cells. *Anal. Biochem.* **269**, 179–188.

Citterio, C., et al. (2008). Unfolded protein response and cell death after depletion of brefeldin A-inhibited guanine nucleotide-exchange protein GBF1. *Proc. Natl. Acad. Sci. USA* **105**, 2877–2882.

Denning, G. M., et al. (1992). Processing of mutant cystic fibrosis transmembrane conductance regulator is temperature-sensitive. *Nature* **358**, 761–764.

Gomes-Alves, P., and Penque, D. (2010). Proteomics uncovering possible key players in F508del-CFTR processing and trafficking. *Expert Rev. Proteomics* **7**, 487–494.

Gomes-Alves, P., et al. (2009). Low temperature restoring effect on F508del-CFTR misprocessing: A proteomic approach. *J Proteomics* **73**, 218–230.

Gomes-Alves, P., et al. (2010a). Rescue of F508del-CFTR by RXR motif inactivation triggers proteome modulation associated with the unfolded protein response. *Biochim. Biophys. Acta* **1804**, 856–865.

Gomes-Alves, P., et al. (2010b). Rescue of F508del-CFTR by RXR motif inactivation triggers proteome modulation associated with the unfolded protein response. *Biochim. Biophys. Acta Proteins Proteomics* **1804**, 856–865.

Gorg, A., et al. (2004). Current two-dimensional electrophoresis technology for proteomics. *Proteomics* **4**, 3665–3685.

Kelsen, S. G., et al. (2008). Cigarette smoke induces an unfolded protein response in the human lung: A proteomic approach. *Am. J. Respir. Cell. Mol. Biol.* **38**, 541–550.

Kerbiriou, M., et al. (2007). Coupling cystic fibrosis to endoplasmic reticulum stress: Differential role of Grp78 and ATF6. *Biochim. Biophys. Acta.* **1772**, 1236–1249.

Kitteringham, N. R., et al. (2009). Multiple reaction monitoring for quantitative biomarker analysis in proteomics and metabolomics. *J. Chromatogr. B Analyt. Technol. Biomed. Life Sci.* **877**, 1229–1239.

Loo, T. W., et al. (2005). Rescue of folding defects in ABC transporters using pharmacological chaperones. *J. Bioenerg. Biomembr.* **37**, 501–507.

Mendes, F., et al. (2003). Unusually common cystic fibrosis mutation in Portugal encodes a misprocessed protein. *Biochem. Biophys. Res. Commun.* **311**, 665–671.

Meunier, B., et al. (2007). Assessment of hierarchical clustering methodologies for proteomic data mining. *J. Proteome Res.* **6**, 358–366.

Minden, J. (2007). Comparative proteomics and difference gel electrophoresis. *Biotechniques* **43**, 739–745.

Penque, D. (2009). Two-dimensional gel electrophoresis and mass spectrometry for biomarker discovery. *Proteomics Clin. Appl.* **3**, 155–172.

Picotti, P., et al. (2010). High-throughput generation of selected reaction-monitoring assays for proteins and proteomes. *Nat. Meth.* **7**, 43–46.

Powers, E. T., et al. (2009). Biological and chemical approaches to diseases of proteostasis deficiency. *Annu. Rev. Biochem.* **78**, 959–991.

Rennolds, J., et al. (2008). Low temperature induces the delivery of mature and immature CFTR to the plasma membrane. *Biochem. Biophys. Res. Commun.* **366**, 1025–1029.

Roxo-Rosa, M., et al. (2004). Proteomics techniques for cystic fibrosis research. *J. Cyst. Fibros.* **3**(Suppl. 2), 85–89.

Roxo-Rosa, M., et al. (2006). Revertant mutants G550E and 4RK rescue cystic fibrosis mutants in the first nucleotide-binding domain of CFTR by different mechanisms. *Proc. Natl. Acad. Sci. USA* **103**, 17891–17896.

Sharma, M., et al. (2001). Conformational and temperature-sensitive stability defects of the delta F508 cystic fibrosis transmembrane conductance regulator in post-endoplasmic reticulum compartments. *J. Biol. Chem.* **276**, 8942–8950.

Singh, O. V., et al. (2008). Chemical rescue of deltaF508-CFTR mimics genetic repair in cystic fibrosis bronchial epithelial cells. *Mol. Cell. Proteomics* **7**, 1099–1110.

Steen, H., and Mann, M. (2004). The ABC's (and XYZ's) of peptide sequencing. *Nat. Rev. Mol. Cell. Biol.* **5**, 699–711.

van Anken, E., and Braakman, I. (2005). Versatility of the endoplasmic reticulum protein folding factory. *Crit. Rev. Biochem. Mol. Biol.* **40**, 191–228.

CHAPTER FOURTEEN

DECREASED SECRETION AND UNFOLDED PROTEIN RESPONSE UPREGULATION

Carissa L. Young,* Theresa Yuraszeck,[†] and Anne S. Robinson*

Contents

1. Introduction	236
2. Heterologous Protein Expression	237
2.1. scFv 4-4-20 as a model scFv	238
3. Quality Control Mechanisms of the Secretory Pathway	238
4. Endoplasmic Reticulum Export and Trafficking	239
5. Experimental Systems to Evaluate Expression, UPR, and Secretory Processing	239
6. *S. cerevisiae* Strains Used for Optimal Expression	240
7. Plasmid Design	241
7.1. scFv 4-4-20 construct	241
7.2. UPR sensors	242
8. Evaluation of Heterologous Protein Expression	243
8.1. Strain growth, expression, and isolation of intracellular heterologous protein	243
8.2. Analysis of GFP fusions by in-gel fluorescence	245
8.3. Experimental results	246
8.4. Pulse-chase and immunoprecipitation protocol	247
9. Statistical Analysis of Microarray Results	248
9.1. Appropriate controls	249
9.2. Selecting differentially expressed genes	251
9.3. Classification	252
9.4. Enrichment of promoter regions in known transcription factor-binding sites	253
9.5. Presence of UPRE-1, UPRE-2, UPRE-3 in the promoter region of differentially expressed genes	254
9.6. Relative quantification by real-time quantitative PCR	254
9.7. Implications for protein secretion and UPR induction	256

* Department of Chemical Engineering, University of Delaware, Newark, Delaware, USA
[†] Department of Chemical Engineering, University of California Santa Barbara, Santa Barbara, California, USA

| Acknowledgment | 257 |
| References | 257 |

Abstract

Recombinant antibody fragments, for example, the classic monovalent single-chain antibody (scFv), are emerging as credible alternatives to monoclonal antibody (mAb) products. scFv fragments maintain a diverse range of potential applications in biotechnology and can be implemented as powerful therapeutic and diagnostic agents. As such, a variety of hosts have been used to produce antibody fragments resulting in varying degrees of success. Yeast, *Saccharomyces cerevisiae*, is an attractive host due to quality control mechanisms of the secretory pathway that ensure secreted proteins are properly folded. However, the expression of a recombinant protein in yeast is not trivial; neither are the quality control mechanisms the cell initiates to respond to overwhelming stress, such as an increased protein load, simplistic. The endoplasmic reticulum (ER) is a dynamic organelle, capable of sensing and adjusting its folding capacity in response to increased demand. When protein abundance or terminally misfolded proteins overwhelm the ER's capacity, the unfolded protein response (UPR) is activated. In the guidelines presented here, we discuss varying aspects of quality control, its modulation, and ways to design appropriate constructs for yeast recombinant protein expression. Furthermore, we have provided protocols and methods to monitor intracellular protein expression and trafficking as well as evaluation of the UPR, with essential controls. The latter part of this chapter will review considerations for the experimental design of microarray and quantitative polymerase chain reaction (q-PCR) techniques while suggesting appropriate means of data analysis.

1. INTRODUCTION

Efficient production of heterologous proteins by eukaryotic hosts, including the yeast *Saccharomyces cerevisiae*, is often hampered by the inefficient secretion of the product. Limitation of protein secretion has often been attributed to a decreased folding rate, leading to novel solutions that have implemented the overexpression of endogenous proteins to support protein folding and maturation (Robinson *et al.*, 1994; Shusta *et al.*, 1998), or the constitutive expression of the unfolded protein response (UPR) transcription factor (TF), Hac1 (Valkonen *et al.*, 2003). Extensive studies conducted with single-chain antibody fragments (scFv) established that overexpression of BiP or PDI increases secretion substantially; and when co-overexpressed, these endoplasmic reticulum (ER)-resident folding assistants act synergistically with the tuning of scFv copy number to increase secretion eightfold (i.e., 20 mg/L; Shusta *et al.*, 1998). Furthermore, our studies have shown that overexpression of PDI or co-overexpression of BiP and PDI significantly reduce the UPR during scFv expression as determined by the canonical UPRE sensor (Mori *et al.*, 1998; Travers *et al.*, 2000). A reduction in the UPR led to an ~3.5-fold increase of secreted scFv (Xu *et al.*, 2005).

The comprehensive characterization of genes required for endogenous protein folding (Jonikas *et al.*, 2009), translational behavior (Arava *et al.*, 2003), and cell growth in *S. cerevisiae* (Brauer *et al.*, 2008) have been described, and are useful for categorizing genes of interest. Therefore, assuming that other components of the secretory pathway affect secretion although are not directly involved in protein folding, one aim of our studies was to evaluate the transcriptome of *S. cerevisiae* during expression of scFv. On a cellular level, we were also interested in understanding the interactions between BiP and PDI with the scFv during expression. We have determined UPR modulation in different cell strains and evaluated the intracellular retention of scFv based on pulse-chase ^{35}S studies to examine trafficking effects.

This chapter focuses on quality control mechanisms of the secretory pathway, specifically impacting UPR upregulation and decreased secretion of scFv. In this regard, we detail experimental methods used to evaluate the UPR in a population, and appropriate means of quantifying the intracellular concentration of a model antibody fragment, scFv 4-4-20, that may be broadly applied to heterologous protein expression and secretion. Rigorous statistical analysis of microarray and quantitative PCR (q-PCR) data is essential when evaluating global data using either a time-course or static experiment. We have carefully outlined methods and caveats in data analysis and interpretation, and utilize our studies of UPR induction by chemical treatment and expression of scFv as case studies.

2. Heterologous Protein Expression

Collectively, heterologous protein secretion involves the coupled processes of protein synthesis, protein folding, and secretory trafficking; thus, a more complete understanding of how these processes interrelate will lead to optimized conditions for scFv expression, secretion, and enhanced activity. In the case of scFv production, there are several reports in literature describing approaches to improve expression: overexpression of folding assistants BiP and PDI (Robinson *et al.*, 1994; Xu *et al.*, 2005) combined with expression level tuning of the heterologous protein (Shusta *et al.*, 1998); modulation of quality control mechanisms including the UPR (Rakestraw and Wittrup, 2006); deletions of vacuolar sorting proteins to improve secretion of heterologous proteins (Zhang *et al.*, 2001); engineered native secretory leader sequences by directed evolution to enhance scFv secretion (Rakestraw et al., 2009); isolation of antibodies for increased stability and affinity using high-throughput yeast surface display (Chao et al., 2006) combined with yeast cDNA overexpression libraries to identify optimal strains leading to scFv secretion (Wentz and Shusta, 2007); and mutational studies to increase the affinity of scFv (Boder et al., 2000; Midelfort and Wittrup, 2006).

2.1. scFv 4-4-20 as a model scFv

To study the effects of scFv expression and trafficking on the yeast UPR, we have focused on a well-characterized construct, scFv 4-4-20, which binds to the ligand fluorescein. This scFv quenches the fluorescence of fluorescein greater than 90% upon binding. Thus, a quick and easy spectrofluorimetric assay is used for the detection of proper folding and activity (Denzin and Voss, 1992). Expression of scFv 4-4-20 in yeast is limited by the activation of quality control machinery, and the subsequent decrease in protein production and secretion (Kauffman et al., 2002).

3. QUALITY CONTROL MECHANISMS OF THE SECRETORY PATHWAY

The secretory pathway of eukaryotic cells is composed primarily of two organelles, the ER and Golgi, responsible for maintaining the fidelity of protein synthesis and maturation. The environment of the ER is specialized to properly fold secretory proteins due to an oxidizing redox potential, appropriate calcium levels, and dedicated enzymes for protein glycosylation and folding (i.e., chaperones and foldases; van Anken and Braakman, 2005). When abnormalities do occur, such as an overwhelming abundance of improperly folded protein retained in the ER, or a decrease in vesicle trafficking from the ER to Golgi, these phenomena, collectively termed "ER stress," upregulate quality control mechanisms to ensure cellular homeostasis. Such stress can be caused by a variety of insults, including nutrient deprivation, pathogenic infection, chemical treatment, and the expression of heterologous protein.

ER-associated degradation (ERAD) and the UPR are two quality control mechanisms that are upregulated upon ER stress. They have several outcomes that occur at various timescales, depend on variations in the spatial organization of organelles, and alter selective protein concentrations and intracellular localization (Brodsky, 2007; McCracken and Brodsky, 2003; Midelfort and Wittrup, 2006; Ng et al., 2000). Removal of misfolded proteins through ERAD occurs by ubiquitination via ER-associated ubiquitin-conjugating enzymes, followed by retrotranslocation, and degradation in the cytoplasm by the proteasome (reviewed in Nakatsukasa and Brodsky, 2008). Local perturbations of unfolded protein levels in the ER activate the UPR, defined as a global cytoprotective signaling cascade that transcriptionally upregulates genes encoding ERAD components, chaperones, and oxidoreductases (Otte and Barlowe, 2004; Travers et al., 2000).

Once the ER's capacity to process nascent proteins is overwhelmed, the ER-resident chaperone BiP is sequestered from the transmembrane kinase protein Ire1p to counteract the stressful condition, which triggers the negative feedback loop known as the UPR. Unbound Ire1p dimerizes, autophosphorylates, and splices an intron from *HAC1* mRNA. The resulting Hac1p transcription factor (TF) binds to the promoter regions of UPR targets, upregulating their expression. However, it must also be noted that unfolded protein may directly initiate the dimerization and activation of Ire1p (Kimata et al., 2007; Oikawa et al., 2007). Either mechanism results in the upregulation of diverse targets that include chaperones and foldases and many elements of the secretory pathway, as well as components of the ERAD machinery. In higher eukaryotes, several additional pathways are activated in response to ER stress, and a general attenuation of translation is observed.

4. ENDOPLASMIC RETICULUM EXPORT AND TRAFFICKING

Secreted heterologous proteins, such as scFv, enter the secretory pathway at the ER and move via vesicular transport to the Golgi. Export from the ER requires specialized ER-resident proteins including COPII machinery, and receptors such as Erv29 help export certain soluble cargo proteins (Otte and Barlowe, 2004). There is increasing evidence that post-ER quality control mechanisms exist and sort defective proteins from the Golgi to the endosomal system for lysosomal/vacuolar degradation (reviewed in Arvan et al., 2002). Interestingly, *vacuolar protein sorting* (*vps*) mutants have resulted in the increased secretion of luminal recombinant proteins (Hong et al., 1996) and membrane proteins (Luo and Chang, 2000).

5. EXPERIMENTAL SYSTEMS TO EVALUATE EXPRESSION, UPR, AND SECRETORY PROCESSING

As a simpler eukaryotic organism, yeast is advantageous to study due to facile genetic manipulation in a sequenced, annotated genome; mammalian-like organelle trafficking of secreted protein with similar mechanisms for protein synthesis, maturation, and secretory trafficking; and microbial growth features. In the following sections, we outline pertinent factors to consider with respect to molecular engineering of constructs for heterologous protein expression and selection of the most appropriate *S. cerevisiae* strain. We also outline experimental protocols and conclude with additional remarks regarding experiments and data analysis.

6. *S. CEREVISIAE* STRAINS USED FOR OPTIMAL EXPRESSION

A yeast strain should be selected based on its suitability for the process being studied, efficiency of transformation, and flexibility with respect to selection. Difficulties associated with the expression level of a recombinant protein, effect of growth rates, and proteases are aspects that should be considered. The choice of an appropriate host strain, induction media, and expression plasmid (i.e., 2 μm, low-copy, or multicopy δ integrating plasmids) can overcome most obstacles. Usually it is desirable to choose a specific parental strain that has been used in previous studies (or industrial applications), therefore allowing direct comparison with established results and not complicating your analysis by differences in strain backgrounds. Additionally, consider strains that carry multiple deletion alleles of auxotrophic markers that will provide flexibility in the future should you choose to introduce episomal plasmids or PCR-based modifications completed by homologous recombination (Brachmann *et al.*, 1998). Deletions of nearly all ∼ 6000 genes have been conducted in multiple *S. cerevisiae* strains (see yeast gene knockout or YKO Collection, Open Biosystems) providing remarkable options. Alternatively, it is rather straightforward to design additional auxotrophic knockouts in your strain of choice (Petracek and Longtine, 2002).

To alleviate the problem of contaminating proteases, a protease-deficient strain (BJ5464 MATα *ura3-52 trp1 leu2Δ his3Δ200 pep4::HIS3 prb1-Δ 1.6R can1* GAL (ATCC 208288)), including mutations in both the *pep4* and *prb1* genes, is recommended (reviewed by Jones, 2002). However, one must keep in mind that all vacuolar proteases increase in concentration as the cells approach stationary phase, and a small increase has been observed at the diauxic plateau; the largest fold increase (i.e., 100× that of log phase) occurs as the cells enter stationary phase (Moehle *et al.*, 1987). Therefore, it is imperative that strain growth and heterologously expressed proteins are monitored by time-course studies. An orthogonal challenge is that pep4 and prb1 are specific precursors for degradation of autophagosomes; therefore, the use of these mutants will potentially lead to an accumulation of terminally misfolded proteins in the cell that may become detrimental to cell growth and homeostasis (Takeshige *et al.*, 1992).

7. Plasmid Design

For the maximal expression of a recombinant protein, the 2 μm is generally considered. Yet, high-level expression will overwhelm the capacity of the secretory pathway; therefore, alternatives include low-copy plasmids (i.e., pRS300 series; Sikorski and Hieter, 1989), direct chromosomal integration, or the flexibility provided by δ integration. Comparison of these molecular engineered constructs are examined in Parekh *et al.* (1996), Shusta *et al.* (1998), and Kauffman *et al.* (2002) by evaluating heterologous protein secretion. When deciding on the appropriate vector, keep in mind that the use of minimal media is essential for auxotrophic markers, resulting in a doubling time of approximately 140 min. This limitation can be avoided by the integration of a PCR-based fragment resulting in a doubling time of 90 min in rich media (Sherman, 2002).

Promoter choices range from constitutive promoters (e.g., PGK and GAPDH) to inducible promoters (e.g., GAL1, GAL1-10, CUP1). Galactose-regulated promoters are induced >1000-fold, yet are strongly repressed by glucose; thus, maximal expression only occurs following glucose depletion (Romanos *et al.*, 2001). For rapid induction, cultures can be grown overnight using a nonrepressing carbon source (e.g., raffinose) and then induced by direct addition of galactose (review protocol in Romanos *et al.*, 2001). Furthermore, regulation of gene expression can be conducted by the tetracycline/tetracycline operator (TetO) systems (Belli *et al.*, 1998) or use of the ADH1 promoter that eliminates the need to shift carbon sources that can alter metabolism and gene expression (Gasch *et al.*, 2000). Expression of scFv in yeast, *S. cerevisiae,* has utilized the GAL1, GAL1-10, and GAPDH promoters (Huang *et al.*, 2008; Shusta *et al.*, 1998; Xu and Robinson, 2009).

7.1. scFv 4-4-20 construct

The autonomous (CEN/ARS) plasmid pRS316-FLAG-4-4-20 consists of an inducible GAL1-10 promoter, synthetic pre-pro sequence (Clements *et al.*, 1991), Glu-Ala-Arg-Pro spacer, FLAG (amino acid sequence Asp-Tyr-Asp-Asp-Asp-Lys) epitope tag, scFv gene, and α-terminator sequence that has been used for pulse-chase studies (Figs. 14.1 and 14.3). Site-directed mutagenesis of two initial stop codons following the scFv 4-4-20 gene restored the original C terminal c-myc epitope (Xu *et al.*, 2005) modified from the construct of Shusta *et al.* (1998). This low-copy plasmid was used in our strain-dependent studies, described in Fig. 14.2, described below. To examine the effects of using a multicopy integration of 4-4-20, the coding sequence described above was subsequently inserted into the δ vector,

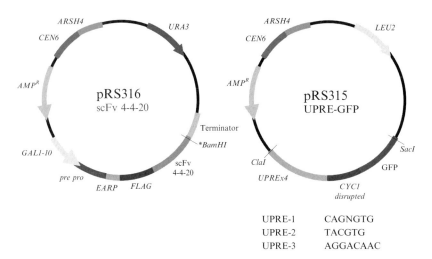

Figure 14.1 Illustration of low-copy plasmids used for heterologous protein expression of scFv 4-4-20 and UPR sensor, UPRE-GFP, whereas any UPR element can be analyzed by fluorescent intensity (Robinson Lab).

pITy-4 (Parekh *et al.*, 1996), and transformed to the BJ5464 parental strain, referred to as high-copy scFv (hcscFv; Kauffman *et al.*, 2002). Integrated hcscFv was grown in 1% raffinose, induced by addition of galactose (1%), and samples were collected at various times ($t = 0, 2, 4, 6, 8,$ and 12 h). Microarray analysis of this data described later in this chapter has identified novel regulation during heterologous recombinant protein expression.

7.2. UPR sensors

The canonical UPR element (UPRE) was originally defined as a 22-bp sequence element of the KAR2/BiP promoter (Mori *et al.*, 1992) and refined to nucleotide precision as a semipalindromic seven-nucleotide consensus, CAGNGTG (Mori *et al.*, 1998), defined as the classical UPR sensor, UPRE-1. A bioinformatics approach conducted by Patil *et al.* (2004) has identified two novel UPRE promoters with the specific nomenclature of UPRE-2 and UPRE-3. In previous studies (Xu *et al.*, 2005), we subcloned a construct containing four tandem repeats of UPRE-1, a disabled *Cyc1* promoter, and green fluorescent protein (GFP) from pKT058 (Travers *et al.*, 2000) into the low-copy pRS300 plasmids of Sikorski and Hieter (1989). Subsequent constructs included four tandem repeats of UPRE-2 and UPRE-3 in the same base vector, thereby providing an appropriate means to compare all three elements and their regulation during UPR induction (Fig. 14.1). UPR elements fused to fluorescent variants have been

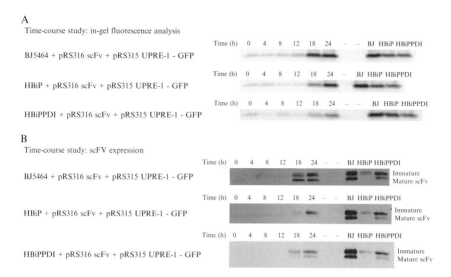

Figure 14.2 Analysis of UPR and intracellular scFv levels following induction of scFv 4-4-20 expression shows UPR initiation and intracellular scFv retention starting at ~18 h. (A) In-gel fluorescence of UPRE-GFP levels in parental strain BJ5464 (top panel) compared to overexpressed BiP (HBiP; middle panel) and co-overexpressed BiP and PDI (HBiPPDI; lower panel). Comparison of each strain at 24 h postinduction of scFv expression, denoted as BJ, HBiP, and HBiPPDI, respectively, was included on each gel. (B) Western analysis using α-FLAG antibodies. Interestingly, BJ5464 maintains the highest level of intracellular expression as compared to HBiP and HBiPPDI strains. Samples of each strain at 24 h postinduction are included on each gel in order to enable a quantitative comparison. Intensities of each Western blot were normalized to the loading control, α-Act1.

integrated in the chromosome of several *S. cerevisiae* strains, resulting in strains capable of growth on rich media and improved fluorescence properties (Robinson Lab, unpublished).

8. EVALUATION OF HETEROLOGOUS PROTEIN EXPRESSION

8.1. Strain growth, expression, and isolation of intracellular heterologous protein

The following time-course protocol is specified for the expression of a heterologous protein and evaluation of the UPR (Fig. 14.2) although it can be modified for any experimental system. Synthetic media has been described elsewhere (Sherman, 2002).

1. Following the restreak of your desired strain onto a selective agar plate (e.g., SD-URA-LEU), resuspend a single colony in 1 mL of appropriate dextrose media and measure the optical density at 600 nm (OD_{600}). Estimate, *a priori*, the volume of the resuspended sample necessary for an overnight culture to achieve a final value of $0.6 \leq OD_{600} \leq 0.8$, assuming a doubling time of 140 min (synthetic media).
2. Grow the desired volume of overnight culture in minimal dextrose media supplemented with the appropriate amino acids (e.g., SD-URA-LEU) at 250 rpm, 30 °C.
3. The following morning, measure the OD_{600}. Calculate the volume to be added for the induction phase, targeting an OD ∼ 0.1–0.3 of the desired total volume of culture.
4. Harvest the necessary cell volume by centrifugation ($2500 \times g$, 5 min, 4 °C), discard the culture supernatant, and keep the cell pellet on ice. Wash away residual glucose by resuspending the pellet in 10 mL of dH_2O. Centrifuge again and repeat.
5. Add the resulting cell pellet to the desired total volume of galactose medium (SG-URA-LEU). For each sample (i.e., desired time points) remove 1 OD_{600} equivalent volume. Spin down ($2500 \times g$, 5 min, 4 °C). Discard the supernatant, resuspend in 0.5 mL dH_2O, and transfer resuspended cells into 1.5-mL eppendorf tubes. Spin down (1 min, 13,000 rcf), remove supernatant, and freeze cell pellet at -80 °C until subsequent analysis.
6. Remove cell pellets from -80 °C and keep on ice from this point forward. Add 100 μL lysis buffer (1% SDS, 50 mM Tris–HCl, pH = 7.4, complete, EDTA-free protease inhibitor cocktail tablets, per manufacturer's instructions (Roche)) per OD_{600}, adapted from Rothblatt and Schekman (1989). Break the cells by adding roughly the same amount of glass beads (0.5 mm Zirconia/Silica beads, Biospec Products, Inc.) as the volume of cell pellet (100 μL equivalent). Vortex on high speed three times for 30 s each and between each cycle chill on ice for a minimum of 30 s.
7. Retrieve the cell lysate by using gel-loading tips, transferring the lysate to a fresh 1.5-mL eppendorf tube. Clarify the extract by centrifugation at top speed for 1 min. Transfer the supernatant to a new tube.
8. Normalize all sample volumes to an equivalent OD_{600} for loading. For measuring intracellular scFv levels, we typically combine 20 μL total lysate with 40 μL $3\times$ SDS loading buffer (150 mM Tris, pH = 6.8; 0.25 mg/mL bromophenol blue; 6% SDS; 30% glycerol; 30 mM dithiothreitol (DTT)). For Western blot analysis, denature samples by heating to 95 °C for 5 min, followed by quenching on ice. A quantitative analysis of the UPRE-GFP sensor can be completed by in-gel fluorescence (see protocol, Section 8.2, below); however, samples should not be heated in order to avoid loss of fluorescence through GFP unfolding.

9. Samples are analyzed by SDS-PAGE, transferred to nitrocellulose paper, and quantified by standard Western blot procedures. Use of primary and secondary antibodies should be optimized according to levels of expressed recombinant protein. See Table 14.1 for more information regarding antibodies currently used in our laboratory. Resulting Western blots are then scanned (Typhoon 9400 Variable Mode Imager, Amersham Biosciences) and quantified using ImageQuant (Molecular Dynamics) or ImageJ (National Institutes of Health) software.

8.2. Analysis of GFP fusions by in-gel fluorescence

Analysis of fusions by in-gel fluorescence was originally described in Newstead *et al.* (2007). Implementation of this technique has proven to be a rapid method to confirm the expression of endogenous and

Table 14.1 Primary and secondary antibody recommendations

Identification	Vendor	Catalog number
Primary antibodies		
Mouse α-FLAG, monoclonal antibody	Sigma-Aldrich®	F1804
Mouse α-myc [9E10], monoclonal antibody	abcam®	ab32
Rabbit α-HA, polyclonal antibody	invitrogen™	71-5500
Rabbit α-GFP, polyclonal antibody[a]	abcam®	ab6556
Rabbit α-RFP, polyclonal antibody	abcam®	ab62341
S. cerevisiae endogenous proteins		
Rabbit α-Kar2/BiP, polyclonal antibody[b]	Robinson Lab	
Loading controls		
Mouse α-beta Actin, monoclonal antibody[c]	abcam®	ab8224
Rabbit α-GAPDH, polyclonal antibody	abcam®	ab9485
Secondary antibodies		
Alexa Fluor® 488 goat α-mouse IgG (H + L)	invitrogen™	A-11029
Alexa Fluor® 633 goat α-rabbit IgG (H + L)	invitrogen™	A-21070
Alexa Fluor® 546 donkey α-rabbit IgG	invitrogen™	A-10040
Alexa Fluor® 633 goat α-mouse IgG (H + L)	invitrogen™	A-21050
Alexa Fluor® 546 donkey α-goat IgG (H + L)	invitrogen™	A-11056

[a] Recommend using the α-GFP rabbit polyclonal antibody ab6556. ab290 and ab6556 are more specific than ab1218; meanwhile, ab6556 is the purified version of ab290.
[b] Rabbit polyclonal antibodies were generated (Capralogics) against a purified MBP–KAR2 fusion protein that was constructed from the C terminal region of KAR2 (*XbaI* site through stop codon) and inserted into pMAL-c2G (New England Biolabs). α-Kar2/BiP was screened for specificity in yeast lysates and purified from serum using a Montage® Antibody Purification kit with PROSEP®-A Media (Millipore) and stored at −80 °C.
[c] ab8224 works well as a loading control for *S. cerevisiae*. We recommend using ab8224 instead of the alternative ab8226.

heterologous fusion proteins in *S. cerevisiae* using multiple GFP and DsRed derivatives Robinson Lab, unpublished.

When using fluorescent protein fusions to quantify protein levels, it is extremely important to normalize protein samples correctly (e.g., to equivalent OD_{600}) while accounting for differences in fluorescence excitation and emission spectra, as well as intensity differences (i.e., quantum yield, extinction coefficient; reviewed in Nagai *et al.*, 2002; Rizzo *et al.*, 2004; Shaner *et al.*, 2004). When quantifying total concentration based on intensity of a fluorescent variant there are a few aspects to consider, including cell preparation (i.e., culture conditions and lysis preparation), storage (e.g., $-80\ °C$ or prepared directly following experiment), and appropriate controls. We also recommend addition of fluorescent standards to determine approximate protein concentration.

Sample preparation is similar to the previous protocol (steps 1–8), although samples should not be heated. Consider the effects of storing your samples at $-80\ °C$ and repeated freeze-thaw cycles, which may result in a significant decrease in fluorescence intensity. Typically, we use a sample load equivalent to $\sim 0.1\ OD_{600}$ and analyze by SDS-PAGE; however, conditions should be optimized for the specific fusion protein of interest. For in-gel fluorescence, protein bands of GFP (or DsRed) variants are imaged using a Typhoon 9400 Variable Mode Imager (Amersham Biosciences) and compared to fluorescent molecular weight ladders at 488 (Benchmark^{TM} Fluorescent Protein Standard, Invitrogen) or 546 nm (DyLight^{TM} 546/649 Fluorescent Protein Molecular Weight Markers, Thermo Fisher Scientific, Inc.). It may be necessary to dilute these ladders in $1\times$ SDS running buffer in order to optimize comparison to protein bands of interest.

8.3. Experimental results

The expression and intracellular retention of scFv 4-4-20 in the parental strain BJ5464, a strain overexpressing BiP herein referred to as HBiP, and a strain co-overexpressing BiP and PDI (i.e., HBiPPDI), were examined as a function of time. UPR induction by in-gel GFP fluorescence (Fig. 14.2), intracellular scFv levels via Western analysis (Fig. 14.2), and BiP/Kar2 protein levels determined by Western blot analysis (data not shown) was measured in all three strains. We confirm that the UPR is significantly upregulated in HBiP as opposed to HBiPPDI (Fig. 14.2A). Interestingly, the UPR response as determined by fluorescence is significantly greater in the parental strain expressing 4-4-20 (i.e., BJ5464); furthermore, when compared to the Western blot analysis of Fig. 14.2B, it is readily apparent that total intracellular scFv protein levels are significantly greater in this parental strain as compared to the strains overexpressing chaperone and/or foldase. In all cases, at 24 h postinduction, the immature scFv species accumulates intracellularly causing an overwhelming stress. Thus, to investigate protein synthesis of this model scFv and concurrent trafficking defects, we recommend taking advantage of the pre-pro leader sequence combined with pulse-chase techniques that will yield valuable insights into quality control mechanisms of the secretory pathway.

8.4. Pulse-chase and immunoprecipitation protocol

To monitor intracellular trafficking of newly synthesized scFv, we have constructed strains in which ER luminal chaperone and foldase, BiP and PDI, respectively, have been overexpressed in order to increase secreted yields of scFv 4-4-20 (Xu et al., 2005). Cells were grown overnight in 5 mL minimal medium (SD-2XCAA; Wittrup and Benig, 1994) supplemented with the appropriate amino acids. The overnight culture was then diluted to an $OD_{600} \sim 0.1$ in fresh SD media for ~ 6 h. In order to continuously provide nutrients to the culture, semicontinuous growth was implemented once expression was induced via galactose (1%). A 0.2 OD_{600} equivalent of cells were resuspended in SG-2XCAA media lacking uracil (SG-URA). Yeast strains were grown semicontinuously in order to maintain the culture between $0.2 \leq OD_{600} \leq 0.3$ by supplementing SG-URA media every 2 h, as needed, for 14 h. Cells were harvested, washed with water to remove methionine, and resuspended at a density of 5 OD_{600} in prewarmed SG-URA minus methionine for 1 h. The culture was then incubated with 240 μCi/OD Trans^{35}S-label (MP Biomedicals, Inc.) for 10 min in an air shaker at 30 °C. Following the pulse, excess cold methionine (100 mM) was added. For each time point, 1 OD_{600} of radiolabeled cells were collected by centrifugation at 14,000 rpm for 2 min. Pellets were then resuspended and incubated in 1 mL chase media consisting of SG media plus 1 mg/mL methionine per OD_{600}.

Radiolabeled scFv samples were immunoprecipitated with cells lysed in lysis buffer (2% SDS, 90 mM HEPES, pH 7.5, 30 mM DTT, and 0.4 mg/mL Pefabloc; Sigma) and denatured at 95 °C for 10 min. Four microliters of the primary antibody (α-FLAG M2 monoclonal, Sigma) was added to each sample for a 1-h incubation at 4 °C, followed by addition of 50 μL of resin (Protein A/G Sepharose Mix, Amersham) and an additional hour incubation at 4 °C. Samples were then centrifuged and the pellets rinsed with cold washing buffer (50 mM Tris, pH 7.5, 250 mM NaCl, 5 mM EDTA, 0.5% NP-40, 0.4 mg/mL Pefabloc, 10 μg/mL Leupeptin, 10 μg/mL Pepstatin, 5 μg/mL Aprotinin) three times prior to heating to 95 °C for 5 min. scFv samples were loaded onto a 11% SDS-PAGE gel; following electrophoresis, the gel was then submerged in Amplify (Amersham) for 30 min, dried, and exposed to a Phosphoimager Scanner SI (Molecular Dynamics). The data was then scanned (Typhoon 9400 Variable Mode Imager, Amersham Biosciences) and quantified using ImageQuant software (Molecular Dynamics).

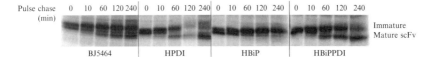

Figure 14.3 ^{35}S pulse-chase analysis of scFv 4-4-20 expression and trafficking effects in *S. cerevisiae*. Data have been modified from Figure 6 of Xu et al. (2005) by permission of Elsevier.

Figure 14.3 depicts typical results obtained from this pulse-chase analysis under conditions in which the culture density was maintained between $0.2 \leq OD_{600} \leq 0.3$ by providing a continuous source of nutrients in semibatch growth. Proper processing of the pre-pro leader sequence should result in a mature scFv species that is secreted to the media. Our results confirm the existence of a mature species in the parental strain (i.e., BJ5464) and HBiPPDI; however, the predominant intracellular scFv species in the HBiP strain is an immature product at all time points. In order to elucidate the specific regulation of protein synthesis, degradation, and ER-to-Golgi trafficking mechanisms, these studies led to further analysis via microarray studies of the global gene expression changes induced by expression of a scFv construct integrated multiple times within the chromosome (i.e., hcscFv) and comparison to the UPR resulting from chemical stress (i.e., DTT addition).

8.4.1. Additional remarks

Accurate quantification of heterologous protein expression and UPR induction requires appropriate normalization procedures. Using these protocols, we normalize to number of cells assayed (via OD_{600}) and compare, by Western blot results, the magnitude of the UPR response and scFv expression to a loading control, Act1. Most importantly, by monitoring time-course expression, we can choose the optimal growth phase to harvest cells in order to achieve reproducible results. It is important to always maintain the appropriate controls, for example, the original plasmid minus the gene of interest, the parental strain without modifications, or multiple integrations of the product compared to a low-copy plasmid or single integration. Examples of these experimental conditions for optimized scFv secretion have been detailed elsewhere (Shusta *et al.*, 1998; Xu *et al.*, 2005). Investigating one strain under one condition with different constructs, and carrying out biological replicates, will ensure reliable results.

9. STATISTICAL ANALYSIS OF MICROARRAY RESULTS

The advent of high-throughput microarray technology to simultaneously measure the expression of thousands of genes has revolutionized the study of biological systems. A single microarray experiment provides an abundance of information from which diverse analyses are possible, from large-scale characterization of the effects of the experimental condition on known cellular processes to network inference efforts in which the regulatory pathways responsible for the system's response to the imposed condition are deduced. However, the obvious benefits of such technology are accompanied by the corresponding challenge of extracting meaningful insight from data that is comprehensive in genomic coverage but generally restricted with respect to aspects such as time-course measurements and biological replication. The challenge is compounded by the lack of a standard approach to microarray data analysis; the appropriate method often depends on the particulars of the experiment that generated the data.

Here we discuss the issues associated with the analysis of microarray data in the context of experiments performed to elucidate the UPR in *S. cerevisiae* (Fig. 14.4). In our experiments, we studied the UPR as activated by chemical treatment with DTT, a well-established UPR inducer, and by heterologous protein expression of hcscFv. RNA isolation methods were performed as described in *Current Protocols in Molecular Biology* (Ausubel *et al.*, 1987). Probe preparation and microarray hybridization were completed by the modified protocol of the UHN Microarray Centre (Canada). Arrays were scanned using ScanArray Express (Microarray Scanner, PerkinElmer) and analyzed by Quantarray software (BioChip Technologies LLC). Additionally, the intensities were normalized by the LOWESS method of TIGR MIDAS (Saeed *et al.*, 2003).

9.1. Appropriate controls

The selection of appropriate controls must be completed prior to the start of the experiment and is dependent upon the nature of the experiment. In the study of the UPR in *S. cerevisiae*, samples were removed from the cell culture at various times following induction. It is desirable, although not strictly necessary, to prepare samples from the treated and control cultures prior to initiating the treatment, at $t = 0$. In our experiment, specifically pertaining to UPR induction by DTT, we account for these possible discrepancies by dividing the original culture at $t = 0$ to separate aliquots (i.e., the negative control and treated sample (i.e., DTT addition)). Minor variation in the conditions of the cultures or even stochastic variation could lead to detectable and significant differences in gene expression; to compensate for these differences and home in on the effect of stress and the ensuing initiation of the UPR, the expression measurements should be normalized to those at $t = 0$. It is not uncommon to assume that the differences between the samples prior to treatment are negligible, but it is prudent to confirm such assumptions.

To further isolate the direct effects of UPR initiation from other influences (such as a metabolic shift), analysis of samples from $\Delta hac1$,

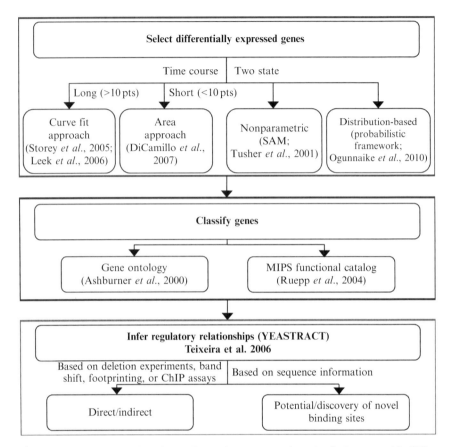

Figure 14.4 Decision chart for analysis of microarray data. Analysis starts with differential expression testing and the method chosen for this analysis depends on the type of experiment. Differentially expressed genes should then be classified according to the Gene Ontology of the Functional Catalog, and then regulatory relationships should be investigated. For *S. cerevisiae*, the YEASTRACT database is a singularly good resource between transcription factors and targets or, in a more discovery-driven effort, on sequence information.

$\Delta ire1$, and/or $\Delta hac1\Delta ire1$ mutants should be considered. Such analysis was undertaken in the canonical work by Travers *et al.* (2000), who induced the UPR by DTT treatment. However, this approach necessarily assumes that the only means by which *S. cerevisiae* responds to ER stress is via the generally accepted IRE1/HAC1 pathway; although pathways paralleling those seen in the higher eukaryotes have not yet been found, such an approach eliminates the possibility.

9.2. Selecting differentially expressed genes

Image analysis and normalization is a fairly straightforward task, with LOW-ESS normalization typically applied to eliminate systematic bias in the data (Quackenbush, 2002). Once these tasks have been completed, differentially expressed genes are selected. Several factors must be considered to select an appropriate method, most importantly whether the study is a time-course or static experiment and the number of available replicates. For a static, two-state experiment, a very popular method is the significance analysis of microarrays (SAM), a modified version of the standard t-test that was developed specifically to handle microarray data (Tusher et al., 2001). This method requires replicates and makes no underlying assumptions about the distribution of the data. It reports the false discovery rate as the metric for determining significant expression, a metric superior to the p-value.

A new mixture model method in which transformed signal intensities are characterized by a set of overlapping beta distributions is also a good choice for selecting differentially expressed genes in a two-state experiment (Ogunnaike et al., 2010). Starting from first principles, the authors showed that this distribution is an adequate descriptor of microarray data and from the set of overlapping fits they calculate the probability that a gene is upregulated, downregulated, or not regulated (Ogunnaike et al., 2010). Replicates can be used to calculate a confidence metric for the reported probabilities (Ogunnaike et al., 2010). This method has the advantages of requiring no replicates and of leveraging all the data in making decisions about the differential expression of each gene, an approach that recognizes gene expression values are not independent (Ogunnaike et al., 2010). Furthermore, it slightly outperformed SAM in a simulation study, recovering 99.6% of differentially expressed genes with five replicates, although SAM's performance is still remarkable with a recovery of 94.7% of the genes (Ogunnaike et al., 2010). A point and click MATLAB implementation is available from the author's Web site (Gelmi).

In the case of a time-course study, such as our microarray analysis of hcscFv expression in *S. cerevisiae*, a method specific to detecting changes over time should be used. Several such methods exist, including the method proposed by Storey et al. (2005), which explicitly accounts for the temporal ordering of the data. In this construction, one is testing whether the entire time-course profile is flat (the null hypothesis). Of course, no profile is ever truly flat, but the variation in some profiles is insignificant and not related to the experimental treatment, but rather a result of limitations of the technology and noise inherent in data collection. To test for differential expression of a given gene, two curves are fit to the expression data for that gene. The first is simply the best-fit curve, which can be fit with a higher order polynomial or cubic splines. The second fitted curve is the best flat line through the data. Then a test statistic that quantifies the relative goodness of fit of the two curves is calculated. To determine significance, a null distribution of test statistics is constructed from the replicate data. Because replicates should, theoretically, be exactly the same value, any discrepancy between them can be attributed to the inherent noise in the experiment that is not related to the biological process under consideration. Null profiles are created from the difference between the replicates, from which the null distribution of test statistics is constructed.

Replicates are not required at every time point nor for every gene, but a sufficient number of replicate permutations must be completed to create a reasonably sized null distribution; 1000 points should be sufficient, although it is often possible to calculate a much larger distribution for little additional computational expense. The main disadvantage of this method is the generally short time courses obtained via microarrays; this method works best when at least 12 points have been gathered, due to overfitting issues. An open-source software package with a user-friendly interface has been developed to complete this differential expression analysis (Leek *et al.*, 2006).

For time-course studies with less than 12 data points, an effective alternative has been proposed (Di Camillo *et al.*, 2007). Rather than fitting curves to the data, the area between the log transformed expression profile and the null profile, a flat line centered at zero, is calculated. Null distributions are again calculated from the replicate data. It was shown that this method was more effective at reducing false positives than the curve-fitting method for shorter time series. We implemented this method in MATLAB (Mathworks, Cambridge, UK), and based on our experience, we suggest this method for time-course studies with few time points.

9.3. Classification

The large amount of data generated in a microarray experiment necessitates a means of consolidating and interpreting that data. Hence the development of structured vocabularies, such as the gene ontology (GO), which provide a means of consolidating information of genes according to the biological process in which they are involved, their molecular function, or their cellular localization (Ashburner *et al.*, 2000). The GO is organized as a directed acyclic graph and its descriptions of categories are very detailed, leading to a large number of terms. The Functional Catalog (FunCat) is an alternative to the GO and does not attempt to exhaustively characterize protein function to its most granular level, but instead classifies proteins according to functional modules (Ruepp *et al.*, 2004). The choice of databases is somewhat arbitrary, as they will provide reasonably consistent results insofar as the categories in each scheme are parallel, but the FunCat's focus on functional modules facilitates data interpretation.

Regardless of the database, two important issues must be considered in classifying interesting genes from microarray experiments. First, the background set against which the interesting set is compared must be established. In most cases, the best choice of a background set is the set of genes measured via microarray; this is not necessarily the set of genes in the genome. If the microarray is enriched in genes associated with metabolism, for example, but the genome is used for the background set, it is likely that the differentially expressed genes will be highly enriched in metabolism. Yet, this result will be a function of the enrichment of the microarray, and not truly a result of the experimental protocol. The second important issue to consider is the statistical significance of the results. Returning to our previous example, if 50% of the genes on the microarray are associated with metabolism, and 50% of the differentially expressed genes are associated with metabolism, the result might simply be a function of random chance rather than associated with the biological phenomena under study. That is not to say that the results are not biologically valid, as our upregulated group was enriched in genes encoding proteins involved in protein folding and stabilization but the conservatively corrected p-value was above the significance threshold. This finding is supported by established knowledge about the UPR, bolstering our confidence in its relevance, but in the absence of other evidence a systematic, well-supported conclusion could not be drawn. Therefore, it is important to evaluate the statistical significance of the results, an easy task to which the hypergeometric distribution function can be applied. The results must also be corrected for the multiple comparisons made in any such classification effort. The Bonferonni correction is typical, although it is very conservative. According to this correction, our upregulated gene group was enriched most notably in genes encoding proteins residing in the ER, with a Bonferonni corrected p-value of 0.02. An alternative approach is to convert the p-values derived from the hypergeometric distribution into q-values, which are the false discovery rates that would be incurred if the associated category were called significant.

9.4. Enrichment of promoter regions in known transcription factor-binding sites

Selecting differentially expressed genes and classifying them according to their biological function provides a great deal of information about the cellular response to ER stress, but it does not illuminate the regulatory pathways by which the response occurs. To address this question, the YEASTRACT database is singularly appropriate; it contains a record of all the documented associations between known *S. cerevisiae* TFs and their target genes, based on deletion experiments, band-shift, footprinting, or ChIP assays (Teixeira *et al.*, 2006). These associations may be direct or indirect. The database also contains information about potential associations based on the presence of known TF-binding site sequences in the promoter regions of *S. cerevisiae* genes. Using sets of differentially expressed genes as inputs, all known TF associations can be mined from this database. One must be careful not to simply assume the TF with the highest number of associations is responsible for the observed phenomena. Rather, evaluations similar to those applied to the classification effort must be completed to find statistically significant associations; again, the hypergeometric distribution is appropriate for this task.

Consider the results we obtained from our microarray experiment in which DTT was used to trigger the UPR. We identified a group of upregulated genes enriched for protein fate, an expected outcome given the knowledge of UPR effects. The TF Ste12p is directly associated with 30% of the genes in this group, while the known UPR regulator Hac1p is only directly associated with 4% of the group. However, when compared against the background set, the Bonferonni corrected p-value associated with Hac1p is $2e^{-4}$, while that associated with Ste12p is 0.5. Only by considering the statistical significance was the known UPR regulator (Hac1p) recovered from this analysis of the data. Discovering new regulators requires the same rigor, lest experimental resources be wasted on unpromising leads.

9.5. Presence of UPRE-1, UPRE-2, UPRE-3 in the promoter region of differentially expressed genes

Greater confidence in new results, such as the novel downregulation that we observe in our experiments to elucidate the UPR, can be obtained by evaluating the upregulated group for consistency with known facts about the UPR. Three TF-binding sites that confer upregulation on UPR targets have been identified (Mori *et al.*, 1998; Patil *et al.*, 2004b; Fig. 14.1); therefore the upregulated group should show significant enrichment in these binding sites. To determine their enrichment, the promoter regions of the upregulated genes must be extracted, necessitating a decision about what constitutes a promoter region; 600–1000 bp upstream of the start codon is a commonly used definition. The degree of allowable mismatch and the presence of the binding sites in the background set must be determined. Then, enrichment and statistical significance can be calculated. In our experiments, we found significant enrichment for these three binding sites; 42% of the upregulated group contained one of the three sites at a 0.03 level of significance.

9.6. Relative quantification by real-time quantitative PCR

Microarray data is inherently noisy and conclusions from these experiments are based on statistical analyses, necessitating further validation that is typically accomplished with q-PCR experiments. Data collected from q-PCR experiments is still widely processed by the so-called $2^{\Delta\Delta Ct}$ method. The full derivation is supplied in Livak and Schmittgen (2001), but the relevant assumption here is that the efficiency with which all RNA is amplified is 100%. However, the fold-change calculation is very sensitive to the amplification efficiency and any violations of this assumption result in significant variability in the results. Bustin and Nolan (2004) observe that the efficiency of reverse transcription is low for genes expressed at low levels and point out that this contributes to the variability and lack of reproducibility of a q-PCR experiment. Ramakers et al. (2003) showed that theoretically, the effect of 20% variability in efficiency can result in fold-change results spanning three orders of magnitude, making it impossible to even determine the direction in which the gene was regulated. In practice, assuming efficiencies are close to 100% can be wildly inaccurate and one can run afoul of the aforementioned problems; in repeated trials, we found that the efficiency with which a canonical UPR gene, KAR2/BiP, was amplified remained consistently at $<80\%$.

Several methods can be used to alleviate problems associated with less-than-ideal efficiencies. The current gold standard is utilizing a standard curve, but this approach has its drawbacks (Rutledge and Stewart, 2008a). Not only is it resource intensive, it also assumes that the efficiency with which any particular sample is amplified is constant (Rutledge and Stewart, 2008a). Although the latter issue may not be significant, as it appears that the variability associated with the sample is due to random error and is relatively less important than the variability associated with the amplicon (Karlen et al., 2007; Nordgård et al., 2006), the former issue remains a problem.

To circumvent this excessive use of resources, methods that determine efficiency directly from the amplification curves can be used. These methods are based on curve fitting and are generally based on either sigmoidal or exponential fits. To calculate fold changes in expression for microarray validation, we suggest the "linear regression of efficiency" (LRE) analysis proposed by Rutledge and Stewart (2008b) to calculate efficiencies for use in Eq. (14.1) which gives the fold change for gene "X."

$$\text{Fold change} = \frac{\left\{\frac{\left(1+E_{hk,treated}\right)^{C_{T,hk,treated}}}{\left(1+E_{x,treated}\right)^{C_{T,x,treated}}}\right\}}{\left\{\frac{\left(1+E_{hk,control}\right)^{C_{T,hk,control}}}{\left(1+E_{x,control}\right)^{C_{T,x,control}}}\right\}} = \frac{\left(1+E_{hk}\right)^{C_{T,hk,treated}-C_{T,hk,control}}}{\left(1+E_{x}\right)^{C_{T,x,treated}-C_{T,x,control}}}$$

(14.1)

where $E_{hk,treated}$, $C_{T,hk,treated}$, $E_{hk,control}$, and $C_{T,hk,control}$ are the efficiencies (E) and C values, respectively, associated with the housekeeping (hk) gene in the treated and control samples, and $E_{x,treated}$, $C_{T,x,treated}$, $E_{x,control}$, and

$C_{T,x,\text{control}}$ are the efficiencies and C_T values associated with the gene "X" for which the fold change is being calculated in the treated and control sample.

Although this method was developed with absolute quantification in mind, it is easily applicable to the relative quantification of RNA. It assumes that fluorescence and amplicon quantity are proportional and has the advantage of accounting for the decline in efficiency over the course of the amplification reaction (Rutledge and Stewart, 2008b); most methods assume that the efficiency is constant in the log-linear region. Based on results showing that the variability in efficiency is random and most strongly influenced by sequence (Karlen et al., 2007; Nordgård et al., 2006), we suggest merging this method with that of Karlen et al. (2007) to provide a robust measure of efficiency. Instead of using sample-specific efficiencies, the efficiency should be taken as the average over all the samples and replicates for an amplicon; such an approach was found to be highly effective (Karlen et al., 2007). Furthermore, it is important to consider the threshold at which the C_T values are calculated. It is common to choose a single value for all the amplicons studied, but this approach does not guarantee that the threshold will fall in the log-linear region of the amplification curve. Thus, it is suggested that the threshold be chosen in an amplicon-specific manner.

Dealing with replicates and establishing error estimates is a subject that is not often addressed in the methodology papers for q-PCR analysis, but is important for establishing the statistical reliability of the validation results. At least three replicates should be used to guarantee statistically significant results and possibly more, depending on the induction ratio of the amplicon being validated and the reproducibility of the amplification reaction curves (Karlen et al., 2007). To estimate the error, standard error propagation formulas should be used (Nordgård et al., 2006).

9.7. Implications for protein secretion and UPR induction

The application of microarray technology and appropriate analysis methods contribute to our understanding of protein secretion in S. cerevisiae and the effect of the UPR on cellular function. In particular, we found that four key trafficking proteins, previously unreported, were upregulated in response to UPR induction by DTT: Sec72, Sil1, Erv29, Erp2. Specifically, Sec72 is a component of the ER translocation complex, Sil1 is a nucleotide exchange factor of chaperone BiP/Kar2, and Erv29 and Erp2 are products of ER-to-Golgi trafficking localized to COPII machinery. Interestingly, Erv29 directly interacts with a specific motif on the pro-signal sequence and is suggested to direct soluble cargo to vesicle components of the secretory pathway (Otte and Barlowe, 2004). Rigorous analysis of microarray data under various environmental stimuli combined with complementary diverse experimental methods has resulted in an improved understanding of scFv expression, trafficking, and ER quality control in yeast, *S. cerevisiae*.

ACKNOWLEDGMENT

The authors were supported in part for the experimental studies by NIH R01 GM 075297.

REFERENCES

Arava, Y., Wang, Y., Storey, J. D., Liu, C. L., Brown, P. O., and Herschlag, D. (2003). Genome-wide analysis of mRNA translation profiles in Saccharomyces cerevisiae. *Proc. Natl. Acad. Sci. USA* **100,** 3889–3894.

Arvan, P., Zhao, X., Ramos-Castaneda, J., and Chang, A. (2002). Secretory pathway quality control operating in Golgi, plasmalemmal, and endosomal systems. *Traffic* **3,** 771–780.

Ashburner, M., Ball, C. A., Blake, J. A., Botstein, D., Butler, H., Cherry, J. M., Davis, A. P., Dolinski, K., Dwight, S. S., Eppig, J. T., Harris, M. A., Hill, D. P., *et al.* (2000). Gene ontology: Tool for the unification of biology. The Gene Ontology Consortium. *Nat. Genet.* **25,** 25–29.

Ausubel, F.M., Brent, R., Kingston, R.E., Moore, D.D., Seidman, J.G., Smith, J.A., and Struhl, K. (1987). Purification and concentration of RNA and dilute solutions of DNA. *Current Protocols in Molecular Biology*, section 2.1.7 (Supplement 59).

Belli, G., Gari, E., Aldea, M., and Herrero, E. (1998). Functional analysis of yeast essential genes using a promoter-substitution cassette and the tetracycline-regulatable dual expression system. *Yeast* **14,** 1127–1138.

Boder, E. T., Midelfort, K. S., and Wittrup, K. D. (2000). Directed evolution of antibody fragments with monovalent femtomolar antigen-binding affinity. *Proc. Natl. Acad. Sci. USA* **97,** 10701–10705.

Brachmann, C. B., Davies, A., Cost, G. J., Caputo, E., Li, J., Hieter, P., and Boeke, J. D. (1998). Designer deletion strains derived from Saccharomyces cerevisiae S288C: A useful set of strains and plasmids for PCR-mediated gene disruption and other applications. *Yeast* **14,** 115–132.

Brauer, M. J., Huttenhower, C., Airoldi, E. M., Rosenstein, R., Matese, J. C., Gresham, D., Boer, V. M., Troyanskaya, O. G., and Botstein, D. (2008). Coordination of growth rate, cell cycle, stress response, and metabolic activity in yeast. *Mol. Biol. Cell* **19,** 352–367.

Brodsky, J. L. (2007). The protective and destructive roles played by molecular chaperones during ERAD (endoplasmic-reticulum-associated degradation). *Biochem. J.* **404,** 353–363.

Bustin, S. A., and Nolan, T. (2004). Pitfalls of quantitative real-time reverse-transcription polymerase chain reaction. *J. Biomol. Tech.* **15,** 155–166.

Chao, G., Lau, W. L., Hackel, B. J., Sazinsky, S. L., Lippow, S. M., and Wittrup, K. D. (2006). Isolating and engineering human antibodies using yeast surface display. *Nat. Protoc.* **1,** 755–768.

Clements, J. M., Catlin, G. H., Price, M. J., and Edwards, R. M. (1991). Secretion of human epidermal growth factor from Saccharomyces cerevisiae using synthetic leader sequences. *Gene* **106,** 267–271.

Denzin, L. K., and Voss, E. W., Jr. (1992). Construction, characterization, and mutagenesis of an anti-fluorescein single chain antibody idiotype family. *J. Biol. Chem.* **267,** 8925–8931.

Di Camillo, B., Toffolo, G., Nair, S. K., Greenlund, L. J., and Cobelli, C. (2007). Significance analysis of microarray transcript levels in time series experiments. *BMC Bioinform.* **8** (Suppl. 1), S10.

Gasch, A. P., Spellman, P. T., Kao, C. M., Carmel-Harel, O., Eisen, M. B., Storz, G., Botstein, D., and Brown, P. O. (2000). Genomic expression programs in the response of yeast cells to environmental changes. *Mol. Biol. Cell* **11**, 4241–4257.

Gelmi, C.A., http://www.che.udel.edu/systems/people/cgelmi/tools.htm.

Hong, E., Davidson, A. R., and Kaiser, C. A. (1996). A pathway for targeting soluble misfolded proteins to the yeast vacuole. *J. Cell Biol.* **135**, 623–633.

Huang, D., Gore, P. R., and Shusta, E. V. (2008). Increasing yeast secretion of heterologous proteins by regulating expression rates and post-secretory loss. *Biotechnol. Bioeng.* **101**, 1264–1275.

Jones, E. W. (2002). Vacuolar proteases and proteolytic artifacts in Saccharomyces cerevisiae. *Methods Enzymol.* **351**, 127–150.

Jonikas, M. C., Collins, S. R., Denic, V., Oh, E., Quan, E. M., Schmid, V., Weibezahn, J., Schwappach, B., Walter, P., Weissman, J. S., and Schuldiner, M. (2009). Comprehensive characterization of genes required for protein folding in the endoplasmic reticulum. *Science* **323**, 1693–1697.

Karlen, Y., McNair, A., Perseguers, S., Mazza, C., and Mermod, N. (2007). Statistical significance of quantitative PCR. *BMC Bioinform.* **8**, 131.

Kauffman, K. J., Pridgen, E. M., Doyle, F. J., III, Dhurjati, P. S., and Robinson, A. S. (2002). Decreased protein expression and intermittent recoveries in BiP levels result from cellular stress during heterologous protein expression in Saccharomyces cerevisiae. *Biotechnol. Prog.* **18**, 942–950.

Kimata, Y., Ishiwata-Kimata, Y., Ito, T., Hirata, A., Suzuki, T., Oikawa, D., Takeuchi, M., and Kohno, K. (2007). Two regulatory steps of ER-stress sensor Ire1 involving its cluster formation and interaction with unfolded proteins. *J. Cell Biol.* **179**, 75–86.

Leek, J. T., Monsen, E., Dabney, A. R., and Storey, J. D. (2006). EDGE: Extraction and analysis of differential gene expression. *Bioinformatics* **22**, 507–508.

Livak, K. J., and Schmittgen, T. D. (2001). Analysis of relative gene expression data using real-time quantitative PCR and the 2-[Delta][Delta]CT method. *Methods* **25**, 402–408.

Luo, W., and Chang, A. (2000). An endosome-to-plasma membrane pathway involved in trafficking of a mutant plasma membrane ATPase in yeast. *Mol. Biol. Cell* **11**, 579–592.

McCracken, A. A., and Brodsky, J. L. (2003). Evolving questions and paradigm shifts in endoplasmic-reticulum-associated degradation (ERAD). *Bioessays* **25**, 868–877.

Midelfort, K. S., and Wittrup, K. D. (2006). Context-dependent mutations predominate in an engineered high-affinity single chain antibody fragment. *Protein Sci.* **15**, 324–334.

Moehle, C. M., Aynardi, M. W., Kolodny, M. R., Park, F. J., and Jones, E. W. (1987). Protease B of Saccharomyces cerevisiae: Isolation and regulation of the PRB1 structural gene. *Genetics* **115**, 255–263.

Mori, K., Sant, A., Kohno, K., Normington, K., Gething, M. J., and Sambrook, J. F. (1992). A 22 bp cis-acting element is necessary and sufficient for the induction of the yeast KAR2 (BiP) gene by unfolded proteins. *EMBO J.* **11**, 2583–2593.

Mori, K., Ogawa, N., Kawahara, T., Yanagi, H., and Yura, T. (1998). Palindrome with spacer of one nucleotide is characteristic of the cis-acting unfolded protein response element in Saccharomyces cerevisiae. *J. Biol. Chem.* **273**(6), 9912–9920.

Nagai, T., Ibata, K., Park, E. S., Kubota, M., Mikoshiba, K., and Miyawaki, A. (2002). A variant of yellow fluorescent protein with fast and efficient maturation for cell-biological applications. *Nat. Biotechnol.* **20**, 87–90.

Nakatsukasa, K., and Brodsky, J. L. (2008). The recognition and retrotranslocation of misfolded proteins from the endoplasmic reticulum. *Traffic* **9**, 861–870.

Newstead, S., Kim, H., von Heijne, G., Iwata, S., and Drew, D. (2007). High-throughput fluorescent-based optimization of eukaryotic membrane protein overexpression and purification in Saccharomyces cerevisiae. *Proc. Natl. Acad. Sci. USA* **104,** 13936–13941.

Ng, D. T., Spear, E. D., and Walter, P. (2000). The unfolded protein response regulates multiple aspects of secretory and membrane protein biogenesis and endoplasmic reticulum quality control. *J. Cell Biol.* **150,** 77–88.

Nordgård, O., Kvaløy, J. T., Farmen, R. K., and Heikkilä, R. (2006). Error propagation in relative real-time reverse transcription polymerase chain reaction quantification models: The balance between accuracy and precision. *Anal. Biochem.* **356,** 182–193.

Ogunnaike, B. A., Gelmi, C. A., and Edwards, J. S. (2010). A probabilistic framework for microarray data analysis: Fundamental probability models and statistical inference. *J. Theor. Biol.* **264,** 211–222.

Oikawa, D., Kimata, Y., and Kohno, K. (2007). Self-association and BiP dissociation are not sufficient for activation of the ER stress sensor Ire1. *J. Cell Sci.* **120,** 1681–1688.

Otte, S., and Barlowe, C. (2004). Sorting signals can direct receptor-mediated export of soluble proteins into COPII vesicles. *Nat. Cell Biol.* **6,** 1189–1194.

Parekh, R. N., Shaw, M. R., and Wittrup, K. D. (1996). An integrating vector for tunable, high copy, stable integration into the dispersed Ty delta sites of Saccharomyces cerevisiae. *Biotechnol. Prog.* **12,** 16–21.

Patil, C. K., Li, H., and Walter, P. (2004). Gcn4p and novel upstream activating sequences regulate targets of the unfolded protein response. *PLoS Biol.* **2,** E246.

Petracek, M. E., and Longtine, M. S. (2002). PCR-based engineering of yeast genome. *Methods Enzymol.* **350,** 445–469.

Quackenbush, J. (2002). Microarray data normalization and transformation. *Nat. Genet.* **32** (Suppl), 496–501.

Rakestraw, A., and Wittrup, K. D. (2006). Contrasting secretory processing of simultaneously expressed heterologous proteins in Saccharomyces cerevisiae. *Biotechnol. Bioeng.* **93,** 896–905.

Rakestraw, J. A., Sazinsky, S. L., Piatesi, A., Antipov, E., and Wittrup, K. D. (2009). Directed evolution of a secretory leader for the improved expression of heterologous proteins and full-length antibodies in Saccharomyces cerevisiae. *Biotechnol. Bioeng.* **103,** 1192–1201.

Ramakers, C., Ruijter, J. M., Deprez, R. H. L., and Moorman, A. F. M. (2003). Assumption-free analysis of quantitative real-time polymerase chain reaction (PCR) data. *Neurosci. Lett.* **339,** 62–66.

Rizzo, M. A., Springer, G. H., Granada, B., and Piston, D. W. (2004). An improved cyan fluorescent protein variant useful for FRET. *Nat. Biotechnol.* **22,** 445–449.

Robinson, A. S., Hines, V., and Wittrup, K. D. (1994). Protein disulfide isomerase overexpression increases secretion of foreign proteins in Saccharomyces cerevisiae. *Biotechnology (NY)* **12,** 381–384.

Romanos, M. A., Clare, J. J., and Brown, C. (2001). Culture of yeast for the production of heterologous proteins. *Curr. Protoc. Protein Sci.* May; Chapter 5, Unit5 8.

Rothblatt, J., and Schekman, R. (1989). A hitchhiker's guide to analysis of the secretory pathway in yeast. *Methods Cell. Biol.* **32,** 3–36.

Ruepp, A., Zollner, A., Maier, D., Albermann, K., Hani, J., Mokrejs, M., Tetko, I., Güldener, U., Mannhaupt, G., Münsterkötter, M., and Mewes, H. W. (2004). The FunCat, a functional annotation scheme for systematic classification of proteins from whole genomes. *Nucleic Acids Res.* **32,** 5539–5545.

Rutledge, R., and Stewart, D. (2008a). Critical evaluation of methods used to determine amplification efficiency refutes the exponential character of real-time PCR. *BMC Mol. Biol.* **9,** 96.

Rutledge, R., and Stewart, D. (2008b). A kinetic-based sigmoidal model for the polymerase chain reaction and its application to high-capacity absolute quantitative real-time PCR. *BMC Biotechnol.* **8,** 47.

Saeed, A. I., Sharov, V., White, J., Li, J., Liang, W., Bhagabati, N., Braisted, J., Klapa, M., Currier, T., Thiagarajan, M., Sturn, A., Snuffin, M., *et al.* (2003). TM4: A free, open-source system for microarray data management and analysis. *Biotechniques* **34,** 374–378.

Shaner, N. C., Campbell, R. E., Steinbach, P. A., Giepmans, B. N., Palmer, A. E., and Tsien, R. Y. (2004). Improved monomeric red, orange and yellow fluorescent proteins derived from Discosoma sp. red fluorescent protein. *Nat. Biotechnol.* **22,** 1567–1572.

Sherman, F. (2002). Getting started with yeast. *Methods Enzymol.* **350,** 3–41.

Shusta, E. V., Raines, R. T., Pluckthun, A., and Wittrup, K. D. (1998). Increasing the secretory capacity of Saccharomyces cerevisiae for production of single-chain antibody fragments. *Nat. Biotechnol.* **16,** 773–777.

Sikorski, R. S., and Hieter, P. (1989). A system of shuttle vectors and yeast host strains designed for efficient manipulation of DNA in Saccharomyces cerevisiae. *Genetics* **122,** 19–27.

Storey, J. D., Wenzhong, X., Leek, J. T., Tompkins, R. G., and Davis, R. W. (2005). Significance analysis of time course microarray experiments. *PNAS* **102,** 12837–12842.

Takeshige, K., Baba, M., Tsuboi, S., Noda, T., and Ohsumi, Y. (1992). Autophagy in yeast demonstrated with proteinase-deficient mutants and conditions for its induction. *J. Cell Biol.* **119,** 301–311.

Teixeira, M. C., Monteiro, P., Jain, P., Tenreiro, S., Fernandes, A. R., Mira, N. P., Alenquer, M., Freitas, A. T., Oliveira, A. L., and Sá-Correia, I. (2006). The YEAS-TRACT database: A tool for the analysis of transcription regulatory associations in Saccharomyces cerevisiae. *Nucleic Acids Res.* **34,** D446–D451.

Travers, K. J., Patil, C. K., Wodicka, L., Lockhart, D. J., Weissman, J. S., and Walter, P. (2000). Functional and genomic analyses reveal an essential coordination between the unfolded protein response and ER-associated degradation. *Cell* **101,** 249–258.

Tusher, V. G., Tibshirani, R., and Chu, G. (2001). Significance analysis of microarrays applied to the ionizing radiation response. *PNAS* **98**(6), 5116–5121.

Valkonen, M., Penttila, M., and Saloheimo, M. (2003). Effects of inactivation and constitutive expression of the unfolded-protein response pathway on protein production in the yeast Saccharomyces cerevisiae. *Appl. Environ. Microbiol.* **69,** 2065–2072.

van Anken, E., and Braakman, I. (2005). Versatility of the endoplasmic reticulum protein folding factory. *Crit. Rev. Biochem. Mol. Biol.* **40,** 191–228.

Wentz, A. E., and Shusta, E. V. (2007). A novel high-throughput screen reveals yeast genes that increase secretion of heterologous proteins. *Appl. Environ. Microbiol.* **73,** 1189–1198.

Wittrup, K. D., and Benig, V. (1994). Optimization of amino-acid supplements for heterologous protein secretion in Saccharomyces cerevisiae. *Biotechnol. Tech.* **8,** 161–166.

Xu, P., and Robinson, A. S. (2009). Decreased secretion and unfolded protein response up-regulation are correlated with intracellular retention for single-chain antibody variants produced in yeast. *Biotechnol. Bioeng.* **104,** 20–29.

Xu, P., Raden, D., Doyle, F. J., III, and Robinson, A. S. (2005). Analysis of unfolded protein response during single-chain antibody expression in Saccharomyces cerevisiae reveals different roles for BiP and PDI in folding. *Metab. Eng.* **7,** 269–279.

Zhang, B., Chang, A., Kjeldsen, T. B., and Arvan, P. (2001). Intracellular retention of newly synthesized insulin in yeast is caused by endoproteolytic processing in the Golgi complex. *J. Cell Biol.* **153,** 1187–1198.

CHAPTER FIFTEEN

Measuring Signaling by the Unfolded Protein Response

David J. Cox,[*] Natalie Strudwick,[*] Ahmed A. Ali,[*,†]
Adrienne W. Paton,[‡] James C. Paton,[‡] and Martin Schröder[*]

Contents

1. Introduction	262
2. Methods to Induce the UPR	266
3. Measuring Signaling by the UPR in *S. cerevisiae*	266
3.1. General guidelines for work with RNA	268
3.2. Required materials	269
3.3. Growth of yeast cultures and UPR induction	271
3.4. Isolation of total RNA from *S. cerevisiae*	271
3.5. Separation of RNA by agarose gel electrophoresis	272
4. Measuring Signaling by the UPR in Mammalian Cells	276
4.1. mRNA methods for analyzing the UPR	277
4.2. Protein methods for analyzing the UPR	281
Acknowledgments	288
References	289

Abstract

The unfolded protein response (UPR) is activated by accumulation of unfolded proteins in the endoplasmic reticulum (ER). The unfolded protein response is associated with many diseases, including cancer, metabolic diseases such as type II diabetes and fatty liver diseases, and neurodegenerative diseases, for example, Alzheimer's disease. The UPR is also activated by numerous toxic chemicals and modulates drug action. Therefore, the UPR becomes increasingly important in toxicological and pharmacological research. In mammals, the UPR is transduced through three parallel signaling pathways originating at the ER-resident transmembrane protein kinase-endoribonucleases (RNase) IRE1, the protein kinase PERK, and a family of type II transmembrane transcription factors, whose most prominent member is ATF6α.

---

[*] School of Biological and Biomedical Sciences, Durham University, Durham, United Kingdom
[†] Department of Molecular Biology, National Research Centre, Cairo, Egypt
[‡] Research Centre for Infectious Diseases, School of Molecular and Biomedical Science, University of Adelaide, Adelaide, Australia

Methods in Enzymology, Volume 491 © 2011 Elsevier Inc.
ISSN 0076-6879, DOI: 10.1016/B978-0-12-385928-0.00015-8 All rights reserved.

We discuss methods to experimentally activate the UPR in the yeast *Saccharomyces cerevisiae* and in cultured mammalian cells. We summarize methods to monitor activation of the three arms of the UPR, while providing detailed protocols for select, reliable assays. To monitor activation of the IRE1 branch, a Northern blotting protocol to monitor splicing of *HAC1* mRNA in yeast and a reverse transcriptase-PCR assay for processing of the IRE1 RNase substrate *XBP1* in mammalian cells are presented. Activation of the IRE1 kinase activity can be assayed by immunoblotting for IRE1 autophosphorylation. Activation of the PERK branch is monitored via phosphorylation of the translation initiation factor eIF2α, induction of CHOP at the mRNA and protein level, and induction of ATF4 at the protein level. Activation of ATF6 is assayed in Western blots through the appearance of its processed 50 kDa soluble cytosolic fragment. We summarize reverse transcriptase-PCR protocols to measure activation of target genes selectively induced by the three branches of the UPR and histological assays for UPR activation in tissue sections. This repertoire of methods will enable the newcomer to the UPR field to comprehensively assess the activation status of the UPR.

1. Introduction

The unfolded protein response (UPR) is a signal transduction pathway that maintains the homeostasis of the endoplasmic reticulum (ER). The UPR plays important roles in many diseases that disrupt ER homeostasis. These diseases include type II diabetes, neurodegenerative diseases, and diseases caused by or associated with mutant secretory cargo proteins, such as α_1-antitrypsin deficiency and cystic fibrosis. Many toxic chemicals and chemotherapeutic drugs evoke ER stress and UPR-mediated apoptosis. The UPR also modulates drug action. For example, induction of the UPR in multiple myeloma treated with the proteasome inhibitor bortezomib not only protects the myeloma from action of the drug through induction of ER-resident chaperones but also mediates the apoptotic effects of the drug (Dong *et al.*, 2009; Neubert *et al.*, 2008). Because of these important roles of the UPR in human diseases, toxicology, and drug development, monitoring the activation status of the UPR has become increasingly important.

The mammalian UPR consists of three principal signaling pathways (Ron and Walter, 2007; Schröder, 2008). Of these, only the pathway engaging the bifunctional ER-resident type I transmembrane protein kinase-endoribonuclease IRE1 is conserved in all eukaryotes (Fig. 15.1). Upon oligomerization and autophosphorylation, IRE1 activates its RNase domain. The RNase activity initiates splicing of mRNAs encoding bZIP transcription factors, Hac1[1] in the yeast *Saccharomyces cerevisiae*, and XBP1 in metazoans. These splicing events remove 252 nt from *HAC1* mRNA and 26 nt from *XBP1* mRNA and can be monitored by Northern blotting or

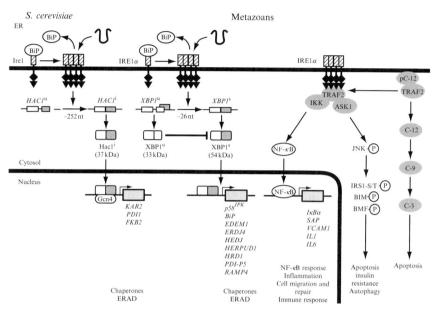

Figure 15.1 The IRE1 pathway in *S. cerevisiae* and mammalian cells. In both *S. cerevisiae* and metazoans, release of the molecular chaperone BiP or direct interaction with unfolded proteins activates Ire1. Upon oligomerization, IRE1 autophosphorylates and activates its RNase domain. Active Ire1 initiates removal of a 252 nt intron harboring a translational attenuator from *HAC1* mRNA. Spliced Hac1 (Hac1^i) induces transcription of genes encoding ER-resident molecular chaperones, such as *KAR2* (BiP), *PDI1*, and *FKB2*, and components of the ER-associated protein degradation (ERAD) machinery. A similar splicing event in metazoan *XBP1* mRNA removes 26 nt, resulting in a significant increase in the size of the XBP1 protein. Spliced XBP1 (XBP1^s) induces expression of ER chaperones and ERAD components. Interaction of active IRE1 with TRAF2 activates the transcription factor NF-κB, the MAP kinase JNK, and a caspase cascade. Abbreviations: C - caspase, pC - procaspase.

reverse transcriptase polymerase chain reaction (PCR). Splicing of *HAC1* mRNA removes a translational attenuator. Therefore, only spliced *HAC1* is translated into a protein that runs in sodium dodecylsulfate (SDS)-polyacrylamide gel electrophoresis (PAGE) gels at ∼37 kDa. Both unspliced and spliced *XBP1* are translated, but due to a frame shift introduced in the splicing reaction, the size of XBP1 increases from ∼33 to ∼54 kDa, producing a significantly more potent transcriptional activation domain in the XBP1 *C*-terminus. Both Hac1^i and XBP1 induce a range of target genes (Fig. 15.1). Mammalian cells possess two IRE1 isoforms, IRE1α and IRE1β, encoded by two separate genes. IRE1α is ubiquitously expressed, whereas expression of IRE1β is restricted to intestinal tissues. In contrast to yeast Ire1, both IRE1α and IRE1β can cleave many other mRNAs encoding secretory proteins, leading to their degradation. IRE1α/β cleavage-mediated decay of mRNAs encoding secretory proteins contributes to decreasing the nascent, unfolded protein load of the ER (Han *et al.*, 2009; Hollien and Weissman, 2006; Hollien *et al.*, 2009). IRE1α activates the transcription factor NF-κB and the MAP kinase JNK via the adaptor protein TRAF2. Sequestration of TRAF2 by IRE1 activates a caspase cascade, leading to proteolytic activation of the proapoptotic executioner caspase, caspase-3. IRE1α and JNK phosphorylation can be monitored in Western blots using phosphoprotein-specific antibodies. NF-κB activation can be monitored using one of many commercial kits.

The second pathway originates at the type I transmembrane protein kinase PERK (Fig. 15.2; Ron and Walter, 2007; Schröder, 2008). After autophosphorylation, PERK attenuates cap-dependent translation by phosphorylating the translation initiation factor eIF2α. This event not only attenuates general cap-dependent translation, but also promotes translation

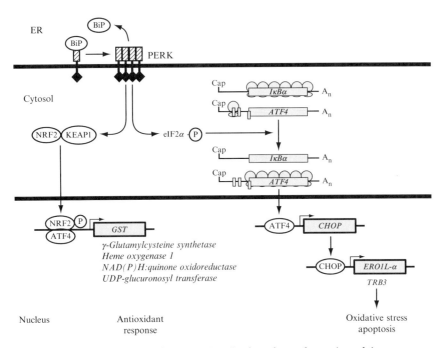

Figure 15.2 The PERK pathway. PERK is thought to be activated in a manner identical to that of IRE1. Phosphorylation of eIF2α by PERK at S51 attenuates general cap-dependent translation, but also promotes translation of mRNAs harboring several short uORFs or IRESs. For example, the mRNA for the bZIP transcription factor ATF4 contains several uORFs. Activation of ATF4 leads to activation of the bZIP transcription factor CHOP, induction of the protein disulfide isomerase ERO1L-α, and ROS generation in ER-stressed cells. This ROS production is countered by PERK-dependent activation of the bZIP transcription factor NRF2, which activates expression genes involved in antioxidant responses.

of mRNAs harboring complex structures in their 5′-untranslated regions (5′-UTRs) such as multiple short upstream open reading frames (uORFs) or internal ribosomal entry sites (IRESs). Phosphorylation of the bZIP transcription factor NRF2 releases NRF2 from cytosolic stores. Active NRF2 mounts an antioxidant response which counters increased reactive oxygen species (ROS) production originating from CHOP-dependent induction of the ER foldase ERO1L-α. Activation of the PERK pathway is most reliably measured through monitoring phosphorylation of eIF2α and induction of CHOP at the mRNA and protein level. Phosphorylation of PERK retards its migration in SDS-PAGE gels considerably, and is also a valid, but more difficult to access tool to monitor activation of PERK.

The third pathway involves translocation of several type II transmembrane bZIP transcription factor precursor proteins from the ER to the Golgi, followed by proteolytic liberation of the soluble, cytosolic transcription factor domain by sequential cleavage by the site 1 and 2 proteases (S1P and S2P) in the Golgi complex (Fig. 15.3; Ron and Walter, 2007; Schröder, 2008). The most prominent member of this family of 7 transcription factors is ATF6α. Conceptually, proteolytic processing of full-length ATF6 (~90 kDa) to its active form (~50 kDa) can be monitored using ATF6-specific antibodies in Western blots. However, success with the currently available anti-ATF6 antibodies is often variable, and therefore, reverse

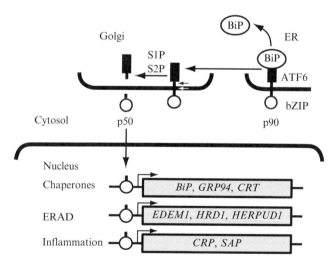

Figure 15.3 The ATF6 pathway. Upon release of BiP from its ER luminal domain, ATF6 translocates to the Golgi complex where sequential cleavage by S1P and S2P processes full-length 90 kDa ATF6 into a soluble 50 kDa species, which can translocate into the nucleus and activate expression of genes encoding ER-resident molecular chaperones, ERAD genes, and several inflammatory genes such as the genes encoding C reactive protein (CRP) and serum amyloid P-component (SAP).

transcriptase PCR monitoring of genes, whose activation during ER stress is dependent on ATF6, is the most robust assay to monitor activity of this UPR branch.

Experimentally, ER stress is often induced by treatment of cells with drugs that inhibit the protein folding machinery of the ER. We will briefly review the mechanism of action and secondary effects of these drugs. We will then describe methods to monitor activation of the three arms of the UPR in yeast and mammalian cells that are currently in use in our group.

2. Methods to Induce the UPR

When measuring signaling by the UPR, positive controls using well-established pharmacological inducers of the UPR should be included in the experimental planning. Drugs to induce the UPR in *S. cerevisiae* have been reviewed (Back *et al.*, 2005). Ca^{2+} ionophores such as ionomycin or A23187 induce the UPR by depleting ER luminal Ca^{2+} in mammalian cells. Many ER-resident chaperones bind Ca^{2+} with high capacity and may require Ca^{2+} to function. Thapsigargin depletes ER Ca^{2+} by irreversibly inhibiting SERCA Ca^{2+} ATPases in the ER membrane. Of the agents depleting ER Ca^{2+} stores, thapsigargin is the most specific inducer of the UPR (Table 15.1). However, thapsigargin also elevates cytosolic Ca^{2+}, which may have profound effects on cell signaling and physiology. Inducers of the UPR with more limited secondary effects are tunicamycin and Shiga toxigenic subtilase AB_5 ($SubAB_5$). Tunicamycin inhibits N-linked glycosylation and thus requires new protein synthesis to act (Back *et al.*, 2005). Several cell lines are resistant to induction of ER stress by tunicamycin. $SubAB_5$ enters the secretory pathway of mammalian cells and cleaves BiP between two leucines at positions 416 and 417 (Paton *et al.*, 2006), which are located in the hinge region connecting the ATPase and peptide-binding domains of the chaperone. $SubAB_5$ is a potent activator of the UPR and probably the elicitor of the UPR in mammalian cells with the least secondary effects SubA is the catalytic subunit of this toxin. A S272A point mutation inactivates the protease activity of the catalytic subunit.

3. Measuring Signaling by the UPR in *S. cerevisiae*

The removal of a 252 nt intron from *HAC1* mRNA is indicative of UPR activation and can be detected by Northern blotting of yeast total RNA. Probes against the 1st *HAC1* exon often also reveal two smaller *HAC1* species, which are the first exon plus the intron and the first exon

Table 15.1 Drugs activating the UPR in mammalian cells

Drug	Working concentration	Stock solution	Primary effects	Secondary effects
Dithiothreitol (DTT)	0.1–10 mM[a]	1 M in H$_2$O. Filter-sterilize. Store in small portions at −20 °C under argon	Reduces disulfide bonds	Metal ion-catalyzed oxidation generates reactive oxygen species[b] Quenches reactive oxygen species Chelates bivalent cations, for example, Zn^{2+}[c,d]
SubAB$_5$	1 µg/ml[e]	1–10 mg/ml	Cleaves BiP in its hinge region	Will affect other processes in which BiP is involved, for example, posttranslational translocation of polypeptide chains into the ER and gating of the SEC61 translocation channel
Thapsigargin	5 nM–1 µM[a]	100 µM in ethanol or DMSO. Store at −20 °C	Irreversible inhibitor of SERCA Ca^{2+} ATPases. Depletes ER Ca^{2+} stores	Elevates cytosolic Ca^{2+} concentrations
Tunicamycin[f]	0.1–10 µg/ml[a]	10–50 mg/ml in DMSO. Store in small portions at −20 °C	Inhibitor of UDP-N-acetylglucosamine-dolichyl-phosphate N-acetylglucosamine phosphotransferase. Inhibits N-linked glycosylation of nascent polypeptide chains translocated into the ER	Inhibits incorporation of D-glucose into D-glucosyl-phosphoryl-dolichol

[a] Rutkowski et al. (2006).
[b] Held et al. (1996).
[c] Cornell and Crivaro (1972).
[d] Krężel et al. (2001).
[e] Wolfson et al. (2008).
[f] The efficacy of tunicamycin strongly depends on the cell line used. Tunicamycin requires new protein synthesis to induce the UPR. Tunicamycin possesses a long fatty acyl moiety which inserts tunicamycin into membranes (Back et al., 2005).

only (Fig. 15.4). Blots can be stripped and reprobed several times for mRNAs encoding ER-resident chaperones, such as *KAR2* or *PDI1*, and a loading control. As loading control, we are using a gel-purified 1.6 kBp *Hin*dIII fragment of plasmid pC4/2 (Law and Segall, 1988). This loading control gives very strong signals. Therefore, it is best to probe for the loading control last.

3.1. General guidelines for work with RNA

When working with RNA, religious adherence to the following general recommendations is strongly suggested. Always wear gloves. Use chemicals, glassware, and plastic materials dedicated for work with RNA. Chemicals should be RNase-free. Sterilize all glassware, all stir bars, and all metal spatulas by baking for ≥ 4 h at 180 °C. Baking destroys RNases more efficiently than autoclaving. Always use sterile plastic materials. Electrophoresis equipment should be thoroughly cleaned with 2% (w/v) SDS and rinsed with type I laboratory water (resistivity ≥ 18 MΩ, total organic carbon ≤ 3.0 ppb) before use, depending on the resistance of the plastic to SDS. Hybridization bottles need to be cleaned and can be sterilized by autoclaving or baking depending on the compatibility of the bottles with these treatments. All solutions should be prepared in type I laboratory water treated with 0.1% (v/v) diethylpyrocarbonate (DEPC). DEPC is a

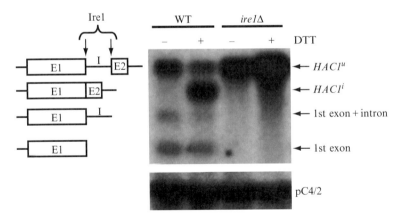

Figure 15.4 Analysis of *HAC1* splicing by Northern blotting. WT and *ire1*Δ *S. cerevisiae* cells were grown to exponential growth phase at 30 °C. To one half of the culture, DTT was added to a final concentration of 2 mM. Both cultures were incubated for another 2 h at 30 °C. Total RNA was extracted as described and analyzed by Northern blotting using a probe against the first exon of *HAC1* and the loading control pC4/2 (Law and Segall, 1988). The diagram to the left of the blot indicates the two *HAC1* exons (E1 and E2) and the intron (I). Unspliced (u), spliced (i), and cleavage intermediates of *HAC1* mRNA are also labeled.

carcinogen and must be used under a fume hood. After addition of DEPC, the solution is stirred vigorously for ≥ 30 min at room temperature (RT) to inactivate RNases. Residual DEPC is destroyed by autoclaving, boiling the solution for 10 min, or incubating the solution overnight at 60 °C. Chemicals containing nucleophilic groups, for example β-mercaptoethanol, EDTA, or tris(hydroxymethyl)aminomethane (Tris), will react with DEPC. Solutions containing these chemicals are therefore prepared in DEPC-treated type I laboratory water and then sterilized depending on the stability and volatility of the dissolved chemicals. Isolation of RNA from yeast immediately after harvesting the cells is most reliable. In our hands, the success of isolating intact RNA from yeast after freezing cells has been variable and dependent on the culture conditions.

3.2. Required materials

3.2.1. Reagents

- YPD-agar (1% (w/v) yeast extract, 2% (w/v) peptone, 2% (w/v) D-glucose, 2% (w/v) agar) or synthetic dextrose (SD) agar Sherman, 1991 plates containing fresh yeast colonies (≤ 1 month old)
- Growth medium such as liquid YPD or SD media (Sherman, 1991)
- Stock solutions of tunicamycin (10 mg/ml in dimethylsulfoxide (DMSO)), DTT (1 M in water; filter-sterilized; store at −20 °C, ideally under argon), or 2-deoxy-D-glucose (1 M in water, filter-sterilized; store at RT)
- Water, treated with DEPC, ice-cold
- 100% and 70% (v/v) ethanol, ice-cold
- *Formamide*: Mix 100 ml formamide with 5 g Dowex MR-3 mixed bed ion exchanger for 1 h at RT. Filter the mixture in a sterile filter unit to separate the Dowex MR-3 mixed bed ion exchanger from the formamide. Store formamide at 4 °C protected from light for up to 1 year.
- Glass beads (Ø ∼ 0.5 mm, BioSpec Products, Bartlesville, OK, USA, cat. no. 11079105), acid-washed. We prefer to buy acid-washed glass beads.
- Phenol:chloroform:isoamyl alcohol (25:24:1, v/v/v), saturated with RNA-buffer, stored at 4 °C protected from light. Once the phenolic phase develops a pink color, discard. Use bottom phase only. To prepare this solution, we mix 1 vol. of commercially available phenol:chloroform: isoamyl alcohol (25:24:1, v/v/v), saturated with 0.1 M Tris–HCl (pH 8.0) with 1 vol. RNA buffer by vortexing, separate the aqueous and organic phases by centrifugation for 5 min at 4,000×g and RT, dispose of the upper aqueous phase according to local waste disposal guidelines, and repeat the extraction once. This procedure should be carried out in a fume hood.
- *RNA buffer*: 0.2 M Tris–HCl (pH 7.5), 0.5 M NaCl, 10 mM EDTA

- DMSO
- *100× Denhardt's solution*: 2% (w/v) Ficoll 400, 2% (w/v) polyvinylpyrrolidone (average molecular weight ~360,000 Da), 2% (w/v) BSA, fraction V. Filter-sterilize and store at 4 °C.
- *Glyoxal*: Mix 50 ml 40% (w/w) glyoxal (=6 M) with 5 g Dowex MR-3 mixed bed ion exchanger for 1 h at RT. Separate the Dowex MR-3 mixed bed ion exchanger from the glyoxal solution by filtration in a sterile filter unit. Store the glyoxal at −20 °C for up to 1 year in 0.5-ml aliquots.
- 10 mM sodium phosphate buffer (pH 7.0)
- 100 mM sodium phosphate buffer (pH 7.0)
- *6× RNA sample loading buffer*: 10 mM sodium phosphate buffer (pH 7.0), 50% (v/v) glycerol, 0.4% (w/v) bromophenol blue
- *20× SSC*: 300 mM sodium citrate, 3 M NaCl
- 0.2× SSC + 0.1% (w/v) SDS, prewarmed to 42 °C
- 20 mM Tris–HCl (pH 8.0), prewarmed to 65 °C
- Random primed ^{32}P labeling kit for DNA probes. We use the Rediprime^{TM}II from GE Healthcare (Little Chalfont, UK)
- Redivue [α-^{32}P]-dCTP, 3,000 Ci/mmol, ~10 µCi/µl
- DNA probe, 25 ng, must be diluted with TE buffer
- Salmon sperm DNA (1 mg/ml, average molecular weight ≤2,000 bp)
- *1× TE Buffer*: 10 mM Tris–HCl (pH 8.0), 1 mM EDTA
- *Hybridization solution*: 50% (v/v) formamide, 5× SSC, 5× Denhardt's solution, 1% (w/v) SDS, 100 µg/ml salmon sperm DNA. To make 20 ml in a 50-ml tube, mix 5 ml 20× SSC, 2 ml 10% (w/v) SDS, and 1 ml 100× Denhardt's solution. A white precipitate may be present in this premix, but will dissolve after addition of the formamide. Warm this premix to 42 °C. In a second 50-ml tube, warm 10 ml formamide to 42 °C. Immediately before use, boil 2 ml 1 mg/ml salmon sperm DNA for 5 min, then snap-cool by placing in an ice water bath. Add the denatured salmon sperm DNA to the prewarmed premix, mix, then add the prewarmed formamide and mix again. At this stage, the white precipitate in the aqueous premix should go into solution.
- Microspin^{TM} S-300 HR column (GE Healthcare)

3.2.2. Disposables

- BioMax MS X-ray film (Eastman Kodak, Rochester, NY, USA, cat. no. Z36006-50EA)
- Cling film
- Ice
- Paper towels
- 5-ml pipette
- Pipette tips, 0.2–2, 2–200, and 200–1,000 µl

- Positively charged Nylon membrane (MP Biomedicals, Illkirch, France, cat. no. MEMP0001)
- 2.0-ml screw-cap microcentrifuge tubes
- Sterile and RNase-free 1.5-ml microcentrifuge tubes
- Sterile 15- and 50-ml tubes
- Whatman 3 MM paper

3.3. Growth of yeast cultures and UPR induction

Always wear gloves and work close to the flame of a Bunsen burner to avoid culture contamination. Prepare an overnight culture by inoculating 2–4 ml of YPD or SD liquid medium with a single colony. Disperse the cells by vortexing for 1 min and incubate at 30 °C for 16–18 h with shaking at ∼250 rpm. This will yield a stationary phase culture ($A_{600} > 2.0$). Determine the A_{600} of the overnight culture using the medium as a blank. A_{600} readings are linear up to $A_{600} \sim 0.6$. Inoculate your experimental cultures (∼20 ml for each experimental condition) at a starting A_{600} of ∼0.01 using the fresh overnight culture and incubate at 30 °C with shaking at ∼250 rpm until the A_{600} reaches 0.4–0.6 indicating that the culture is exponentially growing. Outgrowth of a fresh overnight culture in YPD broth to an A_{600} of ∼0.3 takes ∼6–8 h and in SD medium, it takes ∼12–24 h depending on the genetic background.

Cells for each strain for different experimental conditions are grown to exponential growth phase in the same flask. This culture is split into the desired number of flasks immediately before activation of the UPR. Leave one flask as an untreated control and to the other flasks, add (as desired) 0.5–2 μg/ml tunicamycin, 2–5 mM DTT, or 10–50 mM 2-deoxy-D-glucose. DTT is more effective in SD than in YPD cultures, because its oxidation is catalyzed by monothiols (Cleland, 1964). Since tunicamycin is dissolved in DMSO, an additional control culture to which an equal volume of DMSO is added is required. It is best to process cultures as they reach the exponential growth phase and not to try to synchronize collection times for different cultures. Incubate cultures at 30 °C with shaking at ∼250 rpm for 0.25–2 h (as desired). Collect cells by centrifugation at 4,000×g, 4 °C for 2 min. Aspirate the supernatant and use the cell pellet directly to isolate total RNA.

3.4. Isolation of total RNA from *S. cerevisiae*

Resuspend cells in 1 ml ice cold water and transfer them into a 2.0-ml screw-cap microcentrifuge tube. Centrifuge 10 s at 12,000×g, RT and pipet-out the supernatant. Repeat this step once. Add 300 μl of RNA buffer and resuspend the cell pellet by vortexing. Add 200 μg acid-washed glass beads (assuming that 1 μl of beads weighs 1 μg) and 300 μl phenol:

chloroform:isoamyl alcohol (25:24:1, v/v/v), saturated with RNA buffer. Mix well by flipping the tubes. Lyse cells by vortexing for 3 min at the highest setting. Use of a bead mill will lyse yeast cells faster and more efficiently. We routinely use a Precellys instrument from Bertin Technologies (Saint Quentin en Yvelines, France) at 4 °C using two cycles of 10–15 s at 6500 rpm with a break of 30 s between both cycles. Centrifuge for 1 min at 12,000×g, 4 °C or RT. Transfer the upper phase into a new 1.5-ml tube and add 300 μl phenol:chloroform:isoamyl alcohol (25:24:1, v/v/v), saturated with RNA buffer. Vortex briefly to mix. Centrifuge for 1 min at top speed, 4 °C or RT. Transfer the upper phase into a new 1.5-ml tube, add 900 μl ethanol (ice-cold), and incubate at −80 °C for ≥5 min. The ethanol precipitate is stable for ≥1 month. Centrifuge for 2 min at 4 °C and 12,000×g. Discard the supernatant and add 500 μl 70% (v/v) ethanol (ice-cold) and vortex briefly. Centrifuge for 2 min at 4 °C and 12,000×g. Discard the supernatant and centrifuge for 10 s at 4 °C or RT and 12,000×g. Pipet-out any remaining supernatant and leave the pellet to dry for 5 min at RT. Add 50 μl formamide and dissolve RNA by incubation at RT for 1–2 h. Concentrated samples may require mixing by carefully pipetting up and down several times. Centrifuge for 5 min at 4 °C and 12,000×g and transfer the supernatant into a new 1.5-ml microcentrifuge tube. This step removes particulate material, including partially dissolved RNA and ensures even gel loading. Determine RNA concentration by making a 1:200 dilution of each RNA sample with 2% (w/v) SDS to be in the linear range of the UV spectrophotometer. Include a blank consisting of a 1:200 dilution of formamide in 2% (w/v) SDS. Formamide absorbs in the UV range. For this reason, inclusion of formamide in the blank and a 1:200 dilution of RNA dissolved in formamide are necessary to obtain accurate readings. Determine the RNA concentration and purity by measuring the A_{260} and A_{260}/A_{280} using a UV spectrophotometer and correct for the blank values. One A_{260} absorbance unit equals ∼40 μg RNA. A_{260}/A_{280} ratios between 1.7 and 2.1 indicate pure RNA, ratios of <1.7 contamination with protein, and ratios >2.1 contamination with phenol or dyes absorbing UV light. In the first case, additional phenol:chloroform extractions may be necessary. In the latter case, extractions with 24:1 (v/v) chloroform:isoamyl alcohol will remove residual phenol. Store RNA samples at −80 °C.

3.5. Separation of RNA by agarose gel electrophoresis

3.5.1. Gel preparation

Dissolve the amount of agarose required to make a 1.4% (w/v) solution in an appropriate volume of 10 mM sodium phosphate buffer (pH 7.0) in an Erlenmeyer flask by heating in a microwave. Place the flask into a 50 °C water bath and let the agarose solution equilibrate to 50 °C. Pour the gel and move any air bubbles to the sides of the gel using a glass Pasteur pipette. Allow the gel to solidify (30–60 min). Remove combs and casting dams, and immerse the gel in a 10 mM sodium phosphate buffer (pH 7.0).

3.5.2. Sample preparation

Prepare one sample of 10 μg of a double-stranded DNA marker as a size standard. Denature up to 10 μg total RNA in a sterile 1.5-ml microcentrifuge tube by adding up to 8.09 μl RNA (up to 10 μg/lane), 5.91 μl 6 M glyoxal, 17.50 μl DMSO, and 3.50 μl 100 mM sodium phosphate buffer (pH 7.0). Mix by vortexing and collect the contents by centrifugation for 10 s at 12,000×g and RT. Incubate 1 h at 50 °C. Cool the mixture to RT, add 7 μl 6× RNA sample loading buffer, mix by vortexing, and collect droplets at the bottom of the tube by centrifugation for 10 s at 12,000×g and RT. Samples are now ready for loading onto the gel.

3.5.3. Electrophoresis

Load samples and the DNA size standard onto the gel. Leave at least one lane between the size standard and the samples empty. Electrophorese at 3 V/cm electrode distance with constant buffer recirculation. Buffer recirculation is necessary to prevent a pH gradient forming in the gel. Regularly check that the running front indicated by the bromophenol blue stays blue. A color change to green or yellow indicates insufficient buffer recirculation. Continue electrophoresis until the bromophenol blue migrates ∼80% along the length of the gel. This takes ∼6 h for a gel of a length of ∼10 cm.

3.5.4. Ethidium bromide staining of size standard

Cut off the lane with the size standard using a razor blade and a ruler. Transfer the lane into 0.5 M NH$_4$OAc plus 0.5-μg/ml ethidium bromide. Incubate overnight with slow agitation protected from light. The remainder of the gel is now ready for capillary transfer. Destain the size marker 1–3 times for 10–30 min in water. Photograph the marker alongside a ruler under UV illumination ($\lambda \sim 254$ nm).

3.5.5. Capillary transfer of RNA

Cut eight pieces of Whatman 3 MM filter paper and the membrane to the size of the gel. Cut off the top left corner of the membrane. This edge will serve to orient the membrane. Place four Whatman 3 MM filter papers on top of a stack of paper towels which is ∼5 cm high (Fig. 15.5). The paper towels should be slightly larger than the gel. Place a plastic wrap around the Whatman papers to prevent short-circuiting. Wet a fifth filter paper in 20× SSC and place on top. Remove the air bubbles by gently rolling over the stack with a disposable 5-ml pipette. Wet the membrane briefly in water and then transfer the membrane into 20× SSC. Place the membrane on the top filter paper. Remove the bubbles by rolling a pipette over the surface of the

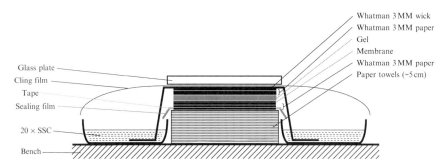

Figure 15.5 Assembly of a capillary transfer of RNA onto a nylon membrane. The assembly of the capillary transfer system is described in the text.

membrane. Pipet 5–10 ml 20× SSC onto the membrane and immediately place the gel onto the membrane. Lay the gel onto the membrane by starting at one side of the stack. Take care to observe the buffer flow below the gel as the gel is placed onto the stack. In case air bubbles are trapped between the gel and the membrane, lift the gel to remove the air bubbles. Soak three Whatman 3 MM papers of the same size as the gel in 20× SSC and place them onto the gel. Remove the air bubbles after laying down each Whatman 3 MM paper. Place a larger Whatman 3 MM paper that has been soaked in 20× SSC on top of the stack with both of its ends submerged in two reservoirs of 20× SSC. Remove the air bubbles. The use of two reservoirs is important to prevent distortion of bands during the transfer. Place a glass plate on top of the assembly. Cover the assembly with a cling film to minimize evaporation. The transfer is complete after ∼1 h. For convenience, we leave the transfer overnight.

Disassemble the transfer unit using forceps. Place the membrane with the nucleic acid side up onto a piece of Whatman 3 MM paper. Immediately, use UV light (120 mJ, $\lambda = 254$ nm) to cross-link the nucleic acid to the damp nylon membrane. Do not let the membrane dry before cross-linking. UV light covalently links the RNA to nylon or positively charged nylon membranes, which will partially protect the RNA samples from degradation by RNases. The conditions for the UV exposure are critical. Too short exposures will result in inefficient cross-linking of the RNA, and too long exposures or exposure of dried membranes will result in too strong cross-linking, decreasing significantly the sensitivity of the hybridization steps. Cover the membrane with a cling film and store at $-20\ ^\circ$C. The membranes can be stored indefinitely at $-20\ ^\circ$C.

3.5.6. Random-primed labeling of DNA probes

Probes are most conveniently prepared by PCR on 0.1–1.0 μg of yeast genomic DNA. For the first exon of *HAC1* mRNA, we use the primers AGCGGATCCATGGAAATGACTGATTTTGAAC and

TCTTGATCACTGTAGTTTCCTGGTCATC, and for *PDI1* CCA-ATCCGGTAAGATTGACG and TTCAGTCATGTCGTGGATGG. A *KAR2* probe is described in Welihinda et al. (1997). The PCR protocol has been described before (Back et al., 2005). Template DNAs of $\sim$500 bp tend to give less background in hybridizations than longer template DNAs. We label 10 ng of DNA probe with 5 μl [α-^{32}P]-dCTP for use in a 5-ml hybridization solution, which is sufficient for a small hybridization bottle (Ø = 35 mm, length $\sim$ 12 cm). For larger hybridization bottles, we label 20 ng of DNA probe with the same amount of [α-^{32}P]-dCTP. We use [α-^{32}P]-dCTP up to an age of $\sim$2 half-lives. Due to the infrequent use of each probe, we do not reuse labeled probes.

Always use screw-cap tubes when using radioactivity. Dilute DNA probe to a concentration of 10 ng in 45 μl of 1× TE buffer (pH 8.0). To obtain high labeling efficiencies, it is important that TE buffer is used and that the final volume used to reconstitute the mix equals 45 μl. Denature the DNA probe by heating for 5 min at 95–100 °C. Place on ice for 1 min and centrifuge to collect the whole content at the bottom of the tube. Add the denatured DNA to a random-primed labeling reaction tube of the Rediprime^{TM}II random-primed labeling kit (GE Healthcare). Dissolve the blue pellet in the reaction tube by repeatedly flipping the tube. Pellets often dissolve readily, but sometimes may take up to several minutes. It is critical that the pellet is completely dissolved before proceeding. Centrifuge for 10 s at RT to collect the whole solution at the bottom of the tube. Add 5 μl Redivue [α-^{32}P]-dCTP and mix by flipping the tube. The denatured DNA solution will change color to purple when the blue pellet has been completely dissolved and mixed with the Redivue. Incubate at 37 °C for 10 min. If desired, the labeling reaction can be allowed to proceed at room temperature, or may also be left to proceed overnight. Labeling at 37 °C for 10 min is sufficient in our experience. Stop the reaction by adding 5 μl of 0.2 M EDTA.

Use a microspin S300 HR column to remove unincorporated nucleotides. Prepare the column by resuspending the resin in the column by vortexing. Loosen the cap of the column and snap-off the bottom closure. Place the column into a 1.5-ml screw-cap microcentrifuge tube for support. Spin the column at 735×g for 1 min. Place the column into a new 1.5-ml tube, remove, but retain, the cap and slowly apply the sample to the top-center of the resin. Be careful not to disturb the bed. Close the tube with the cap, but do not close tightly. Spin the column at 735×g for 2 min. Collect the purified sample at the bottom of the support tube. The sample volume for the Microspin S300 HR columns must be 25–100 μl.

3.5.7. Hybridization

Transfer the membrane into a hybridization bottle. If there is more than one membrane, they should not overlap. Use a nylon mesh to separate several membranes. Incubate the membrane for 15 min at 65 °C in 20 mM

Tris–HCl (pH 8.0). This step removes glyoxal covalently bound to the heterocyclic bases and only needs to be performed before the first hybridization. For subsequent hybridizations, wet the membrane in 6× SSC. For all hybridizations, prehybridize at 42 °C with a 20-ml hybridization solution for ≥3 h. Denature the labeled DNA (2 ng/ml in a 5-ml hybridization solution, cf. Section 3.5.6) by heating to 95–100 °C for 5 min immediately before use, and then place on ice. Centrifuge the tube briefly and mix the contents of the tube well. Add probes to the aqueous premix of the hybridization solution, mix, add the formamide, mix, and add the whole hybridization solution containing the labeled probe to your blot. Hybridize overnight at 42 °C. Wash the membrane 3 × 5 min at RT with ∼200 ml 2× SSC + 0.1% (w/v) SDS, and then wash with ∼200 ml 0.2× SSC + 0.1% (w/v) SDS for 5 min at RT. Wash the membrane for 15 min at 42 °C with ∼50 ml 0.2× SSC + 0.1% (w/v) SDS. Rinse the membrane in 2× SSC and blot-off excess liquid. Wrap the membrane in cling film and expose to Kodak BioMax MS film at −80 °C in an exposure cassette fitted with intensifying screens (e.g., CAWO fast tungstate screens, GRI, Braintree, UK) for 4–6 h. Equilibrate the exposure cassettes to RT and develop the film in a dark room. Perform additional exposures by varying the exposure time by a factor of ∼4 to obtain films showing all *HAC1* species. Blots can be quantified by storage phosphor analysis (Johnston *et al.*, 1990; Reichert *et al.*, 1992; Zouboulis and Tavakkol, 1994) or analyzing films exposed in their linear range with ImageJ (Collins, 2007). Storage phosphor analysis is ∼10 times more sensitive than film exposure and has a ∼1,000 times larger linear range. To rehybridize membranes with other probes (e.g., for the loading control), they need to be stripped first. Incubate blots for 60 min at 65 °C in 50% (v/v) formamide, 1% (w/v) SDS, and 0.1× SSC, replacing the solution once after 30 min. Then, incubate the blots for 10 min at 65 °C in 1% (w/v) SDS, 0.1× SSC. For storage, rinse the membrane in 2× SSC, wrap in a cling film, and store indefinitely at −20 °C. For rehybridization, the protocol is identical to the one described for hybridization above.

4. Measuring Signaling by the UPR in Mammalian Cells

Signaling by the mammalian UPR can be assessed at the mRNA and protein level. At the mRNA level, splicing of *XBP1* mRNA and induction of UPR responsive genes are reliable and specific measurements of UPR activation. At the protein level, processing of ATF6, phosphorylation of IRE1α, PERK, eIF2α, induction of spliced XBP1, CHOP, ATF4, or of ER resident molecular chaperones, such as BiP, may be used to monitor activation of the UPR.

4.1. mRNA methods for analyzing the UPR

4.1.1. XBP1 splicing assay

The splicing of *XBP1* mRNA by IRE1 is a relatively easy way to measure activation of the UPR. After extraction of RNA and cDNA synthesis, the difference in size caused by removal of the 26 nt intron from *XBP1* mRNA can be visualized on a 2% (w/v) agarose gel.

4.1.1.1. Required materials

- Cell culture medium, appropriate for the cell lines of choice
- 10× DNA sample loading buffer (20% (w/v) Ficoll 400, 0.1 M Na$_2$EDTA (pH 8.0), 1% (w/v) SDS, 0.25% (w/v) bromophenol blue, 0.25% (w/v) xylene cyanol)
- 75% (v/v) ethanol
- ER stress-inducing agent (see Section 2). 0.5 μM thapsigargin works well with most cell lines used in our lab.
- 10 mg/ml ethidium bromide, dissolved in water
- EZ-RNA total RNA isolation kit (Geneflow, Fradley, UK, cat. no. K1-0120)
- 20 mg/ml glycogen, aqueous solution, RNA grade (Fermentas, York, UK, R0551)
- 5 U/μl GoTaq® hot start polymerase (Promega, Madison, WI, USA, cat. no. M5005). Supplied with 5× Green GoTaq® flexi buffer.
- 5 U/μl GoTaq® DNA polymerase (Promega, cat. no. M3005). Supplied with GoTaq® buffer.
- Water, DEPC-treated
- Isopropanol
- Liquid nitrogen
- 10 mM dNTP mix in 1 mM Tris–HCl (pH 8.0)
- 2 mM dNTP mix in 1 mM Tris–HCl (pH 8.0)
- Nuclease-free water
- 50 μM Oligo(dT)$_{15}$ primers (Promega, cat. no. C1101)
- Oligonucleotides (Table 15.2)
- 20–40 U/μl RNasin® (Promega, cat. no., N2511)
- 200 U/μl SuperScript® III reverse transcriptase (Invitrogen, Paisley, UK, cat. no. 18080-093)
- 50× TAE Buffer (2 M Tris–HOAc, 0.1 M EDTA, pH ~ 8.5). Dilute to 1× in water before use.

4.1.1.2. Disposables

- 35–60-mm tissue culture dishes
- Cell scraper

Table 15.2 Oligonuclueotides for use in *XBP1* splicing assays

Name	Species	Sequence
XBP1 forward (F2)	*M. musculus*	GATCCTGACGAGGTTCCAGA
XBP1 reverse (R2)	*M. musculus*	ACAGGGTCCAACTTGTCCAG
ACTIN forward	*M. musculus*	AGCCATGTACGTAGCCATCC
ACTIN reverse	*M. musculus*	CTCTCAGCTGTGGTGGTGAA
XBP1 forward (F2)	*H. sapiens*	GAGTTAAGACAGCGCTTGGG
XBP1 reverse (R2)	*H. sapiens*	ACTGGGTCCAAGTTGTCCAG
ACTIN forward	*H. sapiens*	CTGAGCGTGGCTACTCCTTC
ACTIN reverse	*H. sapiens*	GGCATACAGGTCCTTCCTGA

- Nuclease-free pipette tips
- Nuclease-free 1.5-ml microcentrifuge tubes
- Nuclease-free 15-ml tubes
- Nuclease-free PCR tubes
- Serological pipettes

4.1.1.3. Extraction of RNA A 35–60-mm tissue culture dish with adherent cells yields sufficient RNA for measurement of *XBP1* splicing in our hands. Cells should be split one day before the assay with fresh culture medium and allowed to reach 60–80% confluency before the assay. Following treatment of cells with ER stress-inducing agents for the desired amount of time, commercially available kits can be used for extracting RNA from cells. We use the EZ-RNA total isolation kit (Geneflow) which relies on disruption of cells in guanidine thiocyanate/detergent solution, a subsequent phenol:chloroform extraction and precipitation of RNA with alcohol. RNA is isolated according to the manufacturer's instructions. General guidelines for work with RNA should be strictly followed (cf. Section 3.1). Lyse the cells directly in the culture dish by adding 0.5 ml denaturing solution/10 cm^2 of dish area. Use a cell scraper to make sure that cells are completely removed from the base of a dish. Pass the cell lysate through a pipette tip several times to ensure disruption of any cell mass. At this point, it is possible to snap-freeze the cell lysate and then proceed with RNA extraction at a later date. We do this by transferring the cell lysate to a 15-ml falcon tube and submerging it in liquid nitrogen before transferring it to a freezer at $-80\ ^\circ C$ where samples can be stored for up to 1 month, if required, before processing further. Thaw the cell lysate on ice and then leave the tube for 5 min at RT before adding 0.5 ml extraction solution per 0.5 ml denaturing solution. Vigorously shake the tube for 15 s and leave at RT for 10 min. Centrifuge for 15 min at $12,000 \times g$ and $4\ ^\circ C$. Following centrifugation, transfer the upper aqueous phase to a fresh tube. At this point, store the interphase and organic phase at $4\ ^\circ C$ for extracting DNA and protein (for details, refer to the manufacturer's instructions). To the aqueous phase, add isopropanol (0.5 ml/0.5 ml denaturing solution) and leave at RT for 10 min, then centrifuge for 8 min at 4 °C and 12,000×g. If the RNA yield is expected to be low, add 20 µg of glycogen and leave the sample overnight at −20 °C before the centrifugation step. Following centrifugation, remove the supernatant and wash the pellet by vortexing with 1 ml 75% (v/v) ethanol. Centrifuge the sample for 5 min at 4 °C and 7500×g. Remove the supernatant by aspiration and air-dry the pellet for 5 min at RT. Dissolve the RNA pellet in 20–100 µl water. In case of low yields, use less water. Quantify the RNA by measuring the A_{260}. Pure RNA preparations have an A_{260}/A_{280} between 1.7 and 2.1 (cf. Section 3.4).

4.1.1.4. cDNA synthesis We use the Invitrogen SuperScript® III reverse transcriptase with oligo(dT)$_{15}$ primers according to the manufacturer's instructions. We add 5 µg total RNA per reaction. While amounts of total RNA can be reduced, each cDNA synthesis reaction to be compared needs to have the same amount of total RNA. We find it most efficient to perform the heating and cooling steps in a PCR block. In a nuclease-free PCR tube, add the following: 1 µl of 50 µM oligo(dT)$_{15}$, 5 µg total RNA, and 1 µl 10 mM dNTP mix in 1 mM Tris–HCl (pH 8.0). Add water to make a final volume of 13 µl. Briefly centrifuge the mixture, heat to 65 °C for 5 min in a heat block, and cool to 4 °C for at least 1 min. Then, add the following: 4 µl 5× first-strand buffer (250 mM Tris–HCl (pH 8.3), 375 mM KCl, 15 mM MgCl$_2$), 1 µl 0.1 M DTT, 1 µl RNasin®, and 1 µl 200 U/µl SuperScript® III reverse transcriptase. Mix the solution by pipetting up and down and return to the PCR block. Heat to 50 °C for 60 min. Inactivate the reverse transcriptase by heating to 70 °C for 15 min. Store the cDNA at 4 °C or use it immediately in the *XBP1* splicing PCR assay.

4.1.1.5. XBP1 splicing PCR assay Fig. 15.6A shows a schematic representation of the location of primers on the unspliced and spliced forms of *XBP1*. Use of a hot start DNA polymerase, such as hot start GoTaq® from Promega, suppresses unspecific amplification products, which may appear as a ladder throughout the lane. We perform 50 µl reactions and run the full volume on a 2% (w/v) agarose gel. We also routinely perform a further PCR on a reference gene such as *ACTIN*. We use the GoTaq® hot start polymerase with green GoTaq® flexi buffer for easy loading onto the agarose gel following PCR according to the manufacturer's instructions. In a sterile, nuclease-free PCR tube, add the following components: 10 µl 5× Green GoTaq® flexi buffer, 3 µl 25 mM MgCl$_2$, 1 µl 10 mM dNTP mix in 1 mM Tris–HCl (pH 8.0), 5 µl 10 µM forward primer F2, 5 µl 10 µM reverse primer R2, 2.5 µl cDNA directly from the cDNA synthesis reaction, and 0.5 µl 5 U/µl Go*Taq*® hot start polymerase. Add water to a final volume of 50 µl. Run the PCR using the following conditions:

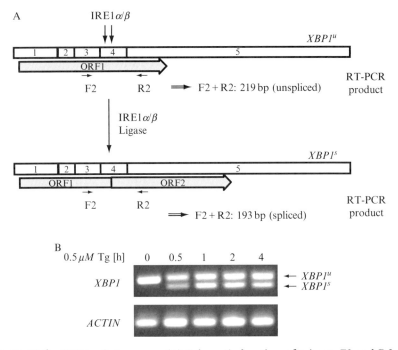

Figure 15.6 *XBP1* splicing assay. (A) Schematic location of primers F2 and R2 on *XBP1* cDNA. The numbers indicate the *XBP1* exons. (B) Resolution of PCR products representing unspliced (u) and spliced (s) *XBP1* mRNAs in a 2% (w/v) agarose gel. ER stress was induced with 0.5 μM thapsigargin for the times indicated.

Initial denaturation	94.0 °C	2 min	
Denaturation	94.0 °C	1 min	
Annealing	59.0 °C	1 min	× 35 cycles
Extension	72.0 °C	30 s	
Final extension	72.0 °C	5 min	
Hold	4.0 °C	Forever	

An *ACTIN* PCR serves as a loading control. A hot start DNA polymerase is not necessary for this PCR. A 25-μl reaction volume is sufficient. We use GoTaq® Green DNA polymerase from Promega. In a sterile, nuclease-free PCR tube, add the following components: 5 μl 5× Green GoTaq® flexi buffer, 1.5 μl 25 mM MgCl$_2$, 2.5 μl 2 mM dNTP mix in 1 mM Tris–HCl (pH 8.0), 2.5 μl 10 μM forward *ACTIN* primer (Table 15.2), 2.5 μl 10 μM reverse *ACTIN* primer (Table 15.2), 1.25 μl cDNA directly from the cDNA synthesis reaction, and 0.25 μl 5 U/μl GoTaq® polymerase. Add water to a final volume of 25 μl. The cycling parameters are the same as for the *XBP1* PCR.

For the *XBP1* PCR, electrophorese the whole 50 μl reaction at 100 V for 1 h on a 2% (w/v) agarose gel with 1× TAE as running buffer containing 1 μg/ml ethidium bromide. The bands for unspliced and spliced *XBP1* will be very close together. Therefore, it is critical to use a high agarose concentration (2% (w/v)) and to run the samples over a sufficient distance. Visualize the bands using UV light ($\lambda = 254$ nm). A typical agarose gel with spliced and unspliced versions of *XBP-1* mRNA is shown in Fig. 15.6B. The *ACTIN* PCR can be run for less time on a smaller gel using 1% (w/v) agarose.

4.1.2. Use of quantitative reverse transcriptase PCR to analyze induction of genes indicative of activation of individual branches of the UPR

Transcriptional activation of a small number of genes has been reported to be nearly exclusively dependent on one of the three branches of the mammalian UPR (Adachi *et al.*, 2008; Harding *et al.*, 2003; Lee *et al.*, 2003; Okada *et al.*, 2002; Wu *et al.*, 2007; Yamamoto *et al.*, 2007). Genes nearly exclusively regulated by the IRE1–XBP1 and the PERK pathway can be readily identified in the literature (Table 15.3). However, transcriptional activation of many chaperone genes by ATF6α (Fig. 15.3) overlaps with IRE1 and PERK signaling activities. A CREB-H·ATF6α heterodimer induces inflammatory genes encoding C reactive protein (*CRP*) and serum amyloid P-component (SAP; Zhang *et al.*, 2006). Induction of these two genes may be indicative of activation of the ATF6 branch of the UPR. Measuring mRNA levels for these genes using Northern blotting or quantitative reverse transcriptase PCR techniques (Bustin, 2002; Bustin *et al.*, 2009; Wilhelm and Pingoud, 2003) will indicate which pathways of the UPR are activated in the experimental system under investigation.

4.2. Protein methods for analyzing the UPR

Measuring UPR activation at the protein level allows direct and reliable assessment of which branches of the UPR are activated in a given experimental setting. Phosphorylation of IRE1α and induction of spliced XBP1 are specific to activation of IRE1α, phosphorylation of PERK is specific to activation of PERK, and processing and nuclear translocation of ATF6 is specific to activation of ATF6. Work with proteins often requires more extensive optimization than work with nucleic acids. Further, cross-reactivities of antibodies may compromise assay reliability. Therefore, other, technically more reliable markers for activation of PERK, such as phosphorylation of eIF2α and induction of CHOP and ATF4, are often used as surrogates for PERK activation. Here, it is important to keep in mind that, in principle,

Table 15.3 Genes indicative of activation of the individual UPR branches

Gene	5′ Oligonucleotide	3′ Oligonucleotide
ATF6 pathway		
CRP[a]		
SAP[a]		
IRE1α pathway		
ERDJ4a[b-e]	TCAGAGAGATTGCAGAAGCG	GACTCCCATTGCCTCTTTGT
HEDJ[c]	TCAGAGAGATTGCAGAAGCG	GACTCCCATTGCCTCTTTGT
RAMP4[c,e]		
PERK pathway		
CHOP[b]	CTGCCTTTCACCTTGGAGAC	CGTTTCCTGGGGATGAGATA
ERO1L-α[f]	TTAAGTCTGCGAGCTACAAGTATTC	AGTAAGTCCACATACTCAGCATCG
GADD34[b,c-g]	ATCTCCTGAACAGAGTCAAGCAGCCCAGAG	TAGCCACCACCTCCCCAAGCCTCTTATCAG
	CCCGAGATTCCTCTAAAAGC	CCAGACAGCAAGGAAATGG
TRB3[b,h]	TCTCCTCCGCAAGGAACCT	TCTCAACCAGGGATGCAAGAG
ER–resident molecular chaperones and oxidoreductases		
BiP[b-d,i,j]	GGTGCAGGAGGACATCAAGTT	CCCACCTCCAATATCAACTTGA
ERO1L-β[i,j]	GGGCCAAGTCATTAAAGGAA	TTTATCGCACCCAACACAGT
GRP94[b,c,i-k]	AATAGAAAGAATGCTTCGCC	TCTTCAGGCTCTTCTTCTGG
p58[IPKb-d,i,j]	TCCTGGTGGACCTGCAGTACG	CTGCGAGTAATTTCTTCCCC

ERAD machinery		
EDEM1[b,d,i]	GCAATGAAGGAGAAGGAGACCC	TAGAAGGCGTGTAGGCAGATGG
HERPUD1[c-e,j]	AGCAGCCGGACAACTCTAAT	CTTGGAAAGTCTGCTGGACA
HRD1[b,d,j]	TGGCTTTGAGTACGCCATTCT	CCACGGAGTGCAGCACATAC

Where available, oligonucleotide sequences for mouse sequences are provided. Human or other species will need alternative primers. Detailed PCR protocols can be found in the references in the footnote to the table.

[a] A CREB-H·ATF6α heterodimer induces *CRP* and *SAP* (Zhang et al., 2006).
[b] Song et al. (2008). *PCR protocol*: Denature at 95 °C for 10 min; then, 40 cycles of denaturation at 95 °C for 15 s, annealing and extension at 59 °C for 1 min.
[c] Lee et al. (2003).
[d] Ma and Hendershot (2004).
[e] Yamamoto et al. (2007).
[f] Marciniak et al. (2004). *PCR protocol*: 40 cycles of denaturation at 95 °C for 1 min, annealing at 72 °C for 1 min, and extension at 72 °C for 1 min.
[g] Ma and Hendershot (2003).
[h] Ohoka et al. (2005).
[i] Wu et al. (2007).
[j] Adachi et al. (2008).
[k] Harding et al. (2003).

activation of other eIF2α kinases, such as GCN2, HRI, or PKR, may have very similar effects to activation of PERK. The same caveat applies to using JNK activation, as monitored by increased dual T-loop threonine and tyrosine phosphorylation of JNK, as a surrogate for IRE1α activation. Finally, immunoblotting to detect increases in ER resident chaperones such as BiP or protein disulfide isomerases may be used as a general indicator of ER stress. Nevertheless, technical limitations may often dictate assay choice. In the following sections, we provide an overview of protein-based assays to monitor activation of the UPR while placing emphasis on assays working reliably in our hands.

4.2.1. Measurement of UPR activation by eIF2α phosphorylation

Signaling through the PERK branch of the UPR results in the phosphorylation of eIF2α at serine 51. An increase in eIF2α phosphorylation in immunoblots is indicative of activation of the UPR. An important control is to reprobe blots with an antibody recognizing total eIF2α to ensure that changes in eIF2α phosphorylation are not solely due to perturbation of the total eIF2α levels or uneven loading of samples. When analyzing eIF2α phosphorylation, the following three points should be kept in mind: (1) eIF2α phosphorylation in the UPR is transient (Marciniak et al., 2004), because a prolonged shutdown of translation interferes with recovery from prolonged ER stress. (2) A basal level of eIF2α phosphorylation will nearly always be detected. Therefore, the presence of eIF2α phosphorylation is not necessarily indicative of PERK activation because of the existence of other eIF2α kinases. (3) Small increases in eIF2α phosphorylation will have significant effects on translation, because S51 phosphorylated eIF2α inhibits the activity of the nucleotide exchange factor eIF2B for the eIF2 complex. Cells possess ~ 10 times more eIF2α than eIF2B (Sudhakar et al., 1999). For less experienced users, we provide detailed protocols for cell lysis, SDS-PAGE, electrotransfer, and immunoblotting conditions. Experienced users are likely to successfully use the recommended antibodies and blocking conditions in adaptations to methods used more routinely in their laboratories.

4.2.1.1. Required materials

- Anti-eIF2α antibody, rabbit polyclonal IgG (Santa Cruz Biotechnology, Santa Cruz, CA, USA, cat. no. sc-11386)
- Anti-phospho-S51 eIF2α antibody, rabbit polyclonal IgG (Cell Signaling Technology, Beverly, MA, USA, cat. no. #9721)
- ECL plus reagents (GE Healthcare)
- Electrophoresis cell and polyacrylamide gels. Both eIF2α and phospho-S51 eIF2α run at around 40 kDa in SDS-PAGE. 12% (w/v) polyacrylamide gels give a good resolution of eIF2α.

- Electrotransfer system. For this assay, we routinely electrotransfer using a semi-dry electrotransfer system. We have also used wet electrotransfer systems with equal success.
- Horseradish peroxidase-conjugated donkey anti-rabbit IgG (Thermo Fisher Scientific, Waltman, MA, USA, cat. no. #31458)
- PBS (4.3 mM Na$_2$HPO$_4$, 1.47 mM KH$_2$PO$_4$, 2.7 mM KCl, 137 mM NaCl (pH 7.4))
- Protein assay. There are numerous assays available to estimate protein concentration. We use the *DC* protein assay from Bio-Rad which is compatible with RIPA lysis buffer. The choice of the protein assay is determined by the compatibility of the assay with interfering chemicals which may be present in buffers or samples. Different protein assays measure different properties of proteins. Therefore, the results obtained with one assay may not be comparable to the results with another assay.
- RIPA buffer (50 mM Tris–HCl (pH 8.0), 150 mM NaCl, 0.5% (w/v) sodium deoxycholate, 0.1% (v/v) Triton X-100, 0.1% (w/v) SDS), ice-cold. Add protease inhibitor just before use. For convenience, we use complete protease inhibitors from Roche Applied Science (Burgess Hill, UK). Protease inhibitors need to be dissolved in the cold.
- 6× SDS-PAGE sample buffer (350 mM Tris–HCl (pH 6.8), 30% (v/v) glycerol, 10% (w/v) SDS, 0.5 g/l bromophenol blue, 2% (v/v) β-mercaptoethanol).
- TBST (20 mM Tris (pH 7.6), 100 mM NaCl, 0.1% (v/v) Tween 20)
- TBST + 5% (w/v) skimmed milk powder. Make fresh before use. Do not heat to dissolve.
- TBST + 5% (w/v) bovine serum albumin (BSA, Sigma-Aldrich Company Ltd., Gillingham, UK, cat. no. A2153). Make fresh before use. Do not heat to dissolve.
- X-ray film (CL-X Posure™ Film, Thermo Fisher Scientific, cat. no. 34091)

4.2.1.2. Disposables

- Cell scraper
- 50-ml centrifuge tubes or plastic hybridization bottles
- 1.5-ml microcentrifuge tubes
- Plastic trays

4.2.1.3. Preparation of lysates

The starting point for preparation of lysates depends on whether tissue, cultured adherent, or suspension cells need to be processed. For cell lines, wash the cells twice with ice-cold PBS. Failure to remove all mediums can interfere with the determination of protein concentration later. Harvest suspension cells by centrifugation and take up the cell pellet in 50–100 μl RIPA buffer per 1×10^6 cells. Scrape adherent cells into RIPA buffer by adding 100–200 μl ice-cold RIPA buffer to a well of a 6-well plate and proportionally more to larger vessels. Detach all the cells using a cell scraper. Cut tissue samples into small pieces using a razor blade and place them into a Dounce homogenizer on ice. Add ice-cold RIPA buffer (~4 vol. compared to the amount of starting tissue) and homogenize the tissue by several strokes with a tight-fitting pestle. Then, leave the homogenized sample to incubate for 15 min on ice. For all types of samples, clear the cell debris from the lysate by a low-speed centrifugation ($500 \times g$ for 10 min at 4 °C).

Lysates may be very viscous due to large amounts of DNA. If this is the case, lysates can be briefly sonicated (10–30 s bursts at 320 W) to decrease their viscosity. Viscous samples will be hard to accurately load onto SDS-PAGE gels. Lysates are then centrifuged at $12,000 \times g$ for 10 min at 4 °C. The supernatant is used for SDS-PAGE. Lysates can be stored at -20 °C. If lysates are going to be used multiple times, it is best to prepare separate aliquots to prevent repeated freeze/thawing, which may denature proteins and inactivate peptide and protein-based protease inhibitors. For the cell lines used in our laboratory, it has not been necessary to use phosphatase inhibitors to preserve S51 phosphorylation of eIF2α. Protein concentration is determined in all samples. Based on these concentrations, between 10 and 25 μg of total protein are processed in a volume that can be confidently loaded into the wells of an SDS-PAGE gel. Sample volumes may be adjusted to the volume of the sample with the lowest protein concentration. To each lysate, 6× SDS-PAGE loading buffer is added to give a final concentration of 1× on ice. The tubes are briefly vortexed, heated at 70 °C for 10 min, samples are cooled for 30 s on ice, centrifuged for 10 s at $12,000 \times g$ and RT, and then kept at RT until they are loaded on the gel. The samples can be stored at -20 °C after heating to 70 °C. It is not necessary to reheat the samples immediately before loading the gels when immunoblotting for eIF2α.

4.2.1.4. SDS-PAGE and immunoblotting 12% (w/v) polyacrylamide gels give a good resolution of eIF2α and phospho-S51 eIF2α. It is preferable to strip membranes and probe for total eIF2α after probing for phosphorylated eIF2α. Stripping of membranes can result in significant protein loss. On the other hand, incomplete stripping may result in misinterpretation of residual signals originating from phospho-eIF2α. In this case, a separate gel may need to be run to blot for total eIF2α. If so, it may be advantageous to run duplicate samples on the two sides of the same gel and to cut the membrane into two halves after electrotransfer. One side is then blotted for phospho-S51 eIF2α and the other side for total eIF2α. We electrotransfer using a semidry electroblotter onto PVDF membrane (Amersham Hybond™-P, pore size 0.45 μm, GE Healthcare, cat. no. RPN303F), using 0.1 M Tris,

0.192 M glycine, and 5% (v/v) methanol as transfer buffer. The system is limited by current, which is set at 2 mA per cm^2 of membrane and run for 60–75 min. After electrotransfer, the membrane is blocked in a large volume of TBST + 5% (w/v) skimmed milk powder for at least 1 h at RT with constant agitation. Blocking can proceed overnight at 4 °C. If the primary antibody is diluted in TBST + 5% (w/v) BSA, the membranes are washed 4× with a large volume of TBST (~100–200 ml) for each 5 min at RT. For incubation with the primary antibody, membranes are transferred to 50-ml plastic tubes. The total eIF2α antibody is diluted 1:500 in TBST + 5% (w/v) skimmed milk powder. The anti-phospho-S51 eIF2α antibody is diluted 1:1,000 in TBST + 5% (w/v) BSA. Milk contains a high amount of phosphopeptides and will interfere with epitope recognition by the anti-phospho-S51 eIF2α antibody. For 50-ml tubes, we use a 3-ml antibody solution, for larger tubes ~10 ml. The 50-ml tubes are placed on a roller and incubated overnight at 4 °C or for 2 h at RT. The membranes are transferred into plastic trays and washed 4× with a large volume of TBST (~100–200 ml) for each 5 min at RT. The membranes are then transferred back into the plastic tubes and incubated with horseradish peroxidase-conjugated secondary antibody diluted 1:20,000 in TBST + 5% (w/v) skimmed milk powder for 1 h at RT with rotation on a roller mixer. The membranes are washed with TBST as described above and developed using ECL plus reagents (GE Healthcare) following the manufacturer's guidelines. Blots are exposed to X-ray film (Fig. 15.7). Exposure times are adjusted on the basis of previous exposures to obtain exposures in the linear range of the X-ray film. Signals can be quantitated with ImageJ (Collins, 2007).

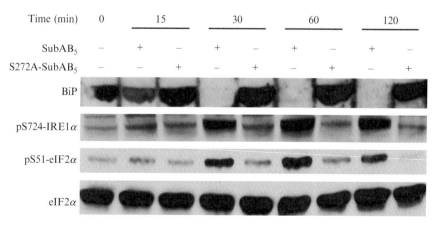

Figure 15.7 Measuring activation of the UPR by immunoblotting for phosphorylation of IRE1α and eIF2α. NS0 cells were treated with 1 µg/ml subtilase AB$_5$ (SubAB$_5$) or catalytically dead mutant S272A-SubAB$_5$ for the indicated times. Cell lysates were prepared and immunoblotted as described in the text.

4.2.2. Other markers of the UPR that can be assayed by immunoblotting

Many other markers of UPR activation can be assessed by immunoblotting, for example, phosphorylation of IRE1α and PERK, processing of ATF6 (Ye et al., 2000; Yoshida et al., 1998), induction of spliced XBP1 (Calfon et al., 2002), CHOP (Wang et al., 1996), ATF4 (Harding et al., 2000), and ER-resident molecular chaperones. An increase in BiP is often used as a marker of UPR activation (Calfon et al., 2002; Haze et al., 1999; Morris et al., 1997). However, increases in BiP at the protein level may be significantly less than increases seen at the mRNA level. An anti-BiP antibody from Sigma-Aldrich Company Ltd. (cat. no. G8918) diluted 1:3,000 in TBST + 5% (w/v) skimmed milk powder (Fig. 15.7) works well in our hands. Phosphorylation of IRE1α can be monitored using anti-phospho-S724 IRE1α antibodies (Abcam, Cambridge, UK, cat. no. 48187; Fig. 15.7). This antibody is diluted 1:1,000 in TBST + 5% (w/v) bovine serum albumin (BSA). Both the antibodies are incubated with the blot overnight at 4 °C. The blots are developed with horseradish peroxidase-conjugated donkey anti-rabbit IgG (Thermo Fisher Scientific) diluted 1:20,000 in TBST + 5% (w/v) skimmed milk powder as secondary antibody for 1 h before ECL plus detection. The anti-phospho-S724 IRE1α antibody displays some cell line-dependent variability in our hands. Antibodies against total IRE1α may require enrichment of ER vesicles or microsomes to allow detection of total IRE1α by immunoblotting. PERK displays a considerable retardation in SDS-PAGE upon activation of the UPR (Harding et al., 1999), which is, however, not seen with all anti-PERK or anti-phospho-PERK antibodies (Martinon et al., 2010).

Many of the markers that can be used to identify activation of the UPR by immunoblotting can also be used to identify its activation in tissues and cells using immunohistochemistry. The phosphorylated forms of PERK, IRE1 and eIF2α (Hoozemans et al., 2009), the upregulation of GRP78 (Brandl et al., 2009), CHOP (Zinszner et al., 1998), ATF6 (Korfei et al., 2008), ATF4 (Korfei et al., 2008), and spliced XBP1 (Maestre et al., 2009) have been used as markers of UPR induction in tissue sections using antibodies against these proteins and a chromogenic detection system. The redistribution of ATF6α from the ER to the nucleus has also been detected by immunocytochemistry (Katayama et al., 2001). Finally, dilation of the ER is a marker of ER stress and can be observed using electron microscopy (Francisco et al., 2010).

ACKNOWLEDGMENTS

This work was supported by the Biotechnology and Biological Sciences Research Council (BB/D01588X/1, BB/D526188/1, BB/E006035/1), Diabetes UK (BDA 09/0003949), the

European Commission (HEALTH-F7-2007-201608), and the Wellcome Trust (079821) to M. S. The research leading to these results has received funding from the European Community's 7th Framework Programme (FP7/2007-2013) under grant agreement n° 201608 acronym "TOBI."

REFERENCES

Adachi, Y., Yamamoto, K., Okada, T., Yoshida, H., Harada, A., and Mori, K. (2008). ATF6 is a transcription factor specializing in the regulation of quality control proteins in the endoplasmic reticulum. *Cell Struct. Funct.* **33**, 75–89.

Back, S. H., Schröder, M., Lee, K., Zhang, K., and Kaufman, R. J. (2005). ER stress signaling by regulated splicing: IRE1/HAC1/XBP1. *Methods* **35**, 395–416.

Brandl, K., Rutschmann, S., Li, X., Du, X., Xiao, N., Schnabl, B., Brenner, D. A., and Beutler, B. (2009). Enhanced sensitivity to DSS colitis caused by a hypomorphic *Mbtps1* mutation disrupting the ATF6-driven unfolded protein response. *Proc. Natl. Acad. Sci. USA* **106**, 3300–3305.

Bustin, S. A. (2002). Quantification of mRNA using real-time reverse transcription PCR (RT-PCR): Trends and problems. *J. Mol. Endocrinol.* **29**, 23–39.

Bustin, S. A., Benes, V., Garson, J. A., Hellemans, J., Huggett, J., Kubista, M., Mueller, R., Nolan, T., Pfaffl, M. W., Shipley, G. L., Vandesompele, J., and Wittwer, C. T. (2009). The MIQE guidelines: Minimum information for publication of quantitative real-time PCR experiments. *Clin. Chem.* **55**, 611–622.

Calfon, M., Zeng, H., Urano, F., Till, J. H., Hubbard, S. R., Harding, H. P., Clark, S. G., and Ron, D. (2002). IRE1 couples endoplasmic reticulum load to secretory capacity by processing the XBP-1 mRNA. *Nature* **415**, 92–96.

Cleland, W. W. (1964). Dithiothreitol, a new protective reagent for SH groups. *Biochemistry* **3**, 480–482.

Collins, T. J. (2007). ImageJ for microscopy. *Biotechniques* **43**, 25–30.

Cornell, N. W., and Crivaro, K. E. (1972). Stability constant for the zinc-dithiothreitol complex. *Anal. Biochem.* **47**, 203–208.

Dong, H., Chen, L., Chen, X., Gu, H., Gao, G., Gao, Y., and Dong, B. (2009). Dysregulation of unfolded protein response partially underlies proapoptotic activity of bortezomib in multiple myeloma cells. *Leuk. Lymphoma* **50**, 974–984.

Francisco, A. B., Singh, R., Li, S., Vani, A. K., Yang, L., Munroe, R. J., Diaferia, G., Cardano, M., Biunno, I., Qi, L., Schimenti, J. C., and Long, Q. (2010). Deficiency of suppressor enhancer lin12 1 like (SEL1L) in mice leads to systemic endoplasmic reticulum stress and embryonic lethality. *J. Biol. Chem.* **285**, 13694–13703.

Han, D., Lerner, A. G., Vande Walle, L., Upton, J.-P., Xu, W., Hagen, A., Backes, B. J., Oakes, S. A., and Papa, F. R. (2009). IRE1α kinase activation modes control alternate endoribonuclease outputs to determine divergent cell fates. *Cell* **138**, 562–575.

Harding, H. P., Zhang, Y., and Ron, D. (1999). Protein translation and folding are coupled by an endoplasmic-reticulum-resident kinase. *Nature* **397**, 271–274.

Harding, H. P., Novoa, I., Zhang, Y., Zeng, H., Wek, R., Schapira, M., and Ron, D. (2000). Regulated translation initiation controls stress-induced gene expression in mammalian cells. *Mol. Cell* **6**, 1099–1108.

Harding, H. P., Zhang, Y., Zeng, H., Novoa, I., Lu, P. D., Calfon, M., Sadri, N., Yun, C., Popko, B., Paules, R., Stojdl, D. F., Bell, J. C., et al. (2003). An integrated stress response regulates amino acid metabolism and resistance to oxidative stress. *Mol. Cell* **11**, 619–633.

Haze, K., Yoshida, H., Yanagi, H., Yura, T., and Mori, K. (1999). Mammalian transcription factor ATF6 is synthesized as a transmembrane protein and activated by proteolysis in response to endoplasmic reticulum stress. *Mol. Biol. Cell* **10**, 3787–3799.

Held, K. D., Sylvester, F. C., Hopcia, K. L., and Biaglow, J. E. (1996). Role of Fenton chemistry in thiol-induced toxicity and apoptosis. *Radiat. Res.* **145,** 542–553.

Hollien, J., and Weissman, J. S. (2006). Decay of endoplasmic reticulum-localized mRNAs during the unfolded protein response. *Science* **313,** 104–107.

Hollien, J., Lin, J. H., Li, H., Stevens, N., Walter, P., and Weissman, J. S. (2009). Regulated Ire1-dependent decay of messenger RNAs in mammalian cells. *J. Cell Biol.* **186,** 323–331.

Hoozemans, J. J. M., van Haastert, E. S., Nijholt, D. A., Rozemuller, A. J. M., Eikelenboom, P., and Scheper, W. (2009). The unfolded protein response is activated in pretangle neurons in Alzheimer's disease hippocampus. *Am. J. Pathol.* **174,** 1241–1251.

Johnston, R. F., Pickett, S. C., and Barker, D. L. (1990). Autoradiography using storage phosphor technology. *Electrophoresis* **11,** 355–360.

Katayama, T., Imaizumi, K., Honda, A., Yoneda, T., Kudo, T., Takeda, M., Mori, K., Rozmahel, R., Fraser, P., George-Hyslop, P. S., and Tohyama, M. (2001). Disturbed activation of endoplasmic reticulum stress transducers by familial Alzheimer's disease-linked presenilin-1 mutations. *J. Biol. Chem.* **276,** 43446–43454.

Korfei, M., Ruppert, C., Mahavadi, P., Henneke, I., Markart, P., Koch, M., Lang, G., Fink, L., Bohle, R. M., Seeger, W., Weaver, T. E., and Guenther, A. (2008). Epithelial endoplasmic reticulum stress and apoptosis in sporadic idiopathic pulmonary fibrosis. *Am. J. Respir. Crit. Care Med.* **178,** 838–846.

Krężel, A., Leśniak, W., Jeżowska-Bojczuk, M., Mlynarz, P., Brasuñ, J., Kozlowski, H., and Bal, W. (2001). Coordination of heavy metals by dithiothreitol, a commonly used thiol group protectant. *J. Inorg. Biochem.* **84,** 77–88.

Law, D. T., and Segall, J. (1988). The *SPS100* gene of *Saccharomyces cerevisiae* is activated late in the sporulation process and contributes to spore wall maturation. *Mol. Cell. Biol.* **8,** 912–922.

Lee, A. H., Iwakoshi, N. N., and Glimcher, L. H. (2003). XBP-1 regulates a subset of endoplasmic reticulum resident chaperone genes in the unfolded protein response. *Mol. Cell. Biol.* **23,** 7448–7459.

Ma, Y., and Hendershot, L. M. (2003). Delineation of a negative feedback regulatory loop that controls protein translation during endoplasmic reticulum stress. *J. Biol. Chem.* **278,** 34864–34873.

Ma, Y., and Hendershot, L. M. (2004). Herp is dually regulated by both the endoplasmic reticulum stress-specific branch of the unfolded protein response and a branch that is shared with other cellular stress pathways. *J. Biol. Chem.* **279,** 13792–13799.

Maestre, L., Tooze, R., Cañamero, M., Montes-Moreno, S., Ramos, R., Doody, G., Boll, M., Barrans, S., Baena, S., Piris, M. A., and Roncador, G. (2009). Expression pattern of XBP1(S) in human B-cell lymphomas. *Haematologica* **94,** 419–422.

Marciniak, S. J., Yun, C. Y., Oyadomari, S., Novoa, I., Zhang, Y., Jungreis, R., Nagata, K., Harding, H. P., and Ron, D. (2004). CHOP induces death by promoting protein synthesis and oxidation in the stressed endoplasmic reticulum. *Genes Dev.* **18,** 3066–3077.

Martinon, F., Chen, X., Lee, A. H., and Glimcher, L. H. (2010). TLR activation of the transcription factor XBP1 regulates innate immune responses in macrophages. *Nat. Immunol.* **11,** 411–418.

Morris, J. A., Dorner, A. J., Edwards, C. A., Hendershot, L. M., and Kaufman, R. J. (1997). Immunoglobulin binding protein (BiP) function is required to protect cells from endoplasmic reticulum stress but is not required for the secretion of selective proteins. *J. Biol. Chem.* **272,** 4327–4334.

Neubert, K., Meister, S., Moser, K., Weisel, F., Maseda, D., Amann, K., Wiethe, C., Winkler, T. H., Kalden, J. R., Manz, R. A., and Voll, R. E. (2008). The proteasome inhibitor bortezomib depletes plasma cells and protects mice with lupus-like disease from nephritis. *Nat. Med.* **14**, 748–755.

Ohoka, N., Yoshii, S., Hattori, T., Onozaki, K., and Hayashi, H. (2005). *TRB3*, a novel ER stress-inducible gene, is induced via ATF4-CHOP pathway and is involved in cell death. *EMBO J.* **24**, 1243–1255.

Okada, T., Yoshida, H., Akazawa, R., Negishi, M., and Mori, K. (2002). Distinct roles of activating transcription factor 6 (ATF6) and double-stranded RNA-activated protein kinase-like endoplasmic reticulum kinase (PERK) in transcription during the mammalian unfolded protein response. *Biochem. J.* **366**, 585–594.

Paton, A. W., Beddoe, T., Thorpe, C. M., Whisstock, J. C., Wilce, M. C., Rossjohn, J., Talbot, U. M., and Paton, J. C. (2006). AB_5 subtilase cytotoxin inactivates the endoplasmic reticulum chaperone BiP. *Nature* **443**, 548–552.

Reichert, W. L., Stein, J. E., French, B., Goodwin, P., and Varanasi, U. (1992). Storage phosphor imaging technique for detection and quantitation of DNA adducts measured by the ^{32}P-postlabeling assay. *Carcinogenesis* **13**, 1475–1479.

Ron, D., and Walter, P. (2007). Signal integration in the endoplasmic reticulum unfolded protein response. *Nat. Rev. Mol. Cell Biol.* **8**, 519–529.

Rutkowski, D. T., Arnold, S. M., Miller, C. N., Wu, J., Li, J., Gunnison, K. M., Mori, K., Akha, A. A., Raden, D., and Kaufman, R. J. (2006). Adaptation to ER stress is mediated by differential stabilities of pro-survival and pro-apoptotic mRNAs and proteins. *PLoS Biol.* **4**, 2024–2041.

Schröder, M. (2008). Endoplasmic reticulum stress responses. *Cell. Mol. Life Sci.* **65**, 862–894.

Sherman, F. (1991). Getting started with yeast. *Methods Enzymol.* **194**, 3–21.

Song, B., Scheuner, D., Ron, D., Pennathur, S., and Kaufman, R. J. (2008). Chop deletion reduces oxidative stress, improves β cell function, and promotes cell survival in multiple mouse models of diabetes. *J. Clin. Invest.* **118**, 3378–3389.

Sudhakar, A., Krishnamoorthy, T., Jain, A., Chatterjee, U., Hasnain, S. E., Kaufman, R. J., and Ramaiah, K. V. (1999). Serine 48 in initiation factor 2α (eIF2α) is required for high-affinity interaction between eIF2α(P) and eIF2B. *Biochemistry* **38**, 15398–15405.

Wang, X. Z., Lawson, B., Brewer, J. W., Zinszner, H., Sanjay, A., Mi, L. J., Boorstein, R., Kreibich, G., Hendershot, L. M., and Ron, D. (1996). Signals from the stressed endoplasmic reticulum induce C/EBP-homologous protein (CHOP/GADD153). *Mol. Cell. Biol.* **16**, 4273–4280.

Welihinda, A. A., Tirasophon, W., Green, S. R., and Kaufman, R. J. (1997). Gene induction in response to unfolded protein in the endoplasmic reticulum is mediated through Ire1p kinase interaction with a transcriptional coactivator complex containing Ada5p. *Proc. Natl. Acad. Sci. USA* **94**, 4289–4294.

Wilhelm, J., and Pingoud, A. (2003). Real-time polymerase chain reaction. *Chembiochem* **4**, 1120–1128.

Wolfson, J. J., May, K. L., Thorpe, C. M., Jandhyala, D. M., Paton, J. C., and Paton, A. W. (2008). Subtilase cytotoxin activates PERK, IRE1 and ATF6 endoplasmic reticulum stress-signalling pathways. *Cell. Microbiol.* **10**, 1775–1786.

Wu, J., Rutkowski, D. T., Dubois, M., Swathirajan, J., Saunders, T., Wang, J., Song, B., Yau, G. D.-Y., and Kaufman, R. J. (2007). ATF6α optimizes long-term endoplasmic reticulum function to protect cells from chronic stress. *Dev. Cell* **13**, 351–364.

Yamamoto, K., Sato, T., Matsui, T., Sato, M., Okada, T., Yoshida, H., Harada, A., and Mori, K. (2007). Transcriptional induction of mammalian ER quality control proteins is mediated by single or combined action of ATF6α and XBP1. *Dev. Cell* **13**, 365–376.

Ye, J., Rawson, R. B., Komuro, R., Chen, X., Dave, U. P., Prywes, R., Brown, M. S., and Goldstein, J. L. (2000). ER stress induces cleavage of membrane-bound ATF6 by the same proteases that process SREBPs. *Mol. Cell* **6**, 1355–1364.

Yoshida, H., Haze, K., Yanagi, H., Yura, T., and Mori, K. (1998). Identification of the *cis*-acting endoplasmic reticulum stress response element responsible for transcriptional induction of mammalian glucose-regulated proteins. Involvement of basic leucine zipper transcription factors. *J. Biol. Chem.* **273,** 33741–33749.

Zhang, K., Shen, X., Wu, J., Sakaki, K., Saunders, T., Rutkowski, D. T., Back, S. H., and Kaufman, R. J. (2006). Endoplasmic reticulum stress activates cleavage of CREBH to induce a systemic inflammatory response. *Cell* **124,** 587–599.

Zinszner, H., Kuroda, M., Wang, X., Batchvarova, N., Lightfoot, R. T., Remotti, H., Stevens, J. L., and Ron, D. (1998). CHOP is implicated in programmed cell death in response to impaired function of the endoplasmic reticulum. *Genes Dev.* **12,** 982–995.

Zouboulis, C. C., and Tavakkol, A. (1994). Storage phosphor imaging technique improves the accuracy of RNA quantitation using ^{32}P-labeled cDNA probes. *Biotechniques* **16** (290–292), 294.

CHAPTER SIXTEEN

Quantitative Measurement of Events in the Mammalian Unfolded Protein Response

Jie Shang

Contents

1. Introduction	294
2. Cell Culture and ER Stress Inducers	295
2.1. Cell culture	295
2.2. Activation of the UPR with ER stress inducers	295
3. Northern Blots for GRP78/BiP and EDEM	296
4. RT-PCR Analysis of XBP1 mRNA Splicing	299
4.1. RT-PCR analysis of XBP1 mRNA splicing	299
4.2. Gel condition determination	300
5. Immunoblotting and Measurement of ATF6	302
6. Measurement of Inhibition of Protein Synthesis	302
6.1. Measurement with L-[^{35}S] methionine	302
6.2. Measurement with [^{3}H] leucine	303
7. Measurement of Stimulation of LLO Extension	303
7.1. Radioactivity measurement with HPLC	303
7.2. Nonradioactivity measurement with FACE	304
8. Calculations of Transcript Accumulation, Signal Activation, and Correlation	304
9. Conclusion	305
Acknowledgments	306
References	306

Abstract

When the homeostasis of endoplasmic reticulum (ER) is disturbed by the accumulation of unfolded or misfolded proteins, a series of signaling responses collectively called the unfolded protein response (UPR) is triggered. UPR transducers IRE1, PERK, ATF6, and UPR-responsive genes such as GRP78/BiP, ERAD genes such as EDEM, and synthesis of the protein N-linked glycosylation donor lipid-linked oligosaccharides (LLOs) are mobilized. This chapter provides

Department of Pharmacology, University of Texas Southwestern Medical Center, Dallas, Texas, USA

Methods in Enzymology, Volume 491 © 2011 Elsevier Inc.
ISSN 0076-6879, DOI: 10.1016/B978-0-12-385928-0.00016-X All rights reserved.

methods used in our laboratory to quantitatively measure the accumulation of mRNAs encoding BiP and EDEM, splicing of XBP1, cleavage of ATF6, inhibition of protein synthesis by PERK, and extension of LLOs under control and stress conditions.

1. Introduction

In eukaryotic cells, all newly synthesized secretory and membrane proteins must first fold properly in the endoplasmic reticulum (ER) before they are delivered to their ultimate destinations (Bernales et al., 2006; Ellgaard and Helenius, 2003). Any circumstances causing the accumulation of misfolded or unfolded proteins in the lumen of the ER (ER stress) would perturb the homeostasis of ER and induce a series of intracellular protein quality control signaling pathways collectively known as the unfolded protein response (UPR), also called the ER stress response (Bernales et al., 2006; Ron and Walter, 2007; Schröder and Kaufman, 2005). During an unfolded protein response, genes encoding ER chaperones (such as GRP78/BiP), ER-resident protein-folding enzymes, and ER-associated degradation (ERAD) proteins (such as EDEM) are upregulated, while lipid-linked oligosaccharides (LLOs) intermediates are extended more efficiently to mature $Glc_3Man_9GlcNAc_2$-P-P-dolichol for protein glycosylation (Lehrman, 2001), to increase protein-folding capacity of ER (Schröder and Kaufman, 2005). Three classes of ER stress transducers typified by IRE1α, PERK, and ATF6α participate in these signaling pathways in metazoan cells, forming an integrated signaling network to mediate the unfolded protein response and participate in many cellular functions, such as lipid synthesis, B cell differentiation, and cell death. (Glimcher, 2010; Ron and Walter, 2007; Schröder and Kaufman, 2005; Todd et al., 2008). IRE1 (inositol-requiring protein-1) is the oldest, most-conserved transducer of the three because it is found in all eukaryotes (Ron and Walter, 2007). There are two types of IRE1 in mammals, IRE1α and IRE1β, IRE1α is expressed ubiquitously, while IRE1β is mainly expressed in pancreatic β cells. IRE1 is a type I transmembrane protein with an N-terminal sensor domain facing ER lumen and kinase and RNase effector domains on the cytoplasmic side (Cox et al., 1993; Mori et al., 1993; Wang et al., 1998). The accumulation of unfolded protein leads to activation, *trans*-autophosphorylation, and oligomerization of IRE1. In metazoans, the activated IRE1 excises a 26-nucleotide intron from XBP1 (X-box binding protein 1) mRNA and results in the unconventional splicing of XBP1 mRNA that encodes a transcription factor binding to the conserved X-box motif in the promoter region to activate UPR genes controlled by UPRE and ERSE promoter elements (Yoshida et al., 2001). PERK (PKR-like ER kinase) is also type I ER transmembrane protein kinase with an N-terminal sensor domain facing ER lumen. The accumulation of unfolded protein in the ER activates PERK, leading to its autophosphorylation and dimerization. Activated PERK attenuates protein translation by phosphorylating and inactivating eukaryotic initiation factor eIF2α and reduces the load of client protein entering the ER (Harding et. al., 1999). ATF6 (activating transcription factor 6) is a type II transmembrane protein. The accumulation of unfolded protein in the ER causes the translocation of ATF6 to Golgi apparatus, where it is proteolytically cleaved by site I-protease (S1P) and site II- protease (S2P; the enzymes processing SREBPs in response to cholesterol deprivation; Ye et al., 2000). This results in release of a soluble 50 kDa N-terminal cytoplasmic fragment, a transcription factor that enters nucleus and activates UPR genes controlled by ERSE promoter elements (Haze et al., 1999). Here I summarize the methods used in our laboratory to quantitatively measure events in mammalian unfolded protein response. These methods were developed for use with human dermal fibroblasts and mouse embryonic fibroblasts. With minor modification, they should be applicable to most metazoan cell cultures.

2. Cell Culture and ER Stress Inducers

2.1. Cell culture

Primary cultures of normal human adult dermal fibroblasts (HADF, ATCC CRL-1904) and mouse embryonic fibroblasts (MEF) were tested. Cultures were grown at 37 °C in the presence of humidified 5% CO_2 in RPMI-1640 medium (HADF) or DMEM medium (MEF; cat. #31800 and #31600, respectively, Invitrogen) supplemented with 10% fetal bovine serum (Atlanta Biologicals, Georgia). Cells were passaged by trypsinization and grown to 70–90% confluence for experiments. Cells were refed within 24 h of analysis to avoid possible stresses such as nutrient deprivation.

2.2. Activation of the UPR with ER stress inducers

The following UPR inducers were used as described in Table 16.1 and the figure legends: DL-dithiothreitol (DTT; cat. #D9779), thapsigargin (TG; cat. #T9033), L-azetidine-2-carboxylic acid (AZC; cat. #A0760), and tunicamycin (TN; cat. #T7765) were from Sigma, St. Louis, MO. Castanospermine (CSN) was from Matreya, LLC, Pleasant Gap, PA (cat. #1800; Shang et al., 2002).

Table 16.1 ER stress inducers and their mechanisms

ER stress inducers	Treatments (human dermal fibroblasts)	Functions
DL-Dithiothreitol (DTT)	0.4 mM, 20 min 2 mM, 20 min	Reduces disulfide bonds
Thapsigargin (TG)	10 nM, 30 min 100 nM, 30 min	Releases ER calcium
L-Azetidine-2-carboxylic acid (AZC)	10 mM, 1 h 60 mM, 1 h	Inhibits prolyl isomerization
Tunicamycin (TN)	5 μg/ml, 1 h 5 μg/ml, 5 h	Inhibits GlcNAc-1-P transferase, blocking N-linked glycosylation
Castanospermine (CSN)	200 μg/ml, 2 h	Inhibits ER glucosidases I and II, blocking calnexin and calreticulin-dependent protein folding

Reproduced and modified with permissions: Shang *et al.* (2002), copyright 2002 Oxford University Press; Shang and Lehrman (2004a), copyright 2004 Elsevier Press.

3. Northern Blots for GRP78/BiP and EDEM

Levels of the UPR-responsive genes GRP78/BiP and EDEM (Hosokawa *et al.*, 2001) were determined by quantitative measurement of their respective mRNAs with Northern blots (Fig. 16.1, panel A), as described below, although more contemporary Q-PCR methods could also be used. Probe templates were prepared from human fibroblast RNA by RT-PCR with primers 5'-TTGCTTATGGCCTGGATAAGAGGG-3' and 5'-TGTACCCTTGTCTTCAGCTGTCAC-3' for human GRP78/BiP (AJ271729), for a 935 bp fragment, and primers 5'-TCATCCGAGTTCCAGAAAGCCGTC-3' and 5'-TTGACATAGAGTGGAGGGTCTCCT-3' for human EDEM (NM_014674), for a 671 bp fragment, and primers 5'-CGGGAAATCGTGCGTGACATTAAG-3' and 5'-TACTCCTGCTTGCTGATCCACATC-3' for human beta actin (BC013380), for a 470 bp fragment (summarized in Table 16.2). Probes were randomly labeled with Rediprime II random prime labeling system (cat. #RPN1633) and Redivue [^{32}P] dCTP (GE Healthcare) and labeled probes were purified with Quick Spin Columns for Radiolabeled DNA Purification Sephadex G-50, fine (cat. #11273973001, Roche Applied Science). Total RNA was isolated from cells 5 h after initiation of treatment with UPR inducers, or from untreated controls with QIAshredder and RNeasy mini kit (cat. #79654 and #74104, respectively, Qiagen, Valencia, CA), and RNA samples were analyzed by 1% formaldehyde

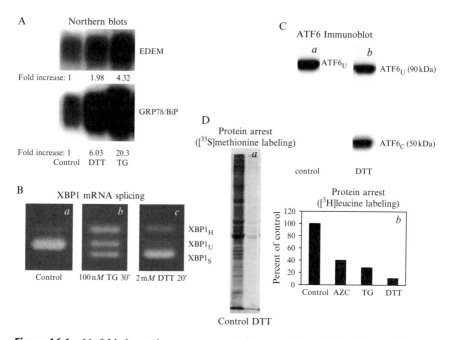

Figure 16.1 Unfolded protein response events in normal dermal fibroblasts and mouse embryonic fibroblasts. (A) Transcription of EDEM and GRP78/BiP. Dermal fibroblasts were left untreated, or treated with either 2 mM DTT for 20 min or 100 nM TG for 30 min. mRNA was harvested to determine the expression of EDEM or GRP78/BiP. Scans of X-ray films of Northern blots are shown. (B) Ire1p-XBP1 signaling in dermal fibroblasts. RT-PCR of XBP1 mRNA shows $XBP1_U$ in control samples, $XBP1_S$ after strong stress with 2 mM DTT for 20 min, and a mixture of $XBP1_U$, $XBP1_S$, and $XBP1_H$ after moderate stress with 100 nM TG for 30 min. (C) ATF6 signaling in dermal fibroblasts. Immunoblot detection of 90 kDa (uncleaved) ATF6 in untreated control cells, and both 90 and 50 kDa (cleaved) ATF6 in cells stressed with 2 mM DTT for 20 min. (D) Translation arrest. (a) Normal dermal fibroblasts were left untreated, or treated with 2 mM DTT for 20 min. Immediately afterward, metabolic labeling with 200 μCi/ml [^{35}S] methionine in complete RPMI 1640 medium was carried out for 20 min. Whole cell lysates were resolved by 10% SDS-PAGE. A phosphorimager scan is shown. (b) Mouse embryonic fibroblasts were left untreated, or treated with 40 mM AZC for 1 h, 100 nM TG for 30 min, or 2 mM DTT for 20 min. Immediately after treatment, cells were labeled in DMEM medium with 0.5 mM glucose, 10% dialyzed FCS, and 5 μCi/ml [^{3}H] leucine for 5 min. Cell lysates were precipitated with 5% cold TCA, washed, and counted by liquid scintillation spectroscopy. Incorporation of [^{3}H] leucine into protein in treated cells was expressed as the percent incorporated into untreated controls, and normalized to total protein. Reproduced and modified with permission, copyright 2004 Elsevier Press, Shang and Lehrman (2004a).

agarose electrophoresis and blotting and washing with standard Northern blot procedures. Damp blots were sealed in a plastic bag and exposed to a phosphorimaging screen for various hours according to the intensity of signals. Then signals were scanned and quantified with a phosphorimager

Table 16.2 RT-PCR primers used to generate northern blot probes and to detect XBP1 splicing

Source	NCBI accession no.	Primer pair sequence (F, forward primer; R, reverse primer)	Amplified fragment length
Human GRP78/Bip	AJ271729	F: 5′-TTGCTTATGGCCTGGATAAGAGGG-3′ R: 5′-TGTACCCTTGTCTTCAGCTGTCAC-3′	935 bp
Human EDEM	NM_014674	F: 5′-TCATCCGAGTTCCAGAAAGCCGTC-3′ R: 5′-TTGACATAGAGTGGAGGGTCTCCT-3′	671 bp
Human Actin	BC013380	F: 5′-CGGGAAATCGTGCGTGACATTAAG-3′ R: 5′-TACTCCTGCTTGCTGATCCACATC-3′	470 bp
Human XBP1	NM_005080	F: 5′-CTGGAACAGCAAGTGGTAGA-3′ R: 5′-CTGGGTCCTTCTGGGTAGAC-3′	Unspliced: 424 bp Spliced: 398 bp
Human XBP1	NM_005080	F: 5′-CCTTGTAGTTGAGAACCAGG-3′ R: 5′-GGGGCTTGGTATATATGTGG-3′	Unspliced: 442 bp Spliced: 416 bp
Mouse XBP1	NM_013842	F: 5′-CTGGAGCAGCAAGTGGTGGATTTG-3′ R: 5′-CTAGGTCCTTCTGGGTAGACCTC-3′	Unspliced: 430 bp Spliced: 404 bp
Rat XBP1	NM_001004210	F: 5′-CTGGAGCAGCAAGTGGTGGA-3′ R: 5′-CTAGGTCCTTCTGGGTAGAC-3′	Unspliced: 430 bp Spliced: 404 bp

(Fuji), and normalized to actin mRNA probed on the same blots after stripping GRP78/Bip or EDEM probes (shake blots in boiling stripping buffer containing 10 mM Na$_2$EDTA, pH 8.0, 0.05× SSC, and 0.1% SDS at room temperature for 20 min, repeated two to three times). For each lane the signal for BiP mRNA or EDEM was divided by the signal for actin. The resulting normalized values were used to calculate the fold enhancements of GRP78 and EDEM signals in treated cells compared to those in untreated controls (Shang and Lehrman, 2004a).

4. RT-PCR Analysis of XBP1 mRNA Splicing

4.1. RT-PCR analysis of XBP1 mRNA splicing

Activation of IRE1 was determined by quantitatively measuring the splicing of its substrate, the mRNA encoding the XBP1 transcription factor (Figure 16.1, panel B). Total RNA was harvested (RNA-Bee RNA Isolation Solvent, TEL-TEST, or RNeasy mini kit, Qiagen) immediately after completion of stress treatments. Typically, 2 μg RNA was used as template in every 20 μl reaction mixture to make first-strand cDNA with Super-Script® First-Strand Synthesis System for RT-PCR kit (cat. #11904-018, Invitrogen, Carlsbad, CA) and 2 μl synthesized first-strand cDNA was used as template in a 50 μl PCR reaction mixture. To amplify human XBP1 (NM_005080), PCR mixture was heated at 94 °C for 2 min, then was for 30 cycles (94 °C for 30 s; 58 °C for 30 s; and 72 °C for 1 min (but 10 min in the final cycle)) using 5′-CTGGAACAGCAAGTGGTAGA-3′ and 5′-CTGGGTCCTTCTGGGTAGAC-3′ (Table 16.2) with Taq DNA Polymerase (cat. #1815105, Roche) or Platinum® Taq DNA Polymerase (cat. #10966-018, Invitrogen). 10 μl PCR products (1/5 total reaction volume) was subjected to electrophoresis with 2% agarose gel (11 × 14 cm, width × length, large gel) until the bromophenol blue dye is 2 cm to the bottom of the gel. Fragments of 398 and 424 bp representing spliced (XBP1$_S$) and unspliced (XBP1$_U$) XBP1, plus a hybrid (XBP1$_H$) migrating as a fragment of approximately 450 bp, were documented after staining with ethidium bromide and scanning photographs or directly scanning gels (Bio-Rad Fluor-S MultiImager). Isolation of XBP$_H$ followed by additional PCR generated all three fragments, but additional PCR of either XBP$_S$ or XBP$_U$ regenerated only the starting fragment. *Apa*LI and *Pst*I, enzymes that cleave within the intron or at the exon border, respectively, cleaved XBP$_U$ but not XBP$_S$ or XBP$_H$. Isolation and sequencing of XBP$_S$ or XBP$_U$ gave the expected sequences lacking or containing intron, but sequencing of XBP$_H$ in either orientation gave composites of the XBP$_S$ and XBP$_U$ sequences (data not shown). XBP1$_H$ was detected with two independent sets of primers, and it was more abundant with milder stresses than robust stresses.

We conclude that XBP_H is a mixture of two hybrid structures, each structure containing one strand from XBP_S and XBP_U that form during annealing in the final PCR step. Thus, XBP_H is greatest with mild stresses because spliced and unspliced strands are both present (Shang and Lehrman, 2004a). Signal activations were determined from Fluor-S readings: percentage activation of XBP1 was calculated as $100 \times [XBP1_S + 0.5\ XBP1_H]/[XBP1_S + XBP1_H + XBP1_U]$. In some experiments fragments were confirmed with a second primer pair, 5′-CCTTGTAGTTGAGAACCAGG-3′ and 5′-GGGGCTTGGTATATATGTGG-3′, which yielded a 442 bp fragment for XBP_U (Yoshida et al., 2001). XBP1 triplet bands ($XBP1_S$, $XBP1_U$, and $XBP1_H$) were also documented with 2% agarose gel electrophoresed XBP1 RT-PCR products amplified with RNA from ER stress induced mouse cell lines (NM_013842) and primers 5′-CTGGAGCAGCAAGTGGTGGATTTG-3′ and 5′-CTAGGTCCTTCTGGGTAGACCTC-3′ (yielding a 430 bp fragment for XBP_U; Fig. 16.2; Table 16.2) or rat cell lines (NM_001004210) and primers 5′-CTGGAGCAGCAAGTGGTGGA-3′ and 5′-CTAGGTCCTTCTGGGTAGAC-3′ (yielding a 430 bp fragment for XBP_U; data not shown; Table 16.2).

4.2. Gel condition determination

The author consistently detected XBP1 RT-PCR triplet bands (spliced XBP1, unspliced XBP1, and hybrid XBP1) amplified with total RNA from various human, mouse, and rat cell lines treated with ER stress inducers. While a number of other studies reported only the spliced and unspliced products, many reported the triplet bands as well (Back et al., 2005, 2006; Hollien et al., 2009; McLaughlin et al., 2007; Penas, et al., 2007; Umareddy et al., 2007; Wilson et al., 2007; Wolfson et al., 2008). In author's hands, all attempts to eliminate XBP_H failed, for example by optimizing PCR conditions, replacing reagents, using more primer, transferring trace amounts of product to new reactions, and refreshing PCRs with polymerase and nucleotides during the final two cycles (unpublished data). As mentioned in 4.1., we conclude that the appearance of XBP_H is simply a PCR artifact that occurs when one spliced and one unspliced strand hybridize; the consumption of primer may be the primary cause of the hybrid. It was suggested that to detect the hybrid XBP1 band, the concentration of agarose in the gel should be 3% or higher and electrophoresis should be as long as possible (Back et al., 2005). In order to clarify the disparity reported among publications, MEF cells were treated with 10 nM TG for 10 min (generating a mixture of all three bands), 30 min (generating mostly the spliced lower band), or left untreated (mostly unspliced middle band). Total RNA isolation, cDNA synthesis, and PCR amplification with mouse (NM_013842) primers 5′-CTGGAGCAGCAAGTGGTGGATTTG-3′ and 5′-CTAGGTCCTTCTGGGTA GACCTC-3′ (Table 16.2) were performed

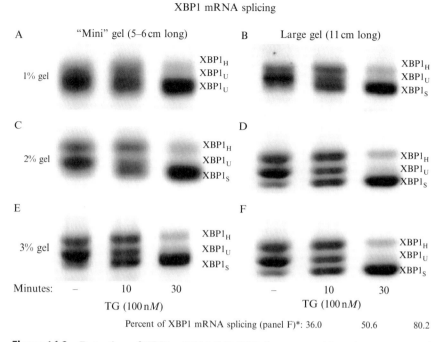

Figure 16.2 Detection of XBP1 mRNA RT-PCR fragments with various agarose gel concentrations and sizes. MEFs were treated with 100 nM TG for 10, 30 min or left untreated as indicated, 10 μl XBP1 RT-PCR products were subjected to electrophoresis with 1% (A and B), 2% (C and D) and 3% agarose gel (E and F) and run 5–6 cm long (A, C, E, "mini" gels) or 11 cm long (B, D, F, large gels). Ethidium bromide stained gels were scanned with a Fluor-S multiImager (Bio-Rad) using the same parameters. For better visualization, panels A and B were expanded. *, the values may vary depending on the cell types.

under the same conditions as mentioned above. 10 μl PCR products amplified with cDNAs from each treatment were subjected to electrophoresis with 1%, 2%, and 3% large agarose gels (11 × 14 cm, width × length). For each gel concentration, gels were stopped either half way (about 5–6 cm from the loading wells, approximately the size of a mini gel) or run until the bromophenol blue dye 2 cm to the bottom of the gel. Ethidium bromide stained gels were scanned (Bio-Rad Fluor-S MultiImager) with the same parameters for direct comparison. As seen in Fig. 16.2, the 1% gel did not separate all three XBP1 bands, while 2% or 3% mini gels were sufficient to separate all three bands. Both the 2% and 3% large gels gave the better separation than their corresponding "mini" gels. Though in author's hands, sometimes 3% large gels tend to produce distorted bands, probably due to the high agarose concentration. Taken together, during electrophoresis of XBP1 mRNA PCR fragments, variations among gel concentration and gel size may contribute to the observation of two or three bands. In conclusion, a 2% mini gel is sufficient to separate all three XBP1 bands, but in order to achieve the best resolution, a 2% large gel or a 3% mini gel is recommended.

5. Immunoblotting and Measurement of ATF6

Activation of ATF6 is determined by quantitative immunoblot analysis of the uncleaved (ATF6$_U$) and cleaved (ATF6$_C$) forms (Fig. 16.1, panel C). Cells were solubilized in 0.5% Triton X-100 with 50 mM Na–HEPES (pH 7.6), 150 mM NaCl, phosphatase inhibitors[1] (100 mM NaF, 1 mM Na$_3$VO$_4$), and protease inhibitors[1] (10.5 μg/ml aprotinin, 1 μg/ml pepstatin A, 1 μg/ml leupeptin, and 1 mM PMSF). Eighty micrograms of cellular protein was resolved by 7.5%, 8%, or 10% SDS-PAGE and transferred to nitrocellulose membranes. ATF6 was identified with a human rabbit polyclonal antibody[2] (gift of Dr. Kazutoshi Mori, Kyoto University, Japan) for the α and β forms, followed by incubation with peroxidase-labeled anti-rabbit secondary antibody (cat. #NA934VS, Amersham, GE Healthcare), detection by chemiluminescence (Amersham ECL^{TM} Western Blotting Detection Reagents, cat. #RPN2109, GE Healthcare), and exposure to KODAK^{TM} X-OMAT^{TM} Blue Film. Signals were measured with a Bio-Rad Fluor-S MultiImager (Shang and Lehrman, 2004a). Signal activations were determined from Fluor-S readings: percentage activation of ATF6 was calculated as 100 × [ATF6$_C$]/[ATF6$_C$ + ATF6$_U$].

6. Measurement of Inhibition of Protein Synthesis

6.1. Measurement with L-[^{35}S] methionine

Activation of PERK is determined by measuring inhibition of total cellular protein synthesis (Fig. 16.1, panel D). Immediately after treatment with UPR inducers, metabolic labeling of fibroblasts with L-[^{35}S] methionine (Amersham Pharmacia Biotech) in complete RPMI 1640 medium was carried out for 20 min as described (Harding et al., 2000). However, to avoid potential stress due to nutrient deprivation, the author omits any preincubation period in methionine-free medium. Whole cell lysates were prepared in RIPA buffer (150 mM NaCl, 1.0% (w/v) Nonidet

---

[1] Can be replaced with PhosSTOP Phosphatase Inhibitor Cocktail Tablets (cat. #04906837001) and complete, Mini Protease Inhibitor Cocktail Tablets (cat. #04693124001) from Roche Applied Science.
[2] Similar human rabbit polyclonal ATF6 antibody is commercially available through http://www.bioacademia.co.jp/en/product_list.php?srch_keyword=73-505, though the author has no experience working with it.

P-40, 0.5% deoxycholate, 0.1% sodium dodecyl sulfate (SDS), 50 mM Tris–Cl (pH 8.0), and protease inhibitors[1] 10 μg/ml aprotinin, 10 μg/ml leupeptin, and 1 mM phenylmethylsulfonyl fluoride (PMSF), all from Sigma) and resolved by SDS-PAGE. Gels were fixed in fixing solution (methanol: acetic acid:H_2O, 40:10:50) for 30 min, followed by soaking and agitating in Amplify™ Fluorographic Reagent (cat. #NAMP100, Amersham, GE Healthcare) for 15–30 min for increased detection efficiency, then placed on top of a piece of Whatman GB002 filter paper and dried at 60–80 °C with a vacuum gel dryer. The dried gel was exposed to a phosphorimaging screen and the total protein radioactivity in each lane was measured with a Phosphorimager (Fuji). The translation inhibition in treated cells was expressed as the percentage decrease of incorporation of L-[^{35}S] methionine into protein compared with untreated controls (Shang et al., 2002).

6.2. Measurement with [^{3}H] leucine

Such assays can also be done by labeling cells with [^{3}H] leucine, and measuring the radioactivity in total precipitated protein. After treatments with potential PERK activators as indicated, cells were labeled for 5 min with 5 μCi/ml [^{3}H] leucine (164 Ci/mmol, Amersham Pharmacia Biotech). Cells were washed three times with ice-cold PBS and lysed with 1 ml of lysis buffer (PBS containing 1% SDS and 10 mM Na_2EDTA). Ice-cold 100% TCA solution was added to lysates to a final concentration of 5% (v/v). Cell lysates were vortexed immediately and kept on ice for 30 min, and precipitates were collected by vacuum-assisted filtration on GF/C glass microfiber filter discs (24 mm Ø, cat. #1822-024, Whatman). Tubes were rinsed once with 1 ml of cold PBS containing 5% TCA, which was applied to the same filter disc. The filter discs were washed four times with 10 ml of ice-cold water containing 5% TCA, and twice with 10 ml ethanol–ether (1:1, v/v) solution prewarmed to 37 °C. Filters were allowed to dry in air, and radioactivity was determined by immersing filters in 0.5 ml water and 6 ml of scintillation fluid (Liquiscint, cat. #LS-121, National Diagnostics) followed by liquid scintillation counting. Radioactivity was normalized to protein in the lysates, which was determined with the Pierce BCA™ Protein Assay Kit (cat. #23225) against bovine serum albumin standard curve (Shang and Lehrman, 2004b).

7. MEASUREMENT OF STIMULATION OF LLO EXTENSION

7.1. Radioactivity measurement with HPLC

LLO extension is assessed by determining the ratios of mature LLOs (with nine mannosyl and three glucosyl residues) to immature LLO intermediates with two to five mannosyl residues. Either during (DTT) or immediately after (AZC, TG, CSN) treatment with UPR inducers, fibroblasts were metabolically labeled with D-[2-^{3}H] mannose (15–20 Ci/mmol, Amersham Pharmacia Biotech) for 20 min in medium with 0.5 mM D-glucose. Isolation of total LLOs, release of ^{3}H-labeled oligosaccharides from the dolichol-P-P carrier with mild acid, and HPLC analyses were carried out as described (Doerrler and Lehrman, 1999). Peak heights for Man$_{2-5}$GlcNAc$_2$ and Glc$_3$Man$_9$GlcNAc$_2$ were measured and then normalized to mannose content to reflect molar quantities of isotope in each oligosaccharide. The molar percentage of Glc$_3$Man$_9$GlcNAc$_2$ in each sample ([mature/[mature + immature]] × 100) was then calculated (Shang et al., 2002).

7.2. Nonradioactivity measurement with FACE

Despite the wide applications of metabolic radiolabeling techniques in the LLO assay, some limitations exist: such as longer labeling periods required for steady-state compositional analysis, challenging labeling in live animals and some difficult cell lines; extra steps need to be taken to determine the actual chemical yields of LLO intermediates due to dilutions of radiolabeled precursors by intracellular precursors; difficulties while handling clinical specimens, and studies of LLO pathways that may block metabolic incorporation (Lehrman, 2007). These limitations can be circumvented by the Fluorophore-Assisted Carbohydrate Electrophoresis (FACE) technique. FACE uses fluorophores to chemically modify released glycans (with exposed reducing termini) from the lipid carrier, followed by electrophoretic separation and fluorescence detection. Two fluorophores are widely used: 7-amino-1, 3-naphthalenedisulfonic acid (ANDS; for oligosaccharides with at least three sugars, Glc$_{0-3}$Man$_{1-9}$GlcNAc$_2$-P-P-Dol) and 2-aminoacridone (AMAC; for monosaccharides and disaccharides, GlcNAc$_{1-2}$-P-P-Dol). FACE is able to detect LLO as low as 1–2 pmol. This technique can be expanded to detect oligosaccharides, monosaccharides, and nucleotide-sugars from diseased or healthy cultured cells or various animal tissues (Cho et al., 2005; Gao and Lehrman, 2002a,b; Gao et al., 2005; Lehrman, 2007; Shang et al., 2007). For detailed protocol of sugar labeling and quantitative measurement with FACE, see Gao and Lehrman (2006).

8. CALCULATIONS OF TRANSCRIPT ACCUMULATION, SIGNAL ACTIVATION, AND CORRELATION

Unfolded protein response signaling activated by accumulation of misfolded proteins within the lumen of the mammalian endoplasmic reticulum (ER) involves cleavage of ATF6, and splicing of XBP1 mRNA initiated by IRE1. Each system responds to diminution of free GRP78/BiP chaperone in the ER lumen (Bertolotti et al., 2000; Shen et al., 2002), so ATF6 cleavage should correlate with XBP1 splicing under a variety of ER stresses. The GRP78/BiP promoter contains an ERSE element responsive to $ATF6_C$ and $XBP1_S$ while the EDEM gene contains a UPRE promoter element responsive to $XBP1_S$ but not $ATF6_C$ (Yoshida et al., 2003). Thus, GRP78/BiP mRNA amounts should correlate with IRE1 and/or ATF6 activation, EDEM mRNA amounts should correlate with IRE1 activation, and amounts of the two mRNAs should correlate with each other (Shang and Lehrman, 2004a). Due to their graded responses to ER stress inducers, dermal fibroblasts were used to test this hypothesis by subjecting them to various dithiothreitol (DTT), thapsigargin (TG), azetidine-2-carboxylic acid (AZC), tunicamycin (TN), and castanospermine (CSN) treatments, followed by quantitative analysis with aforementioned methods. For single ER stress inducers at variable concentrations, there were good correlations between activation of stress signal transducers and transcription of mRNA targets. Further, comparing different ER stresses, we found strong correlations between XBP-1 and ATF6 activation, and between GRP78/Bip and EDEM transcription. However, comparing different ER stresses there was surprisingly no correlation between activation of either of the signal transducers with transcription of either of the mRNAs. Thus, ATF6 and IRE1/XBP1 signaling may be necessary for gene activation, but they do not define the magnitude of UPR-dependent mRNA increases, which appear dependent upon the exact form of ER stress by mechanisms which remain unknown.

9. CONCLUSION

As shown in Table 16.1 and Fig. 16.1, these methods allowed us to quantitatively measure UPR events in a system with graded responses to ER stress inducers. Using these methods we concluded that there were strong mutual correlations between the two signaling events (ATF6 and IRE1) and between the two target transcripts (GRP78/BiP and EDEM), but there was a lack of correlation between either signaling event with either transcript when different ER stresses were compared (Shang and Lehrman, 2004a). We proposed a model in which an additional stress-dependent factor(s) controls the extent of accumulation of both transcripts. Perhaps, ER stresses might differentially affect at least one component other than GRP78/BiP to signal accumulation of misfolded protein. The signal might then activate a factor that acts similarly upon both targets, by increasing either transcription or mRNA stability. Thus, ATF6 cleavage and XBP1 splicing are considered necessary for transcription, but beyond a certain threshold the actual amounts of $ATF6_C$ and $XBP1_S$ do not necessarily determine the levels of GRP78/BiP and EDEM mRNAs (Shang and Lehrman, 2004a).

ACKNOWLEDGMENTS

The author wishes to thank Dr. Mark A. Lehrman for his guidance and valuable comments on the chapter. Work described in this chapter was supported by NIH Grant GM38545 and Welch Grant I-1168 to M. A. L., M. A. L is a professor at UT-Southwestern Medical Center at Dallas.

REFERENCES

Back, S. H., Schroder, M., Lee, K., Zhang, K., and Kaufman, R. J. (2005). ER stress signaling by regulated splicing: IRE1/HAC1/XBP1. *Methods* **35,** 395–416.

Back, S. H., Lee, K., Vink, E., and Kaufman, R. J. (2006). Cytoplasmic IRE1alpha-mediated XBP1 mRNA splicing in the absence of nuclear processing and endoplasmic reticulum stress. *J. Biol. Chem.* **281,** 18691–18706.

Bernales, S., Papa, F. R., and Walter, P. (2006). Intracellular signaling by the unfolded protein response. *Annu. Rev. Cell Dev. Biol.* **22,** 487–508.

Bertolotti, A., Zhang, Y., Hendershot, L. M., Harding, H. P., and Ron, D. (2000). Dynamic interaction of BiP and ER stress transducers in the unfolded-protein response. *Nat. Cell Biol.* **2,** 326–332.

Cho, S. K., Gao, N., Pearce, D. A., Lehrman, M. A., and Hofmann, S. L. (2005). Characterization of lipid-linked oligosaccharide accumulation in mouse models of Batten disease. *Glycobiology* **15,** 637–648.

Cox, J. S., Shamu, C. E., and Walter, P. (1993). Transcriptional induction of genes encoding endoplasmic reticulum resident proteins requires a transmembrane protein kinase. *Cell* **73,** 1197–1206.

Doerrler, W. T., and Lehrman, M. A. (1999). Regulation of the dolichol pathway in human fibroblasts by the endoplasmic reticulum unfolded protein response. *Proc. Natl. Acad. Sci. USA* **96,** 13050–13055.

Ellgaard, L., and Helenius, A. (2003). Quality control in the endoplasmic reticulum. *Nat. Rev. Mol. Cell Biol.* **4,** 181–191.

Gao, N., and Lehrman, M. A. (2002a). Analyses of dolichol pyrophosphate-linked oligosaccharides in cell cultures and tissues by fluorophore-assisted carbohydrate electrophoresis. *Glycobiology* **12,** 353–360.

Gao, N., and Lehrman, M. A. (2002b). Coupling of the dolichol-P-P-oligosaccharide pathway to translation by perturbation-sensitive regulation of the initiating enzyme, GlcNAc-1-P transferase. *J. Biol. Chem.* **277,** 39425–39435.

Gao, N., and Lehrman, M. A. (2006). Nonradioactive analysis of lipid-linked oligosaccharide compositions by fluorophore-assisted carbohydrate electrophoresis (FACE). *Methods Enzymol.* **415,** 3–20.

Gao, N., Shang, J., and Lehrman, M. A. (2005). Analysis of glycosylation in CDG-Ia fibroblasts by fluorophore-assisted carbohydrate electrophoresis: Implications for extracellular glucose and intracellular mannose-6-phosphate. *J. Biol. Chem.* **280,** 17901–17909.

Glimcher, L. H. (2010). XBP1: The last two decades. *Ann. Rheum. Dis.* **69**(Suppl. 1), i67–i71.

Harding, H. P., Zhang, Y., and Ron, D. (1999). Protein translation and folding are coupled by an endoplasmic-reticulum-resident kinase. *Nature* **397,** 271–274.

Harding, H. P., Zhang, Y., Bertolotti, A., Zeng, H., and Ron, D. (2000). Perk is essential for translational regulation and cell survival during the unfolded protein response. *Mol. Cell* **5,** 897–904.

Haze, K., Yoshida, H., Yanagi, H., Yura, T., and Mori, K. (1999). Mammalian transcription factor ATF6 is synthesized as a transmembrane protein and activated by proteolysis in response to endoplasmic reticulum stress. *Mol. Biol. Cell* **10,** 3787–3799.

Hollien, J., Lin, J. H., Li, H., Stevens, N., Walter, P., and Weissman, J. S. (2009). Regulated Ire1-dependent decay of messenger RNAs in mammalian cells. *J. Cell Biol.* **186,** 323–331.

Hosokawa, N., Wada, I., Hasegawa, K., Yorihuzi, T., Tremblay, L. O., Herscovics, A., and Nagata, K. (2001). A novel ER alpha-mannosidase-like protein accelerates ER-associated degradation. *EMBO Rep.* **2,** 415–422.

Lehrman, M. A. (2001). Oligosaccharide-based information in endoplasmic reticulum quality control and other biological systems. *J. Biol. Chem.* **276,** 8623–8626.

Lehrman, M. A. (2007). Teaching dolichol-linked oligosaccharides more tricks with alternatives to metabolic radiolabeling. *Glycobiology* **17,** 75R–85R.

McLaughlin, M., Karim, S. A., Montague, P., Barrie, J. A., Kirkham, D., Griffiths, I. R., and Edgar, J. M. (2007). Genetic background influences UPR but not PLP processing in the rumpshaker model of PMD/SPG2. *Neurochem. Res.* **32,** 167–176.

Mori, K., Ma, W., Gething, M. J., and Sambrook, J. (1993). A transmembrane protein with a $cdc2^+/CDC28$-related kinase activity is required for signaling from the ER to the nucleus. *Cell* **74,** 743–756.

Penas, C., Guzman, M. S., Verdu, E., Fores, J., Navarro, X., and Casas, C. (2007). Spinal cord injury induces endoplasmic reticulum stress with different cell-type dependent response. *J. Neurochem.* **102,** 1242–1255.

Ron, D., and Walter, P. (2007). Signal integration in the endoplasmic reticulum unfolded protein response. *Nat. Rev. Mol. Cell Biol.* **8,** 519–529.

Schröder, M., and Kaufman, R. J. (2005). ER stress and the unfolded protein response. *Mutat. Res.* **569,** 29–63.

Shang, J., and Lehrman, M. A. (2004a). Discordance of UPR signaling by ATF6 and Ire1p-XBP1 with levels of target transcripts. *Biochem. Biophys. Res. Commun.* **317,** 390–396.

Shang, J., and Lehrman, M. A. (2004b). Inhibition of mammalian RNA synthesis by the cytoplasmic Ca^{2+} buffer BAPTA. Analyses of [^{3}H] uridine incorporation and stress-dependent transcription. *Biochemistry* **43,** 9576–9582.

Shang, J., Körner, C., Freeze, H., and Lehrman, M. A. (2002). Extension of lipid- linked oligosaccharides is a high-priority aspect of the unfolded protein response: Endoplasmic reticulum stress in Type I congenital disorder of glycosylation fibroblasts. *Glycobiology* **12,** 307–317.

Shang, J., Gao, N., Kaufman, R. J., Ron, D., Harding, H. P., and Lehrman, M. A. (2007). Translational balancing by the eIF2α kinase PERK couples ER glycoprotein synthesis to lipid-linked oligosaccharide flux. *J. Cell Biol.* **176,** 605–616.

Shen, J., Chen, X., Hendershot, L., and Prywes, R. (2002). ER stress regulation of ATF6 localization by dissociation of BiP/GRP78 binding and unmasking of Golgi localization signals. *Dev. Cell* **3,** 99–111.

Todd, D. J., Lee, A. H., and Glimcher, L. H. (2008). The endoplasmic reticulum stress response in immunity and autoimmunity. *Nat. Rev. Immunol.* **8,** 663–674.

Umareddy, I., Pluquet, O., Wang, Q. Y., Vasudevan, S. G., Chevet, E., and Gu, F. (2007). Dengue virus serotype infection specifies the activation of the unfolded protein response. *Virol. J.* **4,** 91.

Wang, X. Z., Harding, H. P., Zhang, Y., Jolicoeur, E. M., Kuroda, M., and Ron, D. (1998). Cloning of mammalian Ire1 reveals diversity in the ER stress responses. *EMBO J.* **17,** 5708–5717.

Wilson, S. J., Tsao, E. H., Webb, B. L., Ye, H., Dalton-Griffin, L., Tsantoulas, C., Gale, C. V., Du, M. Q., Whitehouse, A., and Kellam, P. (2007). X box binding protein XBP-1s transactivates the Kaposi's sarcoma-associated herpesvirus (KSHV) ORF50 promoter, linking plasma cell differentiation to KSHV reactivation from latency. *J. Virol.* **81,** 13578–13586.

Wolfson, J. J., May, K. L., Thorpe, C. M., Jandhyala, D. M., Paton, J. C., and Paton, A. W. (2008). Subtilase cytotoxin activates PERK, IRE1 and ATF6 endoplasmic reticulum stress-signalling pathways. *Cell. Microbiol.* **10,** 1775–1786.

Ye, J., Rawson, R. B., Komuro, R., Chen, X., Dave, U. P., Prywes, R., Brown, M. S., and Goldstein, J. L. (2000). ER stress induces cleavage of membrane-bound ATF6 by the same proteases that process SREBPs. *Mol. Cell* **6,** 1355–1364.

Yoshida, H., Matsui, T., Yamamoto, A., Okada, T., and Mori, K. (2001). XBP1 RNA is induced by ATF6 and spliced by IRE1 in response to ER stress to produce a highly active transcription factor. *Cell* **107,** 881–891.

Yoshida, H., Matsui, T., Hosokawa, N., Kaufman, R. J., Nagata, K., and Mori, K. A. (2003). Time-dependent phase shift in the mammalian unfolded protein response. *Dev. Cell* **4,** 265–271.

CHAPTER SEVENTEEN

Regulation of Immunoglobulin Synthesis, Modification, and Trafficking by the Unfolded Protein Response: A Quantitative Approach

Adi Drori *and* Boaz Tirosh

Contents

1. Introduction 310
2. Isolation of Splenic B Cells and Infection Thereof by Retroviruses 312
   2.1. Isolation splenic B cells by magnetic sorting 312
   2.2. Generation of retroviruses 314
   2.3. Retroviral transduction of B cells 315
3. Measurement of Protein Synthesis in Plasma Cells 316
   3.1. Metabolic labeling of B cells 316
   3.2. Cell lysis and immunoprecipitation 317
4. Measurement of Ig Mislocalization in Primary B Cells 320
   4.1. Permeabilization of primary B cells using digitonin and trypsin digestion of cytoplasmic proteins 322
Acknowledgments 324
References 324

Abstract

Plasma cells are professional secretory cells, which function as cellular factories for immunoglobulin synthesis and secretion. Being the sole cell type responsible for antibody secretion they play an essential role in the immune response against a broad spectrum of pathogens. Since plasma cells have a long life span and are able to secrete copious amounts of antibody, their number and repertoire should be tightly regulated. Disruption of their homeostasis may lead to severe diseases, such as immunodeficiency or multiple myeloma. Much of the complications of multiple myeloma are attributed to the antibodies themselves, which accumulate in the bloodstream and lead to kidney and pulmonary insufficiencies. Similar pathologies are common to other

Institute for Drug Research, School of Pharmacy, Faculty of Medicine, The Hebrew University, Jerusalem, Israel

Methods in Enzymology, Volume 491 © 2011 Elsevier Inc.
ISSN 0076-6879, DOI: 10.1016/B978-0-12-385928-0.00017-1 All rights reserved.

plasma cell-related diseases, such as AL amyloidosis and autoimmune diseases, in which Ig molecules accumulate to toxic levels without good means to curtail their production.

The process of plasma cell differentiation and maintenance is poorly understood. The discovery that the IRE1/XBP-1 arm of the unfolded protein response (UPR) is necessary to yield full-fledged plasma cells *in vivo* was a breakthrough in the field. Over the years valuable biochemical information on plasma cell differentiation was obtained by exploring the downstream activities of XBP-1. The most pronounced phenotype of XBP-1 deficiency in plasma cells *in vitro* is the steep reduction in μ chain synthesis albeit similar levels of its mRNA. Remarkably, the defect is specific to Ig heavy chains as synthesis of other glycoproteins remains normal. Furthermore, when XBP-1 is absent or its mRNA splicing is inhibited the efficiency of protein translocation into the ER is severely impaired. Still, fundamental questions remain unanswered, such as what exactly generates the conditions of endoplasmic reticulum (ER) stress that activates the UPR in the developing plasma cells. Another enigma is how lipid biosynthesis and protein synthesis, both dramatically modulated during differentiation, are coordinated.

In this chapter, we will provide detailed methodologies for measurements of Ig synthesis and misinsertion into the ER as readout of ER physiology in the course of plasma cell differentiation.

1. INTRODUCTION

The unfolded protein response (UPR), first discovered in yeast (Cox *et al.*, 1993), plays a cytoprotective role against several chemical stresses. Mouse models lacking key elements of the UPR have demonstrated that beyond the protection against conditions in which protein folding in the endoplasmic reticulum (ER) is perturbed, in mammalians the UPR also evolved to govern pathways of cell fate, differentiation, and function (reviewed recently in Rutkowski and Hegde, 2010). One significant example was the discovery of the role of UPR in the terminal differentiation of B lymphocytes to antibody-secreting plasma cells. The differentiation of B lymphocyte is a multistep process which involves the bone marrow and peripheral lymphoid organs. At the immature state, when the B cell receptor genes have completed their somatic recombination, B cell egress from the bone marrow to populate spleen and lymph nodes. In the periphery, following negative selection and receptor editing if necessary, B cells acquire a mature state, where they await activation by a foreign antigen to become antibody secreting cells (ASCs). This process eventually yields plasma cells, which are capable to synthesize and secrete thousands of Ig molecules per cell per second (Oracki *et al.*, 2010).

By the usage of alternative polyadenylation sites, the genes for Ig heavy chain of the IgM and IgG (μ and γ, respectively) encode the mRNA of a membrane and secretory proteins, distinct only in their last exon 5' and untranslated region (5' UTR). While the synthesis of μ heavy and light chain molecules commence constitutively in mature B cells, IgM is not secreted. Mature B cells synthesize roughly equal levels of the membrane (μm) and secretory (μs) forms of μ. In naïve B cells, μm molecules exit the ER, where they undergo critical posttranslational modifications, and travel to the cell surface, while μs is retained in the ER and eventually degraded by the proteasome (a process referred to as ER associated degradation, ERAD; Amitay et al., 1991; Fra et al., 1993). Upon cell activation μs is assembled into pentameric or hexameric IgM and is secreted from the cell, while μm undergoes ERAD (Tirosh et al., 2005).

The differentiation of B cells into plasma cells involves a remarkable remodeling of the secretory pathway. The ER undergoes a radical expansion, required to accommodate the large quantity of newly synthesized Ig molecules and to ensure the successful maturation and assembly of the monomeric Ig subunits into multimeric complexes. Since this process loads the ER, a state of ER stress ensues and the developing plasma cell requires an intact UPR, a cellular response evolved to counteract this stress (Iwakoshi et al., 2003; Reimold et al., 2001).

In vivo, XBP-1−/− B cells fail to yield plasma cells (Reimold et al., 2001). The spliced form of XBP-1 (XBP-1s), major mediator of the UPR, promotes the expansion of the ER and ensures the synthesis, assembly, and secretion of high levels of antibodies (Shaffer et al., 2004; Sriburi et al., 2004).

A few years ago we demonstrated a new and unexpected requirement for XBP-1 during differentiation, in the maintenance of high levels of μ chain synthesis. At similar levels of the μ chain mRNA, the μ protein synthesis was dramatically reduced in the absence of XBP-1 (Tirosh et al., 2005). This was not due to a general defect in ER function as the synthesis of class I and class II MHC and the synthesis of Ig light chains were not affected by lack of XBP-1, indicating a specific regulation of μ chain synthesis by XBP-1. Recently, by crossing mice which are unable to synthesize μs ($\mu s^{-/-}$) with mice deficient of XBP-1 in their B cells, it was demonstrated that XBP-1 specifically regulates the synthesis of μs, while having much more modest effect on μm (McGehee et al., 2009). This was surprising in light of the fact that the two proteins are highly similar. However, it reinforces the need of XBP-1 for the terminal differentiation, in which the μs production predominates.

Signal peptide-dependent insertion of newly synthesized proteins into the ER is a multi-step process, whose fidelity varies with the identity of the protein and the cell type (Blobel and Dobberstein, 1975; Levine et al., 2005).

In addition, environmental signals and physiological conditions in the cell might also affect this process. One example is acute ER stress which was shown to preemptively prevent the translocation of prion proteins into the ER (Kang et al., 2006). While this has been documented for extreme ER stress conditions imposed by chemicals which interfere with protein folding in the ER, and for a special type of proteins, the impact of chronic ER stress on protein translocation in general has not been well characterized. Therefore, we generated chronic ER stress in LPS-activated B cells either by using XBP-1-null B cells or prolonged treatment with proteasome inhibitor, which prevents the splicing of XBP-1 mRNA (Lee et al., 2003). We have learned that under either of these conditions the translocation of US2, a viral-encoded protein with *a-priori* poor insertion efficiency, was impaired. Using monoclonal antibodies that preferentially recognize ER-misinserted μ chains, we found that these conditions also impaired the translocation of μ heavy chains to the ER (Drori et al., 2010). Our data suggest that antibody secreting cells under prolonged ER stress conditions endure cytoplasmic mislocalization of Ig proteins.

We hereby describe methods and guidelines how to explore the control of glycoprotein synthesis and trafficking in ASCs by the UPR.

2. Isolation of Splenic B Cells and Infection Thereof by Retroviruses

In this part, we will describe a refined procedure of how to maximize the yield of B cell isolation from mouse spleens and to optimize the efficiency of their infection by retroviruses. By utilizing these procedures we managed to lower the costs of B cell purification and minimize the number of mice needed per each experiment. To date there are many options for efficient isolation of B cell; most prevalent are based on binding to solid matrixes, labeling with a fluorophore, and using magnetic beads coupled to antibodies.

2.1. Isolation splenic B cells by magnetic sorting

To isolate large numbers of B cells, we recommend using the MACS CD43 depletion midi method by Miltenyi Biotec (CD43 negative B cells are eluted, while CD43 positive cells bind the MACS beads and retained on the column).

We revised the manufacturer's protocol to obtain approximately fivefold more B cells from a single isolation column.

Required materials

Midi size MACS separator, stand, and LD MACS separation columns.
Red blood cells lysis buffer (Sigma, R7757), MACS buffer (PBS, 2 mM EDTA, 0.5% BSA), B cells growth medium (RPMI, 10% heat

inactivated FCS, 1% pen–strep, 1 mM sodium pyruvate, 1% nonessential amino acids, 50 mM 2-mercaptoethanol), anti-CD43 coupled magnetic beads (#130-049-801), BSA, EDTA.

Disposables

5 ml syringes
70 μm cell strainer mesh (Falcon, 352350)
Sterile 50 ml tubes

Protocol:

- Euthanatize/sacrifice mouse and dissect to remove the spleen.
- Place the spleen in a 10 cm dish containing 10 ml of sterile PBS and crush the organ thoroughly, using the back side of a 5 ml plunger to release individual splenocytes.
- Collect and filter through a 70 μm cell strainer mesh (Falcon, Cat. no. 352350) into 50 ml tube. Wash the plate with additional 5 ml of PBS and filter through the mesh into the collection tube.
- Centrifuge ($1000 \times g$, 5 min, room temperature) and resuspend in 5 ml of red blood cells lysis buffer. (*Note*: if more than one spleen is processed use 10 ml). Incubate at room temperature for 5 min and add 40 ml of PBS to neutralize the lysis buffer.
- Centrifuge ($1000 \times g$, 5 min, room temperature).
- Observe the pellet. If it is red, repeat the lysis process. Once all red blood cells have been removed (and the pellet obtains an off-white hue), suspend the cells in 20 ml of MACS buffer and remove an aliquot for cell count.

A single spleen of an adult mouse contains about $80–120 \times 10^6$ cells. We usually combine 3–4 spleens for a single procedure.

- Resuspend $3–5 \times 10^8$ cells in 1 ml of MACS buffer and add 100 μl of CD43 microbeads. Incubate for 15 min on ice. Shake gently every 2–3 min to ensure homogenous binding. At this time precondition the LD column.
- Column preconditioning—assemble the MACS apparatus and load LD column with 2 ml of MACS buffer. Before proceeding to the next step, make sure that all column beads are wet.
- Wash the cells with 15 ml of MACS buffer, centrifuge ($1000 \times g$, 5 min, room temperature), and resuspend in 1 ml of MACS buffer.
- Place a 50 ml collection tube containing 18 ml of B cell medium supplemented with LPS underneath the LD preconditioned column. It is important to add the LPS as early as possible since B cells undergo atrophy in the medium within a few hours after isolation.

- Load the cell suspension by gently dropping it onto the column. Try to avoid air bubbles as much as possible. Let the first mililiter pass through the column almost completely. Do not allow the column to get dry.
- Elute the cells by washing the column with 1 ml of MACS buffer.
- Repeat washing with additional 2 ml of MACS buffer.

Your collection tube now contains 20 ml of purified splenic B cell suspension. Purity can be measured by flow cytometry using staining for CD19 or B220 as positive markers for B cells. We usually obtain over 90% purity and a total yield of 20%. Thus, if one starts with 500×10^6 cells, it is expected that following the procedure he would be left with $90\text{--}100 \times 10^6$ B cells.

- Centrifuge cell suspension ($1000 \times g$, 5 min, room temperature), wash once with fresh B cells medium containing LPS and count.
- Plate the cells at a density of 10×10^6 cells per 10 cm plate. The medium should be supplemented at all time with 20 µg/ml LPS if differentiation into plasma cells is to be induced.

2.2. Generation of retroviruses

Required materials

Polyethylenimine (PEI) (Cat. no. 408727 Aldrich, average Mw $\sim 25{,}000$ by LS, average Mn $\sim 10{,}000$ by GPC, branched), growth medium for 293T HEK cells (DMEM, 10% FCS, pen-strep, glutamine, and sodium pyruvate).

Disposables

Microcentrifuge tubes (1.5 ml)
Tissue culture dishes

To generate retroviruses we routinely use a triple-cotransfection method. The transfected DNA is a mixture of vectors that encodes the retrovirus Gag-pol, Env, and an MSCV-based retroviral vector in a 1:1:3 ratios (Tirosh et al., 2005).

2.2.1. Transfection of packaging cells

293T HEK cells, used as packaging cells, are plated in 10 cm dishes at 30–40% confluence. Transfection is usually performed the next day, when cells reach 60–80% confluence. We found that transfection of 293T with PEI is cheap, has high efficiency, and yields titers of retrovirus sufficient to infect primary B cells.

PEI preparation for transfection—Dissolve the PEI in water at a concentration of 2 mg/ml at 80 °C. After cooling to room temperature, adjust pH to 7.0 with 5 M HCl. Filter the solution through 2 μm filter, aliquot (1 ml) and store at −80 °C.

- Provide cells with fresh DMEM shortly before transfection.
- Mix 12.5 μg of DNA in serum-free DMEM to a final volume of 750 μl (use 1.5 ml microcentrifuge tubes).
- Vortex.
- Add 25 μl PEI solution.
- Vortex and incubate 5 min at room temperature.
- Add in a drop wise fashion the mixture to the cells. Try to drop the solution evenly on the plate.
- Incubate at 37 °C.
- Change the medium the morning after (or at least 7 h post transfection).
- Transfer the cells to a 30 °C incubator for additional 48 h. This may improve the retrovirus titers.
- Harvest the supernatants and filter through 0.45 μm filters.
- Use fresh supernatant to infect the B cells (see Section 2.3).

2.3. Retroviral transduction of B cells

Required materials

Polybrene (hexadimethrine bromide), CpG 1826 (TIB-MOLBIOL), LPS (L3755, Sigma).

Disposables

Sterile 15 ml tubes
Tissue culture dishes (6-well plate)

B cells should be extracted from spleens and plated (1×10^6 cells/ml) in the presence of CpG (100 nM) and LPS (20 μg/ml) a day before harvesting the viral supernatants.

We found the CpG treatment is critical to enhance the efficiency of viral transduction of primary B cells by threefold, probably due to eliciting robust induction of proliferation.

- Centrifuge B cells ($1000 \times g$, 5 min, room temperature).
- Resuspend 2×10^6 cells in 1 ml of fresh medium containing 2× of LPS and 2× CpG. Transfer to a 6-well plate and add 2 μl of polybrene solution (8 mg/ml).
- Add 1 ml of viral supernatant (prepared as previously described). Shake gently to ensure homogenous spread.
- Centrifuge the plate at $2000 \times g$, 45 min at room temperature.
- Incubate at 37 °C, 5% CO_2 for 24 h.

- Collect the infected B cells and wash once with the growth medium to remove polybrene. Count and plate at 1×10^6 cells/ml for 24–48 h, for following analysis.

3. Measurement of Protein Synthesis in Plasma Cells

3.1. Metabolic labeling of B cells

In this protocol, we describe the metabolic labeling of activated B cells with ^{35}S-methionine. The synthesis trafficking and degradation of specific proteins is followed by immunoprecipitation.

Required materials

B cell growth medium (RPMI, 10% heat inactivated FCS, 1% pen–strep, 1 mM sodium pyruvate, 1% nonessential amino acids, 50 mM 2-mercaptoethanol), trypsin–EDTA, starvation medium (cystein/methionine free DMEM, supplemented with 10% fetal calf serum, 2 mM glutamine, 100 unit/ml penicillin, and 100 μg/ml streptomycin), [^{35}S]-methionine/cysteine (we routinely use NEG-072 of Perkin Elmer), MG132 (proteasome inhibitor-provided by multiple suppliers).

Disposables

5 ml syringes
Microcentrifuge tubes (1.5, 2 ml)
Sterile 50 ml tubes
Gel loading tips
Tissue culture dishes (10 cm, 6-well plates)

Protocol:

- Harvest B cells. Transfer the suspension to 50 ml tube.
- Wash the cells from residual medium—centrifuge ($1000 \times g$, 5 min, room temperature), resuspend in PBS and centrifuge again. Cells are now ready for starvation.
- Resuspend up to 5×10^6 cells in 2 ml of starvation medium.
- Incubate for 45 min at 37 °C, preferably in a water bath. Shake the tube every 5–10 min to avoid cell pelleting for too long. Any other reagents needed for the experiment, such as proteasome inhibitors, can be added to the medium at this stage.
- Centrifuge ($1000 \times g$, 5 min, room temperature).

- Resuspend in starvation medium (up to 2×10^6 cells can be resuspended in 200 μl). Transfer cell suspension into 2 ml microcentrifuge tubes.
- Add [^{35}S]-methionine/cysteine in a 1:20 (v/v) dilution. 10 μl of radioactive material for each 200 μl of starvation medium. Since mislocalized ER proteins might be prone to rapid proteasomal degradation, we usually include a proteasome inhibitor (MG132 25 μM final concentration of 10 mM stock solution) in the pulse labeling mixture.
- Incubate in a water bath at 37 °C. Pulse labeling starts immediately upon addition of the radioactive methionine/cysteine. Duration of pulse is adjusted to the protein half-life. Generally, pulse labeling period should not exceed twice the half-life of the protein. This will later complicate the interpretation of the results.
- To terminate the pulse labeling, add 750 μl of ice-cold PBS and dip the tube in ice until centrifugation.
- Centrifuge (table-top centrifuge, 2500 rpm, 3 min, 4 °C).
- Carefully aspirate the medium (use 23G needle or gel loading tip). Cell pellet can now be stored at -80 °C for further analysis. Remember that [^{35}S]-methionine has a half-life of 80 days.

3.2. Cell lysis and immunoprecipitation

Required materials

SDS, protein A and protein G sepharose beads (we recommend Santa Cruz as a supplier).

4× NP-40 lysis buffer (see recipe below). Normal rabbit serum, BSA, PMSF (100 mM dissolved in ethanol), protease inhibitor cocktail (we use Sigma's P2714, 1 tablet for 50 ml solution).

Disposables

Dolphin tubes (Sorenson Bioscience).

First, we will describe an easy protocol to generate total cell lysates.
Whether using freshly labeled cells or frozen pellets, keep them on ice and make sure you work fast, in order to avoid postlysis protein degradation.

- Add 100 μl of 1% SDS solution to cell pellet.
- Immediately add 2.5 μl of PMSF.
- Vigorously vortex the tube for 15 s. Secure the cup with your finger to avoid radioactive spilling. (Once SDS solution is added to the pellets they should be handled in room temperature.)
- Centrifuge (14,000 rpm, 1 min, room temperature).
- Repeat vortex and centrifuge two more times.
- Dilute lysate in 1.5 ml of fresh, ice-cold lysis buffer (see recipe below).

- Shake for 15 min at 4 °C.
- Centrifuge (14,000 rpm, 5 min, 4 °C).
- Transfer the lysate to a precooled 2 ml dolphin tube.
- *Preclearing*: add 30 µl of washed protein A-conjugated sepharose beads.
- Beads should be washed in advance. Wash twice in 1 ml of cold PBS. Centrifuge at 6000 rpm, 1 min, 4 °C.
- Shake for 4 h to overnight at 4 °C.

This step removes proteins that bind nonspecifically to the sepharose beads and to immune complexes.

- Centrifuge (14,000 rpm, 10 min, 4 °C).
- Transfer the cleared lysate to clean 2 ml dolphin tube.
- Add antibody directed against the protein of interest (we usually use 1 µg of polyclonal antibody).
- Shake for 2 h to overnight at 4 °C.
- *Protein pull down*: add 30 µl of protein A or 15 µl of protein G beads to the lysate according to the antibody type used.
- Shake for 1 h. Extended shaking at this stage might result in higher background.
- Centrifuge (7000 rpm, 1 min, 4 °C).
- Aspirate the liquid using loading tip.
- Wash the beads—add 1 ml of NET buffer (see recipe below), supplemented with 0.1% SDS, Centrifuge (7000 rpm, 1 min, 4 °C) and carefully aspirate the liquid.
- Repeat wash two to three more times.

Your protein of interest and its antibody are now bound to the beads.

- Add 50 µl of 1× protein sample buffer.
- Boil the samples, 95 °C, 3 min.
- Centrifuge (14,000 rpm, 1 min, room temperature).

Samples are now ready for SDS-PAGE analysis and autoradiography. For antibody analysis, we use 12% polyacrylamide gel.

From our experience, µ chain is highly susceptible to postlysis degradation, even when lysis is performed in SDS. This can be circumvented by applying the following protocol. However, this protocol poorly recovers nuclear proteins.

- Add to the cell pellet 1.5 ml of freshly prepared 1× lysis buffer, supplemented with protease inhibitor cocktail.
- Immediately add 15 µl of 100 mM PMSF.
- Vigorously vortex the tube for 15 s. Secure the cup with your finger to avoid radioactive spilling. Incubate on ice while handling other samples.
- Repeat vortex two more times.
- Shake for 15 min at 4 °C.

- Centrifuge (14,000 rpm, 1 min, 4 °C).
- Transfer the lysate to a precooled 2 ml dolphin tube.
- Follow from the preclearing step as described above.

4× NP-40 lysis buffer

50 mM Tris (pH = 8)
200 mM NaCl
20 mM MgCl$_2$
2% NP-40 (Igepal CA-630)

Complete 1× NP-40 lysis buffer

Dilute 4× lysis buffer in ddH$_2$O. Per each milliliter of 1× lysis buffer, add the following:
2 μl normal rabbit serum
10 μl 10% BSA
10 μl 0.1M PMSF

10× NET wash buffer

500 mM Tris (pH = 7.4)
1.5 M NaCl
50 mM EDTA
5% NP-40 (Igepal CA-630)

As a source for XBP-1 KO B cells, we use mice that harbor a homozygous floxed allele for XBP-1 (XBP-1$^{f/f}$). When crossed with CD19-Cre mouse strain (CD19-Cre/XBP-1$^{f/f}$), XBP-1 is conditionally deleted in the B cell compartment.

In Fig. 17.1, we show a typical pulse-labeling experiment of wt and XBP-1 KO B cells extracted from spleens of XBP-1$^{f/f}$ and CD19-Cre/XBP-1$^{f/f}$, respectively, using the aforementioned CD43 depletion protocol. Cells were cultured for 3 days in the presence of LPS. 2 × 10^6 of live cells were pulse labeled with ^{35}S-methionine for 5, 10, and 20 min. Cells were lysed in 1% SDS; lysates were diluted in reducing sample buffer, boiled, and analyzed by SDS-PAGE followed by fluorography. An aliquot of the lysate was analyzed by trichloroacetic acid (TCA) precipitation to account for the total level of protein synthesis. Note the stark difference in μ chain synthesis between the wt and XBP-1 deficient B cells. Synthesis of other proteins seems similar.

US2 is an HCMV encoded glycoprotein equipped with a noncleavable signal peptide and a single N-linked glycosylation site. When expressed in cells, the mRNA of US2 yields two polypeptides, one is N-linked glycosylated and localized to the ER, while the second is not glycosylated and

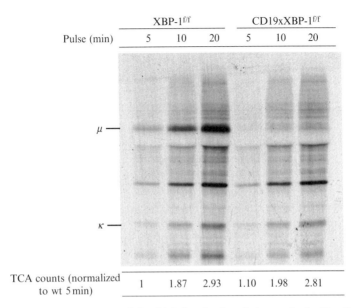

Figure 17.1 XBP-1 is required for high level of μ chain synthesis. B cells were extracted from spleens of XBP-1$^{f/f}$ and CD19-Cre/XBP-1$^{f/f}$ mice. Cells were stimulated for 3 days with LPS and pulse-labeled with ^{35}S-methionine for the indicated time. Total cell extract was prepared and analyzed by SDS-PAGE (12%). Gels were processed for fluorography.

resides in the cytoplasm. These two polypeptides can be easily distinguished, on SDS-PAGE, since the glycosylated form is heavier. Utilizing a short metabolic labeling with ^{35}S-methionine in the presence of a proteasome inhibitor one can score for the efficiency of ER insertion by calculating the ratio between the ER inserted and mislocalized US2. When introduced into wt and XBP-1 deficient B cells by retroviral infection as delineated here, we observed a lower ER insertion efficiency in the XBP-1 KO cells (Fig. 17.2).

4. Measurement of Ig Mislocalization in Primary B Cells

This protocol was used by us to recover ER mislocalized μ chains. However, it can be applied to many other glycoproteins, for which the glycosylation sites are mapped. Our detection protocol relies on the premise that cytoplasmic mislocalized glycoproteins lack N-linked glycosylation. Thus, if antibodies (monoclonal or polyclonal) are raised against the

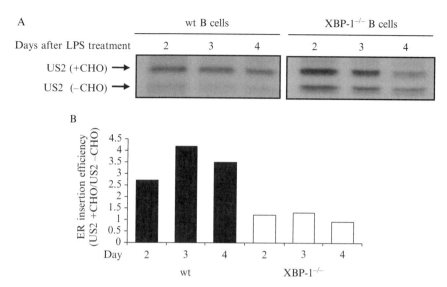

Figure 17.2 XBP-1 is required for efficient insertion of US2 into the ER. (A) Splenic B cells isolated from wt or CD19-CRExXBP-1$^{f/f}$ mice were transduced with a US2 expressing retrovirus a day after stimulation and cultured in the presence of LPS (20 μg/ml) for up to 3 days post infection. Cells were pulse-labeled with ^{35}S-methionine for 15 min in the presence of MG132, and US2 was immunoprecipitated. Immunoprecipitants were resolved on 12% acrylamide gel. Shown is a typical result of three repetitions. (B) Autoradiograms were quantified by phosphoimaging using ScionImage program and the ER-insertion efficiency of US2 was calculated as the ratio between glycosylated US2 and nonglycosylated US2.

glycosylation region they may interact preferentially with the noninserted glycoprotein, where the epitope is unmasked. To prove the cytoplasmic localization of the misinserted glycoprotein, we used a trypsin digestion technique in permeabilized cells. This protocol obviates subcellular fractionation, which is laborious and not always credible.

We applied this approach to μ heavy chains. These Ig molecules in their mature, secreted form are decorated with five glycans. We raised monoclonal antibodies against the peptide sequences that flank glycosylation sites 4 and 5, as these sites are the most conserved. Using these reagents we found that although μ is targeted to the ER with high efficiency, prolonged treatment with proteasome inhibitor dramatically reduces the fidelity of ER insertion and allows detection of nonglycosylated μ chains in the cytoplasm (Drori *et al.*, 2010).

The first step is generating antibodies specific to the glycosylation site. In this chapter, we will not describe in details the protocols for generating the monoclonal antibodies. Suffice it to mention that we used classical protocols. We vaccinated in the base of the tail balb/c mice with KLH-coupled peptides that span the glycosylated site emulsified in complete Freund's adjuvant. This was followed by two i.p. vaccinations in a weekly interval with the KLH-peptides in incomplete Freund's adjuvant. After fusion and selection, we screened the hybridomas by ELISA to the peptides coupled to BSA. For each peptide, we screened approximately 1000 clones and obtained a handful of positive readouts. These clones were further analyzed for immunoprecipitation.

To validate that the antibodies indeed recognize the nonglycosylated protein, we recommend *in vitro* translating (IVT) the protein in reticulocyte lysate (use Promega's TNT® Coupled Reticulocyte Lysate Systems). This requires the cloning of the glycoprotein of interest into a vector appropriate for *in vitro* translation. We routinely use pcDNA3, which is suitable for T7 RNA polymerization.

Applying Promega's specifications, generate the protein by IVT labeled radioactively. The reaction should be subjected to immunoprecipitation with the custom made antibodies and analyzed against a direct loading from the reaction mixture. If tagged by an epitope tag, one can use the appropriate anti-tag antibodies as a positive control (refer to figure 4 in Drori et al., 2010). Only if found efficient, we recommend the following protocol to analyze mislocalization in living cells.

4.1. Permeabilization of primary B cells using digitonin and trypsin digestion of cytoplasmic proteins

Required materials

PBS, KH buffer (100 mM potassium acetate, 20 mM HEPES, pH = 7.2), Digitonin (Roche), MG132 (proteasome inhibitor provided by multiple suppliers).

First, determine the minimal concentration of digitonin needed for efficient permeabilization. Prepare a stock solution of 2% digitonin in water. Dissolve the digitonin by heating it to boiling. The easiest way is by using a heat gun. Once boiled, leave the solution to cool at room temperature, then store at 4 °C. Over time a whitish precipitate will appear in the tube. Avoid using it.

To determine the working dilution of the digitonin wash the cells in PBS and resuspend them in permeabilization buffer (KH buffer, see recipe above). Place the cells on ice and add digitonin. A reasonable range is 0.004–0.025%. The permeabilization occurs very fast. Incubate on ice for 2–3 min and stain the cells with trypan blue. We recommend using the minimal concentration that yields 95% of trypan blue positive cells.

Typical protocol for LPS-activated B cells:

- Use freshly pulse-labeled B cells as described above. Do not freeze the cells as this will lyse a portion of them and will contaminate the sample.
- Wash the pellet with 1.5 ml of ice-cold PBS.
- Centrifuge (2500 rpm, 3 min, 4 °C)
- Resuspend 3×10^6 cells in 200 µl of KH buffer containing 0.008% digitonin, supplemented with 10 µM MG132.
- Add trypsin. The concentration of trypsin should also be adjusted according to the sensitivity and abundance of the protein. For m chains 0.5 µg/ml was sufficient.
- Incubate for 30 min on ice.
- Terminate the reaction by adding 5 mg/ml of soybean trypsin inhibitor.
- Incubate for 5 min at room temperature.
- Add 1.2 ml of ice-cold NP-40 lysis buffer and vortex vigorously.
- Immunoprecipitate the protein of interest as described above.

In Fig. 17.3, we demonstrate ER mislocalization of µ chains in LPS-activated B cells. We generated monoclonal antibodies directed against the

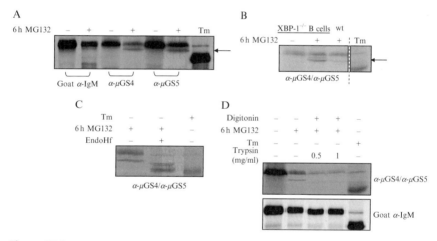

Figure 17.3 Prolonged ER stress and absence of XBP-1 promote µ chain mislocalization in plasmablasts. Wt (A) and CD19-Cre/XBP1f/f (B) primary B cells were pretreated with MG132 for 6 h where indicated to abolish XBP1 splicing. Regardless the pretreatment, cells were incubated with MG132 for 15 min and pulse-labeled with ^{35}S-methionine in the presence of the proteasome inhibitor. Cells were lysed in NP-40 lysis buffer. µ chains were immunoprecipitated using a mixture of anti-GS4/anti-GS5 to increase avidity, or goat α-IgM, as indicated. (C) wt B cells were treated with MG132 for 6 h and lysed using NP-40. Following lysis and µ chain immunoprecipitation, the immunoprecipitates were subjected to EndoHf treatment where indicated. (D) wt B cells, pretreated with MG132 for 6 h, were permeabilized using 0.008% digitonin, and subjected to trypsin digestion at 4 °C (lanes 3 and 4). Trypsin inhibitor was added, cells were lysed and µ chains were immunoprecipitated using anti-GS4$^+$, anti-GS5 (upper panel), or goat anti-IgM (lower panel).

glycosylation site 4 and 5 of μ chains (termed anti-GS4 and anti-GS5, respectively). These antibodies were efficient in immunoprecipitating *in vitro* translated μ. However, they also precipitated the fully glycosylated μ chains. In this experiment, we pulse-labeled LPS-activated B cells with ^{35}S-methionine. Total cell lysates were immunoprecipitated with the commercial polyclonal anti-μ antibody as well as with anti-GS4 and anti-GS5. Only when the cells were treated for 6 h with proteasome inhibitor prior to the pulse-labeling a second polypeptide was seen. This polypeptide was not glycosylated as it was not affected by EndoH digestion. It was also heavier than the μ chain synthesized in the presence of tunicamycin, most likely due to the signal peptide, which was not cleaved, since the polypeptide did not enter the ER. Moreover, these μ chains were susceptible to trypsin digestion in permeabilized cells, further supporting an aborted insertion of μ chain into the ER upon prolonged treatment with proteasome inhibitor. Interestingly, when mislocalization of μ was compared between XBP-1 KO and wt B cells, we observed reduced ER targeting in the KO cells. Lack of XBP-1, same as treatment with proteasome inhibitor, leads to conditions of ER stress. We reason that under these conditions the defect in ER insertion is most likely general.

ACKNOWLEDGMENTS

BT is affiliated with the David R. Bloom Center for Pharmacy at the Hebrew University (Jerusalem, Israel), and with the Dr. Adolf and Klara Brettler Centre for Research in Molecular Pharmacology and Therapeutics at the Hebrew University. The authors declare no financial conflict of interest. Research was funded by grants from David R. Bloom Center for Pharmacy, the Rosetrees fund, the Lower Saxony Research Fund, and Israel Science Foundation grant no. 78/09.

REFERENCES

Amitay, R., Bar-Nun, S., Haimovich, J., Rabinovich, E., and Shachar, I. (1991). Post-translational regulation of IgM expression in B lymphocytes. Selective nonlysosomal degradation of assembled secretory IgM is temperature-dependent and occurs prior to the trans-Golgi. *J. Biol. Chem.* **266,** 12568–12573.

Blobel, G., and Dobberstein, B. (1975). Transfer of proteins across membranes. I. Presence of proteolytically processed and unprocessed nascent immunoglobulin light chains on membrane-bound ribosomes of murine myeloma. *J. Cell. Biol.* **67,** 835–851.

Cox, J. S., Shamu, C. E., and Walter, P. (1993). Transcriptional induction of genes encoding endoplasmic reticulum resident proteins requires a transmembrane protein kinase. *Cell* **73,** 1197–1206.

Drori, A., Misaghi, S., Haimovich, J., Messerle, M., and Tirosh, B. (2010). Prolonged endoplasmic reticulum stress promotes mislocalization of immunoglobulins to the cytoplasm. *Mol. Immunol.* **47,** 1719–1727.

Fra, A. M., Fagioli, C., Finazzi, D., Sitia, R., and Alberini, C. M. (1993). Quality control of ER synthesized proteins: An exposed thiol group as a three-way switch mediating assembly, retention and degradation. *EMBO J.* **12,** 4755–4761.

Iwakoshi, N. N., Lee, A. H., Vallabhajosyula, P., Otipoby, K. L., Rajewsky, K., and Glimcher, L. H. (2003). Plasma cell differentiation and the unfolded protein response intersect at the transcription factor XBP-1. *Nat. Immunol.* **4,** 321–329.

Kang, S. W., Rane, N. S., Kim, S. J., Garrison, J. L., Taunton, J., and Hegde, R. S. (2006). Substrate-specific translocational attenuation during ER stress defines a pre-emptive quality control pathway. *Cell* **127,** 999–1013.

Lee, A. H., Iwakoshi, N. N., Anderson, K. C., and Glimcher, L. H. (2003). Proteasome inhibitors disrupt the unfolded protein response in myeloma cells. *Proc. Natl. Acad. Sci. USA* **100,** 9946–9951.

Levine, C. G., Mitra, D., Sharma, A., Smith, C. L., and Hegde, R. S. (2005). The efficiency of protein compartmentalization into the secretory pathway. *Mol. Biol. Cell* **16,** 279–291.

McGehee, A. M., Dougan, S. K., Klemm, E. J., Shui, G., Park, B., Kim, Y. M., Watson, N., Wenk, M. R., Ploegh, H. L., and Hu, C. C. (2009). XBP-1-deficient plasmablasts show normal protein folding but altered glycosylation and lipid synthesis. *J. Immunol.* **183,** 3690–3699.

Oracki, S. A., Walker, J. A., Hibbs, M. L., Corcoran, L. M., and Tarlinton, D. M. (2010). Plasma cell development and survival. *Immunol. Rev.* **237,** 140–159.

Reimold, A. M., Iwakoshi, N. N., Manis, J., Vallabhajosyula, P., Szomolanyi-Tsuda, E., Gravallese, E. M., Friend, D., Grusby, M. J., Alt, F., and Glimcher, L. H. (2001). Plasma cell differentiation requires the transcription factor XBP-1. *Nature* **412,** 300–307.

Rutkowski, D. T., and Hegde, R. S. (2010). Regulation of basal cellular physiology by the homeostatic unfolded protein response. *J. Cell. Biol.* **189,** 783–794.

Shaffer, A. L., Shapiro-Shelef, M., Iwakoshi, N. N., Lee, A. H., Qian, S. B., Zhao, H., Yu, X., Yang, L., Tan, B. K., Rosenwald, A., Hurt, E. M., Petroulakis, E., *et al.* (2004). XBP1, downstream of Blimp-1, expands the secretory apparatus and other organelles, and increases protein synthesis in plasma cell differentiation. *Immunity* **21,** 81–93.

Sriburi, R., Jackowski, S., Mori, K., and Brewer, J. W. (2004). XBP1: A link between the unfolded protein response, lipid biosynthesis, and biogenesis of the endoplasmic reticulum. *J. Cell Biol.* **167,** 35–41.

Tirosh, B., Iwakoshi, N. N., Glimcher, L. H., and Ploegh, H. L. (2005). XBP-1 specifically promotes IgM synthesis and secretion, but is dispensable for degradation of glycoproteins in primary B cells. *J. Exp. Med.* **202,** 505–516.

CHAPTER EIGHTEEN

USE OF CHEMICAL GENOMICS IN ASSESSMENT OF THE UPR

Sakae Saito *and* Akihiro Tomida

Contents

1. Introduction 328
2. Assessment of the Activation of UPR Transcriptional Program in Cancer Cell 329
   2.1. Microarray expression profiles 330
   2.2. Gene expression signature of glucose deprivation 332
3. Gene Expression Signature-Based Identification of UPR Modulators 334
   3.1. Characterization of UPR modulators 334
   3.2. The effect of UPR modulators on the UPR signal pathway 336
4. Future Perspective of Chemical Genomics in UPR Research 338
Acknowledgments 339
References 339

Abstract

Glucose deprivation, one of the major physiological conditions in solid tumor, leads to activation of the unfolded protein response (UPR) in cancer cells. The UPR occurs through the transcriptional and translational regulatory mechanisms that improve the capacity of the endoplasmic reticulum (ER) to fold and traffic proteins and allows the cell to survive under stress conditions. We previously reported that the macrocyclic compound versipelostatin and the antidiabetic biguanides metformin, buformin, and phenformin could inhibit the UPR during glucose deprivation as well as induce the UPR by treatment of cells with 2-deoxy-D-glucose (2DG), a glycolysis inhibitor. Versipelostatin and biguanides show highly selective cytotoxicity to glucose-deprived tumor cells and exert *in vivo* antitumor activity; thus, these compounds would be interesting anticancer agent candidates. By microarray analysis, we demonstrated that cancer cells under glucose deprivation conditions caused activation of the UPR transcription program, which was suppressed broadly by versipelostatin and biguanides. We also identified the drug-driven gene signatures that can be used to discover pharmacologic UPR modulators. Indeed, we found several bioactive drugs, such as pyrvinium pamoate, valinomycin, and rottlerin, that selectively suppressed 2DG-induced GRP78 promoter activity as versipelostatin and biguanide did. Together with growing bioinformatics databases and analytical tools, our approach could provide a chemical genomic basis for developing UPR-targeting drugs against solid tumors.

1. Introduction

In recent years, evidence from both basic science and clinical research indicates that the unfolded protein response (UPR) associates with immunity and several human diseases, including cancer, diabetes, metabolic disease, and tissue ischemia. In solid tumors, the excess glucose metabolism and proliferative status of cancer cells along with poor vascularization create a unique microenvironment with low oxygen (hypoxia), low nutrition supply, and low pH. One of the important responses to such a microenvironment for tumor development is the UPR, which is thought to protect tumor cells from the stressful conditions of glucose deprivation and hypoxia as well as from immune surveillance (Ron and Walter, 2007; Schroder and Kaufman, 2005). Activation of the UPR has been observed in a range of human solid tumors including breast, lung, gastric cancers, glioma, and melanoma. The elevated levels of the UPR marker glucose-regulated protein 78 (GRP78) correlate with lowered chemosensitivity and poor clinical outcome in breast and lung cancers (Lee *et al.*, 2006; Uramoto *et al.*, 2005). The elevated GRP78 expression also correlates with a higher rate of metastasis to lymph nodes and reduced survival in gastric cancer (Zhang *et al.*, 2006). Thus, developing genetic and chemical interventions in the UPR in tumor cells may be an effective approach to improving cancer chemotherapy.

The UPR is a signaling response activated by unfolded or misfolded proteins in the endoplasmic reticulum (ER). It reduces the ER stress by enhancing the folding and secretory capacity of ER and by diminishing the ability of global translation in cells (Ron and Walter, 2007; Schroder and Kaufman, 2005). In mammalian cells, UPR signaling is initiated mainly through the ER-localized stress sensors activating transcription factor 6 (ATF6), PKR-like ER kinase (PERK/EIF2AK3), and inositol-requiring 1 (IRE1/ERN1). These ER stress sensor proteins produce several different active transcription factors. ATF6 becomes an active transcription factor by proteolytic cleavage whereas IRE1 mediates the unconventional splicing of X-box binding protein 1 (XBP1) mRNA, thereby converting it to a potent UPR transcriptional activator. PERK phosphorylates eukaryotic initiation factor 2 subunit α (eIF2α), which transiently leads to inhibition of general protein translation but, paradoxically, causes selective translation of activating transcription factor 4 (ATF4). These transcription factors, ATF4, XBP1, and ATF6, lead to coordinated induction of divergent UPR target genes, such as the ER-resident molecular chaperones GRP78 and glucose-regulated protein 94 (GRP94). Activation of the UPR executes both a transcriptional and a translational regulatory program to relieve ER stress for cell survival. However, in the case of intolerable levels of ER stress, UPR signaling causes a shift from a prosurvival to a proapoptotic program and induces apoptosis (Ron and Walter, 2007; Schroder and Kaufman, 2005).

Gene expression profiling is a useful method to describe biological states of cells, such as cellular response induced by physiological stimuli, chemicals, and disease genes. In recent studies, the UPR in cell lines defective in such ER stress-signaling genes as PERK, eIF2α, ATF4, ATF6, and XBP1 have been characterized using microarray (Harding *et al.*, 2003; Lee *et al.*, 2003; Scheuner *et al.*, 2001; Wu *et al.*, 2007). Interestingly, each defect has a different effect on the UPR transcription program as well as on cell survival during ER stress, suggesting that gene expression profiling would be useful to define the nature of the perturbation induced by UPR-modulating compounds. Meanwhile, other studies revealed that gene expression profiling can also be used in drug discovery. Notably, the Connectivity Map system has successfully identified functional similarity between seemingly diverse compounds by detecting similarities among gene expression profiles of cellular responses to a large number of bioactive compounds (Lamb *et al.*, 2006). Thus, gene expression profiling with UPR-modulating compounds may provide useful information for drug discovery.

2. Assessment of the Activation of UPR Transcriptional Program in Cancer Cell

In this section, we describe the optimal conditions and procedures for characterizing the UPR transcriptional program in human tumor cells using whole-genome transcriptome analysis. Genome-scale analytical platforms have become widely available and less expensive in the past decade as a result of substantial progress in "omics" technologies. Gene expression profiling is one of the popular techniques to measure the differential expression of thousands of genes at once, thereby creating a global picture of biological states and functions of cells. One of the major physiological conditions of solid tumor cells leading to UPR is glucose deprivation. To characterize UPR activation in cancer cells during glucose deprivation, we first identified differentially expressed genes in cells under ER stress using microarray.

The microarray analysis can also be used to assess drug effects on a genomic scale to validate and identify drug targets. We previously reported that the macrocyclic compound versipelostatin and the antidiabetic biguanides metformin, buformin, and phenformin could inhibit GRP78 protein accumulation and production of the UPR transcription activators XBP1 and ATF4 during glucose deprivation (Park et al., 2004; Saito et al., 2009). Versipelostatin and biguanides show highly selective cytotoxicity to glucose-deprived tumor cells and exert in vivo antitumor activity; thus, these compounds would be interesting anticancer agent candidates. We also obtained gene expression profiles from cells treated with versipelostatin and biguanides under glucose deprivation conditions. Our data indicated that these compounds similarly and broadly modulate the transcription program of the stress response, especially the UPR during glucose deprivation. Our procedures and principal results are shown in Section 2.1.

2.1. Microarray expression profiles
2.1.1. Cell lines and cell treatments
2.1.1.1. Human cancer cell lines Cancer cell lines (human colon cancer HT-29, fibrosarcoma HT1080, and epithelial carcinoma HeLa) can be obtained from the American Type Culture Collection (ATCC). Stomach cancer cell line MKN74 can be obtained from the RIKEN BioResource Center. Cells were maintained in RPMI 1640 medium (for HT-29, HT1080, and MKN74) or in Dulbecco's modified Eagle's medium (DMEM; for HeLa). Both media, containing 2 mg/mL of glucose, were supplemented with 10% heat-inactivated fetal bovine serum (FBS) and 100 µg/mL of kanamycin. All cell lines were cultured at 37 °C in a humidified atmosphere containing 5% CO_2, as the normal growth condition.

2.1.1.2. Glucose deprivation and other stress conditions to induce UPR To induce the UPR, we treated cells under ER stress conditions by replacing the medium with glucose-free medium or by adding 10 mM 2-deoxy-D-glucose (2DG) to glucose-containing medium. The glucose-free RPMI 1640 and DMEM medium were supplemented with 10% heat-inactivated FBS (Ogiso et al., 2000). 2DG is known as a potent glucose metabolism inhibitor and is used as the hypoglycemia-mimicking agent. 2DG was purchased from Sigma-Aldrich (MO, USA), dissolved in sterilized distilled water at stock concentrations of 2 M, and stored at -20 °C. As a different type of chemical stressor, the N-glycosylation inhibitor tunicamycin was also used to induce the UPR. Tunicamycin (Nacalai Tesque, Kyoto, Japan) was dissolved in dimethyl sulfoxide (DMSO) at stock concentrations of 4 mg/mL and stored at -20 °C.

2.1.1.3. UPR-modulating compounds Versipelostatin is a novel compound isolated from *Streptomyces versipellis* and is not commercially available. The detailed method of versipelostatin isolation was described previously

(Park et al., 2002). Versipelostatin was prepared as a stock solution of 10 mM in DMSO. Metformin (Sigma), phenformin (Sigma), and buformin (Wako, Osaka, Japan) were dissolved in sterilized distilled water at stock concentrations of 1 M, 100 mM, and 100 mM, respectively. All stock solutions were stored at -20 °C.

To examine whether the compounds affected the glucose deprivation-induced UPR, the cells were treated for from 15 to 18 h with stressors (glucose withdrawal or 2DG) and UPR modulator compounds (e.g., versipelostatin and biguanides). Each UPR modulator was added at various final concentrations immediately after cells were placed in glucose-free medium or just before the chemical stressors were added to glucose-containing culture medium. The exposure time periods were predetermined to be sufficient to activate the UPR but not long enough to cause apparent, overt cell death, even in the presence of the UPR modulators.

2.1.2. Microarray analysis

Microarray platforms for the whole-genome transcriptome analysis are provided from several suppliers, such as Affymetrix, Agilent Technologies, Illumina, and Nimblegen Systems, Inc. We performed the microarray analysis using Affymetrix GeneChip® systems according to standard Affymetrix protocols.

Briefly, total RNA from cultured cells was isolated using the RNeasy RNA purification kit (Qiagen, Hilden, Germany). The quality of total RNA was analyzed using RNA 6000 Nano LabChip kit on a 2100 Bioanalyzer (Agilent Technologies, CA, USA). The complementary RNA (cRNA) targets for hybridization to GeneChip were prepared by reverse transcription from 5 μg of total RNA. Targets were then labeled with biotin before fragmentation using the GeneChip® One-Cycle Target Labeling Kit (Affymetrix). Hybridization to GeneChip® Human Genome U133 Plus 2.0 arrays (Affymetrix) was carried out using Fulid Station 450 and Scanner 3000 (Affymetrix).

We note that the latest GeneChip® 3' IVT Expression Kit (Affymetrix) is now provided for RNA target preparation, and it produced comparable antisense RNA (aRNA) using only 50–100 ng of input total RNA. We obtained essentially the same gene expression profiles in cancer cells under ER stress conditions using the 3' IVT Expression Kit (data not shown).

2.1.3. Statistical analysis

For genomic data analysis, several statistical methodologies were proposed, and statistical tools are available. Bioconductor is a free, open source, and open development software project for the analysis and comprehension of genomic data generated by wet-lab experiments. Bioconductor packages are distributed from the project website (http://www.bioconductor.org/). Bioconductor is based on the statistical R programming language.

An installer or source codes of R can be downloaded from Comprehensive R Archive Network (CRAN; http://www.r-project.org/).

Normalization of microarray data was carried out by the robust multichip average (RMA) method using the RMAExpress version 1.0 (Irizarry et al., 2003). All microarray data were normalized within each set prior to proceeding to significance analysis of microarray (SAM; Tusher et al., 2001). To select significant genes based on differential expression between the sets of interest, two class paired SAM analysis were performed using the R package samr. A cutoff of false discovery rate (FDR) is applied to identify the signature genes by determining the optimal delta value. Categorical annotation of genes was performed by gene ontology, and significant enrichment for specific biological functional categories was identified using DAVID (Dennis et al., 2003).

To analyze and visualize the results of microarray experiments, hierarchical clustering of signature genes was performed using Cluster 3.0 software (Eisen et al., 1998) and was visualized with Java TreeView. The version of this software that runs on Microsoft Windows, Mac OS, and Linux/Unix can be downloaded at http://rana.lbl.gov/EisenSoftware.htm.

2.2. Gene expression signature of glucose deprivation

2.2.1. Identification of the glucose deprivation signature

To identify an expression signature that captured the cellular response upon glucose deprivation, we obtained gene expression profiles from four different cancer cell lines (HT1080, HT-29, HeLa, and MKN74) that had been cultured under glucose deprivation conditions (glucose withdrawal or 2DG) for 15 or 18 h. From the gene expression data obtained under normal growth and glucose deprivation conditions, 2253 probe sets were selected as differentially expressed genes using SAM analysis, with FDR of 0.5%. Among them, we identified 246 of significant probe sets (178 genes) with the fold change cutoff of >2-fold up and down, and termed this the glucose deprivation signature. We have previously reported a list of the glucose deprivation signature genes, which can be obtained from the *Cancer Research* website as supplementary data (Saito et al., 2009).

Figure 18.1 shows the heat map of the glucose deprivation signature genes, which were either up- (97 genes or 148 probe sets) or downregulated (81 genes or 98 probe sets) by glucose withdrawal and 2DG. Categorical analysis of the upregulated genes using DAVID revealed enrichment of stress response and UPR-related categories, such as endoplasmic reticulum, unfolded protein binding, disulphide isomerase, and chaperone (Saito et al., 2009). The 97 upregulated genes also contained the UPR marker genes GRP78 and GRP94 and transcription factor XBP1 with relatively high scores. Categorical analysis of the 81 downregulated genes revealed enriched cell cycle, cell division, and DNA replication categories.

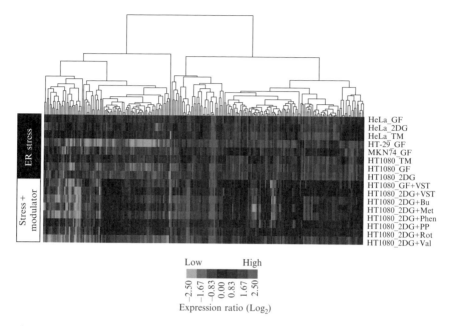

Figure 18.1 Gene expression profile of the human cancer cells under ER stress conditions by the glucose deprivation signature. Glucose deprivation signature genes including 246 probe sets (X axis) sorted by cluster analysis displayed with 16 samples (Y axis). The cells were cultured for 15–18 h under normal growth condition (control), ER stress conditions or ER stress conditions with UPR modulators. The log ratio for each gene was calculated by setting the expression level as 0 (Log_2 1) in an appropriate control sample. GF, glucose-free medium; 2DG, 10 mM 2-deoxyglucose; TM, 5 μg/mL tunicamycin; VST; 10 μM versipelostatin; Bu, 300 μM buformin; Met, 10 mM metformin; Phen, 100 μM phenformin; PP, 0.1 μM pyrvinium pamoate; Rot, 6 μM rottlerin; Val, 10 nM valinomycin. (See Color Insert.)

The glucose deprivation signature genes in cells treated with tunicamycin, another UPR inducer, showed expression change almost similar to cells that were glucose deprived (Fig. 18.1). These results suggest that the cells under glucose deprivation conditions caused activation of the UPR transcriptional program, including some UPR-related genes.

2.2.2. The effect of UPR modulators on the glucose deprivation signature

As shown in the heat map, induction of the glucose deprivation signature genes was largely prevented when versipelostatin and biguanides were added under 2DG stress or glucose withdrawal conditions (Fig. 18.1). In the 148 upregulated probe sets, induction of 81 probe sets (54.7%) was effectively inhibited (inhibition efficiency > 50%) in the presence of versipelostatin and biguanide. By contrast, in the 98 downregulated probe sets, as many as 92 probe sets (93.9%) were hardly affected when the these compounds were added. Taken together, the results demonstrate that versipelostatin and biguanides similarly and broadly modulate the stress response transcription program, especially the UPR during glucose deprivation.

3. Gene Expression Signature-Based Identification of UPR Modulators

To further screen UPR-modulating compounds using our gene expression data, we employed the Connectivity Map, a software tool that searches for similarities between the expression signature of interest and a reference collection of expression profiles obtained from cell lines treated with various drugs (Lamb et al., 2006). Based on the results of the Connectivity Map analysis, we could select as candidate compounds of UPR modulators some bioactive drugs that showed a high-scored match for versipelostatin and biguanide. Then, we tested whether these candidate compounds prevented the UPR and exerted selective toxicity in cancer cells during glucose deprivation.

In this section, we describe our recent results on an anthelmintic pyrvinium pamoate, a potassium ionophore valinomycin, and a PKC inhibitor rottlerin, which were high-scored candidates for UPR-modulating compounds identified by the Connectivity Map analysis.

3.1. Characterization of UPR modulators

3.1.1. Gene expression-based screening

To identify a gene expression signature that can be used for the Connectivity Map analysis, we determined the transcriptional change of cells treated with versipelostatin or biguanides under normal growth conditions for 18 h using microarray. As differentially expressed genes between normal growth cells and versipelostatin- and biguanide-treated cells, 118 probe sets were selected by the SAM analysis with FDR of 20%. As the versipelostatin/biguanide signature, 25 of significant probe sets were selected with the fold change cutoff of >2-fold up and >1.67-fold down. We had to set the fold change cutoff at 1.67 for the downregulated genes because we could not obtain the probe sets (down tags) necessary for Connectivity Map analysis at twofold. We have previously reported a list of the versipelostatin/biguanide signature genes, which can be obtained at the journal's website as supplementary data (Saito et al., 2009). To screen of UPR modulators, pattern-matching analysis of gene expression signatures was performed using the Connectivity Map database build 02 (Lamb et al., 2006). This software tool is available at the project website (http://www.broadinstitute.org/cmap/).

Detailed result of our pattern-matching analysis with the versipelostatin/biguanide signature was previously reported (Saito et al., 2009). From the results of the Connectivity Map analysis, we selected three bioactive drugs, pyrvinium pamoate, valinomycin, and rottlerin, which showed high-scored matches for the versipelostatin/biguanide signature as candidate compounds to modulate UPR.

3.1.2. Reporter assay

3.1.2.1. Chemicals Pyrvinium pamoate (MP Biomedicals, OH, USA), valinomycin (Sigma), and rottlerin (Wako) were dissolved in DMSO at stock concentrations of 10 mM. All the stock solutions were stored at $-20\ ^\circ$C.

3.1.2.2. Plasmid To examine the effect of pyrvinium pamoate, valinomycin, and rottlerin on the UPR, we performed the luciferase reporter assay using GRP78 promoter. The pGRP78pro160-Luc plasmid was created by cloning the human GRP78 promoter region (nucleotide -160 to $+7$ relative to the start of transcription) into the *Kpn*I/*Hin*dIII site of the pGL-Basic vector (Promega, WI, USA), which contains the firefly luciferase gene (Park et al., 2004). This promoter region contains the *cis*-acting ER stress response element (ERSE), which is required for transcriptional activation in response to ER stress (Yoshida et al., 1998).

We also examined the effect of the compounds on XBP1 splicing by IRE1 in HT1080 cells under ER stress conditions. The FLAG-tagged XBP1-Luc plasmid was generated by ligating a FLAG-tag coding sequence and human XBP1 splicing region into the *Hin*dIII/*Apa*I site of the pGL4.13 plasmid vector (Promega) upstream of luciferase (Saito et al., 2009). The XBP1 splicing region encompassed nucleotides 410–633 relative to the start of transcription of the human XBP1 cDNA, including the 26-nt ER stress-specific intron (Yoshida et al., 2001). Cells were transiently transfected with a reporter plasmid encoding a fusion protein of a XBP-1 fragment and luciferase, which produces luciferase activity only when the 26 bp of the XBP-1 fragment mRNA is spliced out by IRE1 during ER stress.

3.1.2.3. Transfection Transient transfections were performed using the FuGENE6 Transfection Reagent (Roche Applied Science, Penzberg, Germany) with antibiotic-free RPMI 1640 medium supplemented with 5% FBS, according to the manufacturer's protocol.

HT1080 cells were cultured overnight in a 12-well plate (3×10^5 cells/well), and transfection was performed. The cells were incubated for 8 h with a transfection mixture containing 500 ng of firefly luciferase-containing reporter plasmids (pGRP78pro160-Luc or FLAG-tagged XBP1-Luc) and 1 ng of renilla luciferase-containing plasmid phRL-CMV (Promega) as an internal control. The medium was then replaced with fresh growth medium, and the cells were incubated for another 4 h. Subsequently, the cells were reseeded in a 96-well plate (5×10^3 cells/well), cultured overnight, and treated for 18 h with various concentrations of versipelostatin, biguanides, or pyrvinium pamoate, with or without 10 mM 2DG or 5 μg/mL of tunicamycin. Relative activity of firefly luciferase to renilla luciferase (mean ± S.D. of triplicate determinations) was determined using the Dual-Glo Luciferase Assay System (Promega).

3.2. The effect of UPR modulators on the UPR signal pathway

To examine the effectiveness of the Connectivity Map-based screening, we tested whether candidate UPR-modulating compounds pyrvinium pamoate, valinomycin, and rottlerin prevented the UPR signaling pathway during glucose deprivation.

As shown in Fig. 18.2, treatments of the pGRP78pro160-Luc transfected cells with 2DG and tunicamycin increased GRP78 promoter activity by approximately five- and sixfold, respectively. In agreement with our previous study (Park *et al.*, 2004; Saito *et al.*, 2009), versipelostatin and biguanide metformin suppressed GRP78 promoter activity under 2DG stress conditions. Similarly, pyrvinium pamoate, valinomycin, and rottlerin suppressed 2DG-induced GRP78 promoter activity in a dose-dependent manner but had little effect on tunicamycin-induced GRP78 promoter activity. The UPR-inhibitory activities of these compounds were also confirmed by immunoblot analysis of endogenous GRP78 (data not shown).

We next examined whether biguanides inhibited IRE1-XBP1 signaling pathways. As shown in Fig. 18.3, treatments of the FLAG-tagged XBP1-Luc transfected cells with 2DG and tunicamycin increased luciferase activity from the fusion reporter by ∼11- and 7-fold, respectively. Adding pyrvinium pamoate and rottlerin as well as versipelostatin and metformin effectively suppressed XBP1 reporter activity, but valinomycin had little effect on XBP1 reporter activity during 2DG stress. In contrast, these compounds did not suppress tunicamycin-induced reporter activity. Interestingly, rottlerin slightly induced XBP1 reporter activity under normal growth conditions. Collectively, these results indicated that pyrvinium pamoate and rottlerin, but not valinomycin, exhibited similar effects to versipelostatin and biguanides on IRE1-BP1 pathways of the UPR, depending on the glucose deprivation conditions.

Consistently, the effects of pyrvinium pamoate and rottlerin on the glucose deprivation signature were quite similar to versipelostatin and biguanide in 2DG-stressed HT1080 cells. In spite of differences in chemical structure and known pharmacological actions, these compounds similarly and broadly modulate the transcription program of 2DG-induced UPR, as

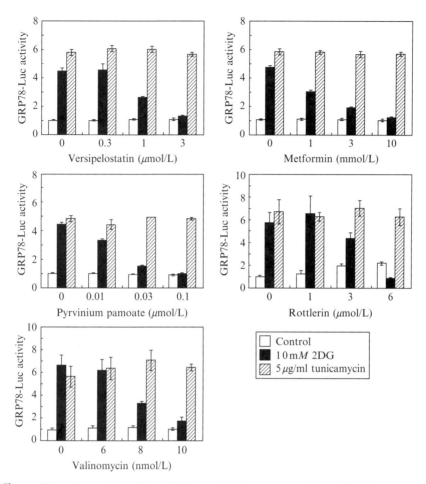

Figure 18.2 Suppressive effect of UPR-modulating compounds on GRP78 promoter activity. HT1080 cells were transfected with pGRP78pro160-Luc and exposed to UPR-inducible stress for 18 h with various concentrations of UPR modulators. Results shown are the mean values of quadruplicate determinations ±S.D.

seen in the glucose deprivation signature (Fig. 18.1). Though valinomycin could inhibit the glucose deprivation-induced GRP78 promoter activity, it had little effect on the glucose deprivation signature in HT1080 cells during 2DG stress. This effect on the glucose deprivation signature correlated with the weak inhibition ability of valinomycin on the IRE1–XBP1 pathway.

We noted that pyrvinium pamoate showed selective cytotoxicity to 2DG-stressed HeLa cells, indicating that modulation of the glucose deprivation signature is associated with cell death as a result of the drug treatment (Saito et al., 2009). In agreement with our findings, pyrvinium pamoate has been

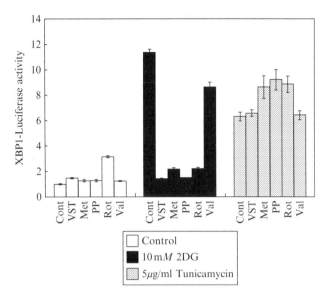

Figure 18.3 Effects of UPR-modulating compounds on the IRE1–XBP1 signaling pathway. HT1080 cells were transfected with FLAG-tagged XBP1-Luc and exposed to UPR-inducible stress for 18 h with UPR modulators. Results shown are the mean values of quadruplicate determinations ±S.D. VST, 10 μM versipelostatin; Met, 10 mM metformin; PP, 0.1 μM pyrvinium pamoate; Rot, 6 μM rottlerin; Val, 10 nM valinomycin.

reported to be more toxic under glucose starvation than under normal growth conditions and to exert antitumor activity against pancreatic cancer xenografts (Esumi et al., 2004). In addition, valinomycin was shown to be toxic to 2DG-treated cancer cells through downregulation of GRP78 and GRP94 (Ryoo et al., 2006), and rottlerin was shown to suppress ER stress-induced autophagy, which is required for survival in response to ER stress (Ozpolat et al., 2007). These consistent observations demonstrated that gene expression-based screening according to Connectivity Map analysis was successfully used to identify UPR-modulating compounds. Our findings demonstrate that disrupting the UPR during glucose deprivation could be an attractive approach for selective cancer cell killing and could provide a chemical genomic basis for developing UPR-targeting drugs against solid tumors.

4. Future Perspective of Chemical Genomics in UPR Research

For chemical genomics research, there are many publicly available, small-molecule databases, such as Chemical Entities of Biological Interest (ChEBI), DrugBank, ChemBank, and Kyoto Encyclopedia of Genes and

Genomes (KEGG). ChemBank houses experimental results from many high-throughput biological assays and small-molecule microarray through a collaboration between Massachusetts Institute of Technology and the Chemical Biology Program and Platform at the Broad Institute of Harvard. To identify and characterize compounds that specifically inhibit the UPR pathway, several cell-based assays have been proposed and performed. The microarray data in relation with the UPR are also available from public repositories, such as the NCBI Gene Expression Omnibus (http://www.ncbi.nlm.nih.gov/projects/geo/).

UPR-modulating compounds have been identified from multiple sources. For example, salubrinal can inhibit eIF2α dephosphorylation and reduces ER stress and cell death (Boyce et al., 2005). It is also known that 4-(2-aminoethyl)-benzenesulfonyl fluoride (AEBSF) inhibits ER stress-induced processing of ATF6 via inhibition of S1P (Okada et al., 2003). Genistein and (−)-epigallocatechin-3-gallate (EGCG) could inhibit GRP78 accumulation, but the specific molecular targets of these compounds are unknown (Lee, 2007). We previously reported that the dysfunction of mitochondria, such as treatment with the electron transport chain blockers rotenone and antimycin A, leads to suppressed UPR activation in cancer cells during glucose deprivation (Haga et al., 2010). These compounds are useful to elucidate the molecular mechanism of UPR, and further identification of the UPR-modulating compounds also advances the chemical genetics approach, which could reveal novel regulatory mechanisms of the UPR.

ACKNOWLEDGMENTS

This work was supported in part by a Grant-in-Aid for scientific research (B) (A. T.); a Grant-in-Aid for Young Scientists (B) (20790217; S. S.) from the Ministry of Education, Culture, Sports, Science and Technology of Japan; a Grant-in-Aid for Cancer Research (21-3-1) from the Ministry of Health, Labour and Welfare (A. T.); and Kobayashi Foundation for Cancer Research (A.T.).

REFERENCES

Boyce, M., Bryant, K. F., Jousse, C., Long, K., Harding, H. P., Scheuner, D., Kaufman, R. J., Ma, D., Coen, D. M., Ron, D., and Yuan, J. (2005). A selective inhibitor of eIF2alpha dephosphorylation protects cells from ER stress. *Science* **307,** 935–939.

Dennis, G., Jr., Sherman, B. T., Hosack, D. A., Yang, J., Gao, W., Lane, H. C., and Lempicki, R. A. (2003). DAVID: Database for annotation, visualization, and integrated discovery. *Genome Biol.* **4,** P3.

Eisen, M. B., Spellman, P. T., Brown, P. O., and Botstein, D. (1998). Cluster analysis and display of genome-wide expression patterns. *Proc. Natl. Acad. Sci. USA* **95**, 14863–14868.

Esumi, H., Lu, J., Kurashima, Y., and Hanaoka, T. (2004). Antitumor activity of pyrvinium pamoate, 6-(dimethylamino)-2-[2-(2, 5-dimethyl-1-phenyl-1H-pyrrol-3-yl)ethenyl]-1-me thyl-quinolinium pamoate salt, showing preferential cytotoxicity during glucose starvation. *Cancer Sci.* **95**, 685–690.

Haga, N., Saito, S., Tsukumo, Y., Sakurai, J., Furuno, A., Tsuruo, T., and Tomida, A. (2010). Mitochondria regulate the unfolded protein response leading to cancer cell survival under glucose deprivation conditions. *Cancer Sci.* **101**, 1125–1132.

Harding, H. P., Zhang, Y., Zeng, H., Novoa, I., Lu, P. D., Calfon, M., Sadri, N., Yun, C., Popko, B., Paules, R., Stojdl, D. F., Bell, J. C., et al. (2003). An integrated stress response regulates amino acid metabolism and resistance to oxidative stress. *Mol. Cell* **11**, 619–633.

Irizarry, R. A., Hobbs, B., Collin, F., Beazer-Barclay, Y. D., Antonellis, K. J., Scherf, U., and Speed, T. P. (2003). Exploration, normalization, and summaries of high density oligonucleotide array probe level data. *Biostatistics* **4**, 249–264.

Lamb, J., Crawford, E. D., Peck, D., Modell, J. W., Blat, I. C., Wrobel, M. J., Lerner, J., Brunet, J. P., Subramanian, A., Ross, K. N., Reich, M., Hieronymus, H., et al. (2006). The Connectivity Map: Using gene-expression signatures to connect small molecules, genes, and disease. *Science* **313**, 1929–1935.

Lee, A. S. (2007). GRP78 induction in cancer: Therapeutic and prognostic implications. *Cancer Res.* **67**, 3496–3499.

Lee, A. H., Iwakoshi, N. N., and Glimcher, L. H. (2003). XBP-1 regulates a subset of endoplasmic reticulum resident chaperone genes in the unfolded protein response. *Mol. Cell. Biol.* **23**, 7448–7459.

Lee, E., Nichols, P., Spicer, D., Groshen, S., Yu, M. C., and Lee, A. S. (2006). GRP78 as a novel predictor of responsiveness to chemotherapy in breast cancer. *Cancer Res.* **66**, 7849–7853.

Ogiso, Y., Tomida, A., Lei, S., Omura, S., and Tsuruo, T. (2000). Proteasome inhibition circumvents solid tumor resistance to topoisomerase II-directed drugs. *Cancer Res.* **60**, 2429–2434.

Okada, T., Haze, K., Nadanaka, S., Yoshida, H., Seidah, N. G., Hirano, Y., Sato, R., Negishi, M., and Mori, K. (2003). A serine protease inhibitor prevents endoplasmic reticulum stress-induced cleavage but not transport of the membrane-bound transcription factor ATF6. *J. Biol. Chem.* **278**, 31024–31032.

Ozpolat, B., Akar, U., Mehta, K., and Lopez-Berestein, G. (2007). PKC delta and tissue transglutaminase are novel inhibitors of autophagy in pancreatic cancer cells. *Autophagy* **3**, 480–483.

Park, H. R., Furihata, K., Hayakawa, Y., and Shin-ya, K. (2002). Versipelostatin, a novel GRP78/Bip molecular chaperone down-regulator of microbial origin. *Tetrahedron Lett.* **43**, 6941–6945.

Park, H. R., Tomida, A., Sato, S., Tsukumo, Y., Yun, J., Yamori, T., Hayakawa, Y., Tsuruo, T., and Shin-ya, K. (2004). Effect on tumor cells of blocking survival response to glucose deprivation. *J. Natl. Cancer Inst.* **96**, 1300–1310.

Ron, D., and Walter, P. (2007). Signal integration in the endoplasmic reticulum unfolded protein response. *Nat. Rev. Mol. Cell Biol.* **8**, 519–529.

Ryoo, I. J., Park, H. R., Choo, S. J., Hwang, J. H., Park, Y. M., Bae, K. H., Shin-Ya, K., and Yoo, I. D. (2006). Selective cytotoxic activity of valinomycin against HT-29 Human colon carcinoma cells via down-regulation of GRP78. *Biol. Pharm. Bull.* **29**, 817–820.

Saito, S., Furuno, A., Sakurai, J., Sakamoto, A., Park, H. R., Shin-Ya, K., Tsuruo, T., and Tomida, A. (2009). Chemical genomics identifies the unfolded protein response as a target for selective cancer cell killing during glucose deprivation. *Cancer Res.* **69,** 4225–4234.

Scheuner, D., Song, B., McEwen, E., Liu, C., Laybutt, R., Gillespie, P., Saunders, T., Bonner-Weir, S., and Kaufman, R. J. (2001). Translational control is required for the unfolded protein response and in vivo glucose homeostasis. *Mol. Cell* **7,** 1165–1176.

Schroder, M., and Kaufman, R. J. (2005). The mammalian unfolded protein response. *Annu. Rev. Biochem.* **74,** 739–789.

Tusher, V. G., Tibshirani, R., and Chu, G. (2001). Significance analysis of microarrays applied to the ionizing radiation response. *Proc. Natl. Acad. Sci. USA* **98,** 5116–5121.

Uramoto, H., Sugio, K., Oyama, T., Nakata, S., Ono, K., Yoshimastu, T., Morita, M., and Yasumoto, K. (2005). Expression of endoplasmic reticulum molecular chaperone Grp78 in human lung cancer and its clinical significance. *Lung Cancer* **49,** 55–62.

Wu, J., Rutkowski, D. T., Dubois, M., Swathirajan, J., Saunders, T., Wang, J., Song, B., Yau, G. D., and Kaufman, R. J. (2007). ATF6alpha optimizes long-term endoplasmic reticulum function to protect cells from chronic stress. *Dev. Cell* **13,** 351–364.

Yoshida, H., Haze, K., Yanagi, H., Yura, T., and Mori, K. (1998). Identification of the cis-acting endoplasmic reticulum stress response element responsible for transcriptional induction of mammalian glucose-regulated proteins. Involvement of basic leucine zipper transcription factors. *J. Biol. Chem.* **273,** 33741–33749.

Yoshida, H., Matsui, T., Yamamoto, A., Okada, T., and Mori, K. (2001). XBP1 mRNA is induced by ATF6 and spliced by IRE1 in response to ER stress to produce a highly active transcription factor. *Cell* **107,** 881–891.

Zhang, J., Jiang, Y., Jia, Z., Li, Q., Gong, W., Wang, L., Wei, D., Yao, J., Fang, S., and Xie, K. (2006). Association of elevated GRP78 expression with increased lymph node metastasis and poor prognosis in patients with gastric cancer. *Clin. Exp. Metastasis* **23,** 401–410.

CHAPTER NINETEEN

SMALL GTPASE SIGNALING AND THE UNFOLDED PROTEIN RESPONSE

Marion Bouchecareilh,[*,†] Esther Marza,[*,†] Marie-Elaine Caruso,[*,†] and Eric Chevet[*,†]

Contents

1. Introduction	344
2. Materials and Methods	346
3. Monitoring the Unfolded Protein Response	348
3.1. Evaluation of the "endogenous" UPR	348
3.2. Recombinant strategies	350
4. Monitoring Small GTPase Activity	353
4.1. Immunofluorescence-based visualization of actin remodeling upon ER stress	353
4.2. GTPase pull-down assay	355
5. Pharmacological and Genetic Modulation of UPR and GTPase Signaling	356
5.1. Pharmacological modulation	356
5.2. Genetic modulation by RNA interference	356
Acknowledgments	358
References	358

Abstract

The endoplasmic reticulum (ER), first compartment of the secretory pathway, is mainly involved in calcium sequestration and lipid biosynthesis and in the translation, folding, and transport of secretory proteins. Under some physiological and physiopathological situations, secretory proteins do not acquire their folded conformation and accumulate in the ER. An adaptive response named the UPR is then triggered from this compartment to restore its homeostasis. In the past few years, interconnections between the UPR and small GTPase signaling have been established. In an attempt to further investigate these novel signaling networks, we hereby provide a detailed description of experimental strategies available. We describe in detail methods to monitor both UPR and small GTPase signaling and the outcomes of such approaches in

[*] Avenir, Inserm U889, Bordeaux, France
[†] Université Bordeaux 2, Bordeaux, France

Methods in Enzymology, Volume 491
ISSN 0076-6879, DOI: 10.1016/B978-0-12-385928-0.00019-5

© 2011 Elsevier Inc.
All rights reserved.

the identification of new links between those signaling pathways using pharmacological and genetic screens. In physiopathological contexts, the guidelines herein should enable researchers in the field to establish essential means for determination of functional interactions between those pathways.

1. INTRODUCTION

The endoplasmic reticulum (ER) is an organelle found in all eukaryotic cells. This organelle, which is the first compartment of the secretory pathway, is mainly involved in calcium sequestration and lipid biosynthesis and in the translation, folding, and transport of secretory proteins (Chevet et al., 2001). These functions require specialized and integrated molecular machines (Caruso and Chevet, 2007). Most of the proteins distributed in organelles of the secretory pathway, expressed at the cell surface or secreted, transit through the ER before reaching their final destination. Indeed, polypeptide chains, translated by ER membrane-bound ribosomes, are first translocated into the ER lumen via the translocon and then processed through the ER folding machineries, which include a chaperone component (e.g., BiP, GRP94, ORP150), a post-translational modification machinery (e.g., N-glycosylation with the oligosaccharyl transferase complex, S–S bond formation with the PDIs or prolyl–proline isomerization), and a quality control component (e.g., calnexin, UGGT). Proteins that do not acquire a correct conformation are retained in the ER by ER-specific quality control mechanisms and are consequently granted further folding attempts. If this fails again, terminally misfolded proteins are degraded via the ER-associated degradation (Wu et al., 2010) machinery (Chevet et al., 2001).

Under basal conditions, these integrated mechanisms maintain the ER protein load in equilibrium with its folding and export capacities, thus reaching homeostasis. However, if one of those components is dysfunctional, the entire chain reaction is perturbed and ER homeostasis is disrupted. This increases the amount of improperly folded proteins that accumulate within the ER lumen. This overload of ER folding capacity triggers an adaptive mechanism to overcome this phenomenon, named the Unfolded Protein Response (UPR), which aims at restoring ER homeostasis (Ron and Walter, 2007; Rutkowski and Hegde, 2010; Tsang et al., 2010) by (1) attenuating protein translation, (2) increasing ER folding capacity, (3) increasing ERAD capacity, and (4) triggering cell death if ER homeostasis is not restored and the cells cannot cope with the initial challenge. This stress response is regulated by the activation of three major ER-resident proximal sensors, which constitute the three arms of the UPR: the Inositol-Requiring Enzyme-1 (IRE-1), the Activating Transcription

Factor-6 (ATF-6), and the Protein Kinase RNA (PKR)-like ER kinase (PERK) (Ron and Walter, 2007; Rutkowski and Hegde, 2010; Tsang et al., 2010). Upon accumulation of misfolded proteins in the ER, these sensors elicit specific signaling cascades mainly through post-translational modification of signaling intermediates (phosphorylation, proteolysis) and RNA modification (through IRE1 endoribonuclease activity). This collectively leads to the regulation of gene expression programs aiming to restore ER normal functions (Ron and Walter, 2007; Rutkowski and Hegde, 2010; Tsang et al., 2010).

The activation of the UPR has been linked to small GTPase signalling. For instance, activation of the UPR was shown to induce the expression of the small G protein SAR-1, a member of the COPII protein complex (Memon, 2004; Saloheimo et al., 2004; Takeuchi et al., 2000). Moreover, expression of a dominant negative form of SAR-1 leads to the accumulation of proteins in the ER (Memon, 2004; Saloheimo et al., 2004; Takeuchi et al., 2000), probably resulting in the saturation of ER folding capacity. In addition to the role of SAR-1 in the regulation of ER protein load, other GTPases of the RAS superfamily were found to play a role in either organelle maintenance/biogenesis (Altan-Bonnet et al., 2003, 2004; Lazarow, 2003) or endomembrane signaling (Philips, 2005; Quatela and Philips, 2006). These observations led us to investigate the role played by GTP-binding proteins of the Ras superfamily as master regulatory components of ER homeostasis (Caruso et al., 2008). Thus far, most of the work was carried out using *Caenorhabditis elegans* as an experimental model (Caruso et al., 2008), and more recent data are now appearing in mammalian culture systems.

Our data indicated a specific role of members of the Rho family of GTPases in the transcriptional activation of UPR target genes. We demonstrated that this possibly occurred through the presence of the Cdc42-related protein CRP-1 in complex with the AAA^+ ATPase CDC-48. We demonstrated physical and genetic interactions between CRP-1 and CDC-48 and consequently delineated a novel signaling component physically linking the ER stress and transcriptional regulation in metazoans (Caruso et al., 2008). Recent reports also suggest the implication of small GTPase signaling in the regulation of ER stress. Indeed, it has been shown that Ras signaling inhibited ER stress in human cancer cells with amplified Myc (Yaari-Stark et al., 2010) or that local Nox4-derived H_2O_2-dependent Ras activation mediated ER signaling (Wu et al., 2010). More indirectly, the Rho-ROCK-LIMK-cofilin pathway has been shown to regulate shear stress-mediated activation of sterol regulatory element-binding proteins (Lin et al., 2003). Similarly, TSC1 and TSC2, which may present GTPase-activating properties toward Rheb, are required for proper ER stress response (Kang et al., 2010). These findings clearly support a connection between small GTPase signaling and the UPR, and this connection remains to be further investigated.

2. Materials and Methods

All the parameters were measured in cell extracts from various cell types from rat and human origin, namely, rat FR3T3 fibroblasts, U87 glioma, HuH7 hepatocellular carcinoma, or HeLa. Cells were cultured in DMEM supplemented with 10% fetal bovine serum (FBS). Once the cells reached 70–80% confluency, they were treated for various periods of time with ER stress chemical inducers, namely, the N-glycosylation antibiotic inhibitor tunicamycin (TM), the proline analog Azetidine-2 carboxilic acid, and the reducing agent dithiothreitol (DTT). Following treatment, the cells were washed with ice-cold phosphate-buffered saline (PBS) and lyzed for either immunoblot or PCR analyses. Lysis was carried out using 30 mM TrisE–HCl pH 7.5, 150 mM NaCl, 1% TX-100, 5 mM NaF, 2 mM Na$_3$VO$_4$, and protease inhibitors for 30 min on ice, and RNA was extracted using Trizol (Invitrogen, Carlsbad, CA, USA), as recommended by the manufacturer. The total RNA was reverse-transcribed into cDNA using the SuperScriptIII cDNA synthesis kit (Invitrogen, Carlsbad, CA, USA). Immunoblot was carried out as previously described (Caruso et al., 2008). Luciferase reporter assays were carried out using the Dual Reporter Assay system (Promega, Madison, WI, USA) and the pGL4 vector backbone. C. elegans strains used are all derived from the reference wild-type bristol N2 strain. C. elegans animals were maintained at 20 °C under standard culture conditions using OP-50 as the food source (Brenner, 1974).

Tunicamycin was purchased from EMD Biosciences/Merck (Darmstadt, Germany), and Azetidine-2 carboxylic acid and DTT were purchased from Sigma (St Louis, MI, USA). Dulbecco's modified Eagle's medium (DMEM) was purchased from invitrogen (Carlsbad, CA, USA). Antibodies and oligonucleotide primers used are detailed in Tables 19.1 and 19.2.

Table 19.1 Antibodies available for monitoring the UPR

Antigen	Antibody type	Species	Company	Reference
PERK	Polyclonal	Rabbit	Santa Cruz Biotech	
P-PERK	Monoclonal	Rabbit	Cell Signaling	Harding et al. (2001)
IRE1	Polyclonal	Rabbit	Santa Cruz Biotech	Drogat et al. (2007)
P-IRE	Polyclonal	Rabbit	Abcam Inc.	Lipson et al. (2006)
ATF6	Monoclonal	Mouse	BioAcademia	Yamamoto et al. (2007)
eIF2α	Monoclonal	Mouse	Cell Signaling	Zamanian-Daryoush et al. (2000)
P-eIF2α	Monoclonal	Rabbit	Invitrogen	Chen et al. (2010)
sXBP1	Monoclonal	Mouse		In-house
RhoA	Polyclonal	Rabbit	Santa Cruz Biotech	Tatin et al. (2006)

Table 19.2 Oligonucleotide primers for PCR analysis

	Forward primer (5′–3′)	Reverse primer (5′–3′)
Chop	ATTGACCGAATGGTGAATCTGC	AGCTGAGACCTTTCCTTTGTCTA
ERdj4	TGGTGGTTCCAGTAGACAAAGG	CTTCGTTGAGTGACAGTCCTGC
Gadd34	GTGGAAGCAGTAAAAGGAGCAG	CAGCAACTCCCTCTTCCTCG
Orp150	GAAGATGCAGAGCCCATTTC	TCTGCTCCAGGACCTCCTAA
Xbp1	GGAACAGCAAGTGGTAGA	CTGAGGGGTGACAAC
Gapdh	AAGGTGAAGGTCGGAGTCAA	CATGGGTGGAATCATATTGG

3. Monitoring the Unfolded Protein Response

The first step in evaluating the role of small GTPase in UPR signaling is to develop standard methodologies to monitor the UPR in cultured cells or in model systems. To this end, various approaches and numerous tools were developed to either measure endogenous UPR parameters or use UPR reporters.

3.1. Evaluation of the "endogenous" UPR

In this part, we describe the tools and methods available for measuring UPR parameters in cultured cells. Several molecular tools are currently available either commercially or developed in-house (see Section 2).

3.1.1. Activation of the PERK pathway and phosphorylation of eIF2α

Upon UPR induction, PERK autophosphorylates and gets activated allowing the phosphorylation of its major substrate, the translation initiation factor eIF2α. PERK pathway activation was monitored by treating HuH7 cells with TM (5 µg/ml) for 0, 30, and 120 min. Cell extracts were clarified by centrifugation, resolved by SDS-PAGE, and transferred to nitrocellulose membranes. The membranes were then blocked using PBS containing 0.1% Tween 20 (PBST) and 3% Bovine Serum Albumin (BSA) for 45 min at room temperature. The primary antibody (anti-P-PERK or P-eIF2α) was incubated (at the right dilution) for periods of time ranging from 2 h at room temperature to 16 h at 4 °C. Following this, the membranes were washed 3 to 5 times with PBST and incubated for 45 min at room temperature with the secondary antibody coupled to horseradish peroxidase. After extensive washing with PBST, immunoreactive bands were revealed using enhanced chemiluminescence. Here, we show the immunoblots performed using anti-P-eIF2α (Fig. 19.1A, top blot) and anti-eIF2α (Fig. 19.1A, bottom blot). Increased eIF2α phosphorylation is shown upon TM treatment in HuH7 human hepatocellular carcinoma cells.

3.1.2. Splicing of XBP1 mRNA

XBP1 mRNA splicing is an event downstream of IRE1α activation. Indeed, upon accumulation of misfolded proteins in the ER, IRE1α oligomerizes and *trans*-autophosphorylates, resulting in the activation of its endoribonuclease activity. IRE1α endoribonuclease activity is involved in two major mechanisms, which are the unconventional splicing of XBP1 mRNA and the degradation of mRNA substrates (Hollien and Weissman, 2006; Hollien et al., 2009; Ron and Walter, 2007). The monitoring of XBP1 mRNA splicing can be performed using RT-PCR, as illustrated in Fig. 19.1B. To this end,

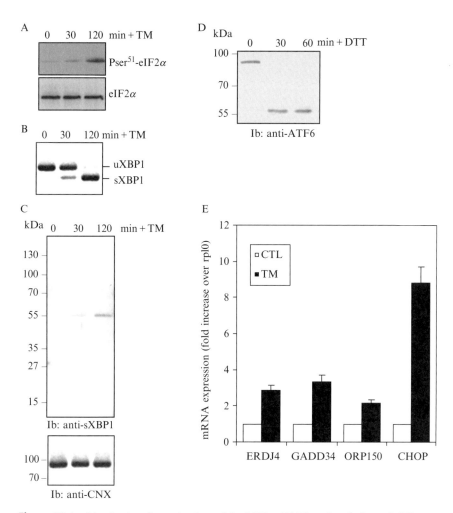

Figure 19.1 Monitoring the activation of the UPR. (A) Phosphorylation of eIF2α on ser51 analyzed by immunoblot on HuH7 cell extracts following TM treatment. (B) XBP1 mRNA splicing as analyzed by RT-PCR on mRNA isolated from HuH7 cells treated with TM. (C) Immunoblot analysis using anti-sXBP1 and CNX antibodies on cellular extracts from HuH7 cells treated with TM. (D) Immunoblot analysis using anti-ATF6 antibodies on cellular extracts from HeLa cells treated with DTT. (E) Q-RT-PCR analysis of ER stress target genes carried out on mRNA isolated from HuH7 cells treated with TM.

cDNA from cells subjected or not to TM treatment were incubated with the primer pair (described in Table 19.2) flanking XBP1 splicing site and subjected to PCR amplification using Taq Polymerase (30 cycles, denaturation 1 min at 94 °C, hybridization 1 min at 60 °C, and elongation 1 min at 72 °C). PCR products were then resolved by agarose gel electrophoresis and visualized using ethidium bromide and UV transillumination. Upon treatment with TM, basal XBP1 mRNA (Fig. 19.1B, top band, uXBP1) is subjected to IRE1α-mediated cleavage of a 26-nucleotides long nonconventional intron (in human) and subsequent ligation (splicing) to result in a shorter mRNA form (Fig. 19.1B, bottom band, sXBP1) with a different open reading frame. The protein product translated from this newly formed mRNA can be detected using antibodies specific to the C-terminus of the protein. We raised monoclonal antibodies that recognize the sXBP1 peptide CSVKEEP-VEDDLVPELG and used them in immunoblotting experiments to detect the appearance of the sXBP1 protein in HuH7 cells subjected to TM treatment, as described in Section 3.1.1 (Fig. 19.1C, top panel). Protein loading was controlled by using anti-calnexin (CNX) antibodies (Fig. 19.1C, bottom panel). sXBP1 protein was translated upon TM treatment in HuH7 cells, which correlated with the splicing of XBP1 mRNA observed in Fig. 19.1B.

3.1.3. Activation of ATF6α

Under ER stress and UPR, ATF6α is translocated to the Golgi where it is cleaved by site 1 and site 2 proteases, releasing the 50 kDa cytosolic domain that will move to the nucleus and play its transcription factor function. ATF6α activation was monitored by immunoblotting using antibodies raised against the cytosolic domain of human ATF6α. HeLa cells were subjected to DTT treatment and cell extracts were prepared, as described in Section 2, and processed for immunoblot, as described in Section 3.1.1. As illustrated in Fig. 19.1D, ATF6α was cleaved upon DTT treatment.

3.1.4. UPR target gene expression

Monitoring of the expression level of mRNA encoding for UPR targets can be achieved using quantitative RT-PCR. Amplification products were generated using primers described in Table 19.2. These primers were designed to amplify products that report for target genes of each of the three arms of the UPR, namely, GADD34 and CHOP as PERK targets, ORP150 as ATF6α, target, and ERDJ4 as IRE1α target. Quantitative PCR was performed using an Applied Biosystems StepOne Real-Time PCR System. *GAPDH* mRNA levels served as internal normalization standards. Data presented in Fig. 19.1E show that upon treatment for 16 h with TM, the expression of ERDJ4, ORP150, GADD34, and mainly CHOP, was enhanced by 2–8-fold in HuH7 cells.

3.2. Recombinant strategies

In this part, we describe how the activation of gene transcription through reporter assays can be monitored in two experimental models, namely, mammalian cells in culture and *C. elegans*.

3.2.1. Luciferase reporter assays in cultured mammalian cells

This was carried out using an ERSE reporter. This construct has been designed to measure the activity of endoplasmic reticulum (ER) stress signaling. Indeed, it is a critical *cis*-acting motif, which mediates the transcriptional response to ER stress (Abdelrahim *et al.*, 2005; Lee *et al.*, 2003). The consensus ERSE contains binding sites for the transcription factors NF-Y and YY1 and can also bind protein complexes containing ATF6α and TFII-I. HeLa cells were transfected using lipofectamine (Invitrogen, Carlsbad, CA, USA) following the manufacturer's recommendations with the ERSE reporter mix containing an ERSE-responsive Firefly luciferase construct and a constitutively expressing Renilla luciferase construct (35/40:1/2). The ERSE-responsive luciferase construct encodes the firefly luciferase reporter gene under the control of a minimal CMV promoter and six repeats of the ERSE transcriptional response element in tandem. The constitutively expressing Renilla luciferase construct encodes the Renilla luciferase reporter gene under the control of a CMV promoter and acts as an internal control for normalizing transfection efficiency (Fig. 19.2A, top panel). Twenty hours posttransfection, the medium was changed to a growth medium without antibiotics. After 30 h of transfection, the cells were treated with either 5 μg/ml of TM, 5 mM Azc, 0.1 mM DTT, or 100 nM thapsigargin for 16 h. Dual Luciferase assay (Promega, Madison, WI, USA) was performed 48 h after transfection, and promoter activity values were expressed as arbitrary units using the Renilla reporter for internal normalization. Experiments were done in triplicates, and the standard deviation is indicated (Fig. 19.2A, bottom panel).

3.2.2. GFP reporter assays in *C. elegans*

To measure the activation of UPR-responsive genes in *C. elegans*, we have used *C. elegans*-based GFP reporter driven by UPR-responsive genes promoters (Caruso *et al.*, 2008). We selected these promoters on their previously reported activation upon ER stress (Calfon *et al.*, 2002; Urano *et al.*, 2002). A 2.5-kilobase-long putative promoter region cloned upstream of the start codon of each target gene was fused to the green fluorescent protein (*gfp*) cDNA. The resulting constructs were used to generate transgenic worm lines expressing GFP under the control of the above-mentioned promoters, as previously described (Dupuy *et al.*, 2007; Hunt-Newbury *et al.*, 2007; McKay *et al.*, 2003a,b). The most commonly used UPR target genes are described elsewhere (Caruso *et al.*, 2008), and here, we show two additional lines expressing GFP under the control of *pdi* and *hsp-3* promoters. *Pdi* is *C. elegans* homolog of the mammalian protein disulfide isomerase and *hsp-3* is *C. elegans* homolog of the mammalian ER chaperone GRP78. To test whether or not these transgenic lines were able to activate the selected GFP reporters upon ER stress, we treated adult hermaphrodites with TM (Fig. 19.2B, left panel) and analyzed them using

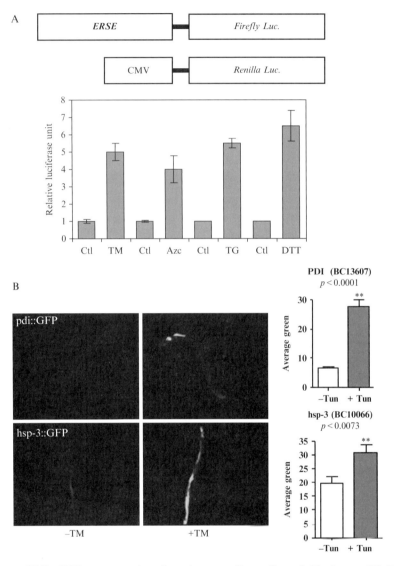

Figure 19.2 UPR reporters in cultured mammalian cells and *C. elegans*. (A) Top panel: schematic representation of the reporter genes used in mammalian cells. Bottom panel: typical reporter gene activation experiment carried out in HeLa cells treated with various ER stressors, TM, Azc, TG, DTT. (B) Left panel: fluorescence images of *pdi*::GFP and *hsp-3*::GFP reporter strains subjected or not to TM treatment. Right panel: quantitation of the previously mentioned experiments using the Union Biometrica COPAS Biosorter (white bars, control; gray bars, TM). More than 100 worms were counted in triplicates for each strain and condition.

the Union Biometrica COPAS Biosort (Union Biometrica) (Dupuy et al., 2007). This instrument is designed to perform multiparametric analyses of micrometric particles in a semiautomated manner. It allowed us to measure the average fluorescence emitted by a large population of transgenic worms (approx. 1000 animals per conditions) in response to TM treatment and to estimate their size. Indeed, the COPAS worm sorter is equipped with two lasers: a red diode excitation laser that is used to analyze the physical parameters of the organism, referred to as time of flight (TOF) and extinction (EXT), and a multiline argon laser that is used to excite various fluorophores. TOF is a measure of the relative length of each animal, and EXT provides a measure of its optical density. We used the 499-nm laser to excite the GFP (Dupuy et al., 2007). Before treatment, worm populations were synchronized as L1 following hypochlorite treatment and 16 h incubation in sterile M9. They were grown to L4 stage on normal nematode growth medium (NGM) plates and transferred to NGM plates containing either 5 μg/ml TM or DMSO as a control for 5 h at 20 °C. The drug was added to the NGM medium before pouring the plates. Worms were then processed through the COPAS, and the two parameters, TOF and EXT, were measured along with the GFP fluorescence. A gating region was defined as an EXT versus TOF dot-plot and helped us to eliminate dead eggs and debris. A TOF count of 100 corresponds to an individual of ∼0.24 mm long. Adults are about 1 mm long and are consequently characterized by a TOF value comprised between 400 and 500. GFP emission of *pdi*::GFP and *hsp-3*::GFP adults was measured in each condition, and TM-treated worms fluorescence was normalized to the GFP emission obtained in DMSO-treated populations. This way, the average GFP emission and the mean value were calculated for each worm population (Fig. 19.2B, right panel). At least 100 worms were counted per strains and conditions in triplicates.

4. Monitoring Small GTPase Activity

Small GTPases of the Ras superfamily comprise more than 80 members conserved across species. Thus far, the Ras and the Rho subfamilies have been clearly involved in the regulation of ER stress signaling (Caruso et al., 2008; Denoyelle et al., 2006; Lin et al., 2003; Yaari-Stark et al., 2010). In the present chapter, we focus on the impact of ER stress on the activation of Rho GTPases.

4.1. Immunofluorescence-based visualization of actin remodeling upon ER stress

One major function of Rho GTPases is to promote the remodeling of actin cytoskeleton. As a consequence, we tested whether chemically induced ER stress promoted actin-remodeling events in rat fibroblasts in culture. To this end, FR3T3 fibroblasts were grown on a 22-mm coverslip (Fisher) and treated, as indicated in Fig. 19.3A. Following ER stress-inducer treatment, cells were washed with PBS, fixed with 3% formaldehyde in PBS for 10 min, permeabilized with 0.2% Triton X-100 in PBS for 5 min, and treated with 50 mM NH$_4$Cl in PBS for 10 min. Cells were then incubated at room temperature with phalloidin-FITC for 1 h, followed by extensive

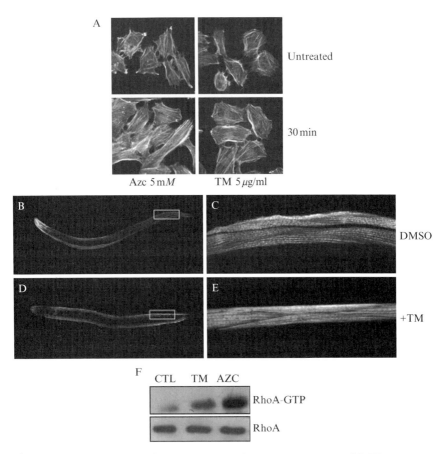

Figure 19.3 Monitoring Rho-GTPase activation upon treatment with ER stressors. (A) Phalloidin staining performed on FR3T3 cells treated with Azc or TM for the indicated period of time and at the indicated concentrations. (B–E) Similar experiment carried out on N2 worms treated (D–E) or not (B–C) with TM. (C) and (E) are enlargements of the yellow box region from (B) and (D). (F) Rho-GTPase activity assay using a pull-down Rhotekin RBD domain fused to GST followed by immunoblot using anti-RhoA antibodies. The experiment was carried out in FR3T3 cell extracts following treatment with Azc or TM at the same concentrations as in (A) for 30 min. (For interpretation of the references to color in this figure legend, the reader is referred to the Web version of this chapter.)

PBS washing. Filamentous actin stained with phalloidin-FITC was visualized using fluorescence microscopy (Leica Microsystems, Mannheim, Germany). Cells subjected to short-term treatment (less than an hour) with chemical ER stressors show significant changes in the architecture of their actin cytoskeleton mainly through the appearance of stress fibers (Fig. 19.3A), thus suggesting the involvement of the Rho member of the Rho subfamily. Similar experiments were performed in *C. elegans*. Adult worms expressing the UPR marker choline/ethanolamine kinase *ckb-2* promoter fused with GFP (*pckb-2::gfp*, strain BC14636) were treated with 5 μg/ml TM or DMSO for 5 h at 20 °C as previously described (Caruso *et al.*, 2008). UPR induction in these worms was first observed under a fluorescent microscope by looking at the GFP fluorescence levels, which increased in TM-treated animals. Actin was then stained using Alexa Fluor 594 Phalloidin (Molecular probes), as previously described (Liu *et al.*, 2010.). Briefly, worms were fixed with 3.7% formaldehyde in PBS for 10 min at RT, washed in PBS, soaked in acetone for 5 min at -20 °C, washed in PBS, and stained with 2 U/ml phalloidin in PBS for 30 min in the dark at RT. After washes in PBS complemented with 0.5% BSA and 0.5% Tween 20, worms were mounted on slides and kept overnight at -20 °C before being observed under a SP5 confocal microscope (Leica). Phalloidin staining in muscle cells appeared different and more intense in TM-treated animals. Fluorescence was increased between the muscle fibers, which appeared less clearly and less punctuated (Fig. 19.3B–F), suggesting a disorganization of actin in these cells. This result is in agreement with our previous finding that small G protein CRP-1 belonging to the Rho family is an essential key player in the induction of the UPR (Caruso *et al.*, 2008). Altogether, these results suggest a link between UPR and the Rho GTPases.

4.2. GTPase pull-down assay

To confirm that the observation made in Section 4.1 was indeed due to an activation of Rho, the proportion of GTP-bound Rho was determined in FR3T3 cells treated for 0 or 30 min with 5 mM Azc or 5 μg/ml TM by performing a GTPase pull-down assay, as previously described (Ren *et al.*, 1999). The Rho-binding domain of Rhotekin (amino acids 7–89), RBD, was used as a bait to fish Rho. Briefly, pGEX-2T constructs containing RBD were transformed into DH5α bacteria, and the GST–RBD produced as a recombinant fusion protein was purified and used to coat beads. Following ER stress-inducing treatment, FR3T3 cells were washed with ice-cold PBS and lysed in 50 mM Tris, pH 7.2, 1% Triton X-100, 0.5% sodium deoxycholate, 0.1% SDS, 500 mM NaCl, 10 mM MgCl$_2$, leupeptin, aprotinin, and 1 mM PMSF. Cell lysates were clarified by centrifugation at $13,000\times g$ at 4 °C for 10 min, and equal volumes of lysates were incubated with GST–RBD (10 μg) coated beads at 4 °C for 45 min. The beads were washed 4 times with 50 mM Tris, pH 7.2, containing 1% Triton X-100, 150 mM NaCl, 10 mM MgCl$_2$, leupeptin, aprotinin, and 0.1 mM PMSF. Bound Rho proteins were detected by immunoblotting using an antibody raised against RhoA (Santa Cruz Biotechnology, see Table 19.1). The amount of RBD-bound Rho was normalized to the total amount of Rho found in cell lysates and GTP-bound Rho levels were determined by immunoblot for the comparison of Rho activity in different samples (Fig. 19.3F). These results confirmed those obtained in Section 4.1 by demonstrating an increased amount of Rho-GTP upon short-term treatment with 5 μg/ml TM or 5 mM Azc.

5. Pharmacological and Genetic Modulation of UPR and GTPase Signaling

Both UPR and small GTPase signaling can be artificially modulated (either activated or inhibited). Based on this observation and the existence of the assays and experimental systems described above, it becomes possible to map the existence of links between small GTPase and ER stress signaling.

5.1. Pharmacological modulation

Pharmacological molecules exist that either activate or inhibit the UPR. Using cultured cells or *C. elegans*, it is possible to perform traditional pharmacological studies by combining both activators (see Table 19.3) and selective inhibitors and then to monitor UPR parameters (as described in Section 3) or physiological outcomes such as cell proliferation, survival, apoptosis, or even necrosis. Moreover, selective inhibitors of Rho GTPases have also been described (Table 19.3). Again, pharmacological approaches combining Rho-GTPase inhibitors and UPR activators may reveal new signaling networks connecting the two pathways.

5.2. Genetic modulation by RNA interference

The other alternative to modulate GTPase or ER stress signaling is to specifically attenuate the expression of specific actors of these respective signaling pathways using RNA interference. This can be carried out both in cultured mammalian cells through si or shRNA or in *C. elegans* using RNAi. We have performed such experimental approaches in both systems to identify new modulators of the UPR. In *C. elegans*, functional screens were carried out on various UPR target genes GFP reporters using a COPAS Biosort-based approach. Average GFP emissions from animals

Table 19.3 Pharmacological modulators of UPR and GTPase signaling (nonexhaustive list)

Class	Name	Target	Reference
UPR activator	Tunicamycin	N-glycosylation	
	BFA	sec7/GEF-ARF1	
	Thapsigargin	SERCA pumps	
	DTT	Disulfide bonds	
	Azc	Proline analog	
PERK	Salubrinal	PP1	Boyce et al. (2005)
IRE1	IREstatin	IRE1	Feldman and Koong (2007)
	Quercetin	IRE1	Wiseman et al. (2010)
ATF6	AEBF	S1P, S2P	Okada et al. (2003)
GTPase inhibitor	CT04	RhoA/B/C	Cytoskeleton Inc.
	ETH 1864	Rac1	Shutes et al. (2007)
	NSC23766	Rac1	Gao et al. (2004)
	Secramine	Cdc42	Pelish et al. (2006)

treated by RNAi against *ire-1*, *atf-6*, and *xbp-1* can be compared to a reference experiment. This comparison enables the calculation of a variation factor β. This variation factor is used to normalize results obtained for each RNAi series to a reference experiment. Consequently, this helps to reduce the fluctuation of GFP expression resulting from environmental variability. Comparison of GFP expression measured in basal condition and upon TM treatment enables the identification of at least three classes of genes: (i) the genes whose silencing does not affect *pdi*::GFP or *hsp-3*::GFP expression (Saloheimo et al., 2004); (ii) the genes whose silencing prevents TM-induced *pdi*::GFP or *hsp-3*::GFP expression (activators); (iii) the genes whose silencing leads to increased basal *pdi*::GFP or *hsp-3*::GFP expression (repressors); and (iv) genes whose silencing differentially alters the expression of *pdi*::GFP or *hsp-3*::GFP (selective modulators). Activators encode proteins whose functional alteration would have a direct effect on ER homeostasis. In opposition, repressors encode products, which are directly involved in the mediation of the UPR signaling pathways toward *pdi* and/or *hsp-3* promoters. Such an approach was successfully utilized using both *hsp-4* and *ckb2* reporter strains and systematic silencing of small GTPases of the Ras superfamily (Caruso et al., 2008).

Similar approaches can be performed in mammalian cells using siRNA. We performed a systematic siRNA screen targeting both small GTPases of the Ras superfamily and ER molecular machines (translocation, folding, quality control, ER-associated degradation) and monitored the activation of ATF6 using a localization reporter (FLAG-ATF6α; Higa-Nishiyama and Chevet unpublished results). Briefly, siRNAs were obtained from RNAi

Co. (Tokyo, Japan) and Ambion (Austin, TX, USA) and targeted specifically the functional groups described above. SiRNA was delivered into HeLa cells by reverse transfection using Lipofectamine RNAiMAX (Invitrogen, Carlsbad, CA, USA) at a siRNA concentration of 12.5 or 25 nM. Twenty-four or 48 h after transfection, cells were subsequently transiently transfected with FLAG-tagged ATF6α using Lipofectamine and PLUS reagents (Invitrogen, Carlsbad, CA, USA) according to the manufacturer's protocols. The localization of FLAG-ATF6α was then monitored using confocal microscopy. These experiments led us to identify new components of ATF6 activation network, which selectively modulate the UPR.

In conclusion, these experimental systems and approaches allow for the identification of the connections between small GTPases and ER stress signaling networks. More specifically, the combination of mammalian and *C. elegans* models, as well as that of pharmacological and genetic approaches, will permit the delineation of specific signaling pathways connecting the UPR to small GTPases of the Rho subfamily.

ACKNOWLEDGMENTS

This work was funded by grants from INSERM-INCa to E. C. E. M. was supported by a postdoctoral fellowship from ARC.

REFERENCES

Abdelrahim, M., et al. (2005). Induction of endoplasmic reticulum-induced stress genes in Panc-1 pancreatic cancer cells is dependent on Sp proteins. *J. Biol. Chem.* **280**, 16508–16513.

Altan-Bonnet, N., et al. (2003). A role for Arf1 in mitotic Golgi disassembly, chromosome segregation, and cytokinesis. *Proc. Natl. Acad. Sci. USA* **100**, 13314–13319.

Altan-Bonnet, N., et al. (2004). Molecular basis for Golgi maintenance and biogenesis. *Curr. Opin. Cell Biol.* **16**, 364–372.

Boyce, M., et al. (2005). A selective inhibitor of eIF2alpha dephosphorylation protects cells from ER stress. *Science* **307**, 935–939.

Brenner, S. (1974). The genetics of *Caenorhabditis elegans*. *Genetics* **77**, 71–94.

Calfon, M., et al. (2002). IRE1 couples endoplasmic reticulum load to secretory capacity by processing the XBP-1 mRNA. *Nature* **415**, 92–96.

Caruso, M. E., and Chevet, E. (2007). Systems biology of the endoplasmic reticulum stress response. *Subcell. Biochem.* **43**, 277–298.

Caruso, M. E., et al. (2008). GTPase-mediated regulation of the unfolded protein response in *Caenorhabditis elegans* is dependent on the AAA+ ATPase CDC-48. *Mol. Cell. Biol.* **28**, 4261–4274.

Chen, Y. J., et al. (2010). Differential regulation of CHOP translation by phosphorylated eIF4E under stress conditions. *Nucleic Acids Res.* **38**, 764–777.

Chevet, E., et al. (2001). The endoplasmic reticulum: Integration of protein folding, quality control, signaling and degradation. *Curr. Opin. Struct. Biol.* **11**, 120–124.

Denoyelle, C., et al. (2006). Anti-oncogenic role of the endoplasmic reticulum differentially activated by mutations in the MAPK pathway. *Nat. Cell Biol.* **8,** 1053–1063.

Drogat, B., et al. (2007). IRE1 signaling is essential for ischemia-induced vascular endothelial growth factor-A expression and contributes to angiogenesis and tumor growth in vivo. *Cancer Res.* **67,** 6700–6707.

Dupuy, D., et al. (2007). Genome-scale analysis of in vivo spatiotemporal promoter activity in *Caenorhabditis elegans*. *Nat. Biotechnol.* **25,** 663–668.

Feldman, D., and Koong, A. C. (2007). Irestatin, a potent inhibitor of IRE1 and the unfolded protein response, is a hypoxia-selective cytotoxin and impairs tumor growth. *J. Clin. Oncol.* **25**(18S), 3514, (June 20 Supplement).

Gao, Y., et al. (2004). Rational design and characterization of a Rac GTPase-specific small molecule inhibitor. *Proc. Natl. Acad. Sci. USA* **101,** 7618–7623.

Harding, H. P., et al. (2001). Diabetes mellitus and exocrine pancreatic dysfunction in perk-/- mice reveals a role for translational control in secretory cell survival. *Mol. Cell* **7,** 1153–1163.

Hollien, J., and Weissman, J. S. (2006). Decay of endoplasmic reticulum-localized mRNAs during the unfolded protein response. *Science* **313,** 104–107.

Hollien, J., et al. (2009). Regulated Ire1-dependent decay of messenger RNAs in mammalian cells. *J. Cell Biol.* **186,** 323–331.

Hunt-Newbury, R., et al. (2007). High-throughput in vivo analysis of gene expression in *Caenorhabditis elegans*. *PLoS Biol.* **5,** e237.

Kang, Y. J., et al. (2010). The TSC1 and TSC2 tumor suppressors are required for proper ER stress response and protect cells from ER stress-induced apoptosis. *Cell Death Differ.* doi: 10.1038/cdd.2010.82.

Lazarow, P. B. (2003). Peroxisome biogenesis: Advances and conundrums. *Curr. Opin. Cell Biol.* **15,** 489–497.

Lee, A. H., et al. (2003). XBP-1 regulates a subset of endoplasmic reticulum resident chaperone genes in the unfolded protein response. *Mol. Cell. Biol.* **23,** 7448–7459.

Lin, T., et al. (2003). Rho-ROCK-LIMK-cofilin pathway regulates shear stress activation of sterol regulatory element binding proteins. *Circ. Res.* **92,** 1296–1304.

Lipson, K. L., et al. (2006). Regulation of insulin biosynthesis in pancreatic beta cells by an endoplasmic reticulum-resident protein kinase IRE1. *Cell Metab.* **4,** 245–254.

Liu, Q., et al. (2010). Facioscapulohumeral muscular dystrophy region gene-1 (FRG-1) is an actin-bundling protein associated with muscle-attachment sites. *J. Cell Sci.* **123,** 1116–1123.

McKay, R. M., et al. (2003a). C elegans: A model for exploring the genetics of fat storage. *Dev. Cell* **4,** 131–142.

McKay, S. J., et al. (2003b). Gene expression profiling of cells, tissues, and developmental stages of the nematode *C. elegans*. *Cold Spring Harb. Symp. Quant. Biol.* **68,** 159–169.

Memon, A. R. (2004). The role of ADP-ribosylation factor and SAR1 in vesicular trafficking in plants. *Biochim. Biophys. Acta* **1664,** 9–30.

Okada, T., et al. (2003). A serine protease inhibitor prevents endoplasmic reticulum stress-induced cleavage but not transport of the membrane-bound transcription factor ATF6. *J. Biol. Chem.* **278,** 31024–31032.

Pelish, H. E., et al. (2006). Secramine inhibits Cdc42-dependent functions in cells and Cdc42 activation in vitro. *Nat. Chem. Biol.* **2,** 39–46.

Philips, M. R. (2005). Compartmentalized signalling of Ras. *Biochem. Soc. Trans.* **33,** 657–661.

Quatela, S. E., and Philips, M. R. (2006). Ras signaling on the Golgi. *Curr. Opin. Cell Biol.* **18,** 162–167.

Ren, X. D., et al. (1999). Regulation of the small GTP-binding protein Rho by cell adhesion and the cytoskeleton. *EMBO J.* **18,** 578–585.

Ron, D., and Walter, P. (2007). Signal integration in the endoplasmic reticulum unfolded protein response. *Nat. Rev. Mol. Cell Biol.* **8,** 519–529.

Rutkowski, D. T., and Hegde, R. S. (2010). Regulation of basal cellular physiology by the homeostatic unfolded protein response. *J. Cell Biol.* **189,** 783–794.

Saloheimo, M., et al. (2004). Characterization of secretory genes ypt1/yptA and nsf1/nsfA from two filamentous fungi: Induction of secretory pathway genes of Trichoderma reesei under secretion stress conditions. *Appl. Environ. Microbiol.* **70,** 459–467.

Shutes, A., et al. (2007). Specificity and mechanism of action of EHT 1864, a novel small molecule inhibitor of Rac family small GTPases. *J. Biol. Chem.* **282,** 35666–35678.

Takeuchi, M., et al. (2000). A dominant negative mutant of sar1 GTPase inhibits protein transport from the endoplasmic reticulum to the Golgi apparatus in tobacco and Arabidopsis cultured cells. *Plant J.* **23,** 517–525.

Tatin, F., et al. (2006). A signalling cascade involving PKC, Src and Cdc42 regulates podosome assembly in cultured endothelial cells in response to phorbol ester. *J. Cell Sci.* **119,** 769–781.

Tsang, K. Y., et al. (2010). In vivo cellular adaptation to ER stress: Survival strategies with double-edged consequences. *J. Cell Sci.* **123,** 2145–2154.

Urano, F., et al. (2002). A survival pathway for *Caenorhabditis elegans* with a blocked unfolded protein response. *J. Cell Biol.* **158,** 639–646.

Wiseman, R. L., et al. (2010). Flavonol activation defines an unanticipated ligand-binding site in the kinase-RNase domain of IRE1. *Mol. Cell* **38,** 291–304.

Wu, R. F., et al. (2010). Nox4-derived H2O2 mediates endoplasmic reticulum signaling through local Ras activation. *Mol. Cell. Biol.* **30,** 3553–3568.

Yaari-Stark, S., et al. (2010). Ras inhibits endoplasmic reticulum stress in human cancer cells with amplified Myc. *Int. J. Cancer* **126,** 2268–2281.

Yamamoto, K., et al. (2007). Transcriptional induction of mammalian ER quality control proteins is mediated by single or combined action of ATF6alpha and XBP1. *Dev. Cell* **13,** 365–376.

Zamanian-Daryoush, M., et al. (2000). NF-kappaB activation by double-stranded-RNA-activated protein kinase (PKR) is mediated through NF-kappaB-inducing kinase and IkappaB kinase. *Mol. Cell. Biol.* **20,** 1278–1290.

CHAPTER TWENTY

Inhibitors of Advanced Glycation and Endoplasmic Reticulum Stress

Reiko Inagi

Contents

1. Introduction	362
2. Advanced Glycation and Its Pathophysiology	363
3. Measurement of Advanced Glycation	364
3.1. HPLC and GC/MS	364
3.2. Immunohistochemistry	365
3.3. Western blot analysis	367
3.4. ELISA	368
4. Link Between Advanced Glycation and ER Stress	369
4.1. Direct induction of ER stress by advanced glycation	370
4.2. Comorbid induction of ER stress and advanced glycation	370
5. Effects of Advanced Glycation Inhibitor Against ER Stress	373
5.1. Antioxidants	373
5.2. Antihypertensive drugs	376
6. Conclusion	377
Acknowledgments	377
References	377

Abstract

Advanced glycation is one of the major pathophysiological posttranslational modifications. Under hyperglycemic conditions or oxidative stress, proteins and DNA are nonenzymatically modified by oxidative glycation and converted to advanced glycation endproducts (AGEs), which induce the loss of protein functions or apoptosis. This conversion to AGEs leads to the disease progression of hyperglycemia- or oxidative stress-related diseases, such as diabetic mellitus and its complications, neurodegenerative disease, atherosclerosis, and kidney disease. Further, recent evidence indicates that advanced glycation is initiated not only by oxidative stress but also by hypoxia, suggesting its pathogenesis across a wide range of diseases associated with aberrant oxygen tension as well as glucose metabolism. In addition to their role in triggering advanced

Division of Nephrology and Endocrinology, University of Tokyo School of Medicine, Tokyo, Japan glycation, these disturbances are also well known as initiators of endoplasmic reticulum (ER) stress, and of the consequent unfolded protein response (UPR). These findings strongly indicate the presence of cross talk between advanced glycation and ER stress in disease progression.

In this chapter, I focus on the link between advanced glycation and ER stress, and the potential use of inhibitors of AGE formation as modulators of ER stress.

1. INTRODUCTION

Advanced glycation is a posttranslational modification of proteins and DNA by highly reactive glucose-derived dicarbonyl compounds. Referred to as advanced glycation endproduct (AGE) precursors, these include glyoxal and methylglyoxal and are mainly generated by glycolysis or oxidation of carbohydrate. AGE precursors are not only cytotoxic intermediates themselves, but also rapidly bind to proteins and DNA to form AGEs, in turn leading to the loss of protein function or apoptosis. Further, AGEs induce cellular signaling by binding to the receptor for AGE (RAGE) and thereby exert pathogenic activities. Advanced glycation occurs more rapidly when the formation of AGE precursors is accelerated, the main triggers of which are reactive oxygen species (ROS) generated by hyperglycemic conditions, hypoxia, or oxidative stress (Gao and Mann, 2009; Inagi et al., 2010). This characteristic suggests that advanced glycation plays a significant role on the pathophysiology of various diseases related to these pathological conditions, including diabetes mellitus and its complications (diabetic nephropathy and retinopathy), ischemic cardiovascular diseases, and inflammation.

ER stress exacerbates the progression of conformational diseases caused by the accumulation of malfolded proteins in the ER due to misfolding-associated mutations (Kuznetsov and Nigam, 1998). In addition to misfolding mutations, other factors that adversely affect ER function include hypoxia and oxidative stress, which lead to pathogenic ER stress in various diseases. These include neurodegenerative and cardiovascular diseases, diabetes mellitus, and kidney diseases (Chiang and Inagi, 2010; Hotamisligil, 2010; Inagi, 2010; Kim et al., 2008). Recent studies have focused on the pathophysiological contribution of the ER stress signal, namely, the unfolded protein response (UPR), in these diseases and the cross talk of stress-induced cellular responses by hypoxia, oxidative stress, and ER stress.

As described above, advanced glycation and ER stress are often observed under the same situations, suggesting that advanced glycation and ER stress pathophysiologically orchestrate the progression of diseases associated with hypoxic and oxidative stress.

2. Advanced Glycation and Its Pathophysiology

While advanced glycation barely occurs under normal conditions, it is markedly enhanced under hyperglycemia. The intermediates of advanced glycation are highly reactive dicarbonyl compounds (α-oxoaldehyde), such as glyoxal and methylglyoxal, which are generated from excess amounts of carbohydrate by glycolysis or hyperglycemia-mediated oxidative stress (oxidation). These bind to lysine or arginine residues of proteins or to DNA and form AGEs, examples of which include pentosidine and N^ε-carboxymethyllysine (CML). These serial reactions are often referred to as the Maillard reaction. AGE formation leads to the conformational modification of proteins, which results in the loss of their original physiological activities, and often also the gain of pathogenic functions. AGEs also exert their aberrant effects by binding to the receptor for AGEs (RAGE). Like AGEs, AGE precursor dicarbonyl compounds also exert a pathogenic effect via the induction of ROS generation, inflammation, or glycation-mediated apoptosis. Of note, AGE precursors modify a wide variety of intracellular and extracellular proteins, including transcription factors and matrix proteins, and thereby perturb cellular homeostasis, suggesting the potent pathophysiological effects of advanced glycation.

In addition to its marked enhancement under hyperglycemia, advanced glycation is also markedly induced by hypoxia or oxidative stress under normoglycemic conditions. While previous studies have characterized oxidative stress as the excess production of ROS under hyperoxic conditions, the current consensus is that oxidative stress is observed whenever oxygen homeostasis is disrupted and can be demonstrated in both hyper- and hypoxia. Thus, under hypoxia and oxidative stress, both of which are associated with pathogenic ROS generation, noxious AGE precursors are formed by glycoxidation in an ROS-dependent manner, and the consequent AGEs contribute to the development of the various manifestations of hypoxia or oxidative stress-related diseases. In fact, advanced glycation under hypoxia is triggered by hypoxia-induced mitochondrial dysfunction, with a subsequent increase in ROS production and alteration of adaptive hypoxic response. These functional abnormalities accelerate the vicious cycle of oxidative stress and hypoxia, and thereby further exacerbate advanced glycation.

In the pathogenesis of advanced glycation, much attention has been focused on methylglyoxal, a major AGE precursor. Like other AGE precursors, methylglyoxal is a highly reactive dicarbonyl compound, binding to both proteins and DNA and subsequently modifying them to form AGEs. Of note, in addition to the harmful effect of methylglyoxal as a precursor of pathogenic AGEs, methylglyoxal itself is cytotoxic and proapoptotic.

Methylglyoxal selectively inhibits mitochondrial respiration and glycolysis in the cells (Biswas *et al.*, 1997; de Arriba *et al.*, 2007). Related to this, methylglyoxal has also been shown to inactivate membrane ATPases and glyceraldehyde-3-phosphate dehydrogenase, a key enzyme in the glycolytic pathway (Halder *et al.*, 1993). Methylglyoxal also induces apoptosis via c-Jun N-terminal kinase (JNK) activation and through the activation of protein kinase Cδ (Chan *et al.*, 2007; Godbout *et al.*, 2002). Cotreatment of cultured human endothelial cells with concentrations of methylglyoxal and glucose comparable to those seen in the blood of patients with diabetes causes cell apoptosis or necrosis with an increase in ROS content (Chan and Wu, 2008).

Previous reports have implicated methylglyoxal in the etiology of various diseases related to oxidative stress, including diabetes mellitus, diabetic complications such as vascular and renal diseases, chronic renal failure, neuronal disease, hypertension, and aging. More recent studies have further demonstrated that methylglyoxal and methylglyoxal-arginine adducts are accumulated in the ischemia–reperfusion-damaged heart and brain, respectively (Bucciarelli *et al.*, 2006; Oya *et al.*, 1999). Methylglyoxal may make a diverse pathogenic contribution to not only chronic diseases but also to acute diseases related to hypoxia.

3. Measurement of Advanced Glycation

The state of advanced glycation is generally assessed using AGEs, including pentosidine and CML or AGE precursors, with the inclusion of glyoxal and methylglyoxal as markers. Because of the variety of late-stage pathways in AGE formation, AGE structure is diverse, and the measurement of total AGE content is accordingly difficult. Measurement of AGE precursors, such as methylglyoxal, is not always suitable due to their high reactivity. The major AGEs can be detected by high performance liquid chromatography (HPLC), gas chromatography/mass spectrometry (GC/MS), immunohistochemistry (IHC), Western blot analysis (WB), or enzyme-linked immunosorbent assay (ELISA). Among these, HPLC and GC/MS are expensive and time-consuming, whereas the others are relatively insensitive and lack specificity. The hyperglycemic status of diabetic patients is determined from the levels of the glycated hemoglobin HbA1c, which is routinely measured by HPLC or immunoassay.

3.1. HPLC and GC/MS

Serum or tissue levels of one of the major AGEs, pentosidine or CML, are quantitatively detected by HPLC or GC/MS (Inagi *et al.*, 2006). Although these methods are suitable for quantitative analysis and have specificity, they are expensive and require many pretreatment steps for the reduction and acid hydrolysis of proteins in the test samples. Chemically synthesized pentosidine and CML are also required as reference standards. The sensitivity of this method depends on the AGE content of the test sample.

Specifically, reduction and hydrolysis of the samples is achieved as follows: renal tissue (~ 100 mg) is reduced for 4 h at room temperature by the addition of an excess of $NaBH_4$ in 0.2 M borate buffer (pH 9.1), precipitated by an equal volume of 20% trichloroacetic acid (TCA), and centrifuged at $2000 \times g$ for 10 min. After the supernatant is discarded, the pellet is washed with 1 ml of 10% TCA, dried under vacuum, and acid-hydrolyzed in 500 µl of 6 N HCl for 16 h at 110 °C in screw-cap tubes which are purged with nitrogen. The hydrolysate is then dried under vacuum, rehydrated in water, and used to measure pentosidine or CML.

The pentosidine content of the samples is analyzed on reverse-phase HPLC. After dilution with PBS, the test sample (20 µl) is injected into an HPLC system and separated on a C18 reverse-phase column (Waters, Tokyo, Japan). The effluent is monitored with a fluorescence detector (RF-10A, Shimadzu, Kyoto, Japan) at an excitation–emission wavelength of 335/385 nm. Synthetic pentosidine is used as a standard and to check sensitivity. The detection limit of pentosidine by HPLC is 0.1 pmol/mg protein.

The CML content of the test samples is measured as its N,O-trifluoroacetyl methyl esters by selected-ion monitoring GC/MS. Before the samples are applied to the GC/MS system, synthetic CML (d4-CML) is added to each sample as an internal control. The detection limit of CML content by GC/MS is 1.0 pmol/mg protein.

Detection of AGE precursor in plasma samples, including glyoxal and methylglyoxal, is also carried out by HPLC (Inagi *et al.*, 2002). Two hundred microliters of plasma ultrafiltrate is reacted with an equal volume of 2,4-dinitrophenylhydrazin (0.025% DNPH in 0.5 N HCl) at 30 °C for 30 min, and subsequently incubated with acetone (1 M, 20 µl) to remove excess DNPH at 30 °C for 10 min. These steps are performed in the dark. The reaction mixture is washed in 200-µl hexane three times, and the aqueous layer is extracted with 200-µl octanol, diluted with acetonitrile, and hydrazone derivatives are analyzed by reversed-phase HPCL utilizing a C18 column as described above. AGE precursor is detected by absorbance at 360 nm. Pure chemical glyoxal or methylglyoxal is used as a standard for calibration.

3.2. Immunohistochemistry

IHC utilizing anti-AGE antibodies is useful in identifying the localization of advanced glycation in various tissues, including brain, heart, aorta, and kidney. Among commercially available monoclonal antibodies to major AGE structures, such as pentosidine, CML, CEL (N^{ε}-carboxyethyllysine), imidazolone, and pyrraline, anti-CML or anti-CEL antibody is widely used, possibly because the absolute content of CML or CEL in the tissues is higher than that of pentosidine or pyrraline, and because the antigen determinants of CML or CEL are structurally accessible by the antibodies. For IHC detection of CML or CEL, unfixed frozen tissues embedded in OCT compound are recommended over paraffin-embedded tissues because these avoid the formation of AGEs which occurs with heating during the paraffin embedding process.

3.2.1. IHC for the detection of CEL in the kidney with ischemia–reperfusion

Advanced glycation level can be estimated by IHC in the detection of methylglyoxal-lysine adducts utilizing anti-CEL monoclonal antibody. Renal cortexes from normal rats or rats with ischemia–reperfusion injury are embedded in OCT compound (Sakura Fine Technical, Tokyo, Japan), frozen in liquid nitrogen, and stored at $-80\ ^\circ$C until use. The sections (4 µm) are washed twice in PBS containing 0.05% Tween 20 (PBS-T) at room temperature for 10 min each and fixed with cold acetone/methanol (1:1) at $-20\ ^\circ$C for 10 min. The slides are then washed twice in PBS-T, incubated with H_2O_2/MeOH (1:99) for 20 min at room temperature to inactivate endogenous peroxidase, blocked with PBS-T containing 2% BSA (bovine serum albumin) at room temperature for 2 h or at 4 $^\circ$C overnight, and the sections are incubated with monoclonal anti-CEL antibody (CEL-SP, 10 µg/ml) (Nagai et al., 2008). The slides are then washed with PBS-T 3 times and reacted with a biotinylated antimouse IgG (1:400; Vector Laboratories, Burlingame, CA) as a second antibody at room temperature for 2 h. After the slides are washed in PBS-T thrice, they are reacted with horseradish peroxidase (HRP)–avidin (1:1000; Vector) at room temperature for 30 min. The slides are washed in PBS-T twice and in 0.05 M Tris–HCl (pH 7.6) at 37 $^\circ$C for 10 min, and then incubated with DAB solution (0.05 M Tris–HCl, pH 7.6, containing 1% H_2O_2 and 0.2 mg/ml 3,3′-diaminobenzidine,tetrahydrochloride: DAB, Wako, Osaka, Japan) at 37 $^\circ$C for 3–10 min for visualization. The enzymatic reaction is stopped by washing in distilled water. In some cases, the use of a blocking kit (Vector Laboratories, Burlingame, CA) to block nonspecific binding of biotin/avidin system reagents is recommended. Nuclear staining with hematoxylin as a counter stain is performed if necessary.

To confirm the specificity of the positive signal, monoclonal anti-CEL antibody (10 µg/ml) is preincubated with CEL-modified bovine serum albumin (CEL-BSA, 0.1 mg/ml) or CML-modified BSA (CML-BSA, 0.1 mg/ml), which is another AGE generated by glyoxal but not methylglyoxal, for 60 min at room temperature and then used.

The results show that CEL content is significantly increased in interstitial lesions of renal cortex in rats with ischemia–reperfusion as compared with that in normal rats. The CEL-positive signal is undetectable when the

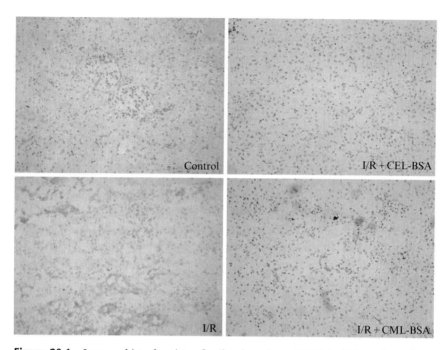

Figure 20.1 Immunohistochemistry for the detection of CEL in kidney tissue from normal rats and rats with ischemia–reperfusion injury. Right nephrectomy was performed 1 week before the induction of ischemic injury by clamping the left renal artery for 45 min. Core body temperature was maintained at 37 °C during surgery using a homeothermic table. Twenty-four hours after the induction of injury, kidney samples were obtained and subjected to immunohistochemical analysis for detection of CEL. Renal ischemia–reperfusion induced the accumulation of CEL. The CEL-positive signal was undetectable when the monoclonal anti-CEL antibody was preincubated with CEL-BSA, but not CML-BSA, another AGE, indicating that the positive signal is specific for CEL. Control, sham-operated kidney incubated with anti-CEL antibody; I/R, kidney with ischemia–reperfusion incubated with anti-CEL antibody; CEL-BSA, immunohistochemistry utilizing anti-CEL antibody preincubated with CEL-BSA; CML-BSA, immunohistochemistry utilizing anti-CEL antibody preincubated with CML-BSA. Original magnification, 200×.

monoclonal anti-CEL antibody is preincubated with CEL-BSA but not with CML-BSA, which is another AGE, indicating that the positive signal is specific for CEL (Fig. 20.1).

3.3. Western blot analysis

As with IHC, the status of advanced glycation is often evaluated with WB utilizing anti-AGE antibodies. This system is helpful in semiquantitatively evaluating the status of advanced glycation by WB, followed by densitometry of the stained bands using NIH ImageJ (http://rsb.info.nih.gov/ij/).

3.3.1. WB for detection of CEL in rat-cultured proximal tubular cells exposed to hypoxia–reoxygenation

Immortalized rat proximal tubular cells (IRPTC) are exposed to hypoxia–reoxygenation as follows: Using an Anaerocult A mini system (Merck, Darmstadt, Germany), which reduces oxygen content to 0.2%, hypoxic stimulation is initiated at 70% confluence of IRPTC cultured in DMEM medium containing 1% FBS. After hypoxia for 24 h, the cells are subsequently cultured under normoxia for 2 h and then collected. The cells are homogenized in a sample buffer (0.35 M Tris–HCl, pH 6.8, 10% SDS, 36% glycerol, 5% β-mercaptoethanol, and 0.012% bromophenol blue), and the cell lysate (20 μg) is electrophoresed on 10–15% SDS-PAGE and transferred to a polyvinylidene fluoride membrane (Amersham Biosciences, Piscataway, NJ). The membrane is washed in TBS-T (25 mM Tris–HCl, pH 7.6, 0.15 M NaCl, 0.05% Tween 20) and blocked with 5% BSA in TBS-T.

For detection of CEL, mouse anti-CEL monoclonal antibody (CEL-SP, 2.5 μg/ml; Nagai *et al.*, 2008, or KH025, 0.5 μg/ml; TransGenic, Inc., Kumamoto, Japan) is used as the first antibody. The membrane is incubated with the antibody at 4 °C overnight, washed in TBS-T 3 times, and subsequently incubated with horseradish peroxidase (HRP)-conjugated antimouse IgG antibody (10,000×; Bio-Rad Laboratories Hercules, CA) for 2 h at room temperature. After the membrane is washed in TBS-T thrice, the bands are detected by an enhanced chemiluminescence system (Amersham Biosciences) and subjected to quantitative densitometry using NIH ImageJ. To normalize the band intensity of CEL by actin, rabbit antiactin polyclonal antibody (1:1000; Sigma Chemical Co., St. Louis, MO, USA) and HRP-conjugated antirabbit IgG antibody (10,000×; Bio-Rad Laboratories) are used as the first and second antibodies, respectively.

The results show that CEL content is significantly increased in IRPC exposed to hypoxia–reoxygenation, but is not increased in untreated IRPTC. In particular, intracellular accumulation of CEL with a molecular weight of 48, 24–30, or 14 kDa is remarkable (Fig. 20.2).

3.4. ELISA

Some anti-AGE antibodies can be used for ELISA. ELISA kits for the detection of AGEs in serum, plasma, or cell culture supernatant are also commercially available, with different sensitivity among them. They include kits for the broad detection of AGEs, including CML, pentosidine, and other AGE structures (Cell Biolabs, Inc., San Diego, CA, with sensitivity more than 250 ng/ml; or Trans Genic, Inc., Kumamoto, Japan, with a sensitivity of more than 7.8 ng/ml). In addition, ELISA for the specific detection of CML is also available (MBL, Nagoya, Japan, with a sensitivity of more than 80 ng/ml; or Cell Biolabs, Inc., San Diego, CA, with sensitivity more than 3 ng/ml).

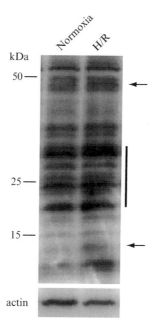

Figure 20.2 Western blot analysis for CEL detection in rat proximal tubular cells (IRPTC) with hypoxia–reoxygenation. The cells were exposed to hypoxia–reoxygenation (0.2% O_2 for 24 h followed by reoxygenation for 2 h) and subjected to Western blot analysis for the detection of CEL utilizing anti-CEL antibody (KH025), followed by densitometry. CEL content was significantly increased in IRPTC exposed to hypoxia–reoxygenation, but not in untreated IRPTC under normoxia. In particular, intracellular accumulation of CEL with a molecular weight of 48, 24–30, or 14 kDa was remarkable (relative increase in CEL/actin, 3.8-, 1.8-, and 4.2-fold, respectively). Representative bands, whose intensity was enhanced by hypoxia–reoxygenation, are indicated by a black bar and arrows. H/R, IRPTC expose to hypoxia–reoxygenation.

While easy to perform, the specificity of this system is lower than that of other systems, particularly when tissue samples are used, with samples often showing nonspecific or false-positive signals caused by a wide variety of unknown cross-reactive factors. Confirmation of ELISA data using the other methods above is advisable.

4. LINK BETWEEN ADVANCED GLYCATION AND ER STRESS

Advanced glycation occurs not only under hyperglycemic conditions but also in oxidative stress and hypoxia, mainly due to increased ROS generation. Functional abnormalities of glycated proteins perturb cellular homeostasis and increase mitochondrial ROS generation, accelerating the vicious cycle of oxidative stress and hypoxia and thereby enhancing subsequent glycation modification.

ER stress is predominantly induced by hypoxic or oxidative stress, and it is reasonable to speculate on the presence of a link between advanced glycation and ER stress on the disease development and progression caused by these disturbances. This chapter summarizes recent evidence for the presence of pathophysiological interaction between advanced glycation and the ER stress response.

4.1. Direct induction of ER stress by advanced glycation

It has been reported that AGEs induce ER stress directly. Among findings, glycated serum albumin (AGE-bovine serum albumin) induces ER stress, as estimated by GRP78 expression, and apoptosis in a dose- and time- dependent manner via an increase in intracellular Ca^{2+} concentration in mouse podocytes. ER stress inhibitor taurine-conjugated ursodeoxycholic acid (TUDCA), which acts as a chaperone that promotes the folding and trafficking of unfolded or malfolded proteins, prevents AGE-induced apoptosis, suggesting that this apoptosis is mediated by the apoptotic UPR pathway rather than by signaling through the receptor for AGE (Chen et al., 2008). Exposure of bovine aortic endothelial cells to clinically relevant concentrations of glycated LDL (low-density lipoprotein) also causes aberrant ER stress, as estimated by increased GRP78, PERK phosphorylation, or ATF6 activation, and this is followed by endothelial dysfunction, such as impaired endothelium-dependent vasorelaxation (Dong et al., 2010). These in vitro results were confirmed by in vivo studies utilizing isolated aorta from mice fed an atherogenic diet.

Like glycated albumin and LDL, AGE precursor also leads to ER stress-induced apoptosis. Extracellular matrix is frequently modified by AGE precursors in the skin of diabetic patients. Type I collagen modified by an AGE precursor such as methylglyoxal or 3-deoxyglucosone causes ER stress-mediated apoptosis via ROS-mediated CHOP activation in dermal fibroblasts, suggesting a pathophysiological role for the link between advanced glycation and ER stress in diabetic wounds (Loughlin and Artlett, 2010).

4.2. Comorbid induction of ER stress and advanced glycation

4.2.1. Hypoxia

The rapid induction of ER stress and subsequent UPR under hypoxia is well known. Hypoxic conditions and the associated loss of energy result in an imbalance between protein-folding capacity and protein-folding load, which in turn results in the accumulation of unfolded proteins in the ER lumen and ER stress.

The proapoptotic UPR is induced in ischemic or hypoxia–reperfusion damage in the heart, brain, and kidney, and the accumulation of AGEs is also observed in these lesions in parallel, suggesting that both advanced glycation and ER stress orchestrate the pathogenesis of hypoxic injury. Previous papers, including ours, have shown an intricate link between advanced glycation and ER stress by the amelioration of ischemia- or hypoxia–reperfusion-induced kidney damage. This kidney damages was ameliorated by inhibition of the UPR pathway by GRP78 overexpression, or by inhibition of advanced glycation by overexpression of glyoxalase I (GLO1), which detoxifies AGE precursors and subsequently inhibits advanced glycation (Kumagai et al., 2009; Prachasilchai et al., 2009).

In a wide range of solid tumors, malignant cells suffer from glucose starvation and hypoxia caused by poor vascularization, which brings about physiological ER stress due to the loss of energy for protein folding. The hypoxia-induced adaptive UPR is crucial to the survival of tumor cells (Koumenis, 2006). ER-resident chaperones, such as GRP78, for example, are induced to enhance protein-folding capacity in a wide range of human cancers, and expression levels correlate with tumor progression, metastasis, and drug resistance (Lee, 2007). Advanced glycation is also elevated and associated with increased metastasis and poorer outcomes in a wide variety of tumors (Logsdon et al., 2007), suggesting the orchestration of advanced glycation and ER stress responses in malignancy.

4.2.2. ROS (oxidative stress)

ROS produced by oxidative stress not only damages protein function by enhancing advanced glycation but also interferes with protein-folding capacity, including protein disulfide bonding, ultimately resulting in protein misfolding in the ER. Studies utilizing overexpression of antioxidative stress enzymes have emphasized the linkage of oxidative stress with advanced glycation and ER stress. For example, the proapoptotic UPR contributed to ischemic neuronal cell death and was markedly less pronounced in animals overexpressing copper/zinc superoxide dismutase (Cu/Zn-SOD), suggesting that ROS play a role in this pathological process (Hayashi et al., 2005). Further, cadmium caused the generation of ROS with subsequent induction of ER stress, estimated by GRP78/94 expression, activation of XBP1, ATF6, or CHOP, in a cultured renal proximal tubular cell line, which in turn led to ER stress-induced apoptosis; this cadmium-induced ER stress and apoptosis were significantly attenuated by transfection with manganese SOD (Yokouchi et al., 2008). Moreover, manganese SOD also attenuated advanced glycation (methylglyoxal-adduct formation) by reducing mitochondrial ROS generation in human aortic endothelial cells under hyperglycemic conditions (Yao and Brownlee, 2010). Thus, pathogenic ROS generation induced by any of the various disturbances is likely to be a key mediator of ER stress and advanced glycation, and SOD induces tolerance against pathogenic ER stress as well as prevents advanced glycation under oxidative stress.

4.2.3. Hyperglycemia

ER stress is induced not only by hypoxia, but also by starvation such as low glucose conditions: loss of energy due to metabolic disorder significantly alters protein-folding capacity, resulting in the accumulation of unfolded proteins in the ER. However, pathologically high glucose conditions also induce ER stress, which is associated with disease progression in diabetes mellitus and its complications, including diabetic nephropathy and retinopathy. From the fact that high glucose causes an increase in ROS production, followed by an increase in AGE precursor formation, and thereby significantly accelerates protein glycation modification, it is easy to speculate that both advanced glycation and ER stress response are initiated by high glucose-induced ROS and together orchestrate the manifestations associated with hyperglycemia. ER-resident proteins, which regulate protein-folding capacity, are also susceptible to oxidation by ROS, namely, glycation or lipoxidation, and exhibit the loss of physiological balance in UPR activity in cells under various stresses (Hayashi *et al.*, 2003; Naidoo, 2009).

4.2.4. Receptor for AGE, RAGE

RAGE is an important molecule in the pathogenesis of advanced glycation via the RAGE-AGE signal pathway. The link between advanced glycation and ER stress may also be mediated by RAGE. In patients with infantile neuronal ceroid lipofuscinosis, a devastating childhood neurodegenerative storage disorder, the neuronal apoptosis is mediated by ER stress. In addition, elevated ER stress stimulates expression in the brain of a Ca^{2+}-binding protein, S100B, and its receptor, RAGE, which results in proinflammatory cytokine production via RAGE signaling in astroglial cells (Saha *et al.*, 2008). These results strongly suggest that ER stress also enhances activation of RAGE-AGE signal pathway and exacerbates advanced glycation-related pathophysiology, for example, by increasing inflammation or ROS production.

4.2.5. Aging

Investigations on longevity have also emphasized the biological significance of interaction between advanced glycation and ER stress pathways. Aging is associated with the accumulation of AGEs, which subsequently contribute to the development of senescence and age-related diseases as well as the chronic complications of diabetes. The UPR molecule GRP78 is also dominantly glycated and enhances the alteration of UPR by aging (Naidoo, 2009).

Studies using *C. elegans* have further emphasized a role of ER stress and advanced glycation in aging. *C. elegans* deficient in HIF-1, a transcription factor for hypoxic inducible pathway, showed an extended life span in a UPR transducer IRE-1-dependent manner and was associated with lower levels of ER stress (Chen et al., 2009). *C. elegans* overexpressed human glyoxalase I homolog, which detoxifies AGE precursors and alters advanced glycation, and also contributes to the extension of the life span by reducing ROS production from mitochondria (Morcos et al., 2008). The appropriate balance between the stress signal pathway derived from ER stress and advanced glycation might be important in cell homeostasis and thereby contribute to longevity.

5. Effects of Advanced Glycation Inhibitor Against ER Stress

Accumulating evidence has emphasized the pathophysiological contribution of ER stress in various diseases, including conformational disease, metabolic disorders, ischemic disease, and oxidative stress-related disease. In parallel, chemical compounds that modulate UPR activation have opened new avenues for therapeutic approaches targeting ER stress in these diseases. Among these compounds, some advanced glycation inhibitors are also characterized as potent ER stress modulators. Some of the ER-resident proteins, including GRP78, GRP94, calreticulin, and PDI (protein disulfide isomerase), are highly glycated and subsequently lose their functions under oxidative stress, in association with an aberrant UPR response (Naidoo, 2009). These findings support the beneficial effects of advanced glycation inhibitors in maintaining ER homeostasis with a physiological UPR level. In this section, I summarize the antiglycation compounds that might suppress pathogenic UPR activation.

5.1. Antioxidants

Antioxidant compounds, acting as ROS scavengers, inhibit the advanced glycation modification mediated by ROS. How do antioxidants affect advanced glycation? ROS initiate the significant generation of advanced glycation precursors by glycoxidation. Antioxidants thus effectively inhibit advanced glycation by scavenging glycation initiators. Apart from this effect on advanced glycation, the ROS scavenging activity of antioxidant compounds also modulates the UPR response. Reagents in this category include catalase *N*-acetyl cysteine, curcumin, tempol, and TM2002, as described below (Table 20.1).

Table 20.1 Modalities targeting ER stress by inhibitors of advanced glycation

Inhibitor of advanced glycation	Site (cause of ER stress)	Consequences (s)	Reference
Antioxidant enzymes			
Catalase	Human endothelial cells (tunicamycin, HIV Tat)	GRP78↓, p-PERK↓	Wu et al. (2010)
Manganese SOD	Porcine renal proximal tubular cells (cadmium)	XBP1↓, ATF6↓, CHOP↓	Yokouchi et al. (2008)
Antioxidants			
N-Acetyl cysteine	Human renal tubular cells (uremic toxin indoxyl sulfate)	CHOP↓	Kawakami et al. (2010)
	Human endothelial cells (oxidized LDL)	IRE1↓, p-PERK↓, ATF6↓	Sanson et al. (2009)
Curcumin	Murine myoblast cells (hydrogen peroxide)	GRP94↑, Caspase 12↓	Pizzo et al. (2010)
	Bovine endothelial cells (lead)	GRP78↓, GRP94↓	Shinkai et al. (2010)
Compounds for ROS scavenging			
Tempol	Bovine endothelial cells (glycated LDL)	GRP78↓, p-PERK↓, ATF6↓	Dong et al. (2010)
TM2002	Rat brain (cerebral ischemia)	ORP150↓	Takizawa et al. (2007)
Antihypertensive drugs			
Olmesartan	Murine heart (aortic constriction)	CHOP↓	Okada et al. (2004)

To my knowledge, the effects of the representative antioxidants vitamins C and E on ER stress have not been reported. One study reported that hyperglycemia induced both ER stress and oxidative stress in human endothelial cells, and that while oxidative stress was reduced with α-tocopherol and ascorbic acid, ER stress was not (Sheikh-Ali et al., 2010). As oxidation of vitamin C also generates AGE precursor and accelerates the generation of carbohydrate and pentosidine, it may be paradoxical to attempt to ameliorate AGE formation using Vitamin C.

5.1.1. The antioxidant enzyme, catalase

Catalase, one of the major antioxidant enzymes, converts hydrogen peroxide into water and oxygen gas. The accumulation of AGEs in cardiomyocytes with aging observed in wild-type mice was significantly reduced in cardiac-specific catalase transgenic mice (Ren et al., 2007). In addition to inhibition of advanced glycation, ER-targeted catalase effectively blocks the UPR activation induced by the ER stress inducer tunicamycin or by HIV Tat in human endothelial cells. The molecular mechanism of this effect involves the scavenging of ER H_2O_2 generated by NOX4 (oxidase expressed on ER) (Wu et al., 2010).

5.1.2. N-acetyl cysteine

N-Acetyl cysteine is a precursor in the formation of the antioxidant glutathione, which inhibits ER stress-induced apoptosis. The uremic toxin indoxyl sulfate, which accumulates in the serum in patients with end-stage renal failure, induces ER stress-induced (CHOP-mediated) apoptosis in tubular cells. N-Acetyl cysteine remarkably suppresses this CHOP activation by indoxyl sulfate (Kawakami et al., 2010) and thereby maintains ER homeostasis and cell survival. N-Acetyl cysteine also prevents ER stress induced by lipoxidation products, including oxidized low-density lipoproteins (LDL), in human endothelial cells. The key modulators of the UPR (IRE1, PERK, and ATF6) and their downstream pathways are markedly activated by lipoxidation products (Sanson et al., 2009). Lipoxidation is somewhat similar to advanced glycation (glycoxidation): lipoxidation precursors are generated from lipids under oxidative stress and readily react with protein to form advanced lipoxidation endproducts (ALEs), which are often colocalized with AGEs in lesions in oxidative stress-related diseases. Note also, however, that the cytotoxic level of N-acetyl cysteine itself will induce proapoptotic UPR-induced cell death in human cancer cells (Guan et al., 2010).

5.1.3. Curcumin

Curcumin is a nontoxic polyphenol which acts as a scavenger of superoxide anion and hydroxyl radicals. It exerts antioxidant defense activity against ROS-induced myogenic cell damage via selective expression of the ER stress-inducible chaperon GRP94 (Pizzo et al., 2010). GRP94 induced by curcumin maintains calcium homeostasis and thereby protects the cell from apoptosis, suggesting that the antioxidant activity of curcumin acts via the UPR pathway. In addition, pretreatment with curcumin significantly abolishes lead-induced expression of GRP78 and GRP94 proteins in a dose-dependent manner in bovine aortic endothelial cells (Shinkai et al., 2010). The ER stress induced by lead is mediated by the JNK-AP-1 pathway, and curcumin inhibits UPR induction by inhibiting transcription factor AP-1.

5.1.4. Low molecular compounds for ROS scavenging

5.1.4.1. Tempol Tempol is a potent radical scavenger and superoxide dismutase mimetic drug, which is widely used as an antioxidant in both *in vivo* and *in vitro* studies. Tempol inhibits the glycation modification of LDL and subsequently blocks the glycated LDL-induced aberrant ER stress associated with endothelial dysfunction (Dong et al., 2010).

5.1.4.2. TM2002 TM2002 is an edaravone (Radicut) derivative. Edaravone is a potent-free radical scavenger, which quenches hydroxyl radical, and is used as a neuroprotective agent for neurological recovery following acute brain ischemia and subsequent cerebral infarction. It also has an antiadvanced glycation effect. The clinical use of this compound is tempered by its severe side effects. Based on the structure–function analysis of edaravone derivatives, 1-(5-hydroxy-3-methyl-1-phenyl-1H-pyrazol-4-yl)-6-methyl-1,3-dihydro-furo[3,4-c]pyridine-7-ol, referred to as TM2002, was synthesized (Izuhara et al., 2008). TM2002 is a nontoxic inhibitor of advanced glycation and oxidative stress which decreased the severity of infarction in permanent focal cerebral ischemia in association with a reduction in both ER stress-induced cell death and advanced glycation, as estimated by the accumulation of AGEs (Takizawa et al., 2007).

5.2. Antihypertensive drugs

The renin–angiotensin system (RAS) is a hormonal system that regulates blood pressure. Pathogenic activation of RAS enhances ROS generation and thereby induces oxidative stress. Angiotensin II type 1 receptor blockers (ARB) and angiotensin converting enzyme inhibitor (ACEi) therefore exhibit antioxidant effects via blocking of the RAS pathway. Of interest, in addition to their antihypertensive effect, they also serve as ROS scavengers to hydroxyl radicals and as inhibitor of transition metals-induced oxidation (Fenton reaction), which generates ROS (Izuhara et al., 2005). Not surprisingly, these oxidative stress-lowering effects significantly inhibit advanced glycation (Nangaku et al., 2003). On the other hand, ER stress is induced via an angiotensin II type 1 receptor-dependent pathway, although the molecular mechanisms of how ER stress is inducted by angiotensin II is still unclear. It is therefore reasonable to consider that the hypertrophy and heart failure induced by aortic constriction is associated with ER stress-induced cardiac myocyte apoptosis during disease progression (Okada *et al.*, 2004). Olmesartan (CS-866), an ARB, prevented upregulation of ER stress associated with proapoptotic UPR activation and thereby ameliorated the heart damage in this model.

6. Conclusion

The pathogenic UPR pathway is induced by AGEs or AGE precursors, and vice versa. This mutual induction suggests that ER stress and advanced glycation influence and interact with each other in the pathogenesis of various diseases. One key mediator of this crosstalk might be ROS, namely, oxidative stress, which is generated from ER or mitochondria. Evidence for this is consistent with the finding that some inhibitors of advanced glycation, which also have antioxidant effects, decrease UPR activation in parallel. It is reasonable to speculate that ER stress inhibitors might modulate the UPR together with advanced glycation and thereby synergistically mitigate disease progression.

Further investigation of the mechanism of this crosstalk will open new avenues in understanding the common pathophysiology of various diseases related to ER stress, advanced glycation, or oxidative stress, and provide exciting hints for the establishment of novel therapeutic strategies.

ACKNOWLEDGMENTS

This work was supported by Grants-in-Aid for Scientific Research from Japan Society for the Promotion of Science (22590880) and a grant from The Kidney Foundation, Japan (JKFB09-2). I am grateful to Masaomi Nangaku, MD, PhD, of Tokyo University School of Medicine, for critically reviewing the chapter.

REFERENCES

Biswas, S., Ray, M., Misra, S., *et al.* (1997). Selective inhibition of mitochondrial respiration and glycolysis in human leukaemic leucocytes by methyl-glyoxal. *Biochem. J.* **323**, 343–348.

Bucciarelli, L. G., Kaneko, M., Ananthakrishnan, R., *et al.* (2006). Receptor for advanced glycation end products: Key modulator of myocardial ischemic injury. *Circulation* **113**, 1226–1234.

Chan, W. H., and Wu, H. J. (2008). Methylglyoxal and high glucose co-treatment induces apoptosis or necrosis in human umbilical vein endothelial cells. *J. Cell. Biochem.* **103,** 1144–1157.

Chan, W. H., Wu, H. J., and Shiao, N. H. (2007). Apoptotic signaling in methylglyoxal-treated human osteoblasts involves oxidative stress, c-Jun N-terminal kinase, caspase-3, and p21-activated kinase 2. *J. Cell. Biochem.* **100,** 1056–1069.

Chen, Y., Liu, C. P., Xu, K. F., Mao, X. D., Lu, Y. B., Fang, L., Yang, J. W., and Liu, C. (2008). Effect of taurine-conjugated ursodeoxycholic acid on endoplasmic reticulum stress and apoptosis induced by advanced glycation end products in cultured mouse podocytes. *Am. J. Nephrol.* **28,** 1014–1022.

Chen, D., Thomas, E. L., and Kapahi, P. (2009). HIF-1 modulates dietary restriction-mediated lifespan extension via IRE-1 in Caenorhabditis elegans. *PLoS Genet.* **5,** e1000486.

Chiang, C. K., and Inagi, R. (2010). Glomerular diseases: genetic causes and future therapeutics. *Nat. Rev. Nephrol.* **6,** 539–554.

de Arriba, S. G., Stuchbury, G., Yarin, J., et al. (2007). Methylglyoxal impairs glucose metabolism and leads to energy depletion in neuronal cells–protection by carbonyl scavengers. *Neurobiol. Aging* **28,** 1044–1050.

Dong, Y., Zhang, M., Wang, S., Liang, B., Zhao, Z., Liu, C., Wu, M., Choi, H. C., Lyons, T. J., and Zou, M. H. (2010). Activation of AMP-activated protein kinase inhibits oxidized LDL-triggered endoplasmic reticulum stress in vivo. *Diabetes* **59,** 1386–1396.

Gao, L., and Mann, G. E. (2009). Vascular NAD(P)H oxidase activation in diabetes: A double-edged sword in redox signalling. *Cardiovasc. Res.* **82,** 9–20.

Godbout, J. P., Pesavento, J., Hartman, M. E., et al. (2002). Methylglyoxal enhances cisplatin-induced cytotoxicity by activating protein kinase C delta. *J. Biol. Chem.* **277,** 2554–2561.

Guan, D., Xu, Y., Yang, M., Wang, H., Wang, X., and Shen, Z. (2010). N-Acetyl cysteine and penicillamine induce apoptosis via the ER stress response-signaling pathway. *Mol. Carcinog.* **49,** 68–74.

Halder, J., Ray, M., and Ray, S. (1993). Inhibition of glycolysis and mitochondrial respiration of Ehrlich ascites carcinoma cells by methylglyoxal. *Int. J. Cancer* **54,** 443–449.

Hayashi, T., Saito, A., Okuno, S., Ferrand-Drake, M., Dodd, R. L., Nishi, T., Maier, C. M., Kinouchi, H., and Chan, P. H. (2003). Oxidative damage to the endoplasmic reticulum is implicated in ischemic neuronal cell death. *J. Cereb. Blood Flow Metab.* **23,** 1117–1128.

Hayashi, T., Saito, A., Okuno, S., Ferrand-Drake, M., Dodd, R. L., and Chan, P. H. (2005). Damage to the endoplasmic reticulum and activation of apoptotic machinery by oxidative stress in ischemic neurons. *J. Cereb. Blood Flow Metab.* **25,** 41–53.

Hotamisligil, G. S. (2010). Endoplasmic reticulum stress and the inflammatory basis of metabolic disease. *Cell* **140,** 900–917, (review).

Inagi, R. (2010). Endoplasmic reticulum stress as a progression factor for kidney injury. *Curr. Opin. Pharmacol.* **10,** 156–165, (review).

Inagi, R., Miyata, T., Ueda, Y., Yoshino, A., Nangaku, M., van Ypersele de Strihou, C., and Kurokawa, K. (2002). Efficient in vitro lowering of carbonyl stress by the glyoxalase system in conventional glucose peritoneal dialysis fluid. *Kidney Int.* **62,** 679–687.

Inagi, R., Yamamoto, Y., Nangaku, M., Usuda, N., Okamato, H., Kurokawa, K., van Ypersele de Strihou, C., Yamamoto, H., and Miyata, T. (2006). A severe diabetic nephropathy model with early development of nodule-like lesions induced by megsin overexpression in RAGE/iNOS transgenic mice. *Diabetes* **55,** 356–366.

Inagi, R., Kumagai, T., Fujita, T., and Nangaku, M. (2010). The role of glyoxalase system in renal hypoxia. *Adv. Exp. Med. Biol.* **662,** 49–55.

Izuhara, Y., Nangaku, M., Inagi, R., Tominaga, N., Aizawa, T., Kurokawa, K., van Ypersele de Strihou, C., and Miyata, T. (2005). Renoprotective properties of angiotensin receptor blockers beyond blood pressure lowering. *J. Am. Soc. Nephrol.* **16**, 3631–3641.

Izuhara, Y., Nangaku, M., Takizawa, S., Takahashi, S., Shao, J., Oishi, H., Kobayashi, H., van Ypersele de Strihou, C., and Miyata, T. (2008). A novel class of advanced glycation inhibitors ameliorates renal and cardiovascular damage in experimental rat models. *Nephrol .Dial. Transplant.* **23**, 497–509.

Kawakami, T., Inagi, R., Wada, T., Tanaka, T., Fujita, T., and Nangaku, M. (2010). Indoxyl sulfate inhibits proliferation of human proximal tubular cells via endoplasmic reticulum stress. *Am. J. Physiol. Renal. Physiol.* **299**, F568–F576.

Kim, I., Xu, W., and Reed, J. C. (2008). Cell death and endoplasmic reticulum stress: Disease relevance and therapeutic opportunities. *Nat. Rev. Drug Discov.* **7**, 1013–1030, (review).

Koumenis, C. (2006). ER stress, hypoxia tolerance and tumor progression. *Curr. Mol. Med.* **6**, 55–69.

Kumagai, T., Nangaku, M., Kojima, I., Nagai, R., Ingelfinger, J. R., Miyata, T., Fujita, T., and Inagi, R. (2009). Glyoxalase I overexpression ameliorates renal ischemia-reperfusion injury in rats. *Am. J. Physiol. Renal. Physiol.* **296**, F912–F921.

Kuznetsov, G., and Nigam, S. K. (1998). Folding of secretory and membrane proteins. *N. Engl. J. Med.* **339**, 1688–1695(review).

Lee, A. S. (2007). GRP78 induction in cancer: Therapeutic and prognostic implications. *Cancer Res.* **67**, 3496–3499.

Logsdon, C. D., Fuentes, M. K., Huang, E. H., and Arumugam, T. (2007). RAGE and RAGE ligands in cancer. *Curr. Mol. Med.* **7**, 777–789, (review).

Loughlin, D. T., and Artlett, C. M. (2010). Precursor of advanced glycation end products mediates ER-stress-induced caspase-3 activation of human dermal fibroblasts through NAD(P)H oxidase 4. *PLoS ONE* **5**, e11093.

Morcos, M., Du, X., Pfisterer, F., Hutter, H., Sayed, A. A., Thornalley, P., Ahmed, N., Baynes, J., Thorpe, S., Kukudov, G., Schlotterer, A., Bozorgmehr, F., et al. (2008). Glyoxalase-1 prevents mitochondrial protein modification and enhances lifespan in *Caenorhabditis elegans*. *Aging Cell* **7**, 260–269.

Nagai, R., Fujiwara, Y., Mera, K., Yamagata, K., Sakashita, N., and Takeya, M. (2008). Immunochemical detection of Nepsilon-(carboxyethyl)lysine using a specific antibody. *J. Immunol. Methods* **332**, 112–120.

Naidoo, N. (2009). The endoplasmic reticulum stress response and aging. *Rev. Neurosci.* **20**, 23–37, (review).

Nangaku, M., Miyata, T., Sada, T., Mizuno, M., Inagi, R., Ueda, Y., Ishikawa, N., Yuzawa, H., Koike, H., van Ypersele de Strihou, C., and Kurokawa, K. (2003). Anti-hypertensive agents inhibit in vivo the formation of advanced glycation end products and improve renal damage in a type 2 diabetic nephropathy rat model. *J. Am. Soc. Nephrol.* **14**, 1212–1222.

Okada, K., Minamino, T., Tsukamoto, Y., Liao, Y., Tsukamoto, O., Takashima, S., Hirata, A., Fujita, M., Nagamachi, Y., Nakatani, T., Yutani, C., Ozawa, K., et al. (2004). Prolonged endoplasmic reticulum stress in hypertrophic and failing heart after aortic constriction: Possible contribution of endoplasmic reticulum stress to cardiac myocyte apoptosis. *Circulation* **110**, 705–712.

Oya, T., Hattori, N., Mizuno, Y., et al. (1999). Methylglyoxal modification of protein. Chemical and immunochemical characterization of methylglyoxal-arginine adducts. *J. Biol. Chem.* **274**, 18492–18502.

Pizzo, P., Scapin, C., Vitadello, M., Florean, C., and Gorza, L. (2010). Grp94 acts as a mediator of curcumin-induced antioxidant defence in myogenic cells. *J. Cell. Mol. Med.* **14**, 970–981.

Prachasilchai, W., Sonoda, H., Yokota-Ikeda, N., Ito, K., Kudo, T., Imaizumi, K., and Ikeda, M. (2009). The protective effect of a newly developed molecular chaperone-inducer against mouse ischemic acute kidney injury. *J. Pharmacol. Sci.* **109,** 311–314.

Ren, J., Li, Q., Wu, S., Li, S. Y., and Babcock, S. A. (2007). Cardiac overexpression of antioxidant catalase attenuates aging-induced cardiomyocyte relaxation dysfunction. *Mech. Ageing Dev.* **128,** 276–285.

Saha, A., Kim, S. J., Zhang, Z., Lee, Y. C., Sarkar, C., Tsai, P. C., and Mukherjee, A. B. (2008). RAGE signaling contributes to neuroinflammation in infantile neuronal ceroid lipofuscinosis. *FEBS Lett.* **582,** 3823–3831.

Sanson, M., Auge, N., Vindis, C., Muller, C., Bando, Y., Thiers, J. C., Marachet, M. A., Zarkovic, K., Sawa, Y., Salvayre, R., and Negre-Salvayre, A. (2009). Oxidized low-density lipoproteins trigger endoplasmic reticulum stress in vascular cells: Prevention by oxygen-regulated protein 150 expression. *Circ. Res.* **104,** 328–336.

Sheikh-Ali, M., Sultan, S., Alamir, A. R., Haas, M. J., and Mooradian, A. D. (2010). Effects of antioxidants on glucose-induced oxidative stress and endoplasmic reticulum stress in endothelial cells. *Diabetes Res. Clin. Pract.* **87,** 161–166.

Shinkai, Y., Yamamoto, C., and Kaji, T. (2010). Lead induces the expression of endoplasmic reticulum chaperones GRP78 and GRP94 in vascular endothelial cells via the JNK-AP-1 pathway. *Toxicol. Sci.* **114,** 378–386.

Takizawa, S., Izuhara, Y., Kitao, Y., Hori, O., Ogawa, S., Morita, Y., Takagi, S., van Ypersele de Strihou, C., and Miyata, T. (2007). A novel inhibitor of advanced glycation and endoplasmic reticulum stress reduces infarct volume in rat focal cerebral ischemia. *Brain Res.* **1183,** 124–137.

Wu, R. F., Ma, Z., Liu, Z., and Terada, L. S. (2010). Nox4-derived H2O2 mediates endoplasmic reticulum signaling through local Ras activation. *Mol. Cell. Biol.* **30,** 3553–3568.

Yao, D., and Brownlee, M. (2010). Hyperglycemia-induced reactive oxygen species increase expression of the receptor for advanced glycation end products (RAGE) and RAGE ligands. *Diabetes* **59,** 249–255.

Yokouchi, M., Hiramatsu, N., Hayakawa, K., Okamura, M., Du, S., Kasai, A., Takano, Y., Shitamura, A., Shimada, T., Yao, J., and Kitamura, M. (2008). Involvement of selective reactive oxygen species upstream of proapoptotic branches of unfolded protein response. *J. Biol. Chem.* **283,** 4252–4260.

Author Index

A

Abdelrahim, M., 351
Abe, S., 167–168
Abu-Asab, M. S., 128–129, 132
Abuelo, D., 147
Aburaya, M., 41
Achsel, T., 149
Ackerman, E. J., 41
Adachi, Y., 281
Adam, A. P., 193
Adhami, V. M., 34
Aebersold, R., 224
Aebi, M., 164, 169, 171
Aguirre-Ghiso, J. A., 193
Ahmed, M. S., 128
Ahmed, N., 373
Airoldi, E. M., 241
Aizawa, T., 376
Akai, R., 189
Akar, U., 338
Akazawa, R., 194, 283
Akha, A. A., 267
Alamir, A. R., 375
Alberini, C. M., 311
Albermann, K., 250, 252
Aldea, M., 241
Ali, A. A., 261
Alliel, P. M., 149
Allison, J., 153
Altan-Bonnet, N., 345
Alt, F., 311
Altmann, F., 171
Amann, K., 262
Amar, L., 149
Ambler, M., 147
Ames, B. N., 75–76
Amitay, R., 311
Ananthakrishnan, R., 364
Anantharamaiah, G. M., 77
Anderson, K. C., 192, 201, 312
Anderson, T. J., 148–149, 156
Andre, B., 201
Andrews, B., 201
Antipov, E., 237–238
Antonellis, K. J., 332
Antonsson, B., 190
Appelt, K., 128
Arai, Y., 41

Arava, Y., 237
Arenzana, N., 189
Arkin, A. P., 201
Arnaud, D., 149
Arnold, S. M., 267
Arsena, R., 75
Artlett, C. M., 370
Arumugam, T., 371
Arvan, P., 237, 239
Ashburner, M., 250, 252
Asim, M., 34
Astromoff, A., 201
Atanasoski, S., 150
Auge, N., 374–375
Ausubel, F. M., 249
Aviles, R. J., 77
Aynardi, M. W., 240
Azar, H. A., 114

B

Babcock, S. A., 375
Backer, J. M., 42–43, 45, 47, 49, 51–52
Backer, M. V., 42–43, 45, 47, 49, 51–52
Backes, B. J., 192, 268
Back, S. H., 75, 77, 187, 192, 281, 288, 300
Badr, C. E., 114–116, 121–123
Bae, K. H., 338
Baena, S., 288
Bailey, S. L., 146
Bajirovic, V., 77
Balasubramanian, D., 77
Ball, C. A., 250, 252
Ballou, B., 114
Bal, W., 267
Bando, Y., 374–375
Barbour, L., 202
Barker, D. L., 276
Barlowe, C., 239
Bar-Nun, S., 311
Barone, M. V., 58, 100
Baron, W., 147
Barrans, S., 288
Barrie, J. A., 148, 300
Bartlett, J. D., 113–114, 122–123
Bartlett, P. F., 153
Bartoszewski, R., 5–7, 9–10, 12–15, 17–18, 218
Bartunkova, S., 152
Baryshnikova, A., 201, 213

Baselga, J., 41–42
Bassik, M. C., 190
Batchvarova, N., 145–146, 288
Bates, S. E., 41
Baudouin-Legros, M., 12
Baumeister, P., 41, 53
Baynes, J. W., 75, 80, 373
Bayyuk, S. I., 114
Bayyuk, W. B., 114
Beal, M. F., 75
Beazer-Barclay, Y. D., 332
Bebok, Z., 4–5, 10, 19
Beckman, K. B., 75
Beddoe, T., 41–42, 266
Bellay, J., 201, 213
Belli, G., 241
Bell, J. C., 281, 283, 288, 329
Belmont, P. J., 194
Belostotsky, D., 10
Benes, V., 281
Benig, V., 237–238
Bensalem, N., 224
Berdiev, B. K., 4
Berenbaum, M. C., 52
Berenbaum, P., 53
Berger, E. G., 171
Berger, P., 150
Bergt, C., 77
Bernales, S., 200, 294
Bernard, O., 153
Bernasconi, P., 190
Berndt, J., 147
Bertolotti, A., 92–93, 97, 146, 190, 302, 305
Best, C. J., 128–129, 132
Beutler, B., 288
Bhagabati, N., 249
Bhamidipati, A., 201
Biaglow, J. E., 267
Bigge, C., 179
Bigge, J. C., 171
Bihani, T., 41
Billings-Gagliardi, S., 149
Bi, M., 104
Bird, T. D., 147, 149
Bishop, A. C., 190
Bishop, J. R., 164
Biswas, S., 364
Biunno, I., 288
Blais, J. D., 104
Blake, J. A., 250, 252
Blakemore, W. F., 157
Blat, I. C., 329, 334
Blethrow, J., 190
Blobel, G., 311
Blount, B. C., 76
Boder, E. T., 241
Boeke, J. D., 241
Boer, V. M., 237

Bohle, R. M., 288
Boise, L. H., 149
Boison, D., 153
Boll, M., 288
Bonner-Weir, S., 329
Boone, C., 201
Boorstein, R., 288
Borgna, J. L., 154
Borojerdi, J., 128–129, 132
Botstein, D., 237, 241, 250, 252, 332
Bouchecareilh, M., 184, 343
Boyadjiev, S., 147
Boyce, M., 339, 357
Bozorgmehr, F., 373
Braakman, I., 218, 238
Brachmann, C. B., 241
Bradbury, N. A., 4
Braisted, J., 249
Brand, H., 42–43, 45, 47, 49, 51–52
Brandl, K., 288
Brandt, G. S., 190
Brasuñ, J., 267
Brauer, M. J., 241
Braun, T., 151
Breakefield, X. O., 114–116, 121–123
Bredesen, D. E., 128
Brennan, M. L., 77
Brenner, D. A., 288
Brenner, S., 346
Brent, R., 249
Brewer, J. W., 288, 311
Brodsky, J. L., 238–239
Brophy, P., 149
Brosius, F. C., 77
Brown, C., 241
Brownlee, M., 75, 371
Brown, M. S., 188, 194, 288, 295
Brown, P. O., 237, 241, 332
Bruce, J. A., 171, 179
Brunet, J. P., 329, 334
Brüning, A., 127–130, 133–135
Brunzell, J., 77
Bryant, K. F., 339
Bucciarelli, L. G., 364
Burda, P., 171
Burger, P., 128–129, 133–135
Burges, A., 128–129, 133–135
Burgess, J. A., 128
Buscemi, S., 75
Bustin, S. A., 255, 281
Butler, H., 250, 252
Butterfield, D. A., 75
Byun, J., 73, 75, 77, 80

C

Calfon, M., 94, 185, 281, 283, 288, 329, 351
Callahan, J., 190, 192

Campanale, K. M., 128
Campbell, R. E., 246
Cañamero, M., 288
Candiano, G., 222
Cantin, A. M., 5
Cao, X. Z., 59
Caputo, E., 241
Cardano, M., 288
Carmel-Harel, O., 241
Caruso, M. E., 343
Carvalho-Oliveira, I., 10
Carvalho, P., 206
Casadevall, A., 222
Casas, C., 300
Castro-Obregon, S., 60
Catlin, G. H., 241
Cavener, D. R., 144
Cecconi, F., 149
Cerasola, G., 75
Cerniglia, G. J., 128
Chait, A., 77
Chambon, P., 150
Chang, A., 237, 239
Chang, X. B., 4, 220
Chan, J., 147
Chan, P. H., 371–372
Chan, W. H., 364
Chao, G., 238
Chapman, R. E., 200
Charles, S. M., 171
Chatterjee, U., 284
Chaudhary, P. M., 58
Chefetz, I., 164
Chen, D., 373
Cheng, S. H., 4
Chen, L., 262
Chen, L. C., 222
Chen, T. C., 26–27, 30, 33, 128–129
Chen, X., 188, 193–194, 262, 288, 295, 305
Chen, Y. J., 346, 370
Cherry, J. M., 250, 252
Chevet, E., 184, 300, 343
Chiang, C. K., 362
Chirgadze, N. Y., 128
Choi, H. C., 370, 374, 376
Choo, S. J., 338
Cho, S. K., 304
Christen, S., 76
Chrong, D. C., 41
Chu, A. M., 201
Chu, G., 250–251, 332
Chung, P., 146, 190
Cidlowski, J. A., 67
Citterio, C., 218
Ciucanu, I., 178
Clare, J. J., 241
Clark, S. G., 185, 288
Clawson, D. K., 128

Cleland, W. W., 271
Clements, J. M., 241
Cobelli, C., 252
Coen, D. M., 339
Cohen, H. R., 128, 135, 187, 190, 192
Collawn, J. F., 4–5
Collin, F., 332
Collins, S. R., 77, 201, 213, 237
Collins, T. J., 276
Connelly, C., 201
Corcoran, L. M., 310
Costa, F., 41
Costanzo, M., 201, 213
Cost, G. J., 241
Cottone, S., 75
Cox, D. J., 261
Cox, J. S., 164, 190, 200–201, 203, 212, 294, 310
Crawford, E. D., 329, 334
Crivaro, K. E., 267
Crowley, J. R., 77
Crozat, A., 58
Currier, T., 249
Curthoys, N. P., 116

D

Dabney, A. R., 239
Dahl, M., 5
Dahms, N. M., 164
Dallner, G., 165
Dalton-Griffin, L., 300
Danda, R., 77, 80
Daneshmand, S., 39
Danish, M., 128–129, 132
D'Antonio, M., 145–146
Dautigny, A., 149
Dave, U. P., 188, 194, 288, 295
Davidson, A. R., 239
Davies, A., 241
Davies, J. F.2nd, 128
Davis, A. P., 250, 252
Davis, K. L., 157
Davis, R. J., 58
Davis, R. W., 201, 250–251
Dean, H. T., 114
de Arriba, S. G., 364
De Clercq, S., 152
De Craene, B., 152
Deegan, S., 135
Degenhardt, K., 59
de Kleine, R., 114, 121
Del Carro, U., 145–146
Deliolanis, N. C., 121
Denic, V., 201, 213, 237
Denmeade, S. R., 50
Denning, G. M., 220
Dennis, G. Jr., 332
Denoyelle, C., 353

Denzin, L. K., 238
Deplazes, P., 176, 178
Deprez, R. H., 255
D'Ercole, A. J., 157
Detorie, N. J., 80
De Vos, I., 152
Dhurjati, P. S., 238
Di, A., 4
Diaferia, G., 288
Diaz, S., 19
di Bello, I. C., 157
DiBenedetto, L. M., 149
Di Camillo, B., 252
Dietz, H. C., 4
Ding, H., 201, 213
Di Sano, F., 149
Dlouhy, S. R., 147
Dmitrovsky, E., 41
Dobberstein, B., 311
Dodd, R. L., 371–372
Doerrler, W. T., 304
Dolinski, K., 250, 252
Dong, B., 262
Dong, D., 40–41, 50, 189
Dong, H., 262
Dong, Y., 370, 374, 376
Doody, G., 288
Dorner, A. J., 288
Dougan, S. K., 311
Dow, S., 201
Doyle, F. J., III, 237–238, 241–242, 247–248
Doyle, M., 190
Drappatz, J., 35
Dressman, B. A., 128
Drew, D., 245
Drogat, B., 152, 346
Drori, A., 309, 312, 321–322
Dubeau, L., 40, 50
Dubois, M., 194, 281, 283, 329
Du, M. Q., 300
Duncan, I. D., 149
Dupuy, D., 351, 353
Du, S., 371
Du, X., 288, 373
Dwek, R. A., 171
Dwight, S. S., 250, 252
Dyer, D. G., 80

E

Eby, M., 58
Eddy, A. A., 77
Edgar, J. M., 148, 300
Edwards, C. A., 288
Edwards, J. S., 250–251
Edwards, R. M., 241
Egea, P. F., 190
Eggens, I., 165

Eikelenboom, P., 288
Eisen, M. B., 241, 332
Ellerby, H. M., 128
Ellgaard, L., 164, 294
Elmberger, P. G., 165
Elvove, E., 114
Eppig, J. T., 250, 252
Ermakova, S. P., 40
Esko, J. D., 164
Esumi, H., 338
Evenchik, B., 59
Everett, E. T., 113

F

Fagioli, C., 311
Fanarraga, M., 149
Fang, L., 370
Fang, S., 328
Farias, E. F., 193
Farmen, R. K., 245
Farnham, P. J., 15
Faulk, M. W., 149
Fay, D. S., 201
Febbraio, M., 77
Feifel, E., 116
Feldman, D., 357
Feldman, E. L., 77
Fels, D. R., 128, 135
Feltri, M. L., 145–146
Fernandes, A. R., 250, 253
Fernandez, J. L., 114, 116, 121
Fernandez, P. M., 39
Ferrand-Drake, M., 371–372
Ferraro, E., 149
Fields, R. D., 157
Finazzi, D., 311
Finer-Moore, J., 190
Fink, G. R., 202, 211
Fink, L., 288
Finley, G., 114
Fisher, J., 190
Fitzgerald, D. J., 50
Fitzgerald, K. D., 155
Fitzgerald, U., 135
Florean, C., 374, 376
Foran, D. R., 153
Fores, J., 300
Fox, P. L., 77
Fra, A. M., 311
Francisco, A. B., 288
Frank, C. G., 171
Franklin, R. J., 157
Fraser, P., 288
Freeze, H., 296, 303–304
French, B., 276
Fribley, A. M., 57, 59
Friedrich, V. L., 147, 152–153, 156

Friend, D., 311
Friese, K., 128–130, 133–135
Frischmeyer, P. A., 4
Fritz, J. E., 128
Fuentes, M. K., 371
Fujita, M., 374, 377
Fujita, T., 362, 371, 374–375
Fujiwara, Y., 366, 368
Fu, L., 4–5
Fung, E., 201
Furihata, K., 331
Furuno, A., 330, 332, 334–337, 339
Fu, X., 77

G

Gale, C. V., 300
Gao, G., 262
Gao, L., 362
Gao, N., 304
Gao, W., 332
Gao, X., 157
Gao, Y., 257, 262
Garbern, J. Y., 145–147, 149
Gardner, E. R., 128–129, 132
Gari, E., 241
Garrison, J. L., 312
Garson, J. A., 281
Gasch, A. P., 241
Gaut, J. P., 80
Geary, R. L., 77
Gehrig, P. M., 176
Gelmi, C. A., 250–251
Gembarska, A., 152
Gencic, S., 147, 149, 152
Genoud, S., 150
George-Hyslop, P. S., 288
Geraci, C., 75
Gething, M. J., 200, 237, 242, 254, 294
Geyer, R., 176
Ghosh, R., 40
Giaever, G., 201
Giepmans, B. N., 246
Gilbert, K., 42–43, 45, 47, 49, 51–52
Gildersleeve, R. D., 75, 77
Gillespie, P., 329
Gillig, T. A., 157
Gills, J. J., 128–129, 132
Gingelmaier, A., 128–130, 133–135
Glavic, A., 190
Glembotski, C. C., 194
Glimcher, L. H., 190, 192, 281, 283, 288, 294, 311–312, 314, 329
Gobe, G., 75
Godbout, J. P., 364
Goder, V., 206
Goldstein, J. L., 188, 194, 288, 295
Gomes-Alves, P., 5, 10, 19, 217–220

Gong, W., 328
Goodwin, P., 276
Goormastic, M., 77
Goossens, S., 152
Gore, P. R., 239
Gorg, A., 222
Gorman, A. M., 128
Gorza, L., 374, 376
Gottstein, B., 176
Goulding, P. N., 171
Gow, A., 143–149, 152–153, 156–157
Gragerov, A., 152–153, 156
Granada, B., 246
Grausenburger, R., 150
Gravallese, E. M., 311
Gray, N. S., 190
Greenblatt, J. F., 201
Greenlund, L. J., 252
Green, S. R., 275
Gresham, D., 241
Griendling, K. K., 75
Griffiths, I. R., 148–149, 156, 300
Groshen, S., 40, 50, 328
Grubenmann, C. E., 171
Grusby, M. J., 311
Gstraunthaler, G., 116
Guan, D., 375
Guarneri, M., 75
Guenet, J. L., 149
Guenther, A., 288
Gu, F., 300
Gu, H., 262
Güldener, U., 250, 252
Gunnison, K. M., 267
Gupta, A. K., 128
Gupta, S., 128, 135
Guptasarma, P., 77
Guthrie, C., 202, 211
Guzman, M. S., 300

H

Haas, M. J., 375
Hackel, B. J., 238
Haenebalcke, L., 152
Haeuptle, M. A., 165, 168
Hafeez, B. B., 34
Haga, N., 339
Hagen, A., 190, 192, 268
Hahn, S. M., 128
Haigh, K., 152
Haimovich, J., 311–312, 321–322
Halder, J., 364
Hamby, C. V., 42–43, 45, 47, 49, 51–52
Hammond, C., 112
Hanaoka, T., 338
Han, D., 190, 192, 268
Hani, J., 250, 252

Han, J., 75, 77
Han, M., 201
Hann, B. C., 206
Harada, A., 194, 281
Harding, H. P., 92–93, 97–98, 100, 145–146, 185, 187, 190, 193–194, 281, 283, 288, 294–295, 302, 304–305, 329, 339, 346
Hardy, M. M., 77
Harris, M. A., 250, 252
Harris, R. J., 164
Hart, E., 171
Hartman, M. E., 364
Harvey, D. J., 171
Hasegawa, K., 296
Hasnain, S. E., 284
Hassan, R., 50
Hatch, S. D., 128
Hattori, N., 364
Hattori, T., 283
Hayakawa, K., 113, 371
Hayakawa, Y., 330–331, 335–336
Hayashi, H., 283
Hayashi, T., 371–372
Hayes, J. M., 77
Haze, K., 93, 188–189, 194, 288, 295, 335, 339
Hazen, S. L., 77
Healy, S. J., 128
He, B., 184
Hecker, L., 75
Hegde, R. S., 310–312, 344–345
Heikkiä, R., 255–256
Heinecke, J. W., 75, 77, 80
Held, K. D., 267
Helenius, A., 112, 164, 169, 294
Hellemans, J., 281
Heller, J. I., 77, 80
Hendershot, L. M., 98, 100, 190, 193, 283, 288, 305
Henneke, I., 288
Hennet, T., 163, 165, 168–169, 171–172, 176, 178
Herrero, E., 241
Herschlag, D., 237
Herscovics, A., 296
Hetschko, H., 58
Hetmann, T., 93, 100
Hetz, C., 93, 96–97, 190
Hewett, J. W., 114–116, 121–123
Hibbs, M. L., 310
Hieronymus, H., 329, 334
Hieter, P., 241–242
Hill, D. P., 250, 252
Hines, V., 236–237, 241
Hinuma, S., 151
Hiramatsu, N., 113, 371
Hirano, Y., 339
Hirata, A., 239, 374, 377
Hitomi, H., 75

Hobbs, B., 332
Hodes, M. E., 147
Hoekstra, D., 147
Hoffman, C. S., 210
Hoffman, E. P., 149
Hofman, F. M., 26, 30, 33
Hofmann, S. L., 304
Ho, H., 146
Holcomb, T., 116
Hollander, M. C., 128–129, 132
Hollien, J., 187, 190, 192, 264, 300, 348
Holzman, C., 50
Honda, A., 288
Hong, E., 239
Hong, M., 6
Hoozemans, J. J., 288
Hopcia, K. L., 267
Hori, O., 374, 376
Horowitz, J. C., 75
Hosack, D. A., 332
Hoshino, T., 41
Hosokawa, N., 296, 305
Hotamisligil, G. S., 184, 362
Houstis, N., 75
Hovnanian, A., 27
Hsu, F. F., 80
Huang, D., 239
Huang, E. H., 371
Huang, T. H., 15
Hubbard, S. R., 185, 288
Hu, C. C., 311
Hudson, L. D., 147, 149, 152
Huggett, J., 281
Huggins, T. G., 80
Hülsmeier, A. J., 163, 165, 168–169, 171–172, 176, 178
Hung, J. H., 6
Hunt-Newbury, R., 351
Hurst, S., 147
Hurt, E. M., 311
Huttenhower, C., 241
Hutter, H., 373
Hwang, J. H., 338

I

Ibata, K., 246
Ichinose, H., 124
Iden, D. L., 148
Ido, Y., 77, 80
Iglesias, B. V., 193
Ihmels, J., 201
Iida, K., 98
Ikeda, M., 371
Imaizumi, K., 288, 371
Inagi, R., 361–362, 364–365, 371, 374–376
Ingelfinger, J. R., 371
Inoue, K., 147

Inouye, S., 122, 124
Ip, C., 41
Irizarry, R. A., 332
Irvine, K. D., 164
Isaacs, J. T., 50
Ischiropoulos, H., 77
Ishihara, T., 41
Ishikawa, N., 376
Ishiwata-Kimata, Y., 239
Ismailov, I. I., 4
Issid, M., 114
Ito, K., 371
Ito, T., 239
Iwakoshi, N. N., 96, 190, 192, 281, 283, 311–312, 314, 329
Iwata, S., 245
Iwawaki, T., 103, 189
Izuhara, Y., 374, 376

J

Jackowski, S., 311
Jackson-Lewis, V., 77, 80
Jaenisch, R., 151
Jagirdar, R., 75
Jain, A., 284
Jain, P., 250, 253
Jamora, C., 38–39
Jandhyala, D. M., 267, 300
Jandhyyala, D. M., 41
Jang, Y., 75
Jaques, A., 179
Jay, D., 75
Jeżowska-Bojczuk, M., 267
Jiang, H., 147, 149
Jiang, W., 145–147
Jiang, Y., 328
Jia, Z., 328
Johnston, R. F., 276
Jolicoeur, E. M., 294
Jones, E. W., 240
Jones, T., 75
Jonikas, M. C., 201, 213, 237
Jousse, C., 193, 339
Jung, M., 149
Jungreis, R., 288
Jungries, R., 146

K

Kahali, S., 53
Kaiser, C. A., 239
Kaji, T., 374, 376
Kalai, M., 149
Kalden, J. R., 262
Kaldor, S. W., 128
Kalish, V. J., 128
Kamholz, J. A:, 149
Kaneko, M., 364

Kang, S. W., 312
Kang, Y. J., 345
Kao, C. M., 241
Kapahi, P., 373
Kardosh, A., 27, 128–129
Karim, S. A., 148, 300
Karlen, Y., 255–256
Karthigasan, J., 149
Kasai, A., 113, 371
Kaser, A., 96
Katayama, K., 167–168
Katayama, T., 288
Kauffman, K. J., 238
Kaufman, R. J., 5, 9, 38, 75, 77, 94, 112, 144, 185, 187, 189, 192–194, 266–267, 275, 281, 283–284, 288, 294, 300, 304–305, 328–329, 339
Kawabata, S., 128–129, 132
Kawaguchi, Y., 53
Kawahara, T., 237, 242, 254
Kawakami, T., 374–375
Keenan, S., 201
Kellam, P., 300
Kelley, K. A., 152–153, 156
Kellogg, A., 77
Kelsen, S. G., 218
Kennedy, P., 149
Kennedy, R. T., 77
Kerbiriou, M., 219
Kerel, F., 178
Kerem, E., 4
Khan, N., 40
Kieran, D., 59
Kimata, Y., 239
Kim, D. E., 114, 116, 121
Kim, H., 245
Kim, I., 362
Kim, S. J., 312, 372
Kim, Y. M., 201, 213, 311
Kingston, R. E., 249
Kinouchi, H., 372
Kirkham, D., 300
Kirschner, D. A., 149
Kitamura, M., 113, 371
Kitao, Y., 374, 376
Kitteringham, N. R., 225
Kjeldsen, T. B., 237
Klapa, M., 249
Klemm, E. J., 311
Klugmann, M., 149, 156
Knauer, R., 171
Ko, B., 41
Kobayashi, H., 149, 376
Koch, M., 288
Koh, J. L., 201, 213
Köhler, P., 176, 178
Kohno, K., 189, 237, 239, 242, 254
Koike, H., 376

Koizumi, A., 58
Kojima, E., 93, 102
Kojima, I., 371
Kollmeyer, J., 77
Kolodny, M. R., 240
Komaki, H., 149
Komuro, R., 188, 194, 288, 295
Koong, A. C., 357
Korennykh, A. V., 190
Korfei, M., 288
Körner, C., 296, 303–304
Kornfeld, S., 164
Korostelev, A. A., 190
Korsmeyer, S. J., 190
Koumenis, C., 104, 128, 135, 371
Kozlowski, H., 267
Kozutsumi, Y., 92
Kraig, R. P., 146
Kramer, E. M., 148
Kredel, A., 288
Kreibich, G., 288
Kreitman, R. J., 41, 50
Kress, C., 190
Kretzler, M., 77
Krishnamoorthy, T., 284
Krivoshein, A., 42–43, 45, 47, 49, 51–52
Krogan, N. J., 201
Kubista, M., 281
Kubota, K., 113
Kubota, M., 246
Kudo, T., 288, 371
Kukudov, G., 373
Kultz, D., 5
Kumagai, T., 362, 371
Kunkler, P. E., 146
Kurashima, Y., 338
Kuroda, M., 58, 145–146, 288, 294
Kuroda, Y., 114
Kurokawa, K., 364–365, 376
Kuznetsov, G., 362
Kvaloy, J. T., 245

L

Ladiges, W. C., 93, 102
Lai, C., 147
Lamb, J., 329, 334
Lamkanfi, M., 149
Landauer, F., 116
Lander, E. S., 75
Lane, H. C., 332
Lane, P., 149
Lang, G., 288
Langridge, J. I., 171
Lau, W. L., 238
Lavail, M. M., 128, 135, 187, 190, 192
Law, D. T., 268
Lawson, B., 288

Laybutt, R., 329
Lazarow, P. B., 345
Lazzarini, R. A., 147
Lee, A. H., 33, 95, 98, 102, 104, 190, 192, 268, 288, 294, 311–312, 329, 351
Lee, A. S., 9, 39, 41, 46, 189, 328, 339, 371
Lee, D. H., 113
Lee, E., 39, 328
Lee, H. H., 76
Lee, H. K., 38–39
Lee, K., 94, 185, 187, 192, 281, 300
Leek, J. T., 239, 250–251
Leeuwenburgh, C., 77, 80
Lee, Y. C., 372
Lehle, L., 171
Lehrman, M. A., 294, 296–297, 299–300, 302–305
Lei, L., 164
Lei, S., 330
LeKniak, W., 267
Lempicki, R. A., 332
Lenhard, M., 128–129, 133–135
Lenoir, D., 147
Leone, D. P., 150
Lerner, A. G., 192, 268
Lerner, J., 329, 334
Levine, C. G., 311
Liang, B., 370, 374, 376
Liang, W., 249
Liao, Y., 374, 377
Li, B., 128
Li, E., 200
Li, G., 101
Lightfoot, R. T., 145–146, 288
Li, H., 128, 135, 144, 187, 190, 192–193, 242, 254, 264, 300
Li, J., 40, 50, 241, 249, 267
Li, L., 152
Lin, A., 41
Lin, J. H., 128, 135, 144, 183–184, 187, 190, 192–193, 264, 300
Lin, M. T., 75
Lin, T., 353
Lin, W., 105, 145–146
Lippow, S. M., 238
Lipson, K. L., 346
Li, Q., 328, 375
Li, S., 288
Lisbona, F., 190
Li, S. Y., 375
Little, E., 189
Litwak, K. N., 77
Liu, C., 329, 370, 374, 376
Liu, C. L., 237
Liu, C. P., 370
Liu, Q., 355
Liu, S., 5
Liu, W., 116

Liu, Y. J., 96
Liu, Z., 374–375
Livak, K. J., 255
Li, X.., 288
Lobel, P., 164
Lockhart, D. J., 200, 237, 239, 242, 250
Logsdon, C. D., 371
Lommel, M., 164
Long, K., 339
Long, Q., 288
Loo, T. W., 219
Lopez-Berestein, G., 338
Lopez-Guisa, J. M., 77
Lopiccolo, J., 128–129, 132
Lorenz, M., 77
Lorito, M. C., 75
Loughlin, D. T., 370
Louie, S. G., 26, 30
Lucau-Danila, A., 201
Lucchesi, J. C., 201
Luckhardt, T. R., 75
Lu, J., 338
Lundholt, B. K., 61
Luo, L., 152
Luo, S., 189
Luo, W., 239
Lu, P. D., 187, 193, 281, 283, 288, 329
Lustgarten, M. S., 75
Lu, Y. B., 370
Lyons, T. J., 370, 374, 376

M

Macchi, P., 148
Macklin, W. B., 150
Ma, D., 339
Maestre, L., 288
Maetens, M., 152
Maguire, C. A., 121
Mahavadi, P., 288
Maier, C. M., 372
Maier, D., 250, 252
Maity, A., 128
Malaguti, M., 145–146
Malhotra, J. D., 75, 77
Malik, A., 34
Mani, R., 201
Manis, J., 311
Mann, G. E., 362
Mannhaupt, G., 250, 252
Mann, M., 224
Manthiram, K., 121
Manz, R. A., 262
Mao, C., 40, 50, 103, 189
Mao, X. D., 370
Marachet, M. A., 374–375
Marciniak, S. J., 100–101, 193, 288
Marcinko, M., 194

Marcu, M. G., 190
Margolis, H. C., 114
Markart, P., 288
Markland, F., 41
Marks, H. G., 149
Martindale, J. L., 59
Martinez, F. J., 75
Martinez-Mier, E. A., 113
Martinon, F., 190, 288
Marza, E., 343
Maseda, D., 262
Masud, A., 60
Masuoka, H. C., 100
Matese, J. C., 241
Matheos, D. P., 190
Matsugo, S., 77
Matsui, T., 136, 185, 194, 281, 295, 300, 305, 335
Matsumoto, M., 58, 100
Matsushima, G. K., 157
Matsuura, G., 40
Matsuzawa, A., 59
Mattei, J. F., 149
Mattei, M. G., 149
Matthijs, G., 171
Mattu, T. S., 171
Ma, W., 200, 294
Ma, Y., 98, 100, 283
Mayatepek, E., 171
May, K. L., 41, 267, 300
Maytin, E. V., 100
Ma, Z., 374–375
Mazza, C., 255–256
McCracken, A. A., 239
McCulloch, M. C., 148–149, 156
McCullough, K. D., 59, 100
McDonald, K. L., 200
McDonald, T. O., 77
McEwen, E., 329
McGehee, A. M., 311
McKay, R. M., 351
McKay, S. J., 351
McLaughlin, M., 148, 300
McLean, L. L., 77
McNair, A., 255–256
Meeson, A., 157
Mehta, K., 338
Meister, S., 262
Memon, A. R., 345
Mendelsohn, J., 41
Mendes, F., 220
Mera, K., 366, 368
Mermod, N., 255–256
Merry, A. H., 171, 179
Messerle, M., 312, 321–322
Metzger, D., 150
Meunier, B., 228
Mewes, H. W., 250, 252
Miao, H., 75, 77

Michalak, M., 185
Midelfort, K. S., 238, 241
Mierzwa, A., 149
Mikoshiba, K., 153, 246
Mi, L. J., 288
Miller, C. N., 267
Miller, J.A.F.P., 153
Miller, S. D., 146
Milner, R. J., 147
Mima, S., 41
Minami, M., 58
Minamino, T., 374, 377
Minden, J., 222
Mira, N. P., 250, 253
Misaghi, S., 312, 321–322
Misra, S., 364
Mitra, D., 311
Miura, M., 189
Miyamichi, K., 152
Miyata, T., 364–365, 371, 374, 376
Miyawaki, A., 246
Mizuno, M., 376
Mizuno, Y., 364
Mizushima, T., 41
Mlynarz, P., 267
Modell, J. W., 329, 334
Moehle, C. M., 240
Moenner, M., 184
Mohapatra, A., 60
Mokrejs, M., 250, 252
Molinari, M., 164
Monsen, E., 239
Montague, P., 148–149, 300
Monteiro, P., 250, 253
Montes-Moreno, S., 288
Mooradian, A. D., 375
Moore, D. D., 249
Moore, G. J., 149
Moorman, A. F., 255
Morahan, G., 153
Morcos, M., 373
Morello, D., 149
Moreno, E. C., 114
Morgan, D. O., 190
Mori, C., 41
Mori, K. A., 136, 185, 188–189, 194, 200, 237, 242, 254, 267, 281, 283, 288, 294–295, 300, 305, 311, 335, 339
Morinaga, N., 40
Morishima, N., 59–60, 200
Morita, M., 328
Morita, Y., 374, 376
Morris, J. A., 288
Moser, K., 262
Moss, J., 40
Mostafavi, S., 201, 213
Mousavi-Shafaei, P., 128
Mueller, D. M., 77, 80

Mueller, R., 281
Mukherjee, A. B., 372
Mukhtar, H., 34
Mule, G., 75
Muller, C., 374–375
Muller, F. L., 75
Müller, R., 171
Munroe, R. J., 288
Münsterkötter, M., 250, 252
Murphy, N., 171
Muzumdar, M. D., 152
Mylonas, I., 128, 130

N

Nadanaka, S., 7, 339
Nadon, N. L., 149, 156
Naem, S., 176, 178
Naessens, M., 152
Nagai, H., 59
Nagai, R., 366, 368, 371
Nagai, T., 246
Nagamachi, Y., 374, 377
Nagata, K., 288, 296, 305
Naidoo, N., 372–373
Nair, S. K., 252
Nakagawa, T., 200
Nakanishi, K., 59–60
Nakatani, T., 374, 377
Nakata, S., 328
Nakatsukasa, K., 239
Namba, T., 41
Nanba, E., 149
Nance, M. A., 147
Nangaku, M., 362, 364–365, 371, 374–376
Nardi, E., 75
Naren, A. P., 4
Navarro, X., 300
Nave, K.-A., 147–149, 156
Neckers, L., 190
Negishi, M., 188, 194, 283, 339
Negre-Salvayre, A., 374–375
Neubert, K., 262
Newstead, S., 245
Ng, D. T., 199–203, 205, 207, 245
Nguyen, L., 113
Nichols, P., 328
Niers, J. M., 121
Nigam, S. K., 362
Nijholt, D. A., 288
Ni, L., 201
Ni, M., 40, 50
Nishi, T., 372
Nishitoh, H., 59
Nislow, C., 201
Noda, M., 40
Noguchi, T., 59
Nolan, T., 255, 281

Nomura, F., 40
Nonaka, N., 176, 178
Norden, A. D., 35
Nordestgaard, B. G., 5
Nordgård, O., 255–256
Normington, K., 237, 242, 254
Novoa, I., 93, 98, 102, 193, 281, 283, 288, 329
Nucho, C. K., 114
Nukuna, B., 77
Nyabi, O., 152

O

Oakes, S. A., 190, 192, 268
Obeng, E. A., 149
O'Brien, A. D., 41
O'Brien, K., 77
O'Byrne, P. M., 5
Ogata, J., 147
Ogawa, N., 237, 242, 254
Ogawa, S., 374, 376
Ogiso, Y., 330
Ogunnaike, B. A., 250–251
Oh, E., 201, 213, 237
Oh-ishi, S., 77
Oh, J., 201
Ohoka, N., 59, 283
Oikawa, D., 239
Oishi, H., 376
Okada, K., 374, 377
Okada, T., 136, 185, 188, 194, 266, 281, 295, 300, 335, 339, 357
Okajima, T., 164
Okamato, H., 364
Okamura, D. M., 77
Okamura, M., 371
Okano, H., 153
Okuno, S., 371–372
Omura, S., 330
Ono, K., 328
Onozaki, K., 283
Oracki, S. A., 310
Oral, E., 77
Oram, J. F., 77
Ord, D., 59
Ord, T., 59
Otipoby, K. L., 311
Otte, S., 239
Oyadomari, M., 193–194
Oyadomari, S., 58, 102, 104, 193–194, 288
Oyama, T., 328
Oya, T., 364
Ozawa, K., 374, 377
Ozpolat, B., 338

P

Paesold-Burda, P., 169, 172
Palmer, A. E., 246

Panning, B., 128, 135, 187, 190, 192
Papa, F. R., 190, 192, 268, 294
Parekh, R. B., 171, 179
Park, B., 311
Park, E. S., 246
Park, F. J., 240
Park, H. R., 330–332, 334–338
Park, J. L., 77
Park, J. W., 67
Park, Y. M., 41, 338
Pasichnyk, K., 77
Pastan, I., 50
Pat, B., 75
Patel, K. G., 121
Patel, T. P., 171, 179
Patil, C. K., 200, 237, 239, 242, 250, 254
Paton, A. W., 40–43, 45, 47, 49, 51–52, 261, 266–267, 300
Paton, J. C., 40–43, 45, 47, 49, 51–52, 261, 266–267, 300
Patterson, J. B., 41
Paules, R., 281, 283, 288, 329
Pearce, D. A., 304
Pearsall, G. B., 149
Peck, D., 329, 334
Pelish, H. E., 357
Penas, C., 300
Pen, L., 40, 50
Pennathur, S., 73, 75, 77, 80, 283
Penn, M. S., 77
Pennuto, M., 101, 145–146
Penque, D., 217, 219–223
Penttila, M., 250–251
Percy, C., 75
Pergola, P., 77, 80
Perseguers, S., 255–256
Pesavento, J., 364
Petasis, N. A., 26, 30
Peterson, A. C., 153
Petroulakis, E., 311
Petsch, D. T., 190
Pfaffl, M. W., 281
Pfaller, W., 116
Pfisterer, F., 373
Pham-Dinh, D., 149
Philips, M. R., 345
Piacentini, M., 149
Piatesi, A., 237–238
Pickett, S. C., 276
Picotti, P., 225
Piddlesden, S., 157
Pike, L., 114, 121
Pingoud, A., 275
Pipe, S. W., 75, 77
Piris, M. A., 288
Piston, D. W., 246
Pizzonia, J., 42–43, 45, 47, 49, 51–52
Pizzo, P., 374, 376

Plesken, H., 146
Ploegh, H. L., 311, 314
Pluckthun, A., 236–237, 241, 248
Pluquet, O., 184, 300
Pollock, R., 193
Pootrakul, L., 39
Pop-Busui, R., 77
Popko, B., 105, 145–146, 157, 281, 283, 288, 329
Poronnik, P., 75
Postma, D. S., 5
Potala, S., 50
Powers, E. T., 218
Prachasilchai, W., 371
Pratt, V. M., 147
Price, M. J., 241
Pridgen, E. M., 238
Prinz, J., 201, 213
Proctor, M., 201
Prywes, R., 185, 188–189, 193–194, 288, 295, 305
Przedborski, S., 77, 80
Puckett, C., 147
Puhlhofer, A., 156
Punna, T., 201
Pyrko, P., 27, 33, 39–40, 128–129

Q

Qian, S. B., 311
Qi, L., 288
Quackenbush, J., 251
Quan, E. M., 201, 213, 237
Quant, E., 35
Quatela, S. E., 345
Quattrini, A., 145–146
Quinton, P. M., 4

R

Rab, A., 5–7, 9–10, 12–13, 18–19
Rabinovich, E., 311
Raden, D., 237, 241–242, 247–248, 267
Raffel, D., 77
Rahmeh, M., 128–129, 133–135
Raines, R. T., 236–237, 241, 248
Rajewsky, K., 311
Rakestraw, A., 237–238
Rakestraw, J. A., 237–238
Ramaiah, K. V., 284
Ramakers, C., 255
Ramos-Castaneda, J., 239
Ramos, R., 288
Rane, N. S., 312
Ranganathan, A. C., 193
Rao, R. V., 53, 60, 128
Rapoport, T. A., 206
Raskind, W. H., 147
Rasmussen, J. E., 80
Rathke-Hartlieb, S., 148

Rathnayaka, T., 114
Rawson, R. B., 188, 194, 288, 295
Ray, J. S., 42–43, 45, 47, 49, 51–52
Ray, M., 364
Ray, S., 364
Reed, J. C., 190, 362
Reichert, W. L., 276
Reich, M., 329, 334
Reimold, A. M., 93, 95–96, 311
Remotti, H., 145–146, 288
Ren, J., 375
Rennolds, J., 220
Ren, X. D., 355
Reynolds, R., 157
Ribick, M., 75, 77
Riccobene, R., 75
Richardson, A., 75
Riles, L., 201
Rincon, M., 58
Riordan, J. R., 4
Rivera, V. M., 193
Rizzo, M. A., 246
Robinson, A. S., 235–238, 241–242, 247–248
Rochefort, H., 154
Rodriguez, B. A., 15
Rojas-Rivera, D., 190
Romanos, M. A., 241
Romero-Ramirez, L., 39, 104
Roncador, G., 288
Ron, D., 75, 77, 144–146, 184–185, 187, 190, 193–194, 276, 281, 283, 288, 294–295, 302, 304–305, 328–329, 339, 344–345, 348
Rose, M. D., 190
Rosenbluth, J., 149, 156
Rosen, E. D., 75
Rosenstein, R., 241
Rosenwald, A., 311
Rossjohn, J., 41–42, 266
Ross, K. N., 329, 334
Rothblatt, J., 244
Roth, F. P., 201
Rowe, S. M., 5
Roxo-Rosa, M., 220, 222
Royle, L., 171
Rozemuller, A. J., 288
Rozmahel, R., 288
Rudd, P. M., 171
Rudnicki, M. A., 151
Ruepp, A., 250, 252
Ruijter, J. M., 255
Ruppert, C., 288
Rusch, V., 41
Rutkowski, D. T., 103–104, 144, 194, 267, 281, 283, 288, 310, 329, 344–345
Rutledge, R., 255–256
Rutschmann, S., 288
Ryoo, I. J., 338

Author Index

S

Sabatini, D. D., 146
Sack, R., 176
Sada, T., 376
Sadri, N., 281, 283, 288, 329
Saeed, A. I., 249
Saha, A., 372
Saha, J., 77
Sahara, Y., 122
Sahoo, S. K., 50
Saito, A., 371–372
Saito, I., 77
Saito, S., 327, 330, 332, 334–337, 339
Sakaki, K., 288
Sakamoto, A., 330, 332, 334–337
Sakashita, N., 366, 368
Sakurai, J., 330, 332, 334–337, 339
Saleh, M., 60
Saloheimo, M., 250–251, 345, 357
Salvayre, R., 374–375
Samali, A., 128, 135
Sambrook, J. F., 200, 237, 242, 254, 294
Sanjay, A., 288
Sanson, M., 374–375
Sant, A., 237, 242, 254
Sarfaraz, S., 34
Sarkar, C., 372
Sasaki, M., 149
Sato, M., 194, 281
Sato, R., 339
Sato, S., 330, 335–336
Sato, T., 167–168, 194, 281
Saunders, T., 194, 281, 283, 288, 329
Sawa, Y., 374–375
Saxena, S. K., 41
Sayed, A. A., 373
Sazinsky, S. L., 237–238
Scaltriti, M., 41–42
Scapin, C., 374, 376
Schapira, M., 281, 283, 288
Schardt, J. A., 39
Schekman, R., 244
Scheper, W., 288
Scherf, U., 332
Scheuner, D., 75, 77, 93, 99, 193, 283, 329, 339
Schiff, R., 149, 156
Schimenti, J. C., 288
Schin, M., 77
Schinzel, A., 190
Schlotterer, A., 373
Schmid, V., 201, 213, 237
Schmitt, D., 77
Schmittgen, T. D., 255
Schnabl, B., 288
Schneider, A., 149, 156
Schollen, E., 171
Schönthal, A. H., 26–27, 30, 33, 128–129

Schröder, M., 5, 9, 38, 112, 187, 261–262, 264–265, 294, 300, 328–329
Schuksz, M., 164
Schuldiner, M., 201, 213, 237
Schultz, P. G., 190
Schulz, J. B., 148
Schwab, M. H., 149, 156
Schwappach, B., 201, 213, 237
Schwartzman, R. A., 67
Schwarze, S. R., 192
Schwiebert, E. M., 4
Scriven, P., 39
Scudder, K. M., 61
Seeger, W., 288
Segall, J., 268
Seidah, N. G., 339
Seidman, J. G., 249
Semler, B. L., 155
Sena-Esteves, M., 121
Sequeira, S. J., 193
Sereda, M., 149
Settle, M., 77
Sevier, C. S., 201, 213
Shachar, I., 311
Shaffer, A. L., 92, 311
Shamu, C. E., 94, 164, 190, 201, 203, 212, 294, 310
Shaner, N. C., 246
Shang, J., 293, 296–297, 299–300, 302–305
Shao, J., 376
Shapiro-Shelef, M., 311
Sharma, A., 311
Sharma, M., 220
Sharma, R., 113–114, 122–123, 147, 149, 157
Sharov, V., 249
Shayman, J. A., 77
Sheikh-Ali, M., 375
Shen, J., 93, 188–189, 193, 305
Shen, X., 185, 288
Shen, Z., 375
Sheppard, D. N., 10
Sherman, B. T., 332
Sherman, F., 269, 291
Shetty, B. V., 128
Shiao, N. H., 364
Shigenaga, M. K., 76
Shigeno, E. T., 76
Shimada, T., 371
Shimazu, T., 59
Shimizu, E., 190
Shinkai, Y., 374, 376
Shin-Ya, K., 330–332, 334–338
Shipley, G. L., 281
Shishehbor, M. H., 77
Shitamura, A., 371
Shoemaker, R. H., 128–129, 132
Shokat, K. M., 128, 135, 187, 190, 192
Shuda, M., 104

Shui, G., 311
Shu, L., 77
Shusta, E. V., 236–239, 241, 248
Shutes, A., 357
Shy, M. E., 149
Siddiqui, I. A., 34
Sidrauski, C., 190, 200
Sikorski, R. S., 241–242
Simon, D., 149
Simons, M., 148
Singh, O. V., 219
Singh, P., 77
Singh, R., 288
Sistermans, E. A., 149
Sitia, R., 311
Skobe, Z., 113–114, 122–123
Smith, C. E., 114
Smith, C. L., 311
Smith, J. A., 249
Smith, J. D., 77
Snead, M. L., 113
Snuffin, M., 249
Sohya, S., 114
Sollott, S. J., 59
Song, B., 75, 77, 101, 194, 281, 283, 329
Sonoda, H., 371
Southwood, C. M., 101, 145–147, 152–153, 156
Spear, E. D., 200–203, 205, 207, 213, 245
Speed, T. P., 332
Spellman, M. W., 164
Spellman, P. T., 241, 332
Spicer, D., 328
Spitaler, N., 116
Sprecher, D. L., 77
Sprecher, E., 164
Springer, G. H., 246
Sriburi, R., 92, 311
Srimanote, P., 40
Srivastava, R. K., 59
Stagljar, I., 171
Staton, M. W., 80
Stecca, B., 152–153, 156
Steen, H., 224
Steinbach, P. A., 246
Steinbrecher, U. P., 77
Stein, J. E., 276
Stevens, J. L., 145–146, 288
Stevens, M. J., 77
Stevens, N., 190, 192, 264, 300
Stewart, D. G., 157, 255–256
Stiles, C., 41
Stoffel, W., 153, 156
Stojdl, D. F., 281, 283, 288, 329
St Onge, R. P., 201, 213
Storey, J. D., 237, 239, 250–251
Storz, G., 241
Strahl, S., 164

Stroud, R. M., 190
Strudwick, N., 261
Struhl, K., 249
Stuchbury, G., 364
Sturn, A., 249
Subramanian, A., 329, 334
Sudhakar, A., 284
Sugio, K., 328
Sullivan, K. A., 77
Sultan, S., 375
Sun, M., 77
Sutcliffe, G., 147
Suter, U., 150
Suzuki, T., 124, 167–168, 239
Swartz, J. R., 121
Swathirajan, J., 194, 281, 283, 329
Swenson, S., 41
Sylvester, F. C., 267
Szent-Gyorgyi, C., 114
Szomolanyi-Tsuda, E., 311

T

Takagi, S., 374, 376
Takahashi, S., 376
Takano, Y., 371
Takashima, S., 374, 377
Takeda, M., 288
Takeshita, K., 149
Takeuchi, M., 239, 345
Takeya, M., 366, 368
Takizawa, S., 374, 376
Talbot, U. M., 40–42, 266
Tanaka, K. I., 41
Tanaka, T., 374–375
Tan, B. K., 311
Tannous, B. A., 113–116, 121–123
Tarlinton, D. M., 310
Tasic, B., 152
Tatin, F., 346
Taunton, J., 312
Tavakkol, A., 276
Tawa, M., 114
Taylor, S., 147
te Heesen, S., 171
Teixeira, M. C., 250, 253
Ten Hagen, K. G., 164, 171
Tenreiro, S., 250, 253
Terada, L. S., 374–375
Tetko, I., 250, 252
Thannickal, V. J., 75
Thiagarajan, M., 249
Thielen, P., 190
Thiers, J. C., 374–375
Thomas, E. L., 373
Thompson, N. J., 201
Thomson, C., 149, 156

Thomson, L., 77
Thornalley, P., 373
Thorp, E., 101
Thorpe, C. M., 41–42, 266–267, 300
Thorpe, J. A., 192
Thorpe, S. R., 75, 80, 373
Thuerauf, D. J., 194
Tian, E., 164
Tibshirani, R., 250–251, 332
Tiemeyer, M., 171
Till, J. H., 185, 288
Tinelli, E., 145–146
Tirasophon, W., 94, 185, 189, 275
Tirosh, B., 96, 309, 311–312, 314, 321–322
Tjon-Kon-Fat, L. A., 121
Todd, D. J., 190, 294
Toffolo, G., 252
Tohyama, M., 288
Tomida, A., 330, 332, 334–337, 339
Tominaga, N., 376
Tompkins, R. G., 250–251
Tooze, R., 288
Toufighi, K., 201, 213
Townes, T. M., 100
Tran, H. D., 80
Travers, K. J., 92, 200, 237, 239, 242, 250
Tremblay, L. O., 296
Trepanier, A., 147
Trotter, J., 148
Troyanskaya, O. G., 237
Tsai, P. C., 372
Tsang, K. Y., 344–345
Tsantoulas, C., 300
Tsao, E. H., 300
Tseng, J., 147, 149
Tsien, J. Z., 190
Tsien, R. Y., 246
Tsuchiya, M., 113–114, 122–123
Tsukamoto, O., 374, 377
Tsukamoto, Y., 374, 377
Tsukumo, Y., 330, 335–336, 339
Tsuruo, T., 330, 332, 334–337, 339
Tsurutani, J., 128–129, 132
Tsutsumi, S., 41
Tufi, R., 149
Turnley, A. M., 153
Tusher, V. G., 250–251, 332

U

Ubersax, J. A., 190
Ueda, Y., 365, 376
Umareddy, I., 300
Upton, J. P., 190, 192, 268
Uramoto, H., 328
Urano, F., 59, 113, 185, 190, 288, 351
Usuda, N., 364
Usuda, S., 124

V

Vaillancourt, J. P., 60
Valkonen, M., 250–251
Vallabhajosyula, P., 311
van Anken, E., 218, 238
Vandenabeele, P., 149
van der Deen, M., 5
van der Knaap, M. S., 149
Vandesompele, J., 281
Vande Walle, L., 192, 268
van Haastert, E. S., 288
van Huizen, R., 98
Vani, A. K., 288
Van Remmen, H., 75
van Ypersele de Strihou, C., 364–365, 374, 376
Varanasi, U., 276
Varga, K., 10, 19–20
Varki, A., 19
Vasudevan, S. G., 300
Vattem, K. M., 187
Verdu, E., 300
Verma, R. S., 50
Veronneau, S., 201
Vincent, A. M., 77
Vindis, C., 374–375
Vink, E., 192, 300
Virrey, J. J., 40, 50
Vitadello, M., 374, 376
Vittal, R., 75
Vivekanandan-Giri, A., 73, 75, 77
Vogel, M., 128–129, 133–135
Voll, R. E., 262
von Heijne, G., 245
Voss, E. W., 238
Voss, V., 58

W

Wada, I., 296
Wada, T., 374–375
Wagner, J. D., 77
Wagner, P., 77
Walker, J. A., 310
Walla, M. D., 80
Walter, P., 94, 128, 135, 144, 164, 184, 187, 190, 192–193, 200–203, 205–207, 212–213, 237, 239, 242, 245, 250, 254, 264, 276, 294, 300, 310, 328–329, 344–345, 348
Wang, H. G., 40, 59, 375
Wang, J. H., 75, 77, 194, 281, 283, 329
Wang, L., 328
Wang, M., 40, 50
Wang, Q. Y., 300
Wang, S., 370, 374, 376
Wang, W., 27, 128–129
Wang, X. Z., 58–59, 94, 100, 145–146, 190, 288, 294, 375

Wang, Y., 189, 237
Ward, C. L., 4
Warfel, N. A., 128–129, 132
Watanabe, M., 40
Watson, N., 311
Weaver, T. E., 288
Webb, B. L., 300
Weibezahn, J., 201, 213, 237
Wei, D., 328
Wei, J., 98
Weinmann, A. S., 15
Weisel, F., 262
Weissleder, R., 114, 116, 121
Weissman, J. S., 187, 190, 192, 200–201, 213, 237, 239, 242, 250, 264, 300, 348
Wek, R. C., 144, 187, 281, 283, 288
Welihinda, A. A., 94, 275
Welsh, J. P., 121
Wenk, M. R., 311
Wen, P. Y., 35
Wentz, A. E., 238
Wenzhong, X., 250–251
Werner, H., 149
Whisstock, J. C., 41–42, 266
Whitehouse, A., 300
White, J., 249
Wiethe, C., 262
Wiggin, T. D., 77
Wilce, M. C., 41–42, 266
Wilds, I. B., 149
Wilhelm, J., 275
Williams, C. A., 147
Williamson, J. R., 77, 80
Wilson, S. J., 300
Winkler, T. H., 262
Wiseman, R. L., 357
Wittrup, K. D., 236–238, 241, 248
Wittwer, C. T., 281
Wodicka, L., 200, 237, 239, 242, 250
Wolf, M. K., 149
Wolfson, A., 75, 77
Wolfson, J. J., 41, 267, 300
Wood, J. L., 190
Woods, I., 59
Woo, K. J., 67
Wormald, M., 179
Wrabetz, L., 144–146, 148
Wrobel, M. J., 329, 334
Wu, H. J., 364
Wu, J., 93–95, 194, 267, 281, 283, 288, 329
Wu, M., 370, 374, 376
Wurdinger, T., 114, 121
Wu, R. F., 344–345, 374–375
Wu, S., 375
Wynn, T. A., 75

X

Xiao, N., 288
Xiao, W., 202
Xie, K., 328
Xiong, S., 40, 50
Xiong, W., 27, 128–129
Xu, A., 164
Xu, J., 200
Xu, K. F., 370
Xu, P., 237, 241–242, 247–248
Xu, W., 192, 268, 362
Xu, Y., 375

Y

Yaari-Stark, S., 345, 353
Yahiro, K., 40
Yamada, K., 167–168
Yamagata, K., 366, 368
Yamaguchi, H., 59
Yamamoto, A., 136, 185, 295, 300, 335
Yamamoto, C., 374, 376
Yamamoto, H., 364
Yamamoto, K., 93–95, 103, 194, 266, 281, 346
Yamamoto, T., 149
Yamamoto, Y., 364
Yamori, T., 330, 335–336
Yanagi, H., 188–189, 194, 237, 242, 254, 288, 295, 335
Yang, J. W., 332, 370
Yang, L., 288, 311
Yang, M., 375
Yankner, B. A., 200
Yan, W., 98, 102
Yao, D., 371
Yao, J., 113, 328, 371
Yarden, Y., 41
Yarin, J., 364
Yasumoto, K., 328
Yasumura, D., 128, 135, 187, 190, 192
Yau, G.D.-Y., 194, 281, 283, 329
Ye, H., 300
Ye, J., 93, 188, 194, 288, 295
Yen, T. S., 184
Ye, P., 157
Ye, R., 40, 50
Ye, W., 40, 50
Yip, H., 76
Yohda, M., 114
Yokota-Ikeda, N., 371
Yokouchi, M., 371
Yoneda, T., 288
Yoo, I. D., 338
Yool, D. A., 149, 156
Yorihuzi, T., 296

Yoshida, H., 9, 93–94, 136, 185, 188–189, 194, 200, 266, 281, 283, 288, 295, 300, 305, 335, 339
Yoshii, S., 59, 283
Yoshimastu, T., 328
Yoshino, A., 365
Young, C. L., 235
Young, C. S., 113
Yuan, J., 200, 339
Yu, M. C., 328
Yun, C. Y., 281, 283, 288, 329
Yun, J., 330, 335–336
Yuraszeck, T., 235
Yura, T., 188–189, 194, 237, 242, 254, 288, 295, 335
Yutani, C., 374, 377
Yu, X., 311
Yuzawa, H., 376

Z

Zamanian-Daryoush, M., 346
Zamorano, S., 190
Zarkovic, K., 374–375
Zeng, H., 146, 185, 281, 283, 288, 302, 329
Zhang, B., 237
Zhang, C., 128, 135, 187, 190, 192
Zhang, H., 77, 149
Zhang, J., 328
Zhang, K., 75, 77, 93, 95, 187, 266, 288, 300
Zhang, M., 370, 374, 376
Zhang, P., 93, 98
Zhang, Y., 144, 146, 187, 190, 193–194, 281, 283, 288, 294–295, 302, 305, 329
Zhang, Z., 372
Zhao, H., 311
Zhao, X., 239
Zhao, Z., 370, 374, 376
Zheng, L., 77
Zhong, L., 147, 149
Zhu, H., 200
Zimmermann, F. K., 149, 156, 202
Zinszner, H., 58, 93, 100, 145–146, 288
Zollner, A., 250, 252
Zouboulis, C. C., 276
Zou, C. G., 59
Zou, M. H., 370, 374, 376
Zufferey, R., 171
Zu, K., 41

Subject Index

A

Activating transcription factor 6, 93–94
Advanced glycation
  AGE precursors, 362
  vs. ER stress
    antihypertensive drugs, 376–377
    antioxidant compounds, 373–376
  link with ER stress
    AGE-RAGE signaling pathway, 372
    aging, 372–373
    direct induction, 370
    hyperglycemia, 372
    hypoxia, 370–371
    ROS, 371–372
  measurement
    ELISA, 368–369
    HPLC and GC/MS, 364–365
    immunohistochemistry, 365–367
    western blot analysis, 367–369
  pathophysiology, 363–364
Advanced glycation endproduct (AGE)
  precursors, 362
Amino acid oxidation markers.
  See Oxidative stress
Animal models
  activating transcription factor 6, 93–94
  ATF4, 99–100
  cancer, 104
  CHOP, 100–101
  EIF2α, 98–99
  ERAD, 92
  folding enzymes upregulation, 92
  GADD34, 101–102
  hypoxia, 104
  inflammatory-mediated demyelination, 105
  IRE1/X-box-binding protein-1
    antibody-secreting plasma cells, 96
    BiP depletion, 94
    Crohn's disease, 96
    dendritic cells, 96
    ERAD components, 94–95
    high rate secretory cells, 95
    IgM synthesis, 96
    $Ire1\alpha^{-/-}$ and $Xbp1^{-/-}$, 95
    Lou Gehrig's disease, 97
    Paneth cells, 96–97
    RAG2-deficient blastocysts, 95
    ulcerative colitis, 96

  XBP1 transcriptional activity, 94
  knockout animals, 92
  lipid metabolism, 103–104
  mouse, 92–93
  P58IPK, 102
  PKR-like ER kinase, 97–98
  protein load decrease, 92
  transgenic mouse model, 102–103
Antibody secreting cell (ASC) process, 310
Antioxidant buffer, 78–79
Antioxidant compounds
  catalase, 375
  curcumin, 375–376
  N-acetyl cysteine, 375
  reagents, 373–374
  ROS scavengers, 373
  tempol, 376
  TM2002, 376
Apoptosis. *See* UPR-mediated apoptosis

B

Bioconductor, 331–332

C

N^ε-Carboxyethyllysine (CEL) detection
  IHC, 366–367
  WB analysis, 368–369
Catalase, 375
C/EBP-homologous protein (CHOP), 145, 187
Cell lysis
  cells-to-CT™, 63
  immunoprecipitation
    buffer, 319
    generation, protocol, 317–318
    materials, 317
    SDS, 318
    US2, 319–321
    XBP-1, μ chain synthesis, 319–320
Cells-to-CT™
  cell lysis, 63
  quantitative reverse transcription real-time PCR
    cDNA preparation, 64–65
    data analysis, 66
    plate setup, 65
    TaqMan gene expression assay, 64
    thermocycling, 65

399

Cells-to-CT™ (cont.)
  reverse transcription, 63–64
  XBP1u and XBP1s, 66–67
CFTR. See Cystic fibrosis transmembrane
    conductance regulator
CFTR expression regulation
  cell surface signaling complexes, 4
  cellular stress responses, 5
  chronic airway disorders, 5
  cystic fibrosis, 4–5
  epithelial cell differentiation, 5
  ER stress induction, 6–7
  mutations, 4
  protein levels assessment
    maturation efficiency measurements, 19–20
    western blot, 20–22
  RNA isolation
    binding capacity, 7
    cell confluency, 7
    cell lysate, 8
    materials and equipment, 7
    RNeasy spin columns, 8
  semiquantitative RT-PCR, 8–9
  transcriptional regulation
    chromatin immunoprecipitation assays
        (see Chromatin immunoprecipitation
        assays)
    promoter reporter assay, 14–15
  UPR effects
    mRNA levels measurement, 10
    mRNA stability (see mRNA)
  UPR reporter, 9
Chaperone-mediated protein folding, 112
Chemical–genetic manipulation
  ER stress, 189
  Fv2E–PERK, 191, 193
  IRE1[I642G], 190–192
  isogenic cell line, 189–190
  TET-ATF6, 194
  wild-type ATF6, 193–194
  wild-type IRE1 activation, 190–191
  wild-type PERK, 191, 193
Chemical genomics
  activation, ER, 328
  active transcription factor, 328
  chemical genomics research, 338–339
  GRP78 expression, 328
  modulators
    gene expression-based screening, 334–335
    reporter assay, 335–336
    UPR signal pathway, 336–338
  transcriptional program, cancer cell
    gene expression profiling, 329
    glucose deprivation signature, 332–334
    microarray expression profiles, 330–332
    versipelostatin and biguanides, 330
  tumor development, 328
Chromatin immunoprecipitation assays

buffers, materials and equipment, 15
principle, 15
protocol
  immunoprecipitation and DNA recovery, 18
  input preparation, 16–18
  sample preparation and cross-linking, 16
  sonication and determination, DNA
      content, 16
Connectivity map analysis, 334–335
Curcumin, 375–376
Cystic fibrosis (CF), 4–5
Cystic fibrosis transmembrane conductance
    regulator (CFTR), 218–219
Cytoplasmic protein extraction, 11

D

Delipidation, 79–80
2-Deoxy-D-glucose (2DG), 330
2D gel electrophoresis
  Coomassie and fluorography, 222–223
  2D maps image analysis, 223
  2D-PAGE, 221–222
  sample preparation, 221
  statistical analysis, 223
Diethylpyrocarbonate (DEPC), 268–269
DNA fragmentation analysis, 67–69
Dolichol-linked oligosaccharide (DLO) analysis
  2AB labeling, 171–172
  chemical and materials, 169
  extraction of, 170
  [^{3}H]-labeling, 170
  HPLC analysis, 171–172
  hydrolysis and purification, 170–171
Dolichol phosphate (Dol-P) analysis
  anthryldiazomethane labeling, 167–168
  devices and materials, 165–166
  extraction and purification, 166–167
  HPLC analysis, 168
  negative ion electrospray mass spectrometry,
      168–169

E

Endoplasmic reticulum associated degradation
    (ERAD), 311
  animal models, 92
  secretory pathway, 238–239
Endoplasmic reticulum (ER) stress, 75–76.
    See also Oxidative stress
  advanced glycation
    AGE-RAGE signaling pathway, 372
    aging, 372–373
    antihypertensive drugs, 376–377
    antioxidant compounds, 373–376
    direct induction, 370
    hyperglycemia, 372
    hypoxia, 370–371
    ROS, 371–372

Subject Index

ATF6 levels, 29
GRP78
  glioma cell lines and primary cell culture, 27–28
  glioma tissue, 28–29
GRP78/BiP levels *in vitro*
  cell death assays, 32
  EGCG modulation, 33–34
  siRNA modulation, 33
  TMZ and EGCG, 32
  TMZ *in vitro* treatment, 34
GRP78 levels, 29
immunohistochemistry, 29–30
intermediate markers
  cellular translation, 30
  cycloheximide, 30
  intracytoplasmic calcium, 30–31
in vivo modulation, GRP78
  progression and survival models, 35
  rodent glioma models, 34–35
  tissue analysis, 35–36
late markers, apoptotic, 31
transgenic mouse model, 102–103
UPR activity
  RNA isolation, 7–8
  semiquantitative RT-PCR, 8–9
  stress induction, 6–7
  UPR reporter assessment, 9
Epidermal growth factor receptor (EGFR)
  dimerization and autophosphorylation, 41
  EGF–SubA
    breast cancer cells, 42–43
    cells of interest, 43
    cell susceptibility, 44–46
    Celltiter 96®, 43
    disposables, 43
    MTS, 42
    plate reader, 43
    prostate cancer cells, 42
    tumor cell preparation, 44
  GRP78 cleavage, 42
  STAT3 and E2F1 transcription factors, 42
Escherichia coli, 41

F

FACS. *See* Fluorescence activated cell sorting
False discovery rate (FDR), 332
Flourescence microscopy
  antibodies, 130–132
  glibenclamide, 133–134
  proteins, 132–133
Flouride-induced endoplasmic reticulum stress
  calcium depletion, 112
  chaperone-mediated protein folding, 112
  Golgi transport, 112
  mutant protein synthesis, 113
  SEAP, 113–114
  TEM, 113
  UPR pathway components, 113
Fluorescence activated cell sorting (FACS), 120
Fluorophore-assisted carbohydrate electrophoresis (FACE) technique, 304

G

Gaussia Luciferase (Gluc). *See also* Flouride-induced endoplasmic reticulum stress
materials
  anti-Gluc antibody, 117
  assay plates, 118
  cells, 115–116
  CTZ substrate, 117
  Gluc lentivirus vector, 116–117
  growth medium, 116
  hexadimethrine bromide, 117
  luminometer, 117–118
  microscope, 118
  sodium fluoride, 117
procedure
  FACS, separation, 120
  fluoride, 121–123
  measurement of, 120–121
  separation of, 120
  transduction, 118–120
  UPR activation, 123
properties of, 114–115
Gene ontology, 226–227
Genetic switch technology (GST)
  applications, 156–157
  generalization, 157
  oligodendrocytes (*see* Oligodendrocytes)
  phenotypes, 156
GFP reporter assays, *C. elegans*
  multiparametric analyses, micrometric particles, 353
  strains, 351–352
  TOF and EXT, 353
  transgenic worm lines generation, 351
Glucose-regulated protein 78 (GRP78), 328. *See also* Endoplasmic reticulum (ER) stress
BiP levels *in vitro*
  cell death assays, 32
  EGCG modulation, 33–34
  siRNA modulation, 33
  TMZ and EGCG, 32
  TMZ *in vitro* treatment, 34
early markers
  ATF6 levels, 29
  GRP78 levels, 29
  immunohistochemistry, 29–30
  glioma cell lines and primary cell culture, 27–28
  glioma tissue, 28–29
intermediate markers
  cellular translation, 30

Glucose-regulated protein 78 (GRP78) (cont.)
  cycloheximide, 30
  intracytoplasmic calcium, 30–31
  in vivo modulation
    progression and survival models, 35
    rodent glioma models, 34–35
    tissue analysis, 35–36
  late markers, apoptotic, 31
Glycoprotein maturation and UPR
  endoplasmic reticulum, 163
  FGF23, 164
  GlcNAc$_2$Man$_9$Glc$_3$, 164
  N-glycosylation (see N-Glycosylation)
  O-glycosylation (see O-Glycosylation)
  O-fucosyltransferase (OFUT1) enzyme, 164
N-Glycosylation
  DLO analysis
    2AB labeling, 171–172
    chemical and materials, 169
    extraction of, 170
    [^{3}H]-labeling, 170
    HPLC analysis, 171–172
    hydrolysis and purification, 170–171
  Dol-P analysis
    anthryldiazomethane labeling, 167–168
    devices and materials, 165–166
    extraction and purification, 166–167
    HPLC analysis, 168
    negative ion electrospray mass spectrometry, 168–169
  endoplasmic reticulum, 163–164
  site occupancy
    Asn$_{413}$ and Asn$_{611}$, 172
    calculation of, 174–175
    devices and materials, 173
    human serum transferrin analysis, 172
    liquid chromatography, 174–175
    proteolytic digestion and deglycosylation, 173–174
    transferrin purification, 173
O-Glycosylation
  β-elimination reaction
    devices and materials, 176
    hydrazinolysis, 179
    MALDI-MS, 178–179
    mild alkali treatment, 176–177
    peeling reaction, 176
    permethylation and purification, 178
    sample preparations, 176
  hydrazinolysis
    devices and materials, 179
    fluorescence chromatogram, 179–180
    sample preparation, 179–181
  mucin-type glycosylation, 164, 175
Green fluorescent protein (GFP) fusions, 245–246
Glucose deprivation signature
  cells treatment, tunicamycin, 333
  cellular response, 332
  human cancer cells, ER stress, 332–333
  UPR modulators, 333–334

H

Hela cells, 351
Heterologous protein expression
  experimental results, 246–247
  GFP fusions, 245–246
  pulse-chase and immunoprecipitation protocol, 247–248
  scFv 4-4-20, 238
  scFv secretion, 237–238
  synthetic media, 243–245
Histone deacetylase (HDAC) inhibitors
  clonogenic assay, 54
  materials required, 53–54
Human cancer cell lines, 330
Hyperglycemia, 372
Hypoxia, 370–371

I

Immortalized rat proximal tubular cells (IRPTC), 368
Immunoblot analysis, 134–136
Immunohistochemistry (IHC), 365–367
Immunoprecipitation assay protocol, 208–209
Inositol-requiring protein-1 (IRE1), 294
In vitro translating (IVT) method, 322
IRE1/X-box-binding protein-1
  antibody-secreting plasma cells, 96
  BiP depletion, 94
  Crohn's disease, 96
  dendritic cells, 96
  ERAD components, 94–95
  high rate secretory cells, 95
  IgM synthesis, 96
  Ire1α$^{-/-}$ and Xbp1$^{-/-}$, 95
  Lou Gehrig's disease, 97
  LPS-stimulated WT, 95–96
  Paneth cells, 96–97
  RAG2-deficient blastocysts, 95
  ulcerative colitis, 96
  XBP1 transcriptional activity, 94
Isolation splenic B cells, retroviruses
  magnetic sorting
    disposables, 313
    protocol, 313–314
    purity, 314
    required materials, 312–313
  retroviral transduction, 315–316
  transfection of packaging cells, 314–315

L

Linear regression of efficiency (LRE) analysis, 255–256
Lipid-linked oligosaccharides (LLOs) extension

Subject Index

nonradioactivity measurement, FACE, 304
radioactivity measurement, HPLC, 303–304
Lou Gehrig's disease, 97

M

Maillard reaction, 363
Mammalian cells
 UPR activation, eIF2α phosphorylation
  disposables, 285
  lysates preparation, 285–286
  markers, immunoblotting, 288
  required materials, 284–285
  SDS-PAGE and immunoblotting, 286–287
 XBP1 splicing assay
  cDNA synthesis, 279
  disposables, 277–278
  genes indicative of activation, 281–283
  PCR, 279–281
  required materials, 277
  RNA extraction, 278–279
Mammalian unfolded protein response
 ATF6
  activity, 188
  cleavage, 304–305
  immunoblotting and measurement, 302
 cell culture, 295
 chemical–genetic manipulation
  ER stress, 189
  Fv2E–PERK, 191, 193
  IRE1[I642G], 190–192
  isogenic cell line, 189–190
  TET-ATF6, 194
  wild-type ATF6, 193–194
  wild-type IRE1 activation, 190–191
  wild-type PERK, 191, 193
 ER resident transmembrane proteins, 184
 ER stress inducers, 295–296
 inhibition of protein synthesis
  [^{3}H] leucine, 303
  measurement, L-[^{35}S] methionine, 302–303
 intrinsic function, 184–185
 IRE1, 294
 IRE1 activity, 185–187
 mechanistic link, 184–185
 nonradioactivity measurement, FACE, 304
 Northern blots, GRP78/BiP and EDEM
  fold enhancements, 299
  RT-PCR primers, 296, 298
  transcription, 296–297
 PERK activity, 187–188
 radioactivity measurement, HPLC, 303–304
 UPR elements, 188–189
 XBP1 mRNA splicing
  dermal fibroblasts, 297, 299
  gel condition determination, 300–302
  isolation and sequencing, 299
  RT-PCR fragments, agarose gel concentrations, 300–301
  RT-PCR primers, 298, 300
Methylglyoxal, 364
Morgan Elson chromogen, 176
mRNA
 CFTR mRNA half-life
  equipment and materials, 12
  protocol, 12–13
 CFTR mRNA levels measurement, 10
 cytoplasmic RNase activity
  cytoplasmic protein extraction, 11
  equipment and materials, 11
  RNA extraction, 12
  RNA samples incubation, 11
Multiple reaction monitoring (MRM), 82, 84–86
Multiple sclerosis (MS), 105
Myelin basic protein (MBP), 153

N

Nelfinavir-induced endoplasmic reticulum stress
 analysis of
  fluorescence microscopy (see Fluorescence microscopy)
  phase contrast microscopy, 130
 cell culture experiments, 129
 immunoblot analysis, 134–136
 mean plasma concentration, 129
 Pfizer, 129
 RT-PCR analysis
  cDNA synthesis, 139
  HeLa cells, 136
  new anticancer drug, 135
  PCR and PCR conditions, 139–141
  RNA extraction protocol, 138
  RNA preparation, 138–139
  specific primers availability, 136–137
  XBP1 splicing region, 136
 subcutaneous glioblastoma xenografts, 129–130
Viracept®, 129

O

Oligodendrocytes
 CreER^{T2} coding region, 152
 demyelination/remyelination processes, 154
 design construction, 152–153
 genetic switch detection, 154
 genomic DNA source, 150–151
 induction and labeling, 155
 MBP promoter/enhancer region, 153
 MCreER^{T2}G transgene expression effect, 153–154
 Neo^{msd} allele recombination, 154
 PGKneo cassette, 151
 Quickchange kit, 152
 ROSA26 allele engineered mice, 152

Oligonucleotide primers, 346–347
Oxidative stress, 371–372
　authentic and isotope standards preparation, 80–81
　chronic fibrotic diseases, 75–76
　end-organ damage, 75–76
　isotope dilution MS, 76–77
　LC/MS/MS detection, 82–84
　MRM analysis, 84–86
　oxidized amino acid content, 77–78
　redox signaling, 75
　RNS, 75
　ROS, 75
　sample preparation
　　antioxidant buffer, 78–79
　　delipidation, 79–80
　　delipidation buffer, 79
　　lysates hydrolysis, 79–80
　　protein isolation, 79
　　protein precipitation, 79–80
　　solid-phase extraction, 80
　　tissue collection, 79
　separation and detection, analytes
　　GC/MS, 81–82
　　LC/MS, 82

P

PCR. *See* Polymerase chain reaction
Pelizaeus-Merzbacher disease (PMD), 145
Phase contrast microscopy, 130
PKR-like ER kinase (PERK), 294–295
Plasmablasts, 323–324
Plasma cells
　ASC process, 310
　cell differentiation, 311
　Ig mislocalization, primary B cells
　　glycoproteins, 320–321
　　μ heavy chains, 321
　　nonglycosylated protein recognition, 322
　　permeabilization, 322–324
　　Promega's specifications, 322
　isolation splenic B cells, retroviruses
　　magnetic sorting, 312–314
　　retroviral transduction, 315–316
　　transfection of packaging cells, 314–315
　protein synthesis
　　cell lysis and immunoprecipitation
　　　(*see* Cell lysis, immunoprecipitation)
　　metabolic labeling, 316–317
Plasmid
　design
　　galactose-regulated promoters, 241
　　scFv 4-4-20 construct, 241–242, 248
　　UPR sensor, 242–243
　　reporter assay, 335
PLP1 gene mutation
　missense and nonsense, 147–148
　Pelizaeus-Merzbacher disease, 147, 149
　wild-type PLP1 and DM20, 147–148

Polymerase chain reaction (PCR), 211
Protein identification
　MALDI analysis, 224–225
　protein gel cut and digestion, 224
　Sequest/Mascot, 224
　types, 223
Protein Information and Knowledge Extractor (PIKE), 227–229
Protein isolation, 79
Proteostasis network signaling pathways
　BHK cell line, 219
　biochemical validation, 225–227
　bioinformatics analysis
　　BHK-F508del, 228–230
　　gene ontology, 226–227
　　PIKE, 227–228
　　tools, 226
　cell lines, 220
　CFTR, 218–219
　clustering analysis
　　data mining approach, 229
　　fast hierarchical clustering analysis, 229, 231
　　geWorkbench output, 230–231
　　protein expression analysis, 229
　2-DE, protein expression profile, 219–220
　2D gel electrophoresis
　　Coomassie and fluorography, 222–223
　　2D maps image analysis, 223
　　2D-PAGE, 221–222
　　sample preparation, 221
　　statistical analysis, 223
　F508del mutation, 218
　low temperature treatment and *in vivo* labeling, 220–221
　MIAPE, 231
　MS analysis, protein identification
　　MALDI analysis, 224–225
　　MS/MS spectra data, 224
　　PMF, 223
　　protein gel cut and digestion, 224
　RXR mutagenic repair, 218–219
Pulse-chase protocol, 207–208
Pyrvinium pamoate, 335

Q

Quantitative PCR (qPCR) analysis
　BiP, 188
　CHOP, 187

R

Reactive nitrogen species (RNS), 75
Reactive oxygen species (ROS), 75
Redox signaling, 75
Renin-angiotensin system (RAS), 376
Reverse transcription, 63–64
RNA. *See also* mRNA
　general recommendations, 268–269
　isolation, 271–272

separation, agarose gel electrophoresis
  capillary transfer of RNA, 273–274
  DNA probes labeling, 274–275
  electrophoresis, 273
  ethidium bromide staining, 273
  gel preparation, 272–273
  hybridization, 275–276
  sample preparation, 273
RT-PCR analysis
  cDNA synthesis, 139
  HeLa cells, 136
  new anticancer drug, 135
  PCR and PCR conditions, 139–141
  RNA extraction protocol, 138
  RNA preparation, 138–139
  specific primers availability, 136–137
  XBP1 splicing region, 136

S

Saccharomyces cerevisiae, 200
  growth of yeast cultures, 271
  HAC1 splicing, Northern blotting, 268
  isolation, RNA, 271–272
  required materials
    disposables, 270–271
    reagents, 269–270
  RNA, 268–269
  RNA separation, agarose gel electrophoresis
    capillary transfer of RNA, 273–274
    DNA probes labeling, 274–275
    electrophoresis, 273
    ethidium bromide staining, 273
    gel preparation, 272–273
    hybridization, 275–276
    sample preparation, 273
Secreted alkaline phosphatase (SEAP) reporter, 113–114
Single-chain antibody fragments (scFv), 236
Small GTPase signaling
  adaptive mechanism, 344
  endogenous UPR
    ATF6α activation, 350
    PERK pathway and phosphorylation, eIF2α, 348
    UPR target gene expression, 350
    XBP1 mRNA splicing, 348–350
  ER organelle, 344
  genetic modulation, RNA interference
    cultured mammalian cells, 356
    GFP expression, 357
    mammalian cells, siRNA, 357
  GTPase pull-down assay, 355–356
  immunofluorescence-based visualization, 353–355
  materials and methods, 346–347
  pharmacological modulation, 356
  recombinant strategies
    GFP reporter assays, *C. elegans*, 351–353

luciferase reporter assays, cultured mammalian cells, 351
stress response, 344–345
target genes, transcriptional activation, 345
UPR activation, 345
Streptococcus pneumoniae, 165
Synthetic genetic array (SGA), 201

T

Tamoxifen, 152
Temozolomide (TMZ), 31–32
Time of flight (TOF), 353
Transient transfections, 335–336
Tunicamycin, 346
Two-dimensional gel electrophoresis (2DE), 219–220

U

Unfolded protein response (UPR)
  ATF6 pathway, 265–266
  BiP and PDI, 236–237
  budding yeast, 200, 215
  cancer, 104
    breast cancer models, 39–40
    glioblastoma models, 39
    Grp78-knockdown fibrosarcoma cells, 39
    tumorigenesis, 38–39
    XBP1, 39
  CFTR mRNA levels
    cytoplasmic RNase activity, 11–12
    determination of, 10–11
    measurement of, 10
    mRNA half-life, 12–13
  characteristics, 149–150
  chemical genomics (*see* Chemical genomics)
  CHOP protein, 145
  cloning, complementation, 209–210
  definition, 200
  EGF–SubA with histone deacetylase
    clonogenic assay, 54
    materials required, 53–54
  EGF–SubA with thapsigargin
    assessment, 52
    MDA231luc cell preparation, 52
    with thapsigargin, 50–51
  endoplasmic reticulum export and trafficking, 239
  (−)-epigallocatechin gallate, 40
  β-galactosidase assay, 212–213
  gene characterization, 237
  genetic screen, 201
  GRP78 cleavage
    bacterial toxin SubAB, 40–41
    band normalization analysis, 46
    dimerization and autophosphorylation (*see* Epidermal growth factor receptor)
    disposables, 48

Unfolded protein response (UPR) (cont.)
    EGF–SubA-induced cleavage, 45–46
    ER stress signaling pathways, 41
    Escherichia coli, 41
    reagents, 47–48
    tumor cell exposure, 48–49
    tumor cell preparation, 48
    western blot analysis, 46
GST, 156–157
HAC1 mRNA monitoring, 213
Hac1p, 200
HAC1 splicing, 263
heterologous protein expression
    experimental results, 246–247
    GFP fusions, 245–246
    pulse-chase and immunoprecipitation
        protocol, 247–248
    scFv 4–4–20, 238
    scFv secretion, 237–238
    synthetic media, 243–245
hypoxia, 104
induction, 266–267
inflammatory-mediated demyelination, 105
IRE1 pathway, 262–263
library transformation and plasmid isolation,
    210–211
lipid metabolism, 103–104
mammalian UPR (see Mammalian unfolded
    protein response)
media recipes, 214
mutant gene identification, 211
mutant screening, viability
    colony color phenotype, 202, 204
    colony-color sectoring assay, 202
    immunoprecipitation assay protocol, 208–209
    kill rate, 204–205
    per screening strains, 202–203
    primary genetic screening, 202, 204
    pulse-chase protocol, 207–208
    recessive mutant, 205
    secondary screen, per mutants, 206–207
    tetrad analysis, 206
    UV mutagenesis, 202, 204
    YPD media, 205
NF-κB activation, 264
oligodendrocytes
    CreER^{T2} coding region, 152
    demyelination or remyelination
        processes, 154
    design construction, 152–153
    genetic switch detection, 154
    genomic DNA source, 150–151
    induction and labeling, 155
    MBP promoter/enhancer region, 153
    MCreER^{T2}G transgene expression effect,
        153–154
    Neo^{msd} allele recombination, 154
    PGKneo cassette, 151–152
    Quickchange kit, 152
    ROSA26 allele, 152

PERK signaling pathway
    cap–dependent translation, 264–265
    CHOP function, 146–147
    temporal delay circuit, 144–145
plasma cells (see Plasma cells)
plasmid design
    Galactose-regulated promoters, 241
    scFv 4–4–20 construct, 241–242, 248
    UPR sensor, 242–243
PLP1 gene mutation
    missense and nonsense, 147–148
    Pelizaeus-Merzbacher disease, 147
    rodent models, PMD, 149
    wild-type PLP1 and DM20, 147–148
PMD neurodegenerative disease, 145
quality control mechanism, secretory pathway,
    238–239
regulatory circuit, 214
RNase activity, 262
Saccharomyes cerevisiae strains, optimal
    expression, 240
SGA database, 213–214
signaling measurement
    mammalian cells (see Mammalian cells)
    S. cerevisiae (see Saccharomyces cerevisiae)
signal pathway
    connectivity map-based screening, 336
    IRE1-XBP1 signaling, 336, 338
    pyrvinium pamoate, 336–337
    suppressive effect, GRP78 promoter
        activity, 336–337
    valinomycin, 338
small GTPase signaling (see Small GTPase
    signaling)
statistical analysis
    amplification reaction curve, 256
    appropriate control, 249–250
    classification, 252–253
    $2^{\Delta\Delta}$Ct method, 254–255
    differentially expressed gene selection,
        251–252
    high-throughput microarray technology,
        248–249
    LRE analysis, 255–256
    microarray data analysis chart, 249–250
    promoter regions enrichment, 253–254
    protein secretion and UPR induction
        implication, 256–257
    UPRE 1, UPRE 2, UPRE 3, 254
synthetic lethality, definition, 201
toxic effects, 144
UPR and secretory processing evaluation,
    239–240
yeast, 239–240
UPR. See Unfolded protein response
UPR-mediated apoptosis
    ATF4-mediated induction, 58
    cell death target genes
        cell culture, 62–63
        cell lysis, 63

cells-to-CT™ (see Cells-to-CT™)
gene expression assays, 62
protocol modification, 62
quantitative reverse transcription real-time PCR, 64–66
reverse transcription, 63–64
semiquantitative PCR analysis, 66–67
CHOP, 58–59
DNA fragmentation analysis, 67–69
eIF2α phosphorylation, 58
JNK cell activation, 59
monitoring proliferation and caspase activation
dual γ-axis representation, 61–62
edge effect, 61
luminescent assay, 60–61
mitochondria-ionophore-induced ER stress, 60
PERK activation, 58

V

Versipelostatin, 330–331

W

Western blotting (WB) process
analysis, 246–247, 367–369
eIF2a and ATF4, 187–188
FLAG-ATF6, 188
protein levels assessment, 20–22

X

XBP1 splicing
assay, 185–187
cDNA synthesis, 279
disposables, 277–278
genes indicative of activation, 281–283
PCR, 279–281
required materials, 277
RNA extraction, 278–279
mRNA
dermal fibroblasts, 297, 299
gel condition determination, 300–302
IRE1α endoribonuclease activity, 348
isolation and sequencing, 299
monitoring, 348–349
protein loading, 350
RT-PCR fragments, agarose gel concentrations, 300–301
RT-PCR primers, 298, 300

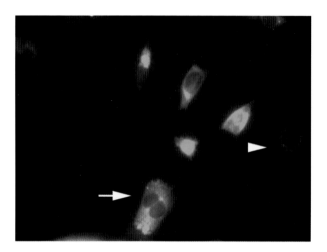

Ramaswamy Sharma et al., Figure 7.2 Identification of transduced cells. LS8 cells were transduced with a lentivirus construct expressing Gluc-YFP fusion. Fluorescence microscopy for YFP shows a successfully transduced cell (long arrow) and a nontransduced cell (short arrow).

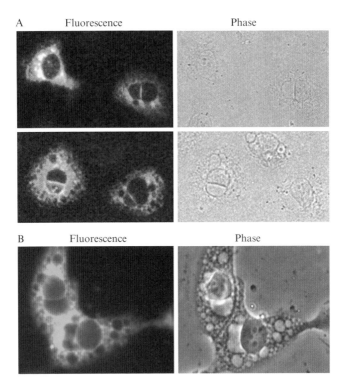

Ansgar Brüning, Figure 8.2 Fluorescence analysis of nelfinavir-treated cancer cells. A specific ER organelle staining in nelfinavir-treated cancer cells can either be achieved by BiP immunostaining (A), or with fluorescent glibenclamide to stain for ER membranes (B; previously published in Brüning et al., 2009). Samples: OVCAR3 cells treated for 24 h with 8 μg/ml nelfinavir.

```
                    →
TGCTGAAGAGGAGGCGGAAGCCAAGGGGAATGAAGTGAGG
CCAGTGGCCGGGTCTGCTGAGTCCG:CAGCACTCAGACTA
CGTGCACCTCTG:CAGCAGGTGCAGGCCCAGTTGTCACCC
CTCCAGAACATCTCCCCATGGATTCTGGCGGTATTGACTC
TTCAGATTCAGAGTCTGATATCCTGTTGGGCATTCTGGAC
                                    ←
```

Ansgar Brüning, Figure 8.6 XBP1 splicing site and primer orientation. The 26 nucleotides that are spliced within an exon of XBP1 (exon 4, marked in blue) are indicated in red. Nucleotides of the adjacent exons are printed in black. Binding sites of PCR primers are underlined and their orientation (5′–3′) is indicated by an arrow. Primers were placed before the exon–exon boundaries to avoid amplification of exon 4 from genomic DNA. A suggested sequence for a TaqMan Probe (e.g., 6-FAM-GCTGAGTCCGCAGCAGGTG-TAMRA) that specifically recognizes the spliced form of XBP1 and would allow quantitative real-time PCR analysis is underlined in blue.

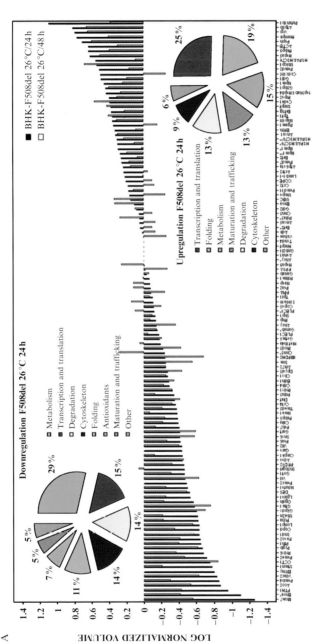

Figure 13.3 (Continued)

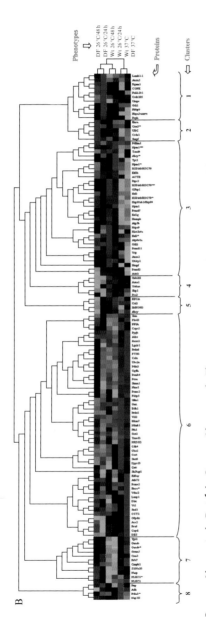

Patrícia Gomes-Alves et al., Patrícia Gomes-Alves et al., Figure 13.3 Graphical representation of the log normalized volume of proteins in BHK-F508del (A). Expression levels of those proteins in BHK-F508del at 37 °C were considered as reference to determine the up- or downregulation of the correspondent protein in the other phenotype. Proteins up- and downregulated in F508del cells were also classified according to their molecular function. Unsupervised hierarchical clustering (Euclidean distances and average linkage) of the expression profiles of the 139 protein spots differentially expressed in the six phenotypes studied (B). Each horizontal row in the heat map represents the expression level for one protein in the different phenotypes (columns). The relative abundance is displayed by color intensity (light red—more abundant; light green—less abundant).

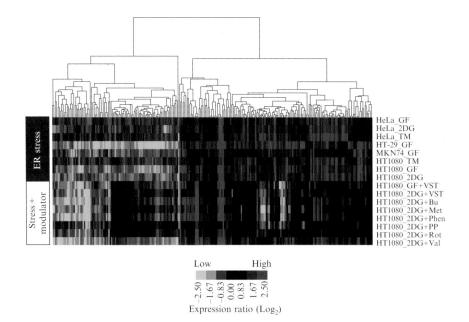

Sakae Saito and Akihiro Tomida, Figure 18.1 Gene expression profile of the human cancer cells under ER stress conditions by the glucose deprivation signature. Glucose deprivation signature genes including 246 probe sets (X axis) sorted by cluster analysis displayed with 16 samples (Y axis). The cells were cultured for 15–18 h under normal growth condition (control), ER stress conditions or ER stress conditions with UPR modulators. The log ratio for each gene was calculated by setting the expression level as 0 (Log_2 1) in an appropriate control sample. GF, glucose-free medium; 2DG, 10 mM 2-deoxyglucose; TM, 5 μg/mL tunicamycin; VST; 10 μM versipelostatin; Bu, 300 μM buformin; Met, 10 mM metformin; Phen, 100 μM phenformin; PP, 0.1 μM pyrvinium pamoate; Rot, 6 μM rottlerin; Val, 10 nM valinomycin.